Lecture Notes in Networks and Systems

Volume 1025

The series "Lecture Notes in Networks and Systems" publishes the latest developments in Networks and Systems—quickly, informally and with high quality. Original research reported in proceedings and post-proceedings represents the core of LNNS.

Volumes published in LNNS embrace all aspects and subfields of, as well as new challenges in, Networks and Systems.

The series contains proceedings and edited volumes in systems and networks, spanning the areas of Cyber-Physical Systems, Autonomous Systems, Sensor Networks, Control Systems, Energy Systems, Automotive Systems, Biological Systems, Vehicular Networking and Connected Vehicles, Aerospace Systems, Automation, Manufacturing, Smart Grids, Nonlinear Systems, Power Systems, Robotics, Social Systems, Economic Systems and other. Of particular value to both the contributors and the readership are the short publication timeframe and the world-wide distribution and exposure which enable both a wide and rapid dissemination of research output.

The series covers the theory, applications, and perspectives on the state of the art and future developments relevant to systems and networks, decision making, control, complex processes and related areas, as embedded in the fields of interdisciplinary and applied sciences, engineering, computer science, physics, economics, social, and life sciences, as well as the paradigms and methodologies behind them.

Indexed by SCOPUS, INSPEC, WTI Frankfurt eG, zbMATH, SCImago.

All books published in the series are submitted for consideration in Web of Science.

For proposals from Asia please contact Aninda Bose (aninda.bose@springer.com).

Soumi Dutta · Abhishek Bhattacharya ·
Celia Shahnaz · Satyajit Chakrabarti
Editors

Cyber Intelligence and Information Retrieval

Proceedings of CIIR 2023, Volume 2

Springer

Editors
Soumi Dutta
Sister Nivedita University
Kolkata, West Bengal, India

Abhishek Bhattacharya
Sister Nivedita University
Kolkata, West Bengal, India

Celia Shahnaz
Bangladesh University of Engineering
and Technology (BUET)
Dhaka, Bangladesh

Satyajit Chakrabarti
University of Engineering and Management
Kolkata, West Bengal, India

ISSN 2367-3370 ISSN 2367-3389 (electronic)
Lecture Notes in Networks and Systems
ISBN 978-981-97-3593-8 ISBN 978-981-97-3594-5 (eBook)
https://doi.org/10.1007/978-981-97-3594-5

This Springer imprint is published by the registered company Springer Nature Singapore Pte Ltd.
The registered company address is: 152 Beach Road, #21-01/04 Gateway East, Singapore 189721, Singapore

Preface

We hereby are delighted to announce that the Institute of Engineering and Management, India, in association with Computer Society of India, Kolkata Section, has hosted the eagerly awaited and much coveted 2nd International Conference on Cyber Intelligence and Information Retrieval (CIIR 2023) in Hybrid Mode on December 14–15, 2023.

The second version of the conference was able to attract a diverse range of engineering practitioners, academicians, scholars, and industry delegates, with the reception of abstracts including more than 1400 authors from different parts of the world. The committee of professionals dedicated toward the conference is striving to achieve a high-quality technical program with tracks on cyber intelligence, advance computing, information retrieval, data science and data analytics and pattern recognition. All the tracks chosen in the conference are interrelated and are very famous among the present-day research community. As such, a great deal of research is being conducted in the aforementioned tracks and their associated sub-areas. More than 370 full-length papers have been received, among which the contributions are focused on theoretical, computer simulation-based research, and laboratory-scale experiments. Among these manuscripts, only 81 papers have been included in the Springer proceedings after a thorough two-stage review and editing process. All the manuscripts submitted to the CIIR 2023 were peer-reviewed by at least two independent reviewers. The comments from the reviewers were communicated to the authors, who incorporated the suggestions in their revised manuscripts. The recommendations from two reviewers were taken into consideration while selecting a manuscript for inclusion in the proceedings. It is an exalting experience to finally see these elite contributions materialize into two book volumes as CIIR 2023 proceedings by Springer entitled *Cyber Intelligence and Information Retrieval*. The articles are organized into two volumes in some broad categories covering subject matters on cyber Intelligence, data mining, information retrieval, networks, soft computing, and cloud computing. CIIR 2023 invited six keynote speakers, who are eminent researchers in the field of computer science and engineering, from different parts of the world. In addition to the keynote sessions on each day of the conference, three concurrent technical sessions are held every day to assure the oral presentation of around 81 accepted papers. Prominent researchers in the session's subject area have served as keynote speakers and session chair(s) for each of the concurrent sessions. The research part of the conference was organized in a total of 11 special sessions. These special sessions and international workshops provided the opportunity for researchers conducting research in specific areas to present their results in a more focused environment. An international conference of such magnitude and release of the CIIR 2023 proceedings by Springer has been the remarkable outcome of the untiring efforts of the entire organizing team. An event's success surely depends on the diligent work of many contributors at various points in time, determined by their commitment and sincerity. Fortunately, CIIR 2023 has benefited from contributions and support from all sides since the start of its journey. We extend our gratitude to everyone who sent their best wishes and made any kind of contribution to the success of CIIR 2023. The tenacity of the steering, advising, and technical program committee members made the edited proceeding volumes published by Springer possible. The CIIR 2023 organizers would like to thank all of the participating writers for their outstanding works and attention. We would like to express our sincere gratitude to the authors, editors, committee members, and production workers whose hard work helped to define the CIIR 2023 proceedings. Finally, we express our gratitude to Springer for agreeing to publish the conference papers of the CIIR 2023. The acquisition senior editor, Mr. Aninda Bose, has been a great help throughout the process.

Kolkata, India Soumi Dutta
Kolkata, India Abhishek Bhattacharya
Dhaka, Bangladesh Celia Shahnaz
Kolkata, India Satyajit Chakrabarti

About the Conference

Welcome to the 2nd International Conference on Cyber Intelligence and Information Retrieval (CIIR 2023) held at the Institute of Engineering and Management, Kolkata, India, in association with CSI Kolkata section on December 14–15, 2023.

As a premier conference in the field, CIIR 2023 provides a highly competitive forum for showcasing the latest developments in the research and application of cyber security and information retrieval. We are pleased to present the proceedings of the conference as its published record. The theme this year is crossroad of cyber security and information retrieval, a topic that is quickly gaining traction in both academic and industrial discussions because cyber intelligence relies heavily on effective information retrieval techniques to collect and analyze data from various sources, such as threat feeds, vulnerability databases, and dark web forums.

CIIR 2023 is a young conference for research in the areas of cyber security, data sciences, information retrieval and data mining. Although 2022 was the debut year for CIIR, it has already witnessed significant growth. As evidence of that, CIIR 2023 received a record 373 submissions. The authors of submitted papers come from 14 countries and regions. Authors of accepted papers are from 11 countries.

We hope that this program will further stimulate research in information security and data mining and provide practitioners with better techniques, algorithms, and tools for deployment. We feel honored and privileged to serve the best recent developments in the field of cyber security to you through this exciting program.

Dr. Abhishek Bhattacharya
Convener, CIIR 2023

About the Editors

Contents

Editors and Contributors

About the Editors

Dr. Soumi Dutta is Associate Professor at Sister Nivedita University, India. She has completed her Ph.D. (CST), IIEST, Shibpur. She received her B.Tech. (IT) and M.Tech. (CSE) securing 1st position (Gold medalist), from MAKAUT. She is certified as Publons Academy Peer Reviewer, 2020, and Certified Microsoft Innovative Educator, 2020. Her research interests are Data Mining, Information Retrieval, Online Social Media Analysis, and Image Processing. She has published 30 conference and journal papers in Springer, IEEE, IGI Global, Taylor & Francis, etc. She has 5 book chapters in Taylor & Francis Group and IGI Global. She is Peer Reviewer and TPC Member in different international journals. She was Editor in CIPR 2020, CIPR2019, IEMIS 2020, CIIR 2021, and IEMIS 2018 special issues in *IJWLTT*. She is Member of several technical functional bodies such as ACM, IEEE, IFERP, MACUL, SDIWC, Internet Society, ICSES, ASR, AIDASCO, USERN, IRAN, and IAENG. She has published 3 patents. Dr. Soumi Dutta has delivered 15 keynote talks in different international conferences. She has been awarded with Rashtriya Shiksha Ratna Award, InSc Research Education Excellence Award, and International Teacher Award 2020–21 by Ministry of MSME, Government of India.

Dr. Abhishek Bhattacharya is Associate Professor at Sister Nivedita University, India. He has completed his Ph.D. (CSE) from BIT, Mesra. He is certified as Publons Academy Peer Reviewer, 2020. His research interests are data mining, cyber security, and mobile computing. She has published 25 conference and journal papers in Springer, IEEE, IGI Global, Taylor & Francis etc. He has 3 book chapters in Taylor & Francis Group EAI. He is peer reviewer and TPC member in different international journals. He was the editor in IEMIS 2020, IEMIS 2018, IEMIS 2022, CIIR 2023, ICIDA 2022 and special issues in *IJWLTT*. He is the member of several technical functional bodies such as IEEE, IFERP, MACUL, SDIWC, Internet Society, ICSES, ASR, AIDASCO, USERN, IAENG. He has published 3 patents.

Celia Shahnaz received Ph.D. degree in electrical and computer engineering from Concordia University, Montreal, QC, Canada, in 2009. Currently she is serving as Professor in the Department of Electrical and Electronic Engineering, BUET, from where she received her B.Sc. and M.Sc. degrees in 2000 and 2002, respectively. Dr. Celia is Senior Member of IEEE, Fellow of IEB, and has published more than 80 international journal and conference papers. She has been appointed as 2017 Communications Chair, IEEE SIGHT steering committee, 2017 Chair, IEEE WIE Workshops Subcommittee and IEEE PES Women in Power (WiP) Region 10 Regional Representative, and 2017 Member, IEEE SSIT WIE and SIT subcommittees. Recently, she has been selected as the recipient of IEEE Member and Geographic Activities (MGA) Leadership award with citation "For leadership in engineering and technology driven innovative IEEE Women in Engineering activities for enhanced membership development and engagement in Region 10 and across the globe". Her research interests include the areas of speech analysis, speech enhancement, digital watermarking, biomedical signal processing, audio-visual recognition for biometric security, pattern recognition, multimedia communication, control system, robotics and signal processing, and pattern recognition for power signals.

Dr. Satyajit Chakrabarti is Pro-Vice Chancellor, University of Engineering and Management, Kolkata and Jaipur Campus, India, and Director of Institute of Engineering and Management, IEM. As Director of one of the most reputed organizations in Engineering and Management in Eastern India, he launched a PGDM Program to run AICTE-approved management courses, Toppers Academy to train students for certificate courses and Software Development in the field of ERP solutions. Dr. Chakrabarti was Project Manager in TELUS, Vancouver, Canada, from February 2006 to September 2009, where he was intensively involved in planning, execution, monitoring, communicating with stakeholders, negotiating with vendors and cross-functional teams, and motivating members. He managed a team of 50 employees and projects with a combined budget of $3 million.

Contributors

Omar Rafat Adnan Department of Computer Science and Engineering, East West University, Dhaka, Bangladesh

Ambuj Kumar Agarwal Department of Computer Science and Engineering, School of Engineering and Technology, Sharda University, Greater Noida, India

Amjed A. Ahmed Department of Computer Techniques Engineering, Imam Al-Kadhum College (IKC), Baghdad, Iraq;
Center for Cyber Security, Faculty of Information Science and Technology, Universiti Kebangsaan Malaysia (UKM), Bangi, Malaysia

Md. Al-Amin CSE, East West University, Dhaka, Bangladesh

Nazib Al-Hoque CSE, East West University, Dhaka, Bangladesh

Jaffar Alkhayer Faculty of Legal Sciences, Shoolini University, Solan, Himachal Pradesh, India

Kavita Arora School of Computer Applications, Manav Rachna International Institute of Research and Studies, Faridabad, Haryana, India

Danish Ather Amity University, Tashkent, Uzbekistan

Baranitharan Balakrishnan Department of Agricultural Engineering, Kalasalingam Academy of Research and Education, Krishnankoil, TN, India

Ayanta Banerjee Institute of Engineering & Management, Kolkata, India

Partho Banerjee Department of Law, University of Petroleum and Energy Studies, Dehradun, India

Arittra Basu Department of Computer Application and Science, Institute of Engineering and Management, Kolkata, India

Arkajyoti Basu Department of Computer Application and Science, Institute of Engineering and Management, Kolkata, India

Tapas Bhadra Aliah University, Kolkata, WB, India

Abhishek Bhattacharya Sister Nivedita University, Chakpachuria, West Bengal, India

Rupam Bhattacharya Department of Computer Application Institute of Engineering and Management, Kolkata, West Bengal, India

Vasundhara Bhatt Law College Dehradun, Uttaranchal University Dehradun, Dehradun, India

Ankan Bhowmik Department of Computer Application and Science, Institute of Engineering and Management, Kolkata, India

V. Bhuvaneswari Law College Dehradun, Uttaranchal University Dehradun, Dehradun, India

Lindita Bicaj Rochester Institute of Technology, Europe, Kosovo

Manashi Chakraborty The Neotia University (TNU), Sarisa, WB, India

Karthik Chandran Department of Robotics and Automation, Jyothi Engineering College, Chelakkara, KL, India

Saumitra Chattopadhyay Graphic Era Hill University, Dehradun, Uttarakhand, India

S. K. Chaturvedi School of Law, H.N.B Garhwal Central University, Tehri Garhwal, India

Tanupriya Choudhury Adjunct Professor, CSE Department, Symbiosis Institute of Technology, Symbiosis International (Deemed University), Pune, Maharashtra, India;
Research Professor, CSE Department, Graphic Era Deemed to be University, Dehradun, Uttarakhand, India

Tashfia Choudhury Department of Computer Science and Engineering, East West University, Dhaka, Bangladesh

Maleha Israt Chowdhury Department of Computer Science and Engineering, East West University, Dhaka, Bangladesh

Md Tanvir Chowdhury Department of Computer Science and Engineering, East West University, Dhaka, Bangladesh

Subrata Chowdhury Department of Computer Science and Engineering, SITAMS (A) Chittoor, Murukambattu, AP, India

Priyanka Das Department of Computer Application Institute of Engineering and Management, Kolkata, West Bengal, India

Tropa Das CSE, East West University, Dhaka, Bangladesh

G. Dharma Teja Department of Computer Science and Engineering, Saveetha School of Engineering, Saveetha Institute of Medical and Technical Sciences, Saveetha University, Chennai, India

Divyansh Department of Computer Science & Engineering, Sharda School of Engineering & Technology, Sharda University, Greater Noida, India

Anil Kumar Dixit Law College Dehradun, Uttaranchal University Dehradun, Dehradun, India

Utkarsh Dixit Ajay Kumar Garg Engineering College, Ghaziabad, India

Soumi Dutta Department of Computer Science, Sister Nivedita University, Kolkata, West Bengal, India

Atul Garg Chitkara University Institute of Engineering and Technology, Chitkara University, Rajpura, Punjab, India

Deepanshi Garg CSE Department, University of Petroleum and Energy Studies (UPES), Dehradun, Uttarakhand, India

Nidhi Bansal Garg Chitkara University Institute of Engineering and Technology, Chitkara University, Rajpura, Punjab, India

Vinay Gautam Chitkara University Institute of Engineering and Technology, Chitkara University, Punjab, India

Soumen Ghosh Haldia Institute of Technology, Haldia, WB, India

Anvi Goel School of Computer Science, University of Petroleum and Energy Studies (UPES), Dehradun, Uttarakhand, India

Chander Mohan Gupta Faculty of Legal Sciences, Shoolini University, Solan, Himachal Pradesh, India

Sonam Gupta Ajay Kumar Garg Engineering College, Ghaziabad, India

Varun Haralalka Department of Computer Application and Science, Institute of Engineering and Management, Kolkata, India

Mahamudul Hasan CSE, East West University, Dhaka, Bangladesh

Md. Miskat Hossain East West University, Dhaka, Bangladesh

Md Jakir Hossain Department of Computer Science and Engineering, East West University, Dhaka, Bangladesh

Md Sabbir Hossain Department of Computer Science and Engineering, East West University, Dhaka, Bangladesh

Md Shohrab Hossain Department of Textile Engineering, Ahsanullah University of Science and Technology, Dhaka, Bangladesh

Izazul Hoq Imran CSE, East West University, Dhaka, Bangladesh

Sajida Islam Ipty Department of Computer Science and Engineering, East West University, Dhaka, Bangladesh

Anurag Jain School of Computer Sciences, University of Petroleum and Energy Studies (UPES), Dehradun, Uttarakhand, India

Vishal Jain Department of Computer Science & Engineering, Sharda School of Engineering & Technology, Sharda University, Greater Noida, India

Rayirth Jaiswal Department of Computer Application and Science, Institute of Engineering and Management, Kolkata, India

Sanjay Jasola Graphic Era Hill University, Dehradun, Uttarakhand, India

Maheshwar Narayan Joshi Law College Dehradun, Uttaranchal University Dehradun, Dehradun, India

Mukesh Joshi School of Computing, Graphic Era Hill University, Dehradun, UK, India

Animesh Kairi Institute of Engineering Management, Kolkata, WB, India

K. Kannabiran Department of Horticulture, Kalasalingam Academy of Research and Education, Krishnankoil, TN, India

Ravi Kant Law College Dehradun, Uttaranchal University Dehradun, Dehradun, India

Gagandeep Kaur Department of Law, School of Law, University of Petroleum & Energy Studies, Dehradun, Uttarakhand, India

Vandana Dixit Kaushik Department of Computer Science and Engineering, Harcourt Butler Technical University, Kanpur, Uttar Pradesh, India

Rubina Liyakat Khan The Applied College, Imam Abdulrahman Bin Faisal University, Dammam, Saudi Arabia

Ketan Kotecha Symbiosis Center for Applied Artificial Intelligence, Symbiosis Institute of Technology, Symbiosis International (Deemed University), Pune, Maharashtra, India

Monika Kothiyal ICFAI Law School Dehradun, Dehradun, India

Raj Kumar Department of Computer Applications, Manav Rachna International Institute of Research and Studies (MRIIRS), Faridabad, India

Maisha Mahajabin East West University, Dhaka, Bangladesh

Nusrat Maisha East West University, Dhaka, Bangladesh

Shreya Majumder Institute of Engineering and Management, Kolkata, India

Momena Khatun Methila Department of Computer Science and Engineering, East West University, Dhaka, Bangladesh

Kalyan Mitra Indian Maritime University - Kolkata Campus, Kolkata, India

Nayantara Mitra Institute of Engineering & Management, Kolkata, India

Ammar Abdulhassan Muhammed Department of Computer Techniques Engineering, Imam Al-Kadhum College (IKC), Baghdad, Iraq

Hassan Jameel Mutasharand Department of Computer Techniques Engineering, Imam Al-Kadhum College (IKC), Baghdad, Iraq

Pranta Nath Nayan East West University, Dhaka, Bangladesh

Ahmed J. Obaid University of Kufa, Kufa, Iraq

Sonal Pathak School of Computer Applications, Manav Rachna International Institute of Research and Studies, Faridabad, Haryana, India

Anish Paul Department of Computer Application and Science, Institute of Engineering and Management, Kolkata, India

Suplab Podder Postdoctoral Researcher, Rochester Institute of Technology, Prishtina, Europe, Kosovo

Tridip Pramanick Department of Computer Application and Science, Institute of Engineering and Management, Kolkata, India

Sabyasachi Pramanik Haldia Institute of Technology (HIT), Haldia, WB, India

Pranjal Department of Computer Science & Engineering, Sharda School of Engineering & Technology, Sharda University, Greater Noida, India

Suhail Javed Qurashi School of Computer Applications, Manav Rachna International Institute of Research and Studies, Faridabad, Haryana, India

Abdur Rahman East West University, Dhaka, Bangladesh

Habibur Rahman Department of Computer Science and Engineering, East West University, Dhaka, Bangladesh

Mayur Rahul Department of Computer Application, School of Engineering and Technology, CSJM University, Kanpur, Uttar Pradesh, India

Dharm Raj Department of Computer Science & Engineering, Sharda School of Engineering & Technology, Sharda University, Greater Noida, India

S. Ramesh Department of Electronics and Communication Engineering, Saveetha School of Engineering, Saveetha Institute of Medical and Technical Sciences, Saveetha University, Chennai, India

Poonam Rawat Law College Dehradun, Uttaranchal University Dehradun, Dehradun, India

Rituja Haldia Institute of Technology, Haldia, WB, India

Arindam Roy Institute of Engineering Management, Kolkata, WB, India

Atanu Roy The Neotia University (TNU), Sarisa, WB, India

Atrri Roy Department of Computer Application and Science, Institute of Engineering and Management, Kolkata, India

Reena Roy Law College Dehradun, Uttaranchal University Dehradun, Dehradun, India

Tanmay Sinha Roy Swami Vivekananda University, Kolkata, India

Raifa Runa Department of Computer Science and Engineering, East West University, Dhaka, Bangladesh

Manjima Saha Institute of Engineering and Management, Kolkata, India

Shubhamita Saha Institute of Engineering and Management, Kolkata, India

Tufan Saha Institute of Engineering Management, Kolkata, WB, India

Ashok Kumar Sahoo Graphic Era Hill University, Dehradun, Uttarakhand, India

Debabrata Samanta Rochester Institute of Technology, Prishtina, Europe, Kosovo

Sabyasachi Samanta Haldia Institute of Technology, Haldia, WB, India

Ayan Sar School of Computer Science, University of Petroleum and Energy Studies (UPES), Dehradun, Uttarakhand, India

Kaustav Sarkar Department of Computer Application and Science, Institute of Engineering and Management, Kolkata, India

Sagar Saxena Law College Dehradun, Uttaranchal University Dehradun, Dehradun, India

A. Selvakumar Department of Computer Science and Engineering, Saveetha School of Engineering, Saveetha Institute of Medical and Technical Sciences, Saveetha University, Chennai, India

Amit Seth Department of Computer Science and Engineering, Harcourt Butler Technical University, Kanpur, Uttar Pradesh, India

Shabrina Akter Shara Department of Computer Science and Engineering, East West University, Dhaka, Bangladesh

Hitesh Kumar Sharma School of Computer Science, University of Petroleum and Energy Studies (UPES) Energy Acres, Bidholi, Uttarakhand, India

Sangeet Sharma Law College Dehradun, Uttaranchal University Dehradun, Dehradun, India

Satyam Sharma Law College Dehradun, Uttaranchal University Dehradun, Dehradun, India

Vishal Shaw Department of Computer Application and Science, Institute of Engineering and Management, Kolkata, India

Shneha Haldia Institute of Technology, Haldia, WB, India

Himanshu Shukla Department of Computer Application, School of Engineering and Technology, CSJM University, Kanpur, Uttar Pradesh, India

Akhilesh Singh Department of Computer Application, School of Engineering and Technology, CSJM University, Kanpur, Uttar Pradesh, India

Anupam Singh Department of Computer Science and Engineering, Graphic Era Hill University, Dehradun, India

Malikhan Singh Computer Science and Engineering, College of Engineering & Technology, Mathura, India

Ramveer Singh Department of CSE, Galgotias College of Engineering & Technology, Greater Noida, India

Rishav Raj Singh Department of Computer Application and Science, Institute of Engineering and Management, Kolkata, India

Thipendra Pal Singh School of Computer Science Engineering and Technology, Bennett University, Greater Noida, UttarPradesh, India

Monjurul Islam Sumon Department of Computer Science and Engineering, East West University, Dhaka, Bangladesh

Tasnim Israk Synthia Department of Computer Science and Engineering, East West University, Dhaka, Bangladesh

Mahir Tagwar Department of Computer Science and Engineering, East West University, Dhaka, Bangladesh

Abu Talha Department of Computer Science and Engineering, East West University, Dhaka, Bangladesh

M. Tharun Kumar Reddy Department of Electronics and Communication, Saveetha School of Engineering, Saveetha Institute of Medical and Technical Sciences, Saveetha University, Chennai, India

Naveen Tiwari School of Computing, Graphic Era Hill University, Bhimtal Campus, Uttrakhand, India

Raj Gaurang Tiwari Chitkara University Institute of Engineering and Technology, Chitkara University, Punjab, India

Naresh Kumar Trivedi Chitkara University Institute of Engineering and Technology, Chitkara University, Punjab, India

Vaishnavi Department of Computer Science & Engineering, Sharda School of Engineering & Technology, Sharda University, Greater Noida, India

Sarvachan Verma Ajay Kumar Garg Engineering College, Ghaziabad, Uttar Pradesh, India

Shekhar Verma Department of Computer Application, School of Engineering and Technology, CSJM University, Kanpur, Uttar Pradesh, India

Richa Vijay Amity University, Uttar Pradesh, Noida, India

Sudhanshu Vishwakarma CSE Department, University of Petroleum and Energy Studies (UPES), Dehradun, Uttarakhand, India

Arun Kumar Yadav National Institute of Technology, Hamirpur, HP, India

Divakar Yadav IGNOU, New Delhi, India

Vikash Yadav Department of Technical Education, Government Polytechnic Bighapur Unnao, Manpur, Uttar Pradesh, India

Alba Ymerhalili Rochester Institute of Technology – RIT Global Campus, N.N., Germia Campus, Prishtina, Europe, Kosovo

Farid ishraq Zarif Department of Computer Science and Engineering, East West University, Dhaka, Bangladesh

Cyber Intelligence

An Innovative Method for DNA Cryptography-Based Digital Signature Verification

Animesh Kairi, Tapas Bhadra, Arindam Roy, and Tufan Saha

Abstract Online services have become increasingly important in our daily lives in recent years. Despite its advantages, the communication network has some security problems. The security components include data integrity, user authentication, and information confidentiality. To comply with all of these standards, the sensitive material must first be signed electronically by the sender and then authenticated by the intended recipient. Therefore, more research on digital signatures is needed to improve data security and transfer data validity. In this study, a secure online signature algorithm is developed. To create secure digital signatures, it is necessary to combine safe hash codes with ElGamal, DNA, and crypto techniques. Then, the crypto time complexity of our study methodology and the current PlayGamal method are compared. By utilizing the SHA algorithm, a safe hash code is produced, and the computational complexity of the encryption method is reduced by hybridization.

Keywords Security · Cryptography · Digital signature · Verification

1 Introduction

Internet access is a need for everyone's daily existence in the current climate. Everyone's life and every industry, whether it is to be education, business, marketing, or entertainment, require internet access. There is an increasing risk of data theft as these standards expand. Therefore, a secure communication route is required to guard against data theft and forgeries. Interception of data as it is being transmitted is one of the security concerns. Therefore, a digital signature [10] is necessary to confirm that the data is coming from an authentic user in order to avert these problems. Figure 1 shows various goals of cryptography.

A. Kairi (✉) · A. Roy · T. Saha
Institute of Engineering Management, Kolkata, WB, India
e-mail: ani.kairi@gmail.com

T. Bhadra
Aliah University, Kolkata, WB, India

S. Dutta et al. (eds.), *Cyber Intelligence and Information Retrieval*, Lecture Notes in Networks and Systems 1025, https://doi.org/10.1007/978-981-97-3594-5_1

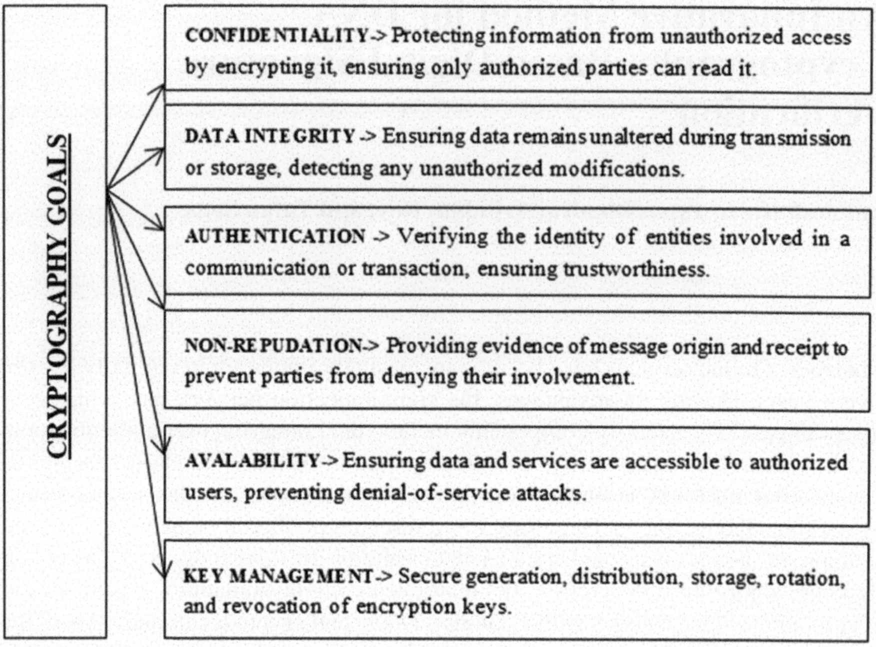

Fig. 1 Goals of cryptography

Verifying a digital signature is necessary to confirm that the correct information is coming from the verified sender and to identify users. Digital signatures are used for non-repudiation, authentication, and integrity verification [5].

The creation of a custom character that verifies the user and data validity is one way to get past the non-repudiation restriction. Data protection techniques have included the use of cryptography [3, 4]. One of the most secure network technologies, digital signature can be used to do this.

Public key algorithms and hashing algorithms are both used concurrently in the execution of digital signatures [6]. A digest of the message of data is created using hashing methods, and it is subsequently utilized to check the precision of the information stored on the server located in the cloud. On the other hand, data that is encoded for transmission via the internet is created using a public key encryption approach [7].

In the first stage, a hash value is produced from the message that is known as the signature. Any asymmetric key technique is then used to encode this hash value before it is sent to the receiver.

In the digital signature mechanism (Fig. 2), the following steps are occurred:

- The message to be signed is input into a hash function, which generates a fixed-size hash value unique to the message.
- The digital signature is created by encrypting the hash value using the private key, sometimes referred to as the signing key.

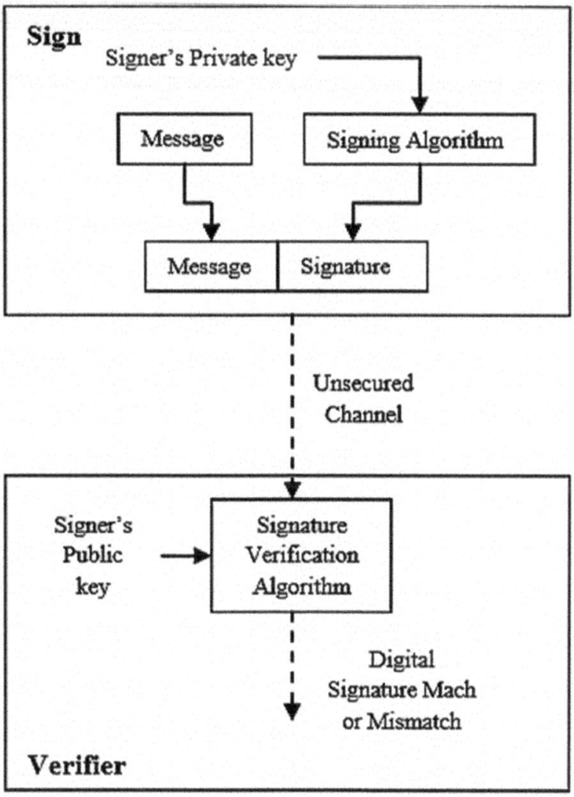

Fig. 2 Mechanism of digital signature

- The digital signature, along with the message, is transmitted or stored securely.
- To verify the digital signature, the recipient or verifier uses the corresponding public key, also known as the verification key.
- The message is again input into the same hash function used during the signing process to generate a hash value.
- The verifier decrypts the digital signature using the public key, which produces a decrypted hash value.
- The decrypted hash value is compared to the newly computed hash value of the received message.
- If the two hash values match, the digital signature is considered valid, indicating that the message has not been tampered with and was indeed signed by the private key holder.

The specific cryptographic algorithms, key management procedures, and additional security measures may vary depending on the chosen digital signature scheme [9]. The receiver checks the authenticity of the signature after receiving the message with the hash value in the send step. If a match is made, it demonstrates the veracity of the sender and, consequently, the message or information. The communication is denied if the signature cannot be validated.

The following categories are used to categorize electronic signature schemes:

A. Digital Signature Using Direct Method.

This mechanism is used directly between the information's source and receiver. The following actions are taken in this method:

i. The sender transmits the publicly available key to the recipient.
ii. The person who sent it then uses their own private key to generate a hash or di- gest, encrypts it using the private key included in the message, and sends it through the communication channel [14].
iii. After acquiring these facts, the receiver verifies them by generating hash from the contents of the message and contrasting them with the hash value received. If there is a match, the user is legitimate [15].

Private key security is the primary area of concern in this form of signature [16].

E. Arbitrated Digital Signature.

On behalf of the recipient, data's digital signature is checked by a third-party auditor (TPA). The digitally signed communication is sent by the sender to the TPA for further validation [17]. When a sender sends knowledge, TPA verifies the information's electronic signature on the receiver's account and provides the confirmation report to the recipient to ensure that the data is authentic and that all security checks have been successfully completed. Validation, authenticity verifying, and sender authentication are carried out following transmission from the sender to the recipient while upholding the idea and requirement of the electronic signature [18].

A digital signature is created for integrity verification, which is an electronic digest that verifies the authenticity of the entire document. This prohibits the processing and encryption of the full document. Digital signature systems are susceptible to various attack types, including:

iv. Attacks where the perpetrator only has knowledge of the public key are one of the numerous types of attacks that digital signature systems are susceptible to.
v. A cyber-attack where the perpetrator has access to several message signatures.
vi. It allows the attacker to view the signatures of any communication of his choosing.

Digitally signed papers are susceptible to additional attacks in addition to the ones outlined above, including existential, selective, and universal counterfeit assaults. Even though there are numerous common e-signature plans, none of the above assaults can effectively defend against them. In reality, additional elements, such as the key's dimensions, the computing strategy, the hashing algorithm, influence the effectiveness of these digital signature methods. The efficacy and usability of many electronic mechanisms are being improved constantly through the development of electronic signature procedures. They will eventually combine with elliptical curved cryptographic methods to create ECDSA from DSA, also known as EC-ElGamal at ElGamal.

2 Related Work

Can et al.'s [13] study examines a novel cone-like digital signature scheme that uses two private keys to make it more difficult to find the signature keys.

Jian-zhi et al. [12] developed a digital signature strategy using the DSA method and the hyper-elliptical curve technique, which results in an elevated level of security and validates the authenticity of the data.

Xuan et al. [11] conducted a comparative investigation of digital signatures generated by the RSA, ECDSA, and DSA methods and found that ECDSA outperforms the others.

Geri et al. [8] developed a digital signature for XML transactions. This study is solely concerned with XML data storage.

Picture data was given a digital signature, and Alam et al. [4] further encrypted picture digital signatures utilizing the RSA method.

Using the ECDSA method, Sergei et al. [3] developed a digital signature method for the marine sector.

Kadek Dwi Budi Utama et al. [2] developed a digital signature algorithm by combining the MAC address with AES-128. The machine address (MAC address) and the algorithm known as AES features are combined to provide an encryption technique that is used after the hash value has been computed using SHA-256.

Using components of the Playfair cipher and the ElGamal algorithm, Abdul Ghofar et al.'s PlayGamal digital signature method was first presented in 2017 [1].

In 2017, Smith et al. [19] introduced DNA sequence encoding of digital signatures and hybrid encryption combining DNA cryptography and traditional cryptographic techniques.

In 2018, Brown et al. [20] introduced DNA encoding and decoding of digital signatures and polymerase chain reaction (PCR) for DNA amplification.

Zhang et al. [21] introduced general DNA sequence encoding of digital signatures and DNA-based encryption and decryption.

Patel et al. [22] introduced hybrid encryption combining DNA cryptography and traditional cryptographic techniques and polymerase chain reaction (PCR) for DNA amplification.

Lee et al. [23] introduced DNA encoding of digital signatures, DNA-based encryption and decryption, security analysis, and performance evaluation.

3 Research Work

The secured digital signature used in this study is formed in three phases as follows:

- Message digest generation using the hashing algorithm.
- The (Decimal Bond DNA cryptography (DBD)Gamal technique is used to encrypt the resulting message digest and submit it as a digital signature.

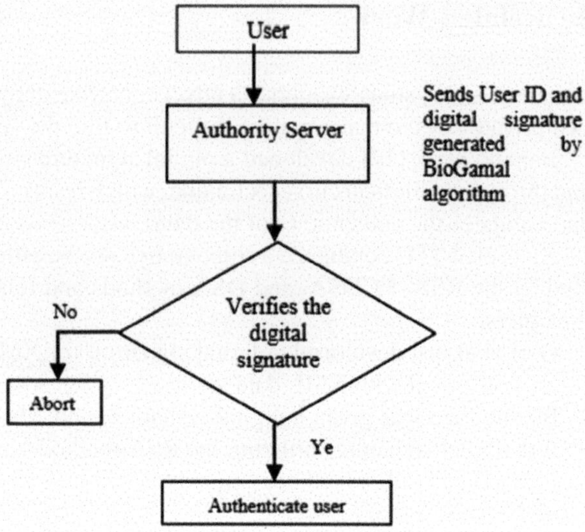

Fig. 3 Research techniques

- The DBDGamal algorithm is used to further decrypt the received message to verify the e-signature.

The communication is from the authorized sender if the material of the communication digests generated at the end of reception, and those obtained there are identical (Fig. 3).

Process Flow of Research Techniques

A digital signature method called DBDGamal is created in this research effort and is used both at the transmitter end and at the receiver end of the communication.

This process includes employing the SHA hashing algorithm to build a message digest and the DBDGamal algorithm to encrypt it. ElGamal and the DNA encryption algorithm are both used in the encryption process. To send a message with a digital signature, the two algorithms' processes are combined. The method of decrypting a digital signature uses the DNA and ElGamal decryption algorithm, as shown in Fig. 4's encryption flow procedure [25].

A. DBDGamal Algorithm

The DNA encryption/decryption method and the ElGamal encryption/decryption method are combined to create the DBDGamal algorithm. The hash code of the data file on the first level is encrypted using DNA cryptography. The binary digital coding used in information science is encoded using two states of 0 or 1 and a combination of 0 and 1. DNA is coded using a biological sequence comprising the bases adenine (A), thymine (T), cytosine (C), and guanine (G). Table 1 shows how to convert the digital DNA bases into binary numbers [24]. Table 1 shows a binary table, Table 2 shows the DNA encoding library, and Fig. 3 shows the ciphertext of the Decimal Bond DNA (DBD) cryptography method [18, 24] (Fig. 5).

B. ElGamal Algorithm

Three steps make up the ElGamal algorithm:

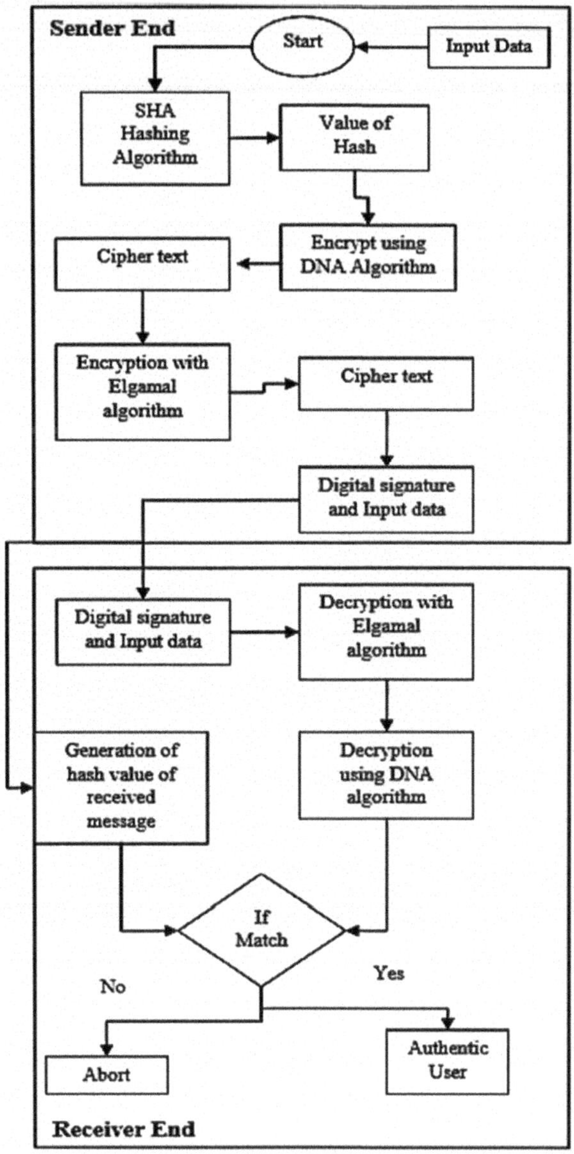

Fig. 4 Research methods flowchart

Table 1 Binary table

Binary value	DNA coding
00	A
01	T
10	G
11	C

Table 2 DNA encoding library

AATC	ACGC	TTTG	ACAG	ACCG
ATCA	GCCT	CTCA	GTAA	ATTG
AGGT	TTAA	GTAT	ACCT	ACGG
CCAG	CCAC	GCCG	GCGC	TCTA
TTAC	TGCC	AAAT	CCGT	GGAG
CCAA	ACAC	TAAC	AGGC	CCGG
GTCA	TGGT	CAGA	ACTC	CACG
TTAT	CGCC	ACTT	CTAT	CTTA
TACA	AGAG	CGGA	GAAC	GCAG
ACAT	CTGT	ACTG	TAAA	TCAA
CGCT	CCTT	TGTT	GTCC	GACC
TCCA	GTAG	GTAC	GCGG	TGAG
TCTG	GGTG	GACA	CAAA	CGAA
GCTG	TGGG	TTGG	CGTC	TCAG
GTTG	AACG	TAGC	AAAC	TTCG
GGTA	AACC	GAAT	TCGC	GGGT
AGTC	TGTG	AGGA	GTGG	TTCA
GTCG	AGCA	TTTT	TGCG	CAAG
GTGA	CATA	CGGG	TTGT	CAGG

```
21230301201233012123030121230123030122301301212301
230120123123023013012230112302301123030121230123023013012301201231
2123030121230123023013012301201231
2012301232301123001232301123012301230301212301230012330123012123023010
12301230123230130123012012330122301123023011
2301201230123230112300123012323011230301212301230230101232301123012303
0121230123012303012301212301
2301201231230230130122301301201231230230112301230123001230123230130122
301012312301 21230230123011230301230120123301201233012301212301
2123023011230123023010123012301230123301230121230301223013012012301232
3011230123023013012301201231230230101231230230101232301123012300123012
3230112302301012312301230301212301230012323011230123001230123012323011
21230230101231230301223011230230123010123012301231
2012312302301012323010123012301231230301223012301123030121230123001233
```

Fig. 5 Ciphertext

i. Formation of Key
ii. Encryption
iii. Decryption.

It is a type of block cryptography that divides the initial message into fragments to allow for further encryption of the data [6]. The original data blocks are processed by the encryption algorithm, which then transforms them into cipher data blocks. The cipher data block is then transferred to the receiving end.

The following is how the ElGamal method is used:

i. Create a primitive group Zp* and a prime number p.
ii. Make a private key and a public key. Create a second primitive component and an element that is free with the values {00, 01, 02,…, p-2} to accomplish it.

iii Three sets of numbers, as follows, make up the public key:

$$\beta = \alpha^a \bmod p$$

where a is the secret key, whose value is unknown. While the sender is uninformed of the confidential key, the receiver participates in the creation of both the public and private keys. Only the recipient's public key for data encryption is known to the sender.

- Encryption method ElGamal employs a public key and a secret integer k (k = 0, 1…, p − 2).
- A distinct k number is used to encrypt every word in the message. Coding that has the shape of a unit made up of the two values (r,t) will be generated from a numerical value representing one character in ASCII.
- A message letter is selected, converted to ASCII code, and encrypted as follows:

  Use the equation to calculate the values of r and t.

$$r = \alpha(\bmod p)$$

$$t = \beta^k(\bmod p)$$

- The ciphertext obtained is (r,t).
- The process is repeated for every message segment in the data.
- Using the public keys (p, α, β), along with the private key (a), the procedure for decryption is carried out. The receiving ciphertext (r,t) is converted into plaintext as follows:

$$M = \left(r^a\right)^{-1} \bmod p$$

M is plaintext with the value r^a $(r^a)^{-1} = r^{p-1-a}$ mod p.

4 Analysis of Result

The proposed DBDGamal and PlayGamal are primarily compared about of performance time required for execution of the decryption and encryption procedures. The PlayGamal algorithm [1] combines characteristics of the Playfair cipher and the

ElGamal algorithms to provide a secure digital signature. Due to the Playfair algorithm's characteristics and the fact that a 6*6 Playfair matrix was used in its creation, the time complexity of the current work rose. DBDGamal is therefore designed to lessen the playgame's time complexity.

Standalone computer (Core i7, 8GB RAM). In terms of encryption and decryption times, the PlayGamal approach and the DBDGamal method are compared and evaluated with numerical, alphanumeric, and alphabetic characters.

Encryption time and decryption time are terms used to describe the entire amount of time required to encrypt and decode data files, respectively.

The calculation of the encryption time for various types of communications is shown in Table 3 and Fig. 6. Based on the findings, it has been determined that the DBDGamal algorithm performs better than the PlayGamal algorithm. PlayGamal's encryption process takes roughly 89% longer than DBDGamal's. Calculations of the decryption times for various types of communications are shown in Table 3 and Fig. 7. Based on the findings, it has been determined that the DBDGamal algorithm performs better than the PlayGamal algorithm. PlayGamal's decryption time is roughly 75% longer than DBDGamal's.

Table 3 Crypto time analysis

Message	PlayGamal [1] encryption time	DBDGamal encryption time	PlayGamal [1] decryption time	DBDGamal decryption time
Numeric (0, 1, 2, 3……)	9,671,490	1,443,711	3,445,705	622,385
Alphanumeric (a1, b2, …)	9,483,318	1,419,259	2,237,733	685,378
Alphabetic (a, A, b, B, …)	20,056,212	2,103,361	4,491,046	1,457,664

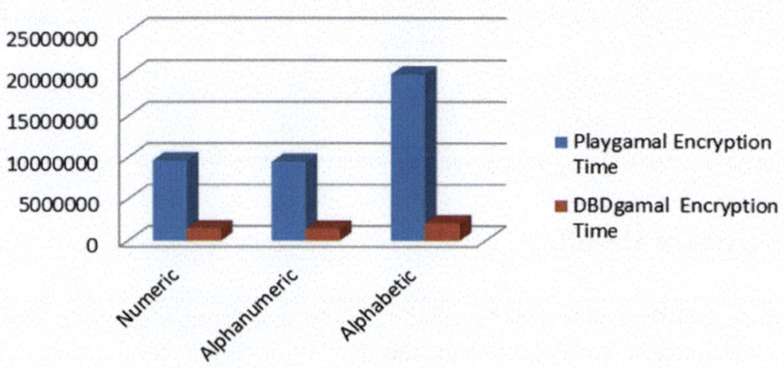

Fig. 6 DBDGamal versus PlayGamal encryption time

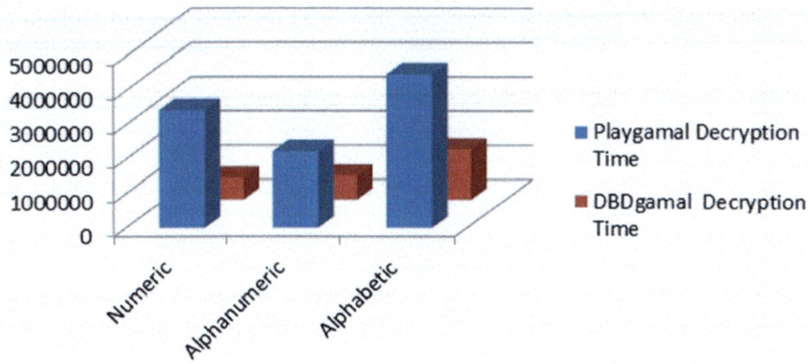

Fig. 7 DBDGamal versus PlayGamal decryption time

5 Conclusion

Data integrity checks and user authentication are both done using digital signatures. The creation of a secure digital signature technique for data files is the aim of this project. For this, the DNA and ElGamal methods are combined to generate the DBDGamal algorithm. ElGamal, a type of asymmetric algorithm, encrypts and decrypts data using a variety of key pairs, whereas the DNA algorithm creates ciphertext by employing biological sequencing. ElGamal and the DNA encryption algorithm are both used in the encryption process. The system's security is improved, and the time complexity is decreased, thanks to the characteristics of these two algorithms. The decryption procedure of the DBDGamal algorithm is used at the receiver end to verify the digital signature that was received. It is between 35 and 45% more efficient than the PlayGamal algorithm in terms of crypto time.

References

1. Ghofar A, Hardi M, Firdaus MN, Shidik GF (2017) Digital signature based on playgamal algorithm. In: International seminar on application for technology of information and communication (iSemantic). IEEE, Oct 2017, pp 58–65
2. Utama KDB, Rizqia Al-Ghazali QM, Irfan Bayu Mahendra L, Fajar Shidik G (2017) Digital signature using MAC address based AES128 and SHA-2 256-bit. In: International seminar on application for technology of information and communication (iSemantic). IEEE, Oct 2017, pp 72–78
3. Chernyi SG, Ali AA, Veselkov VV, Titov IL, Budnik VY (2017) Security of electronic digital signature in maritime industry. In: IEEE conference of Russian young researchers in electrical and electronic engineering (EIConRus), Feb 2017, pp 29–32
4. Alam S, Jamil A, Saldhi A, Ahmad M (2015) Digital image authentication and encryption using digital signature. In: International conference on advances in computer engi neering and applications (ICACEA), Mar 2015, pp 332–336

5. Jaju SA, Chowhan SS (2015) A modified RSA algorithm to enhance security for digital signature. In: International conference and workshop on computing and communication (IEMCON), Oct 2015, pp 1–5
6. Kioon MCA, Wang Z, Das SD (2013) Security analysis of MD5 algorithm in password storage. In: International symposium on computer, communication, control and automation, pp 0706–0709
7. Kuppuswamy P, Mohammad Appa P, Al-Khalidi SQY (2012) A new efficient digital signature scheme algorithm based on block cipher. IOSR J Comput Eng (IOSRJCE) 7(1), 47–52
8. Gerić S, Vidačić T (2012) XML digital signature and its role in information sys- tem security. MIPRO 2012, 21–25 May 2012, Opatija, Croatia
9. Nguyen MH, Hoi DN, Luu DH, Moldovyan AA, Moldovyan NA (2011) On functionality extension of the digital signature standards. In: International Conference on Advanced Technologies for Communications (ATC 2011)
10. Zhang Q, Li Z, Song C (2011) The Improvement of digital signature algorithm Based on elliptic curve cryptography. In: International conference on artificial intelligence, management science and electronic commerce (AIMSEC). IEEE, Aug 2011, pp 1689–1691
11. Xuan Z, Du Z, Chen R (2009) Comparison research on digital signature algorithms in mobile web services. Nat Nat Sci Found China IEEE
12. Jian-zhi D, Xiao-hui C, Qiong G (2009) Design of hyper elliptic curve digital signature. In: International conference on information technology and computer science, July 2009, pp 45–47
13. Can X, Lin Y (2009) A new conic curve digital signature scheme. In: Information assurance and security, Aug 2009, pp 623–626
14. Hai-peng C, Xuan-jing S, Wei W (2009) Digital signature algorithm based on hash round function and self-certified public key system. In: International workshop on educa tion technology and computer science, May 2009, pp 618–624
15. Romney GW (2007) Digital signature signing engine to protect the integrity of digital as- sets. IEEE, Apr 2007
16. Jarusombat S, Kittitornkun S (2006) Digital signature on mobile devices based on location. IEEE
17. Campbell S (2003) Supporting digital signatures in mobile environments. In: IEEE interna tional workshops on enabling technologies: infrastructure for collaborative enterprises (WETICE'03). IEEE
18. Harn L (1994) Group-oriented (t, n) threshold digital signature scheme and digital multi signature. IEEE Proc-Comput Digit Tech 141(5)
19. Smith J, Johnson R, Anderson L (2017) DNA cryptography for digital signature verification. Int J Adv Res Comput Sci Softw Eng
20. Brown A, Wilson B, Thompson C (2018) DNA-based digital signatures: a review of techniques and applications. In: IEEE international conference on computer and information technology (2018)
21. Zhang L, Li S, Chen X (2019) DNA-Based digital signature scheme for secure data communication. IEEE Trans Inf Foren Secur
22. Patel M, Sharma R, Gupta S (2020) Enhancing digital signature verification using DNA cryptography. Int J Sci Res Comput Sci Eng
23. Lee K, Park S, Kim J (2021) DNA-Based authentication scheme for digital signatures. ACM Trans Inf Syst Secur
24. Kairi A, Bhadra T, Decoding the future using a novel DNA-based cryptosystem. In: European chemical bulletin. https://doi.org/10.48047/ecb/2023.12.si10.004132023.30/08/2023
25. Kairi A, Bhadra T, Roy A (2023) Secured healthcare ecosystem using DNA cryptography in 5G network. J China Petroleum Proc Petrochem Technol

A Novel Framework for Facial Emotion Detection Using Deep Learning Algorithms for HCI-Enabled System

Hitesh Kumar Sharma, Ayan Sar, Tanupriya Choudhury, Ketan Kotecha, Soumi Dutta, and Abhishek Bhattacharya

Abstract Deep learning-based facial expression recognition has become increasingly popular in the creation of systems that support Human–Computer Interaction (HCI). The use of techniques under deep learning for facial emotion recognition in HCI applications is briefly discussed in this work. It emphasizes the importance of accurately detecting and comprehending human emotions in HCI systems points out the drawbacks of conventional approaches and investigates the effective usage of deep learning models such as Convolutional Neural Networks (CNNs) for this purpose. To improve user engagement and experience the evaluation also covers the integration of face emotion recognition with HCI technologies. It also discussed the difficulties and potential and effective applications of deep learning in the field of facial emotion recognition in HCI-enabled systems. For researchers and practitioners who are looking to use the concept of deep learning methods to achieve facial emotion recognition in the context of HCI applications this succinct review is an invaluable resource.

Keywords Facial recognition · Deep learning · CNN neural network · Human computer interaction (HCI)

H. K. Sharma · A. Sar
School of Computer Science, University of Petroleum and Energy Studies (UPES) Energy Acres, Bidholi, Uttarakhand, India

T. Choudhury (✉)
Research Professor, CSE Department, Graphic Era (Deemed to be University), Dehradun, India
e-mail: tanupriyachoudhury.cse@geu.ac.in; tanupriya.choudhury@sitpune.edu.in

Adjunct Professor, CSE Department, Symbiosis Institute of Technology, Symbiosis International (Deemed University), Pune, India

K. Kotecha
Symbiosis Center for Applied Artificial Intelligence, Symbiosis Institute of Technology, Symbiosis International (Deemed University), Pune 412115, India
e-mail: Director@sitpune.edu.in

S. Dutta · A. Bhattacharya
Sister Nivedita University, Chakpachuria, India

S. Dutta et al. (eds.), *Cyber Intelligence and Information Retrieval*, Lecture Notes in Networks and Systems 1025, https://doi.org/10.1007/978-981-97-3594-5_2

1 Introduction

The process of mechanically identifying and deciphering the emotions displayed on human faces known as facial emotion detection has received a lot of interest lately. Human–computer interaction (HCI) applications in particular heavily rely on the capacity to comprehend human emotions. To improve user experience and engagement HCI seeks to design systems and technologies that enable seamless and natural interaction between humans and computers.

The constructed features and classifiers used in traditional methods for detecting facial emotions frequently have a limited ability to capture the subtle and complex variations in facial expressions. However, the science of face emotion recognition has undergone a paradigm shift with the introduction of deep learning techniques particularly Convolutional Neural Networks (CNNs). Using deep learning techniques, it is now possible to automatically learn expressive and discriminative representations from unprocessed facial data. Deep learning algorithms can efficiently capture complex patterns and subtleties in facial expressions by utilizing large-scale datasets and potent computer resources enabling more precise and reliable emotion identification. The combination of deep learning-based facial emotion identification and HCI-enabled technology brings up a plethora of opportunities for improving user experiences in many scenarios.

In affective computing deep learning-based facial emotion detection also enables the automatic analysis of massive amounts of emotional data providing insights and opening up new opportunities for applications in fields like sentiment analysis market research and healthcare. An in-depth analysis of face expression recognition using deep learning techniques for HCI-enabled systems is the goal of this article. It will explore typical deep learning-based emotion detection approaches frameworks and datasets. The paper will also go over the difficulties and potential directions in this area highlighting the significance of ethical issues interpretability and user-centric design principles.

HCI-enabled devices have the power to revolutionize how people interact with computers and technology through the effective and ethical utilization of the process of deep learning for the achievement of facial emotion recognition. In-depth discussions of specific methods innovations and developments in deep learning-based facial emotion recognition will be covered in the next sections of this article with an emphasis on how they may affect HCI-enabled systems and what that can mean for improving user experiences.

2 Literature Review

Recent years have seen substantial breakthroughs in facial expression detection for HCI-enabled systems employing deep learning techniques. The existing literature in this field is thoroughly reviewed in this part with emphasis on methodology architectures datasets problems and future perspectives. To detect facial emotions in HCI-enabled devices deep learning models particularly Convolutional Neural Networks (CNNs) have been widely used [1]. To accurately recognize emotions, CNNs have exhibited exceptional success in the automatic learning of certain discriminative features from facial photos and expressions. For this job, a variety of CNN architectures including VGGNet, ResNet, and InceptionNet have been used producing cutting-edge results on benchmark datasets. CNNs have been combined with models like Long Short-Term Memory (LSTM) and Gated Recurrent Units (GRUs) [2] to efficiently handle spatiotemporal facial data and enhance emotion recognition performance. Several benchmark datasets have been extensively utilized to train and assess deep-learning models for facial emotion detection. AffectNet the Extended Cohn-Kanade (CK+) dataset and the Emotion Recognition in the Wild (EmotiW) [3] dataset are further noteworthy datasets. These datasets include a wide variety of facial expressions and are useful training tools. As an illustration auditory qualities like speech intonation and prosody might reveal more details about the user's emotional state [4]. Additional helpful indications for emotion recognition can come from text analysis of supplementary captions or social media posts. Researchers have also looked into including generative models like generative adversarial networks (GANs) [5] for the identification of face expressions. GANs make it possible to create artificial facial expressions enhancing the training set and enhancing the generalization skills of deep learning models. Khajontantichaikun et al. in [6] proposed a method for emotion detection in Thai elderly facial expressions using a hybrid approach using a combination of object detection techniques. The requirement for extensive diversified datasets that accurately represent multiple demographic groupings cultural contexts and environmental factors is a big challenge [7]. Giri et al. [8] introduced a facial emotion detection system leveraging the potential of CNN and OpenCV for facial feature recognition. Azizi et al. in [9] focused on emotion detection through facial expression images employing CNN as the underlying technology. In conclusion, facial emotion detection using deep learning approaches for HCI-enabled systems has witnessed significant advancements. The integration of deep learning models' multimodal data and generative models has propelled the accuracy and robustness of facial emotion recognition.

3 Methodology

There are various crucial phases and elements involved in facial emotion recognition for HCI-enabled systems employing deep learning algorithms. This section describes a typical development and implementation process for a deep learning-based HCI facial emotion recognition system as shown in Fig. 1.

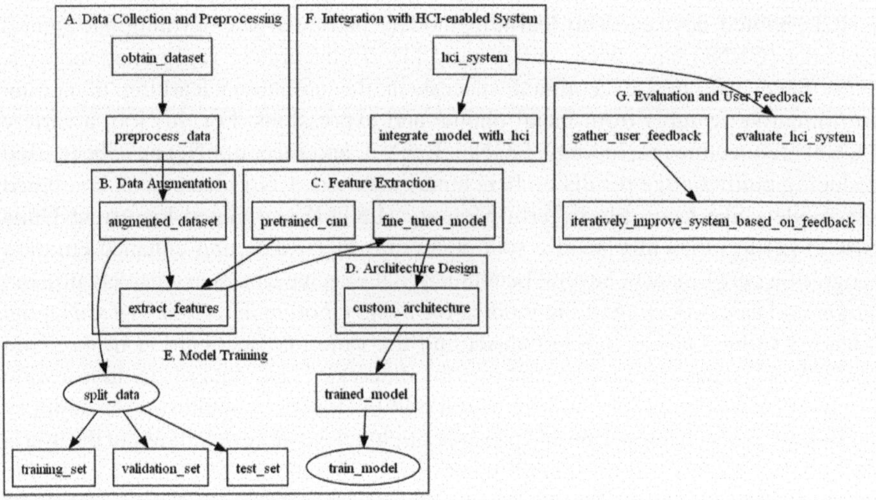

Fig. 1 Methodology flowchart

A. **Data Collection and Preprocessing**

- Obtain a relevant dataset that includes facial photos that have been labeled with the appropriate emotions. FERC CK+ AffectNet and custom datasets gathered especially for the intended HCI application are examples of popular datasets.
- To ensure consistent input dimensions and quality preprocess the facial photos by performing operations like resizing normalization and alignment. To precisely extract facial regions, use face detection and facial landmark localization methods.

B. **Data Augmentation**

- Boost the dataset's size and diversity by adding new data. To create more training samples, use tricks like random translations flip rotations, and lighting changes.
- Create artificial face expressions with the use of generative models like GANs to expand the dataset and solve the problem of insufficient annotated data.

C. **Feature Extraction**

- Utilize a deep learning model typically a pre-trained CNN to extract high-level features from the pre-processed facial images.
- Fine-tune the pre-trained CNN on the emotion recognition task by training the last few layers or specific modules of the network while keeping the initial layers frozen.

- Alternatively, it can be considered a good option to adopt the idea of implementing a pre-trained model in this context and then training it on a large-scale dataset (e.g. ImageNet). Alternatively, fine-tuning the model on the emotion recognition dataset to achieve the expected results.

D. **Architecture Design**

- Create an appropriate deep learning architecture for the particular HCI-enabled system. This could be done by the integration of CNNs with RNNs for the capturing of temporal dependencies or the addition of special attention algorithms to increase the concentration on distinguishable facial regions.
- Improving the functionality of the emotion detection system and experimentation with other various architectural designs which included multiple network depths layer configurations and other activation functions. The methodological flowchart is shown in Fig. 1.

E. **Model Training**

- Create training validation and testing sets from the dataset. Utilize the training set for backpropagation and gradient descent-based optimization of the model.
- Regularise the model by using strategies like drop-out batch normalization or L1/ L2 regularisation to avoid overfitting.
- It is also required to keep a constant check on the performance of the model on the validation set so that necessary arrangements can be made as hyperparameter tuning (such as learning rate and batch size).
- Use the relevant evaluation metrics such as accuracy precision recall and F1 score to assess the model's performance on the testing set.

F. **Integration with HCI-Enabled System**

- Integrate the HCI-enabled system architecture with the learned face emotion recognition model. This can entail creating suitable APIs and interfaces for easy integration.
- Implement face data processing in real-time allowing the system to record and interpret facial expressions as users interact with it.
- Integrate the HCI system's response or behaviour with the identified emotions for example by updating the expressions of virtual characters changing the dynamics of the game or producing the proper feedback.

G. **Evaluation and User Feedback**

- Completely assess the HCI-enabled system taking into account aspects like precision real-time performance user satisfaction and user experience.
- Gather user feedback through qualitative interviews questionnaires or user studies to evaluate the system's efficacy usability and perceived advantages.

- Iteratively improve the system based on user feedback fixing any detected flaws or restrictions.

It is significant to highlight that depending on the needs and restrictions of the HCI-enabled system as well as the resources and data accessible the specific methodology may change. However, this generic methodology offers a framework for creating a system for detecting facial emotions in an HCI setting utilizing deep learning techniques.

4 Dataset Description

For this research project, the dataset that was selected for further processing and training of the model was the Facial Images Dataset taken from Kaggle. It consists of 3950 pictures with various expressions on their faces. The collection primarily included five categories of facial expressions. The plant village dataset that was used for this study is readily accessible.

Table 1 contains the description of the images from the dataset with each number of images present in each class.

In Fig. 2 it is shown that some randomly taken images from the dataset.

4.1 Dataset Standardization

The input photos were standardized using an image generator. It will be used to modify the photos so that the standard deviation becomes 1 and the mean pixel intensity is 0. The formula below will be used to create new values to replace the existing pixel values in an image. Each value in this formula will be subtracted from the mean and the result will be divided by the standard deviation [7].

Table 1 Description of the facial dataset [https://www.kaggle.com/datasets/apollo2506/facial-recognition-dataset]

Emotion class	No. of images
Anger	541
Sadness	891
Happiness	771
Neutral	761
Surprised	991

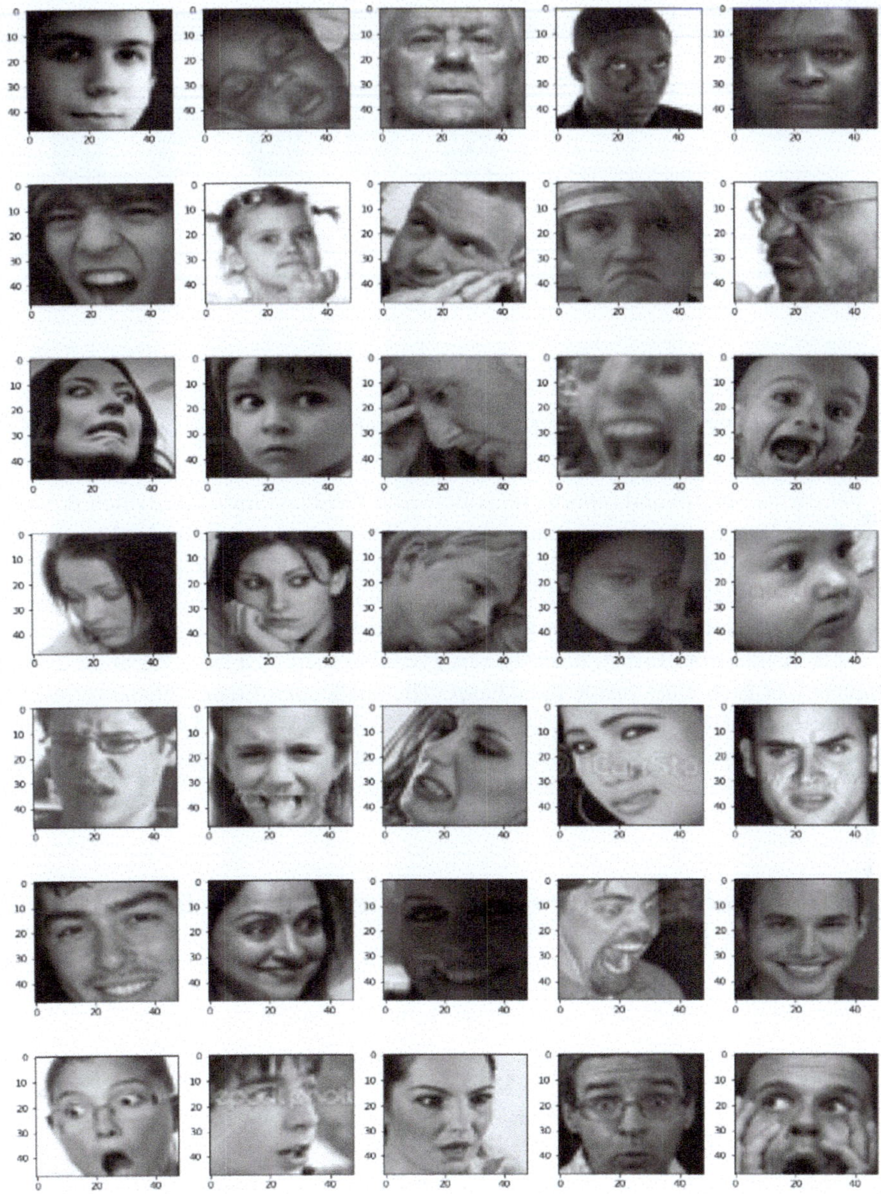

Fig. 2 Dataset sample images

Fig. 3 Class frequency plot

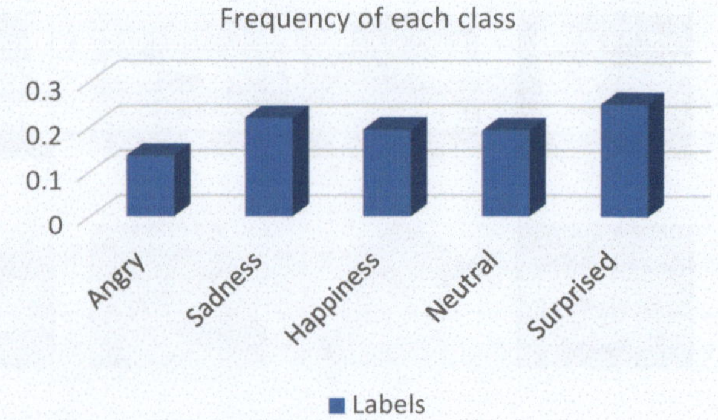

4.2 Imbalance of Classes in the Dataset

As per earlier studies and reports, it is often seen that class imbalance often proved as a threat later, especially in image datasets. This is very common in the case of image datasets, where the images in all the classes are not similar. In these cases, a bias situation occurs for the class which has more images [7].

The original number of images which are present in the dataset is plotted in Fig. 3.

As is visible from Fig. 3, it can be observed that there is a large variation in the frequency of each class inside the dataset. The challenge of data imbalance inside the dataset created a full biasing situation which in turn reduced the performance of the model. Now, to overcome this challenge of data imbalance, the following mathematical formulations and equations below were applied.

4.3 Elimination of the Problem of Class Imbalance Using the Loss Function

The equation for the calculation of the cross-entropy formula is given in Eq. (1).

$$L_{\text{cross entropy}}(X_i) = -(Y_i Log(f(X_i)) + (1 - Y_i)Log(1 - f(X_i)))$$

Cross-entropy loss formula (1)

The output of the model is represented as $f(X_i)$ where X_i and Y_i stand for the input feature and its matching label which is the chance that the outcome will be favorable.

The following equation is the written formula for the total average cross-entropy loss for the entire dataset D of size N as in Eq. (2).

$$L_{\text{cross entropy}}(D) = -\frac{1}{N}\left(\sum_{\text{positive}} Log(f(X_i)) + \sum_{\text{Negative}} Log(1 - f(X_i))\right)$$

Total average cross-entropy loss for the entire dataset (2)

This formulation makes it abundantly evident that if there is a significant amount of imbalance in the dataset and a low frequency of positive labels the loss will be predominated by negative labels in Fig. 4. The formula for the calculation is shown in Eqs. (3) and (4).

$$\text{Freq}_{\text{positive}} = \frac{\text{No. of Positive Samples}}{N}$$

Frequency calculation of positive labels (3)

$$\text{Freq}_{\text{negative}} = \frac{\text{No. of Negative Samples}}{N}$$

Frequency calculation of negative labels (4)

It is clearly visible from Fig. 4 that the positive labels contributed less as compared to the negative ones. However, this contribution ought to be equal to get reliable findings. The only approach to ensure that each class contributes equally to the classification model is to multiply each class frequency by a class-specific weight factor Wpositive and Wnegative.

So, the formulation will be represented as in Eq. (5).

$$W_{positive} \times Freq_{positive} = W_{negative} \times Freq_{negative}$$

Formula for calculation of the contribution of each class to the classification model (5)

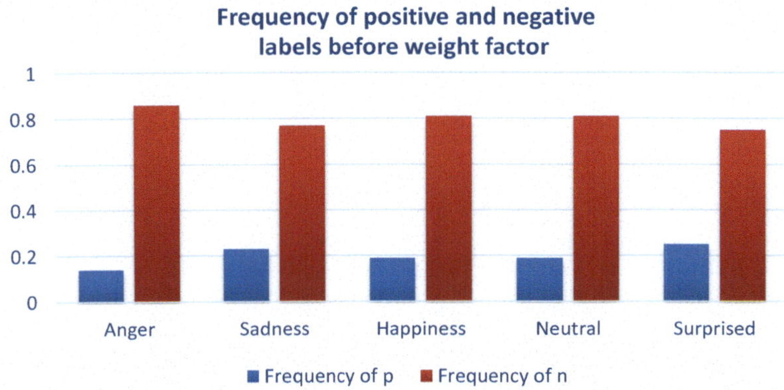

Fig. 4 Frequency of positive and negative labels of 5 classes in the dataset before weight factor

Which we can do simply by taking

$$W_{\text{positive}} = \text{Freq}_{\text{negative}}$$

$$W_{\text{negative}} = \text{Freq}_{\text{positive}}$$

The problem of class imbalance will be addressed using the aforementioned approach. We produced a frequency graph to test the concept and the expected chart is displayed. Figure 5 portrays the balancing of both frequencies as shown in Table 2.

As portrayed in Fig. 5, the correct application of the weight factor function on the positive and negative labels resulted in equal distribution to the loss function. The equation to loss function is given below in Eq. 6.

$$L^{w}_{\text{cross entropy}}(X) = -(W_{\text{positive}} Y Log(f(X)) + W_{\text{negative}}(1 - Y) Log(1 - f(X))) \quad \text{Loss function calculation} \tag{6}$$

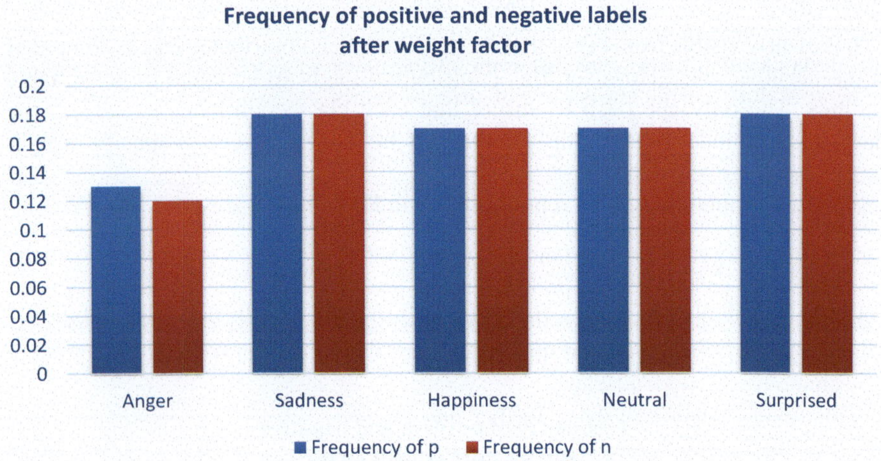

Fig. 5 Frequency of positive and negative labels of 5 classes in the dataset after application of weight factor

Table 2 Computational table for data imbalance

Class labels	Total images	wp	wn	freq_p	freq_n
Anger	541	0.87	0.15	0.13	0.13
Sadness	891	0.78	0.24	0.18	0.18
Happiness	771	0.82	0.18	0.17	0.17
Neutral	761	0.82	0.18	0.17	0.17
Surprised	991	0.76	0.24	0.18	0.18

This total fusion and adjustment of the loss function eliminated the challenge of class misbalancing and bias situations in the dataset.

5 Methodology for Model Creation

The framework proposed in this research to achieve accurate emotion recognition was executed by the creation of a complex convolutional neural network which was proved as a productive model for effective image characterization. CNN models were prepared by the dataset preprocessing, removing the challenge of class imbalance and removing the bias situation. The next step consisted of the creation of a variety of layers which included convolutional ReLU pooling and fully associated layers that were applied to the dataset. These images were captured in groups. The suggested framework specified a group size of 64 with 100 epochs built into the model. Small highlights are first concentrated by the model, but as the readiness developed more distinct highlights will be separated. One of the key advantages of CNN is that the majority of the preprocessing is done in this way. Despite that images for the information were resized. Incorporating additional operations like turn shear and other operations expansion increases the dataset's size. The model discovers highlights and designs during the preparatory interaction and learns them.

The following layer setup is displayed in Table 3.

6 Experimental Results

The proposed CNN model prepared for accurate emotion detection exhibited solid and impressive performance based on various evaluation metrics. During the training of the model, it achieved an accuracy of nearly 95% in Fig. 6, which reflected its ability to correctly classify the emotions in the training dataset. The validation accuracy, which is also a crucial measure to check the robustness of unseen data, stood at nearly 92% in Fig. 6. The training and validation losses of 0.18 and 0.32 respectively in Fig. 7, suggested accurate convergence with minimum overfitting possible. The F1-Score, Precision, and Recall metrics further highlighted the robustness and proficiency of the model, with near values of 91.8% for precision in Fig. 8, 91.9% for recall in Fig. 9, and an exemplary F1-score of 91.8% in Fig. 10. While precision score denoted the accuracy of the positive predictions, recall measured the ability to capture all the positive instances with F1-Score representing a harmonic mean of the precision and recall. The graphs given above represent the evaluation metrics for the proposed model, highlighting its stability and convergence during the training. Overall, these results suggested that the CNN model was well-prepared for accurate detection of emotions.

Table 3 Specifications of layers defined in the proposed CNN model

Specifications of the model	
Layers	Specifications of the layer
Model_Optimizer used	Adam_Optimizer
Metrics of the model	['model_accuracy']
Metrics of the model loss	Categorical
Layer 1_Conv2D	5 × 5 filter size each consisting total 32 filters using ReLU activation
Layer_2_Max-Pool	(Size of Kernal) 2 × 2
Dropout	20%
Layer 3_Conv2D	64 filters of size 3 × 3 filter size (ReLU activation)
Layer_4_Max-Pool	(Size of Kernal) 2 × 2
Layer dropout	20%
Layer 5_Conv2D	128 filters of size 3 × 3 filter size (ReLU activation)
Layer_6_Max-Pool	(Size of Kernal) 2 × 2
Layer dropout	30%
Layer 7_Conv2D	256 filters of size 5 × 5 filter size (ReLU activation)
Layer_8_Max-Pool	(Size of Kernal) 2 × 2
Layer dropout	30%
Output layer	5 classes using softmax

The proposed model contained
Total parameters: 10,750,658
Parameters which are trainable: 10,750,897

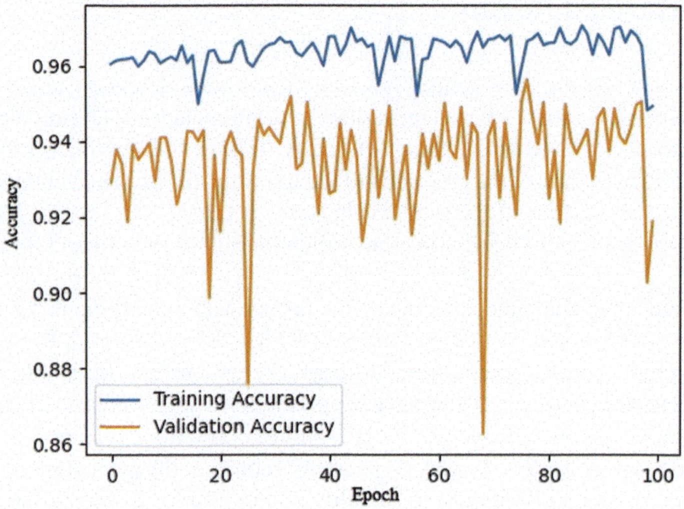

Fig. 6 Model training for the proposed model

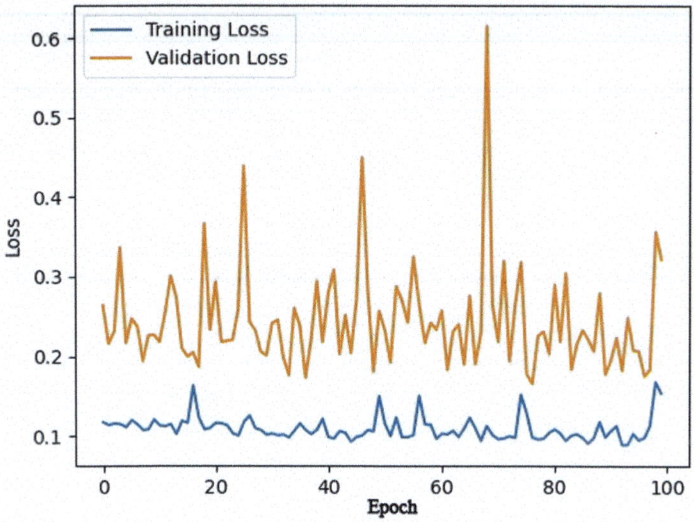

Fig. 7 Model loss during training for the proposed model

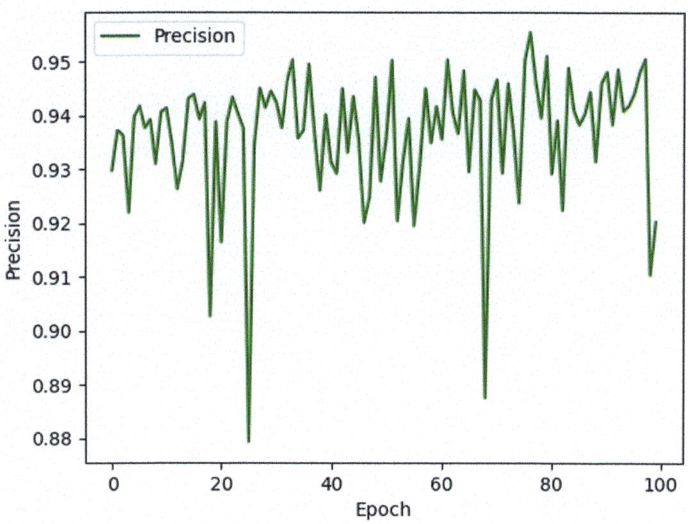

Fig. 8 Precision graph of the model trained

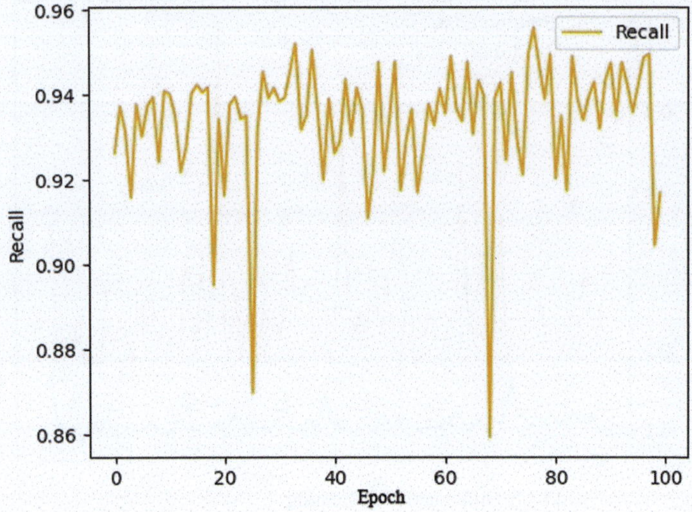

Fig. 9 Recall graph of the trained model

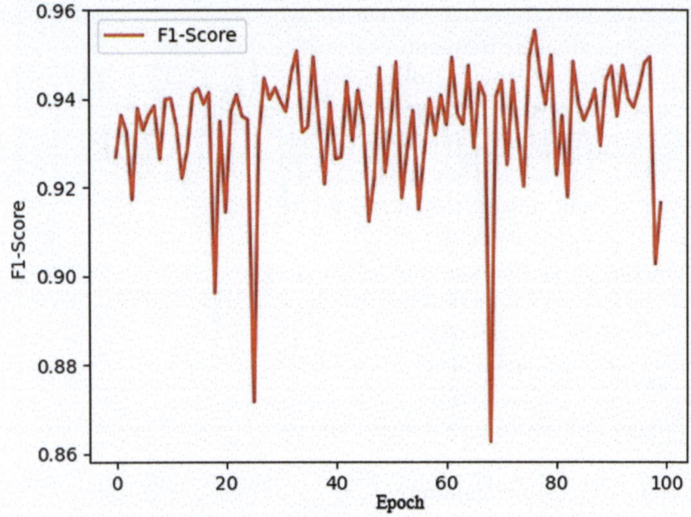

Fig. 10 F1-score graph of the trained model

7 Conclusion

Facial expression recognition in HCI-enabled devices gives up fascinating new opportunities for designing experiences that are more immersive sympathetic and user-centered. About autonomously extracting expressive and discriminative characteristics from facial data deep learning models like CNNs and RNNs have achieved great success. Future research should concentrate on overcoming the issues and investigating fresh ideas to enhance the efficacy interpretability and moral considerations of facial expression recognition for HCI-enabled systems as the area continues to develop.

References

1. Liu Y, Zhang Y (2020) Deep learning for facial emotion recognition: a survey. IEEE Trans Affect Comput 11(4):554–577
2. Zhao R, Liu X (2020) Deep learning for facial expression recognition: a comprehensive review. Neurocomputing 418:100–116
3. Ding X, Liu X, Li L (2020) Deep learning for facial expression recognition: a survey. Wiley Interdiscip Rev Data Mining Knowl Discov 10(5):e1355
4. Chen Y, Jiang H, Li C, Jia X (2016) Profound component extraction and characterization of hyperspectral pictures dependent on convolutional neural network (CNN). IEEE Trans Geosci Remote Sens 54(10):6232–6251
5. Chuang M-C, Hwang J-N, Williams K (2016) A feature learning and object recognition framework for underwater fish images. IEEE Trans Image Process 25(4):1862–1872
6. Khajontantichaikun T, Jaiyen S, Yamsaengsung S, Mongkolnam P, Ninrutsirikun U (2022) Emotion detection of Thai elderly facial expressions using hybrid object detection. In: 2022 26th international computer science and engineering conference (ICSEC), Sakon Nakhon, Thailand, pp 219—223. https://doi.org/10.1109/ICSEC56337.2022.10049334
7. Ren R, Cheng J, Yin Y, Zhan J, Wang L, Li J, Luo C (2018) Deep convolutional neural networks for log event classification on distributed cluster systems. In: 2018 IEEE international conference on big data (big data)
8. Giri S et al (2022) Emotion detection with facial feature recognition using CNN & OpenCV. In: 2022 2nd international conference on advance computing and innovative technologies in engineering (ICACITE), Greater Noida, India, pp 230–232. https://doi.org/10.1109/ICACITE53722.2022.9823786
9. Azizi FN, Kurniawardhani A, Paputungan IV (2022) Facial expression image-based emotion detection using convolutional neural network. In: 2022 IEEE 20th student conference on research and development (SCOReD), Bangi, Malaysia, pp 157–162. https://doi.org/10.1109/SCOReD57082.2022.9974104

A Performance Analysis to Detect Synthetic Images with Deep Learning-Based Image Classification

Vinay Gautam, Mukesh Joshi, Raj Gaurang Tiwari, Ambuj Kumar Agarwal, and Naresh Kumar Trivedi

Abstract This study tackles the growing public concern about the spread of AI-generated images and their impact on public trust in the media and society. The ability to distinguish between true and fraudulent contents has grown increasingly difficult, resulting in the spread of misinformation and the possibility of disinformation operations. Deepfake technology, in particular, offers a huge risk since it generates incredibly realistic manufactured films and images. To address this issue, we suggest the creation of an AI-based system capable of distinguishing between actual and fraudulent photos. We use a deep learning-based convolutional neural network (CNN) architecture, which is well-known for its image classification skills. The research focuses on improving performance by optimising the CNN model and hyperparameters. A 120,000-image CIFAKE dataset is employed, with 60,000 AI-generated synthetic photos and 60,000 real-captured photographs. To test the efficiency of the trained model, performance evaluation metrics such as accuracy, precision, recall, and F1-score are used. With a high accuracy rate of 95.9%, the proposed method achieves remarkable outcomes in properly distinguishing between actual and AI-generated images. Furthermore, the model has high precision, recall, and F1-score values, suggesting its robustness in dealing with both types of images. The study provides light on the issues posed by AI-generated images' rising sophistication and emphasises the necessity for automated solutions to tackle fraudulent

V. Gautam · R. G. Tiwari (✉) · N. K. Trivedi
Chitkara University Institute of Engineering and Technology, Chitkara University, Punjab, India
e-mail: rajgaurang@chitkara.edu.in

V. Gautam
e-mail: vinay.gautam@chitkara.edu.in

N. K. Trivedi
e-mail: nareshk.trivedi@chitkara.edu.in

M. Joshi
School of Computing, Graphic Era Hill University, Bhimtal Campus, Dehradun, UK, India

A. K. Agarwal
Department of Computer Science and Engineering, School of Engineering and Technology, Sharda University, Greater Noida, India
e-mail: ambuj.agarwal@sharda.ac.in

S. Dutta et al. (eds.), *Cyber Intelligence and Information Retrieval*, Lecture Notes in Networks and Systems 1025, https://doi.org/10.1007/978-981-97-3594-5_3

content. This research contributes to sustaining public trust in media and limiting the harmful societal impacts of AI-generated disinformation by attaining a high level of accuracy in deepfake identification.

Keywords AI-generated images · Deepfake detection · Image classification · Deep learning · CNN

1 Introduction

The emergence of images produced by artificial intelligence can now make it difficult to distinguish between actual and unreal contents, weakening public confidence in the media and society. The development of technology has made it simpler for anybody to produce intricate works of art or lifelike visuals without specialised knowledge. As a result, misinformation has grown and is now being used to promote political polarisation. The remarkable potential of AI to imitate and adapt makes it difficult for tech businesses, researchers, picture agencies, and news organisations to create guidelines for material origin and ownership.

AI-generated images can be used to create disinformation campaigns, fake news, and propaganda that can have far-reaching political and social impacts. Additionally, they might be utilised in fraud to make fraudulent papers or identities AI-generated images may also have a substantial impact on the media and advertising industries, as well as uses in law enforcement and security.

For example, deep fakes can be used maliciously, such as distributing fabricated videos of elected officials or public figures making inflammatory statements, eroding public confidence, and severely impacting public dialogue. Furthermore, deep fakes might be used to shame or blackmail anyone with access to confidential material, or they could be used to build fake social media profiles to attract Western sources. Deep fakes, for example, might be used to manufacture incendiary information, such as convincing footage of the US military forces committing war crimes, with the goal of radicalising communities, recruiting terrorists, or inciting violence. Furthermore, the "Liar's Dividend" effect may be at work, in which people dispute the authenticity of real material by arguing that it is a profound fake, throwing doubt and weakening trustworthiness [1].

As the quality and accuracy of AI-generated images are improving over the past few years, it is challenging for humans to determine whether the image is real or fake. Various synthetic image generation algorithms like Generative Adversarial Network (GAN) generate fake images which are impossible to detect with the naked eye. GAN comprises a generator and a discriminator neural network, where the discriminator evaluates the images generated by the generator.

The aim of this research is to create an AI system that can accurately distinguish between real and fake images. Convolutional neural networks (CNNs) [2, 3], a deep learning model for image classification, will be used to train the system. To ensure the system's proper functioning in practical situations, several metrics will be used to assess its performance, including accuracy, precision, recall, and F1-score. Additionally, the research will focus on examining the ideal architecture and hyperparameters for the CNN model to ensure optimal performance.

2 Background

Preprints and recent thesis are the primary formats in which the majority of the cutting-edge research in this area are presented. For instance, researchers in [4] suggested training SDM on medical imaging data, producing higher-quality images that could improve the model's capabilities through data augmentation. Studies [5, 6] have shown that diffusion models can also produce audio in addition to visuals. Furthermore, diffusion models are quite good at producing realistic artwork, even leading human viewers to believe that the paintings were produced by human painters, according to a research by Yi et al. [7] in 2021. This discovery has prompted scholars to think about the possibilities of diffusion models when co-creating with human artists [8].

However, it should be noted that some latent diffusion techniques, as indicated in the DE-FAKE plan put forth by Sha et al. [9], may result in images with observable digital fingerprints that denote their synthetic origin. Although improvements in model quality have considerably decreased visual flaws, it is still possible that computer vision techniques could recognise certain characteristics in images beyond what the human eye is able to. Examples of these digital fingerprints utilising Fourier transformations are given in [10].

According to study [11], optical flow approaches can detect synthetic human faces in the FaceForensics dataset with an accuracy of about 81.61% when used in the context of picture alteration detection. Despite small visual glitches at the pixel level, [12] presents recurrent AI-based video recognition, attaining 97.1% accuracy over 80 frames. With an F1-score of 0.88 and an AUC of 0.95, Wang et al. [13] suggest utilising EfficientNets and Vision Transformers in a system that can identify fake images created by adversarial models.

Convolutional and temporal approaches were also suggested in [14] to detect mixed sets of fake data with accuracy levels ranging from 66.26 to 91.21%. [15] found the chrominance components CbCr in digital photos as a possible method for spotting minute pixel differences that are frequently present in synthetic images. Human identification of picture manipulation is getting harder as a result of improvements in image generating techniques. Humans have trouble identifying images that have been altered using image processing techniques, according to a 2017 study by Nightingale et al. Since then, the field has experienced rapid growth for almost five years.

In 2018, Raturi [16] proposed an architecture to find fake accounts on social media, primarily Facebook. Based on the posts and wall placements of fake accounts, machine learning techniques including Support Vector Machine (SVM) and Complement Naive Bayes (CNB) were applied in predicting them. Through text categorization and data analysis, offensive terms and their frequency were examined to validate content. SVM recognised Bag-of-Words (BOW)-based fake accounts with 97% accuracy, whereas CNB did so with 95% accuracy. The study emphasised that one of the main security concerns in social networks is the improper validation of data before to release.

By comparing 4000 fake and 4000 actual photos using error-level analysis, Kuruvilla et al. [17] successfully trained a neural network. The trained neural network showed an impressive 83% success rate in distinguishing between fake and real photos.

3 Method

This section defines the procedures followed along with their implementation and how they are used. The first section defines the data collection and then image classification steps used are defined in next sections.

3.1 Data Collection

Dataset of 120,000 images which contains 60,000 real-captured images and 60,000 fake AI-generated images is collected from CIFAKE [18]. The real 60,000 images are 32 × 32 in size and are divided into ten classes, and those classes within the dataset are aeroplane, automobile, bird, cat, deer, dog, frog, horse, ship, and truck. Per class, there are 6,000 images. A testing dataset of 16.6% is utilised for each class, with 5,000 photos being used for training and 1,000 for testing. In this study, the positive class "REAL" is trained using all of the photos from the training dataset. Therefore, 10,000 are utilised for testing, and 50,000 are used for training.

3.2 Data Preprocessing and Augmentation

The real and fake image datasets were obtained from separate folders. Each image was converted into a numerical array format. The pixel values of all the images were adjusted to ensure uniformity by dividing them by 255.0, which produced values ranging from 0 to 1. In order to perform additional analysis, we separated these processed photos into distinct lists.

We used data augmentation approaches to improve the model's ability to adapt and successfully handle various variances in the training data. The training images were subjected to a number of modifications using the ImageDataGenerator module from Keras. These adjustments included rotation, width and height changes, shear, zooming, horizontal flipping, and nearest neighbour filling.

3.3 Model Architecture

KNN, Logistic Regression, and two CNN models were used in a thorough training and testing process that involved four deep learning models [19, 20]. Our primary goal was to evaluate how well they did at correctly identifying fake or real images. A total of 100,000 images—50,000 true labels and 50,000 fake labels—made up the training dataset.

TensorFlow within the PyCharm IDE was used to train the CNN models. Additionally, the KNN and Logistic Regression models were trained using the Orange data mining tool. Orange is a flexible tool with a variety of machine learning techniques, which makes it appropriate for our needs.

We utilised a sequential model to construct the image classification model. The model's architecture was made up of a number of convolutional layers, followed by max-pooling layers. The ReLU activation function was incorporated into the convolutional layers to add nonlinearity and capture complex patterns. The model's final layer is a fully connected layer with a sigmoid activation function. We used the binary cross-entropy loss function to measure the model's accuracy, the Adam optimizer to assemble the architecture, and the accuracy metric to determine the model's performance.

3.4 Model Training and Evaluation

To train the model, we utilised the augmented training data obtained from the ImageDataGenerator. The model's performance was evaluated on the validation data after each training period, which comprised several epochs. For analysis and visualisation in the future, we meticulously captured the training history, including the accuracy and loss values.

The trained model was then tested using a dataset that included both 10,000 real images and 10,000 fake ones. The testing images underwent preprocessing steps to assure consistency, just like the training images did. We generated numerous evaluation measures to fully evaluate the model's performance in classifying actual and fake photos. Accuracy, precision, recall, F1-score, and area under the receiver operating characteristic curve (ROC) were some of these measurements [21]. These indicators provided us with useful information about the model's performance.

We also created a confusion matrix to give the classification results a more understandable representation. Understanding the distribution and precision of the model's predictions was made easier by this visual help.

The accuracy and loss data for training and validation were plotted to show how the model learned throughout the course of the epochs. To further illustrate the trade-off between true positive rate and false positive rate, as well as precision and recall, respectively, a receiver operating characteristic (ROC) curve and a precision–recall curve were plotted.

4 Results and Observations

This section provides an overview of the results and outputs obtained during the model training and deployment process. Four deep learning models are contrasted in the following bar graph (Fig. 1). Figures 2 and 3 also show the accuracy and loss graphs for the suggested CNN model. An impressive validation accuracy of 95.9% was attained by the proposed CNN model, demonstrating its superior accuracy.

A number of assessment measures are also used to evaluate the performance of the suggested CNN model in addition to the results given above. The confusion matrix is shown in Figs. 4 and 5 which illustrates the ROC curve, Fig. 6 presents the precision–recall curve, and Fig. 7 showcases the calibration plot graph. These visualisations provide useful information about the model's categorization performance.

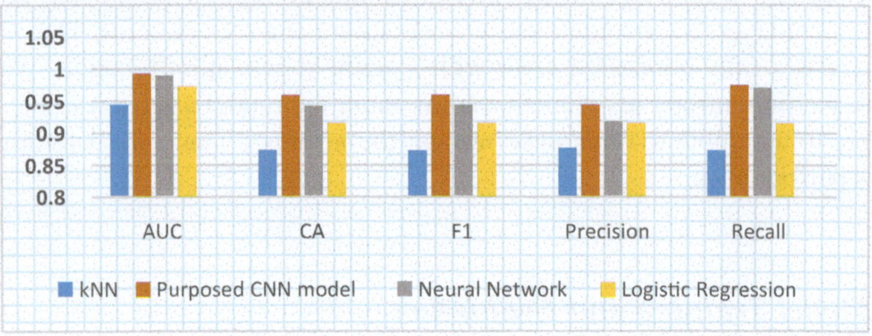

Fig. 1 Performance of four classification models based on evaluated metrics

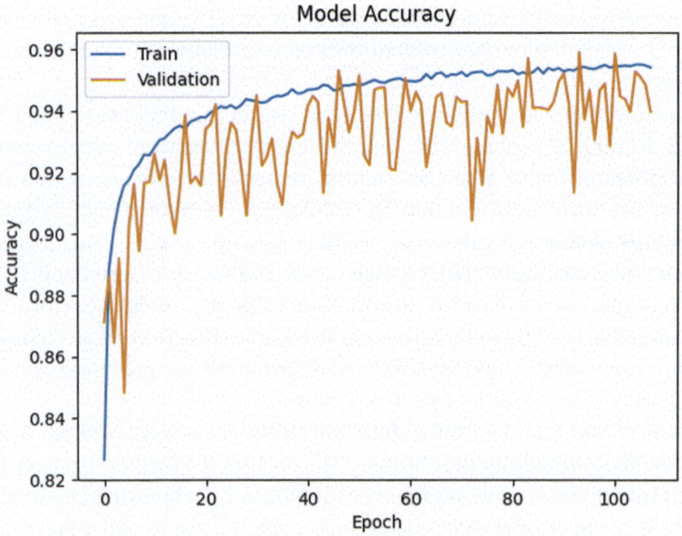

Fig. 2 Accuracy graph of purposed CNN model

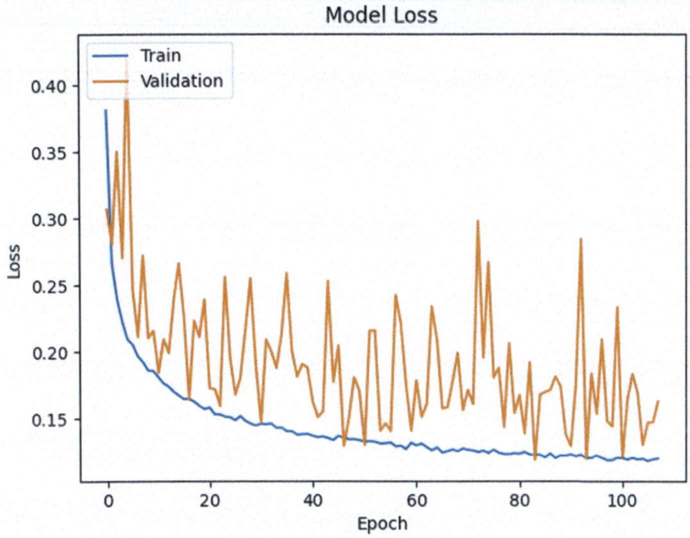

Fig. 3 Loss graph of purposed CNN model

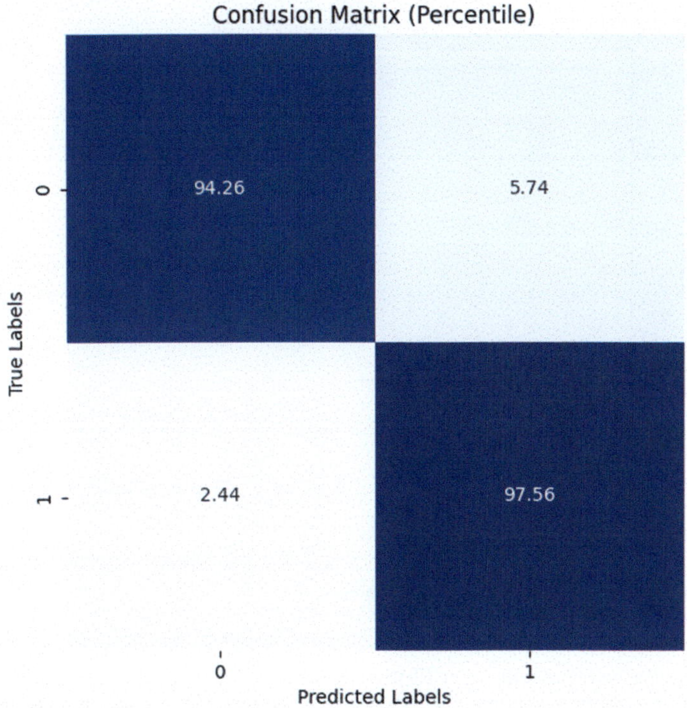

Fig. 4 Confusion matrix of purposed CNN model

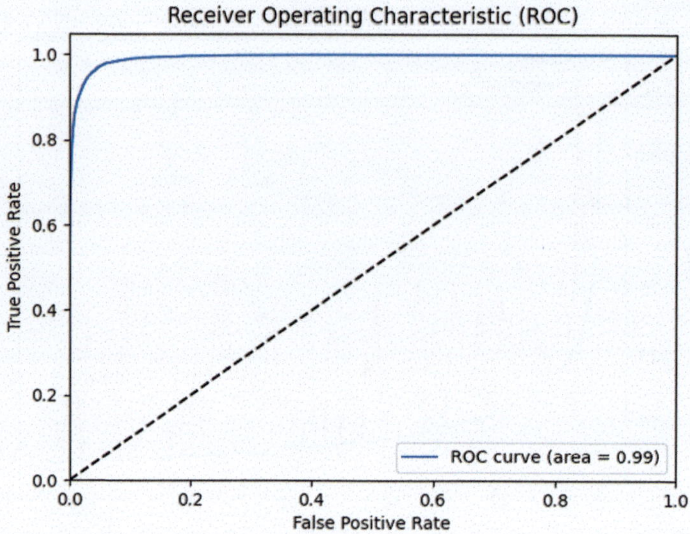

Fig. 5 ROC curve of purposed CNN model

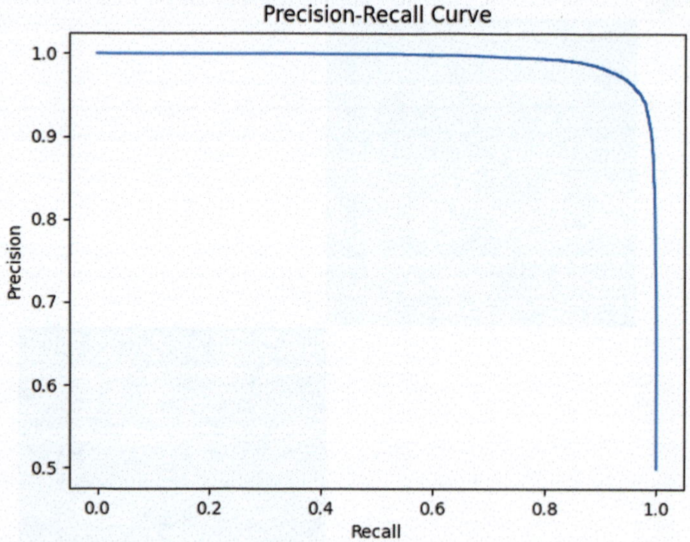

Fig. 6 Precision–recall curve of purposed CNN model

Additionally, screenshots of a Flutter application that uses the suggested CNN model have been shown in Fig. 8. These images show how the model is used in practice within a user-friendly interface.

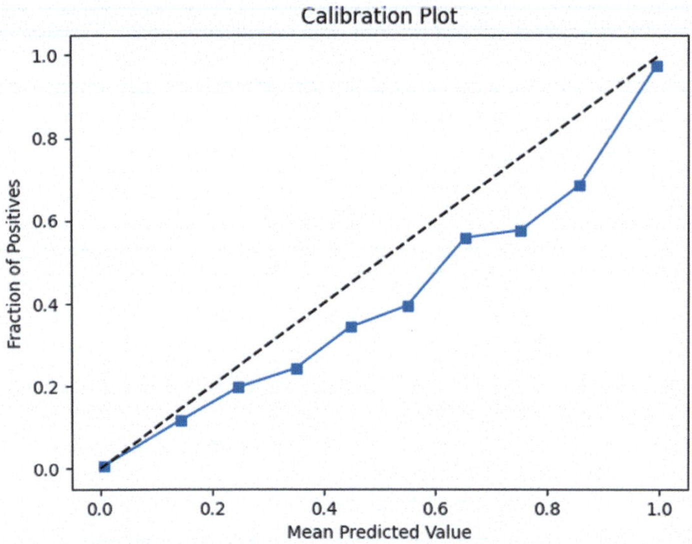

Fig. 7 Calibration plot of purposed CNN model

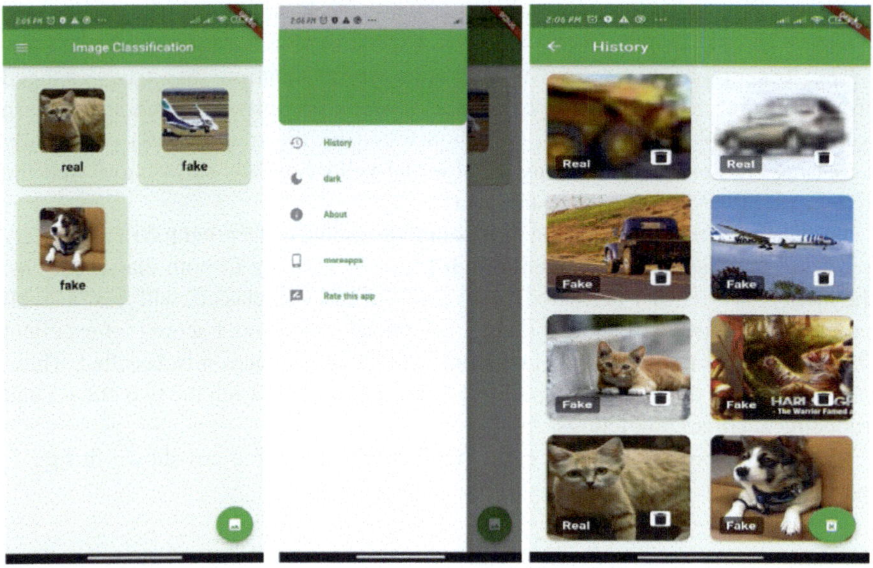

Fig. 8 Screenshots of implemented purposed CNN model in Flutter

The KNN model's AUC of 0.944 indicated remarkable discrimination power. It classified 87.3% of occurrences correctly, with a classification accuracy of 0.873. This model had an F1-score, precision, and recall of 0.873, 0.877, and 0.873, respectively.

With an amazing AUC of 0.99, the proposed CNN model surpassed the other models in terms of performance. Its classification accuracy score of 0.959 demonstrated a high degree of predictability. This model's F1-score, precision, and recall were each 0.959, 0.944, and 0.975, respectively.

With an AUC of 0.983, the neural network model demonstrated promising results. It classified 94.1% of occurrences correctly, according to its classification accuracy score of 0.941. The F1-score, precision, and recall are 0.943, 0.917, and 0.97, respectively.

An AUC of 0.974 for the Logistic Regression model indicated that it had a strong ability to discriminate across classes. It classified 91.7% of occurrences correctly, with a classification accuracy of 0.917. This model had an F1-score, precision, and recall of 0.917, 0.917, and 0.917, respectively.

The accuracy graph shows how training accuracy and validation accuracy have changed throughout the course of the epochs. The training accuracy shows a steady upward trend, rising until the training procedure is completed, when it reaches a staggering 95.51%. However, compared to the training accuracy, the validation accuracy also shows progress. At the epoch number 94, it reaches a noteworthy accuracy of 95.91%.

The loss graph shows how training loss and validation loss have changed throughout the course of the epochs. The model's capacity to successfully learn from the available training data is demonstrated by the training loss continually decreasing. A similar diminishing trend in the validation loss shows how well the model generalises to fresh, untested data.

We can see from the confusion matrix that the model did a good job of accurately predicting the classes. It obtained a high level of accuracy in both classifications. The model successfully predicted 94.26% of the cases in class 0, with just a small percentage of misclassifications (5.74%). Similarly, the model scored an excellent 97.56% accuracy rate for class 1, with only 2.44% of instances misclassified. These results show that the model is capable of differentiating between the two classes and producing accurate predictions.

ROC curve, precision–recall curve, calibration plot curve are shown in Fig. 5, Fig. 6, and Fig. 7, respectively.

5 Conclusion and Future Work

In conclusion, this study analyses the concerning rise of AI-generated images and their potential societal effects, such as the dissemination of false information, political manipulation, and a reduction in public trust. As technology advances, it becomes increasingly difficult to recognise between actual and false images, necessitating the creation of robust systems capable of doing so.

Using convolutional neural networks (CNNs), the study describes a way for developing an AI system that can categorise real and false photos with high accuracy. A dataset of 60,000 actual photos and 60,000 AI-generated images is used to train and assess the system. To evaluate the system's performance, many measures like as accuracy, precision, recall, and F1-score are used.

The suggested CNN model demonstrates the potential of AI in addressing the issue of image categorization by utilising advanced deep learning techniques and data augmentation methodologies. The design of the model, which includes larger image sizes, enhances accuracy by catching more subtle elements and patterns in the photos.

The experimental results show that the suggested technique efficiently distinguishes between actual and AI-generated images, with high precision and recall rates. These discoveries broaden the repertory of image recognition techniques and provide useful insights for applying AI-powered image detection systems in real-world contexts.

References

1. Ahmed SRSEAMRDAD (2022) Analysis survey on deepfake detection and recognition with convolutional neural networks, HORA 2022—4th international congress on human-computer interaction, optimization and robotic applications, proceedings
2. Tiwari RG, Yadav DS, Misra A (2022) Performance evaluation of optimizers in the classification of marble surface quality using CNN. In: Proceedings of third doctoral symposium on computational intelligence: DoSCI 2022. Springer Nature Singapore, Singapore, pp 181–191
3. Dwivedi M, Tiwari RG, Ujjwal N (2022) Deep learning methods for early detection of monkeypox skin lesion. In: 2022 8th international conference on signal processing and communication (ICSC). IEEE, pp 343–348
4. Chambon P, Bluethgen C, Langlotz CP, Chaudhari A (2022) Adapting pretrained vision-language foundational models to medical imaging domains. arXiv:2210.04133
5. Schneider F, Jin Z, Schölkopf B (2023) Moûsai: text-to-music generation with long-context latent diffusion. arXiv:2301.11757
6. Schneider F (2023) Archisound: audio generation with diffusion, Master's thesis, ETH Zurich
7. Yi D, Guo C, Bai T (2021) Exploring painting synthesis with diffusion models. In: 2021 IEEE 1st international conference on digital twins and parallel intelligence (DTPI). IEEE, pp 332–335
8. Guo C, Dou Y, Bai T, Dai X, Wang C, Wen Y (2023) Artverse: a paradigm for parallel human–machine collaborative painting creation in metaverses. IEEE Trans Syst Man Cybern: Syst
9. Sha Z, Li Z, Yu N, Zhang Y (2022) De-fake: Detection and attribution of fake images generated by text-to-image diffusion models. arXiv:2210.06998
10. Corvi R, Cozzolino D, Zingarini G, Poggi G, Nagano K, Verdoliva L (2022) On the detection of synthetic images generated by diffusion models. arXiv:2211.00680
11. Amerini I, Galteri L, Caldelli R, del Bimbo A (2019) Deepfake video detection through optical flow based CNN. In: Proceedings of the IEEE/CVF international conference on computer vision workshops, pp 0–0
12. Güera D, Delp EJ (2018) Deepfake video detection using recurrent neural networks. In: 2018 15th IEEE international conference on advanced video and signal based surveillance (AVSS). IEEE, pp 1–6
13. Wang J, Wu Z, Ouyang W, Han X, Chen J, Jiang Y-G, Li S-N (2022) M2tr: Multi-modal multi-scale transformers for deepfake detection. In: Proceedings of the 2022 international conference on multimedia retrieval, pp 615–623

14. Saikia P, Dholaria D, Yadav P, Patel V, Roy M (2022) A hybrid CNN-LSTM model for video deepfake detection by leveraging optical flow features. In: 2022 international joint conference on neural networks (IJCNN). IEEE, pp 1–7
15. Li H, Li B, Tan S, Huang J (2020) Identification of deep network generated images using disparities in color components. Signal Process 174:107616
16. Raturi R (2018) Machine learning implementation for identifying fake accounts in social network. Int J Pure Appl Math 118(20):4785–4797
17. Villan M, Kuruvilla A, Paul KJ, Elias EP (2017) Fake image detection using machine learning. IRACST—Int J Comput Sci Inf Technol Secur (IJCSITS)
18. Bird JJ, Lotfi A (2023) CIFAKE: image classification and explainable identification of AI-Generated synthetic images. arXiv:2303.14126
19. Ramesh TR, Lilhore UK, Poongodi M, Simaiya S, Kaur A, Hamdi M (2022) Predictive analysis of heart diseases with machine learning approaches. Malaysian J Comput Sci 132–148
20. Lilhore UK, Poongodi M, Kaur A, Simaiya S, Algarni AD, Elmannai H, Vijayakumar V, Tunze GB, Hamdi M (2022) Hybrid model for detection of cervical cancer using causal analysis and machine learning techniques. Comput Math Methods Med
21. Agarwal AK, Kiran V, Jindal RK, Chaudhary D, Tiwari RG (2022) Optimized transfer learning for dog breed classification. Int J Intell Syst Appl Eng 10(1s):18–22

A Study of AI-Based Systems in the Judicial Interpretation of the Law

Sangeet Sharma, Anil Kumar Dixit, Sagar Saxena, Gagandeep Kaur, and V. Bhuvaneswari

Abstract ML, DL, neural networks, NLP, computer vision, and robotics are just a few of the examples of recent breakthroughs in the field of AI, which has seen significant advancements in recent years. These approaches are already being aggressively implemented in the judicial system, and it is anticipated that a significant portion of the legal service operations that are presently being carried out by attorneys will be taken over by AI in the years to come. The possible uses and effectiveness of AI in the execution of justice are investigated in this research. The purpose of this study is twofold: first, to examine the ethical issues that AI faces in the judicial system; second, to highlight the limits of AI in the judiciary via some instances of AI tools that have recently been produced. AI searching tools, for example, are one way that AI algorithms are beginning to assist attorneys and judges. AI algorithms are also beginning to assist court administrations with automated prediction and company analytics based on computational analysis of big data. Legal knowledge-based technologies that use the notion of AI have the potential to speed up the supply of legal services by shifting the focus from the traditional searching of relevant cases to the extraction of specific information in a tailored manner.

Keywords Machine learning · Artificial intelligence · Deep learning · Natural learning process · Commission for the efficiency of justice · European Union

---

S. Sharma · A. K. Dixit · S. Saxena (✉) · V. Bhuvaneswari
Law College Dehradun, Uttaranchal University Dehradun, Dehradun, India
e-mail: adv.sagarsaxena@gmail.com

S. Sharma
e-mail: sharmasangeet852@gmail.com

A. K. Dixit
e-mail: anilkdixit44@gmail.com

V. Bhuvaneswari
e-mail: bhuvaneswarivellayan@gmail.com

G. Kaur
School of Law, University of Petroleum & Energy Studies, Dehradun, India
e-mail: gkaur@ddn.upes.ac.in

S. Dutta et al. (eds.), *Cyber Intelligence and Information Retrieval*, Lecture Notes in Networks and Systems 1025, https://doi.org/10.1007/978-981-97-3594-5_4

1 Introduction

Modern civilization values electronic data due to rapid technological advancement. Digital data allows speedy searching and processing. New technologies are being introduced as the court system digitalizes and the law invades digital social life. Technology and computer computational power drive AI progress. AI is used in many fields. It has not been implemented in court. AI-based IT systems may be used in court law interpretation [1]. Despite not having a constitutional definition, AI has long been discussed in legal discussions. AI requires interpretation; hence there is no single, widely accepted definition. AI has becoming more common in public debate as AI-powered systems, including algorithms we use daily, gain appeal. Artificial intelligence (AI) analyzes its environment and acts autonomously to achieve goals [2]. Another definition of AI is an algorithm developed to interact with humans that, given an intricate objective, engage in the real or virtual world by recognizing their surroundings, understanding the data gathered data, whether it is organized or unorganized, deductive reasoning on the information derived from this data, and determining which course of action to take by predefined variables to achieve the objective [3]. Judicial application of law must be studied before addressing AI in a court system. It means applying a legal rule, not the law. The court's law enforcement affects parties' rights. The law usually applies only to governmental agencies or analogous authorities. A party exercises its right. This rule authorizes a state entity to act. Find two legal uses. The law distinguishes judicial and managerial application [6]. These are not law enforcement terms. Judicial interpretation of the law is confined to establishing facts and setting a norm. State bodies' acts are determined by substantive law rules, but they can state formally and enforceable that a given standard of substantive law applies in certain circumstances, giving them the power to establish a specific individual and concrete norm for a given case. In managerial law implementation, the organ can issue substantive law acts. Legally, justice originates from the Constitution [4].

AI-based systems have piqued attorneys' interest in these challenges. AI-related rules were created as legal disciplines used AI. However, attorneys' interest in AI in judicial interpretation of the law is mostly academic. However, some governments are contemplating AI-based IT systems for their legal systems [5]. Implementing the law with AI would be a technological and conceptual revolution. When discussing the possibility of implementing that can assist or replace the judge in significant situations, it must be noted that advances in technology that allow the analysis of large amounts of data led to the widespread adoption of AI in many areas of life. In several fields, these technologies outperform people in accuracy, speed, and reliability. Courts use two AI models to make substantive judgements. It could be a consultation model or fully computerized. The first involves creating an AI-based judge replacement. The AI would analyze the evidence and follow all processes and justifications to provide an enforceable decision for all parties. This process would be automated and human-free. The advising model aids the human adjudicator rather than replacing him. The technology would assess the dispute and provide a decision to a human adjudicator [6]. These computers would have access to a wealth of data, including national, EU, and international laws and tribunal judgements. This advanced technology could also use legal information systems with a large database of legal scholars and expert opinions. Systems that replace judges raise worries about justice. It must be assessed if one will still view the court in the sense of the Constitution during the adjudication process using AI-based tools and data collecting and analysis. Only judges are independent in adjudication; hence an autonomous court must have individuals. The court makes decisions utilizing an AI-based IT system that is not adjudication-independent [7]. Another difficulty with an AI-only system is the possibility of algorithm intervention affecting rulings. The algorithm developers and overseers of the IT system that produces rulings may influence their substance, even defining the accepted line of judicial action in a certain instance. Tampering with the system's algorithm might change the court's ruling, raising worries about its freedom and liberty. Courts should be independent of other powers. Executive directives to create an IT system increase the potential of the executive influencing judgements, which would be improper intervention in the judicial system.

Thus, the authors aspire to accomplish the following objectives:

a. The present research centers on the implementation of AI-based systems in the legal interpretation of the law, as well as a brief examination of the modern technologies most prominently associated with court proceedings.

b. In addition, the primary focus of this study is the implementation of AI-based IT systems that should assist and recommend a verdict to a human adjudicator.

The remaining research paper is organized as follows: Sect. 2 will include an application of AI-based IT systems in the judicial application of the law; Sect. 3 will go through the works related to machine learning tools for judiciary practices. Next Sect. 4 talks about AI in court practice: ethical principles. Further, next Sect. 5 proposes limits of judicial artificial intelligence. Then a recommendation is presented in Sect. 6. Finally, we concluded this paper with the conclusion in Sect. 7.

2 Implementation of AI-Based Systems in the Judicial Interpretation of the Law

AI will transform the legal industry. AI could streamline most judicial tasks. These include case push, knowledge index query, supplementary conviction, sentencing guidance, evidence assessment, verdict, integrated court support, visual speech reading and modification, handling cases, deviance analysis, judicial and administrative relationship, autonomous record generation, and collaborative proof reading [8] (Fig. 1).

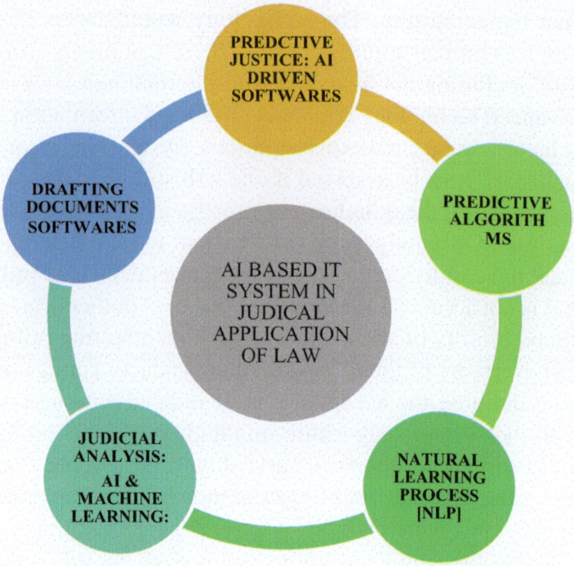

Fig. 1 AI-based IT system in judicial application of law

2.1 *Judgment Prediction*

Prediction of legal judgment has been an enduring subject in both legal theory and practice, with the aim of enhancing judicial uniformity, administrative effectiveness, and access to justice. In order to predict legal judgements, numerous methods and techniques have developed over time, ranging from simple calculative models to extremely sophisticated analytical algorithms. Diverse models have been explored; notably, approaches based on AI have gained traction, particularly since the introduction of AI. AI-powered software can study large numbers of cases, all relevant court documents and judge decisions, and other significant public information.

2.2 *Police Prediction Algorithms*

Police prediction algorithms use crime information to identify trends and predict outcomes. Law enforcement has used predictive software to determine whether a weapon-wielding suspect was the culprit or the victim [9]. Based on occupation, age, and family criminal histories, an algorithm creates a numerical score and classifies people as low, moderate, or high risk of reoffending. The future criminal risk and recidivism of an accused person have also been calculated using similar algorithms [10].

2.3 Natural Language Processing (NLP) for Attorneys and Adjudicators

A system for legal automation the objective of attorneys and advocates has been to forecast verdicts, and ongoing endeavours have been dedicated to improving its precision. It analyses the contents of an immense amount of judgement documents using a specialised algorithm to render decisions that automatically determine a case's outcome. In recent years, there has been a substantial increase in the volume of litigation, which has made it more challenging and time-consuming for human attorneys to render accurate verdicts in court reviews. Subsequently, the emergence of automated legal systems based on big data necessitates the development of systems capable of accurately predicting legal judgements by analysing a vast quantity of legal precedents.

2.4 Court Data Analysis

Software that analyzes a lot of case law in a second to predict a court ruling. The majority of predictive systems are used before litigation. AI-powered software can study large numbers of cases, all relevant court documents and judge decisions, and other significant public information [11].

2.5 Document Processing System

Several legal firms are also beginning to adopt this technology by using automated programs to design legal documents. Plenty of these app developers assert that the last documentation is generated in mere minutes, whereas manual human writing could take days.

3 Computational Learning Methods in the Judiciary

It has so far been noted in both the public and private sectors that AI systems are being used to aid legal practitioners in their work. In the following, you will find a few of the most popular ML applications in the legal system. The following are some of them: [12, 13].

3.1 ROSS Intelligence in the USA

IBM's American legal research service developed this AI-driven tool. IBM's DeepQA research, which generated Watson, an automated system that answers natural language questions, inspired ROSS.

3.2 Predictive in France

It is a tool for predicting the favorable outcome rate of trials using predictive analytics. The creators of predictive assert that the algorithm can analyze thousands of judicial decisions per second, and in the past 2 years they have begun working with more than 400 attorneys.

3.3 Luminance in the UK

The ML system analyzes and stores document data to increase legal intelligence. Over fourteen of the world's 100 major law firms use Luminance since 2016. Its detection of patterns technology, advanced probabilistic analysis, and superintend and not-superintend ML algorithms may find commonalities, distinctions, and abnormalities at all levels of legal record examination for due diligence or conformance evaluation.

3.4 Victor Robot in Brazil

This is an initiative devised by the Supreme Court of Brazil in collaboration with the University of Brasilia. The device is intended to assist the Brazilian Supreme Court by analyzing cases that reach the Court using documentation evaluation and NLP techniques.

3.5 COMPASS Algorithm

It is a risk assessment algorithm used to forecast the likelihood that the accused will commit another crime. Using a proprietary algorithm, it offers the convict the necessary advice, including case-related information that may be expected during punishment.

3.6 *EBrevia*

In conjunction with ML and NLP technology, AI is used to extract pertinent data from legal agreements and other records, bring remarkable precision and rapidity to due diligence, and assist lawyers in their research.

3.7 *Caseiq*

The program is a simulated legal study assistance that autonomously assesses the brief's language and feeds the data to a sophisticated predictive algorithm to identify possible inconsistencies of law or competing arguments (Table 1).

4 Ethics of AI in the Judicial System

How we should engage with AI in the actual world is still debated. Over 25 documents, including IEEE, EU, and International bylaws, have outlined AI ethics in writing. Council of Europe Commission for the Efficiency of Justice (CEPEJ) addressed it. The CEPEJ Working Party on Quality (GTQUAL) developed AI ethics for judicial administration. CEPEJ approved them in December 2018. These five "Ethical Standards" overlap, making thorough and methodical treatment difficult [14].

4.1 *Respect for Basic Human Liberties*

To ensure that the development and operation of AI applications and resources are compatible with fundamental rights such as the right to confidentiality, equal consideration, and an impartial trial.

4.2 *Equal Treatment*

To prevent group bigotry. PREjudice and arbitrary isolation of people and groups are significant threats, as COMPAS shows. The bias may come from the algorithm's data or the approach itself.

Table 1 Technological table

S. no	Technology	Key features	Advantage	Limitation
1	ROSS intelligence	Legal Helix is an AI framework that utilizes technologies like IBM Watson's artificial intelligence capabilities to assist lawyers	Makes use of NLP and ML techniques to locate applicable legal precedents for a given question	Using an algorithm to make decisions that restrict individual freedom
2	Predictive analytics tool	To better construct a case, establish a legal strategy, comprehend the strategies of opposing counsel, evaluate suspects, and forecast case results, predictive analytics can uncover patterns, irregularities, and correlations in massive volumes of data	Regression analytics can predict changes in interactions between variables, whereas ML techniques can analyze data to discover causal connections between factors	Variables are ignored by computers and algorithms
3	Luminance	Functionality for Pattern Recognition, Warnings, Legal Compliance, Time Tracking, Risk Analysis, Database Export, and Monitoring	Legal Inference Transformation Engine (LITE) advanced AI for legal process automation in the world. Luminance's AI integrates several cutting-edge ML techniques, including deduction, DL, NLP, and recognition of trends	Problems with ethics, unavailability of explanations, and the fact that they can only learn from observing natural behavior
4	Victor rRobot	Reasoning in the fields of software engineering, computer science, and law for the Artificial Intelligence Project	Processing, organizing procedural components, and assisting technical team judgments at the Supreme Court all fall under the purview of ML and neural networks	Models for analyzing and assessing legal thinking are constantly being improved upon
5	COMPASS algorithm [Correctional Offender Management Profiling for Alternative Sanctions]	To measure future recidivism risk	Risk assessment tools for general and violent recidivism, as well as pretrial infractions, were developed by Northpointe	The algorithm is susceptible to being biased

(continued)

Table 1 (continued)

S. no	Technology	Key features	Advantage	Limitation
6	eBrevia AI by DFIN	A form of AI that is preoccupied with the investigation of algorithmic structures is known as NLP technology with ML	Make significant changes to the process of contract review, within minutes, you may extract the most important clauses and data points from thousands of papers	Still in the process of being developed
7	CASEIQ: legal research assistance	Method for prediction based on artificial intelligence that draws on the accumulated knowledge of legal professionals	The legal document is automatically analyzed for cited judgments, legislation, and legal defense patterns. In addition, provide suggestions in the form of phrases, laws, or important cases to make your research more exhaustive	Still in the process of being developed

4.3 *Protection of Data*

When analyzing judgments, data that are unalterable and verified information and sources should be used alongside interdisciplinary models in an encrypted technological framework.

4.4 *Transparency*

Data processing methods should be transparent and open to independent evaluation. Legal precedent now demands disclosure. The end user of an algorithm must accurately, quickly, and satisfactorily share their judgments, facts, and hypotheses. This complete, timely, and suitable release should allow judicial review of the decisions and data, reasoning, and assumptions used to provide effective legal defense against decisions based on these decisions, information, logic, and pre-conceptions. The Dutch legal system today follows this example.

4.5 *User-Managed AI*

As the computer cannot independently make decisions, the algorithm cannot be used as a prescription. The actions of the AI must be transparent to the user, who must also feel in control of their own choices. This indicates that consumers should have no difficulty opposing the algorithm's predicted outcome. This matter of human agency was crucial in the Loomis case, which reached the Wisconsin Supreme Court [15]. At stake were:

1. Because the accused cannot independently verify the reliability and validity of the risk evaluation in question, the use of COMPAS in cases where the operation is a company's secret raises the question of whether the accuser's right to an impartial trial has been violated.
2. Does a risk assessment that considers gender and ethnicity violate the right to an impartial trial? The Wisconsin SC rejected Loomis's arguments but said the court should clarify COMPAS use. The US Supreme Court declined to review the case.

The Dutch Council of State suggested redefining sound governance principles like reasoned judgment and due diligence in the perspective of automation [16]. Among other considerations, a judgment must indicate which algorithms were used and which data were shared from other government organizations. This will boost citizen input into automation and chain decision-making; personal reconsideration is advised when objecting to automated decisions. In the end, AI can help courts, judges, litigants, and justice seekers. The task with the most success so far is organizing massive amounts of data that could improve justice. Advice and prophecies are still questioned. Using AI in a judge's decision-making is acceptable if justified. However, courtroom AI has limitations.

5 Limitations of Judicial AI

The current law makes AI replacing judges inevitable. Like nuclear fission, AI can either improve or destroy cities [17]. AI is now used sparingly in the judicial system. In detail, judicial AI can only support a judge's decisions and cannot replace them. For productivity, it may be better for technology and secondary work. In depth, judicial AI has been effective in using competent advanced technologies like facial recognition and speech transformation and in subjects with less technical requirements, sufficient investment, and successfully overcome obstacles like electronic data. However, AI in the judicial system lacks implementation, pragmatism, and real impact [18, 19].

5.1 Legal Authority

The initial query is whether a machine's algorithm or automated system can substitute for a human judge. Concerning a computerized system that makes administrative decisions, "Justice Perry" raises issues regarding who makes the decision and who has the legal authority to do so [20]. Is it the computer programmer, policymaker, or individual who makes decisions, or the machine or computer system?

5.2 *Translating Law into Code*

Analysts wonder how to correctly translate the legislation into computer orders, codes, and operations. Legal language is often complex and context-specific. Computer programmers and IT workers rarely have legal, strategic, or administrative experience. However, these professionals must turn cases and regulations into computer rules and orders for autonomous decision-making. These complex laws work within legislative assumptions and subjective judgments. These intricacies are hard to code into a self-driving mechanism. Due to these constraints, many suggest that other legal domains may be better for computer code [21]. These regulations will need to be updated due to several changes, new legal rulings, and complex transitioning clauses. Self-governing systems must be able to implement the law at different times to ensure that cases are decided based on the laws in effect when the required acts happened.

5.3 *Discretionary Judgments*

Many legal decisions require discretion. Reasoning software processes input data by predefined algorithms to produce a specified result. Such rigor may conflict with judgment. Social principles, party personalities, and other factors may be considered while making discretionary judgments. However, modern bias and judgment are still problematic. As said, judgment can be affected by many variables that would not exist if AI were engaged, but AI procedures may also bias results. According to the access to justice movement, court adjudication can be altered by numerous factors, including representation, assets available to the party in dispute, decision-making quality, and rights context [22].

5.4 *Syntax and Semantics*

AI in law may encounter the philosophical distinction between syntax and semantics. Algorithms have syntax but no semantics. Digital technology processes data as zeros and ones. Technology can process and manipulate symbols, but it does not understand their significance. The machine cannot grasp the data it processes. In contrast, the human intellect can grasp processed information. This issue shows that computer algorithms will soon be able to mimic human thought processes, but it will take time to do so exactly [23].

6 Recommendations

After examining the study or research of distinguished scholars or researchers:

A. First, we would like to recommend that considering the significant advancements in this discipline, members of the judiciary must discuss the difficulties and possibilities related to deploying AI in judicial systems and its consequences for the rights of people and the rule of law.
B. Second, the application of AI in the judicial system is to facilitate the deliverance of services in completely novel ways. AI cannot replace judges instead it can help judges in the matter of legal knowledge and understanding.

7 Conclusion

In many nations, judicial interpretation of the law does not use AI. Legal documents and other relevant publications support the study's assumption. It has been suggested that AI-based systems could help adjudicators in all types of legal cases. This approach is advised due to the growing number of disputes decided by the courts and the growing body of judicial rulings. Thus, the adoption of automated systems with artificial intelligence will require Constitutional amendments and legal code updates to comply with new laws. The right to justice includes the right to have one's case examined by any government entity, but it must also have certain features to function. AI-based computerized courts cannot meet this goal because only judges can adjudicate independently. Given the rapid growth of technology and the changing environment, AI systems will soon be applied in the legal system. Many factors support the use of these systems, including the fact that they reduce court case review time, court costs, and analytic powers that exceed even the most skilled judges. However, such a metamorphosis has flaws. In this last context, the legal system's social functions must be considered. AI in the court system could make it less compassionate. To prevent illegal influence on judgments, such systems must be accessible and secure. Therefore, more study is needed to build an AI-based judicial execution of the legislation in the legislative and administration.

References

1. Nowotko PM (2021) AI in judicial application of law and the right to a court. Procedia Comput Sci 192:2220–2228
2. Plan A (2011) Communication from the commission to the European parliament, the council, the European economic and social committee, and the committee of the regions. European Commiss
3. AI H (2019) High-level expert group on artificial intelligence. Ethics guidelines for trustworthy AI, vol 6

4. Al-Alawi AI, A-Lmansouri AM (2023) Artificial intelligence in the judiciary system of Saudi Arabia: a literature review. In: 2023 international conference on cyber management and engineering (CyMaEn). IEEE, pp 83–87, Jan 2023
5. Partyk A (2021) Artificial intelligence and the administration of justice: remarks on independent judges. Artificial intelligence and the administration of justice: remarks on independent judges, pp 91–106
6. Abiodun OS, Lekan AJ (2020) Exploring the potentials of artificial intelligence in the judiciary. Int J Eng Appl Sci Technol 5(8):23–27
7. Xu Z (2022) Human judges in the era of artificial intelligence: challenges and opportunities. Appl Artif Intell 36(1):2013652
8. Baser P, Saini JR (2022) AI-Based intelligent solution in legal profession. In: ICT systems and sustainability: proceedings of ICT4SD 2022. Springer Nature Singapore, Singapore, pp 75–84
9. Davey M (2016) Chicago police try to predict who may shoot or be shot. New York Times, vol 21
10. Cheng KM (2018) Predictive analytics in the criminal justice system: media depictions and framing
11. Stewart P, Stuhmcke A (2020) Judicial analytics and Australian courts: a call for national ethical guidelines. Alternative Law J 45(2):82–87
12. Anderson T, Torreggiani WC, Munk PL, Mallinson PI (2020) The impact of the introduction of artificial intelligence in radiology and its potential legal implications in the UK and Ireland. BJRl Open, 2, 20200030
13. Sil R, Roy A, Bhushan B, Mazumdar AK (2019) Artificial intelligence and machine learning based legal application: the state-of-the-art and future research trends. In: 2019 international conference on computing, communication, and intelligent systems (ICCCIS). IEEE, Oct 2019, pp 57–62
14. European Commission for the Efficiency of Justice (CEPEJ) (2019) European ethical charter on the use of artificial intelligence in judicial systems and their environment strasbourg, Council of Europe
15. Washington AL (2018) How to argue with an algorithm: lessons from the COMPAS-ProPublica debate. Colo Tech LJ 17:131
16. Reiling AD (2020) Courts and artificial intelligence. In: IJCA, vol 11, p 1
17. Barrat J (2015) Our final invention: artificial intelligence and the end of the human era. St. Martin's Griffin, New York
18. Zuo WM (2018) Some thoughts on the application prospects of legal artificial intelligence in China. Tsinghua Law Science 12(2):108–124
19. Zuo WM (2020) From generalization to specialization: Rethinking the application of artificial intelligence in China's judicial system. Law Forum 35(2):17–23
20. Sourdin T (2018) Judge v Robot?: Artificial intelligence and judicial decision-making. Univ New South Wales Law J 41(4):1114–1133
21. Bathurst TF (2015) Advocate v Rumpole: who will survive? An analysis of advocates' ongoing relevance in the age of technology. Australian BAR Rev 40(3):185–194
22. Sourdin T (2016) The role of the courts in the new justice system. Available at SSRN 2713576
23. Searle J (2002) Can Computers Think? In: Chalmers DJ (ed), Philosophy of mind: classical and contemporary readings, vol 669. Oxford University Press, p 671

Broken Authentication and Its Significance in Protecting Online Applications: An Overview Paper

Md. Miskat Hossain, Mahamudul Hasan, Maisha Mahajabin, Abdur Rahman, Nusrat Maisha, and Pranta Nath Nayan

Abstract Broken authentication vulnerabilities pose significant risks to the security of web applications, potentially leading to unauthorized access, data breaches, and other detrimental consequences. This review paper presents a thorough assessment of the literature on broken authentication, highlighting various methods, constraints, and conclusions. This paper under consideration stresses the significance of fixing session management problems, putting robust authentication procedures in place, and giving priority to the security of login information and session cookies. The use of complicated passwords, avoiding session IDs in URLs, using robust encryption, and wiping session IDs after inactivity are important considerations. This review paper also goes over the value of manual testing and comprehending authentication schemas to spot and stop flawed authentication assaults. Overall, the review paper highlights the vital necessity for strong security measures, meticulous testing, and prioritizing the protection of authentication and session management procedures to maintain the security of online applications.

Keywords Broken authentication · Security · Unauthorized · Vulnerability · Credentials

1 Introduction

In today's digital age, security is of utmost importance for any organization that handles sensitive data. Authentication is a significant element of any secure system since it ensures that only authorized users have access to sensitive resources. Broken authentication, on the other hand, is a common vulnerability caused by authen-

Supported by organization x.

Md. M. Hossain (✉) · M. Hasan (✉) · M. Mahajabin · A. Rahman · N. Maisha · P. N. Nayan
East West University, A/2 Jahurul Islam Ave, Dhaka 1212, Bangladesh
e-mail: miskatbd42@gmail.com

M. Hasan
e-mail: munna09bd@gmail.com

S. Dutta et al. (eds.), *Cyber Intelligence and Information Retrieval*, Lecture Notes in Networks and Systems 1025, https://doi.org/10.1007/978-981-97-3594-5_5

tication processes that are defective, misconfigured, or easily guessable, allowing unauthorized access to user accounts and sensitive data. In web applications, broken authentication refers to inconsistencies in the authentication and session management processes that could let attackers access user accounts or other sensitive data without authorization. The Open Web Application Security Initiative frequently ranks this vulnerability as one of the top ten most critical web application security vulnerabilities (OWASP).

Broken authentication refers to a variety of security flaws that can be utilized by hackers to masquerade as authorized users online. Security professionals are very concerned about the potential of broken authentication attacks because they have been implicated in numerous large data breaches. This includes flaws in the handling of sessions and credentials, which enable attackers to access systems using hacked session IDs or stolen login credentials. Phishing and credentials that have been stolen are the most often used techniques in data breaches, according to Verizon's 2020 Data Breach Report [1]. Broken authentication, in contrast to other security flaws, cannot be found by automatic scanners, necessitating manual testing to assure the security of authentication procedures. The web application provides the user with a session ID for each visit. This ID is required for the program to interact with the user and answer queries. The stated principles can be used by protocol designers to create safe and user-friendly protocols. Forbes.com reported there are 2.5 quintillion bytes of data produced each day on a worldwide scale in addition statista.com also summarizes in January 2021 there were 4.66 billion active internet users worldwide [2]. Several broken instances of authentication attacks are:

(1) Session Hijacking: Checked session IDs can hijack user identities. Suppose a user forgets to log off from a public device. In that case, the same session ID that was previously generated for the original user can be used by some other person to continue the session.
(2) Session ID URL: The Session ID appears in the website URL and can be used by any person who accesses the URL through a wired or wireless network to impersonate the identity of the user.
(3) Credential Stuffing: Hackers access a database containing unencrypted user passwords and use techniques to assess if the passwords are correct and usable and protocols that protect against such attempts must be provided for a stable web application.
(4) Password Spraying: Password spraying refers to the use by hackers attempting to access protected accounts of the most common and poor passwords, such as 'password' or '123456.' Consequently, to prevent such attacks, minimum password specifications have been added.
(5) Phishing Attacks: To get users to reveal their login credentials, hackers give users links to websites that resemble the original web application. However, with due caution and by checking the web application in use, phishing attacks can be easily avoided [3].

The huge growth of information reported is also on par with the number of penetrations and cyber-crime attempts, these penetration attempts have various purposes, several penetrations are meant to take benefit, misuse the data for wrongdoing, and attempt to prove themselves, or have fun with the information [4]. Policies and multifactor authentication are just some of the many security measures that can be implemented to ensure the safety of user accounts and confidential information. However, these measures alone may not be enough to guarantee complete protection against cyber threats. Regular testing and auditing of the authentication process is crucial to identify any weaknesses or vulnerabilities that may exist. This allows for timely remediation and helps prevent potential breaches before they occur. A robust and secure authentication mechanism is essential for protecting against unauthorized access, which can lead to devastating consequences such as data theft or loss. Therefore, it is important for organizations to invest in comprehensive security protocols that are regularly reviewed and updated to stay ahead of evolving threats in today's ever-changing digital landscape.

This review paper will delve further into the idea of broken authentication. We will look at the many types of authentication flaws and how attackers exploit them. We will also look at the impact of broken authentication on web application security as well as the ramifications of these attacks for both businesses and users. Lastly, we will examine effective methods for preventing broken authentication and limiting its effects, such as strong password restrictions, multifactor authentication, and regular monitoring of authentication logs. Ultimately, the goal of this review article is to provide a thorough understanding of broken authentication and its significance in protecting modern web applications.

2 Literature Review

The purpose of the literature review section is to offer a thorough overview of the range of knowledge already available on broken authentication. The protection of sensitive data depends on finding solutions to this vulnerability, which has been a recurring issue in the realm of security. The following section (Table 1) displays the current methods and drawbacks of flawed authentication in online applications.

Based on [6], detect and mitigate cross-site request forgery (CSRF) and broken authentication and session management attacks in web applications. Cross-site request forgery attacks (CSRF) are maliciously used to send requests from an authenticated user to a web application. Since the target cannot see the responses to the modified queries, this is a very difficult attack strategy that works behind the scenes to cause disruptions. Instead of the initial data theft, CSRF attacks concentrate on state changes. CSRF may also lead to the loss of sensitive data. To be able to cope, the authors of this study proposed an effective system to detect CSRF and broken authentication and session management attacks in web applications. The system lowers the risks of unwanted access and data breaches by identifying and stopping these attempts. The authors focus on identifying and countering CSRF attacks. To find malicious scripts in incoming requests, it uses a packet tracker module. If no harmful material is discovered, a request checker policy is used to further evaluate

Table 1 Approach and limitation comparison

Reference	Approach	Limitations
[5]	Implementing in place defenses against attacks that compromise session management and authentication	It's possible that not all security threats have been addressed in full
[6]	With new rule libraries, an efficient test environment, and a simulation environment that detects CSRF vulnerabilities, weak authentication, and session management threats	This paper does not include an extensive analysis of the stated defensive strategy for dealing with CSRF, broken authentication, and session management attacks. Its scope, implementation details, and insights are limited, and it may not be applicable to other sorts of attacks or vulnerabilities
[7]	The attack using session misconfiguration and guessing/cracking weak passwords were shown to be the most effective ways to exploit the web application's Broken Authentication and Session Management vulnerabilities on those domains	In recent years, several tool combinations have been found to improve protection, but up to the publication of this study, it was not possible to combine the various tool types that are now on the market
[8]	This study presented five exploitation strategies that were assessed within the country and also applied the prevention strategies that a developer might use to lessen exploitation	The findings of the study offer a limited scope, lack of comparability, limited insights, and no solution for resolving the web application vulnerabilities that have been found. It concentrates on a specific web application that may or may not be indicative of other web apps or sectors
[9]	In this paper, a tool that can be used to find web application vulnerabilities in website source code is described. This tool aids web application developers in reducing web application vulnerabilities. Following the scanning procedure, it will produce a report listing all vulnerabilities and leaks found by presenting the name of the compromised file, its location, and its description	This paper has limitations in that it only works for asp.net and their supported language
[10]	The paper's technique is based on a dynamic inspection of the web application. The researchers developed a tool that can trawl online sites automatically, detect the session management techniques put to use by the application, and then examine them for vulnerabilities. To find and fix session management vulnerabilities, the tool uses a range of approaches, including pattern matching and state machine analysis. It can detect common vulnerabilities such as session fixation, hijacking, and poisoning of sessions	The tool does not take secure coding standards into account, generates false positives, lacks explanation, and cannot be compared to other automated vulnerability detection tools
[11]	The piece of research examines the performance of many popular online application security scanners, including Acunetix, AppScan, Burp Suite, and OpenVAS, by testing them on vulnerable web apps and assessing their ability to find vulnerabilities	The study simply compares a few web application security scanners, focusing on the OWASP Top 10 security concerns. It does not provide a full analysis of the vulnerabilities found by each scanner, making replication and extension of the investigations difficult for other researchers
[12]	The individual prevention architecture for SQL Injection, cross-site scripting, and Defective Authentication and Session Management is suggested in this work. This model can be used to defend a real-world website against outside threats	This work contributes to some improvement in online security against recent outside attacks. However, this strategy does not account for all online platforms and security issues

the request. However, if malicious information is found, an attack handler module raises an alert and examines the CSRF attack type. While reflected CSRF attacks are examined using the PhishTank database, stored CSRF assaults are examined using the request checker policy. For system enhancement, the detected attack parameters are saved. To improve security and thwart CSRF attacks, the design places a strong emphasis on distinct session IDs and makes use of whitelist rules, the PhishTank database, and whitelisting. The authors carried out experiments on a testbed made up of numerous web apps to determine the usefulness of the system. For the purpose of simulating actual-world situations, they gathered a dataset of both honest and dishonest requests. Following that, the system's detection accuracy, precision, and recall were evaluated using this dataset. The evaluation results showed that the implemented system achieved high accuracy in detecting and mitigating CSRF, broken authentication, and session management assaults. The precision and recall measures showed that, while limiting false positives and false negatives, the system successfully recognized and categorized malicious requests. The results and evaluation of the system showed high accuracy and effectiveness in detecting and classifying malicious requests related to CSRF and broken authentication and session management attacks. The system achieved a high precision and recall rate, indicating its ability to correctly identify and handle attack instances. The performance evaluation demonstrated the system's robustness and efficiency in protecting web applications against these vulnerabilities, outperforming other existing techniques.

The concept discussed in [7] revolves around the issue of broken authentication and session management in web applications, which has led to significant data breaches. Broken authentication refers to vulnerabilities that hackers utilize to pose as authorized users, most commonly due to session management and credential issues. Phishing attacks and the exploitation of stolen credentials have become popular ways to commit data breaches. Because broken authentication assaults are difficult to identify by automatic scanners, it is essential to manually test systems and understand the authentication schema in order to find and stop such attacks. It emphasizes that inadequate online application design, particularly flaws in session Management is a significant contributor to successful attacks. This suggests defensive measures to guard against broken authentication vulnerabilities, including requiring complex passwords, making sure session IDs are hidden in URLs, using strong encryption for data transfer, erasing session IDs after a certain amount of inactivity, and displaying generic error messages for failed authentication.

The use of various manipulation techniques to attack session misconfiguration, weak passwords through guessing or cracking, authentication issues, and decoding insufficient encryption in order to exploit these vulnerabilities. The importance of strong security mechanisms, careful testing, and prioritizing the protection of credentials and session cookies in order to reduce the danger of broken authentication and session management vulnerabilities in online applications. Figure 1 shows the authentication and session management process, Fig. 2 shows the exploitation of generally broken authentication and session management.

The paper [9] addresses how the World Wide Web has developed into a platform for web apps and emphasizes how popular web applications are growing because

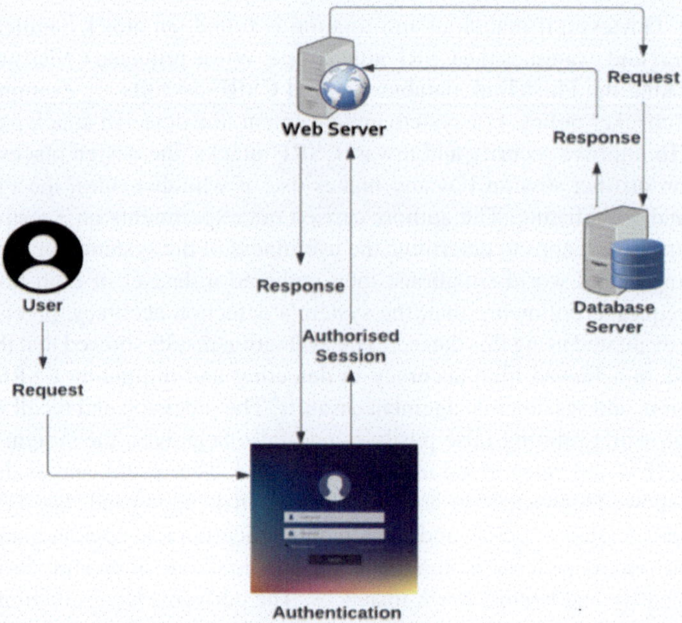

Fig. 1 Authentication and session management process [7]

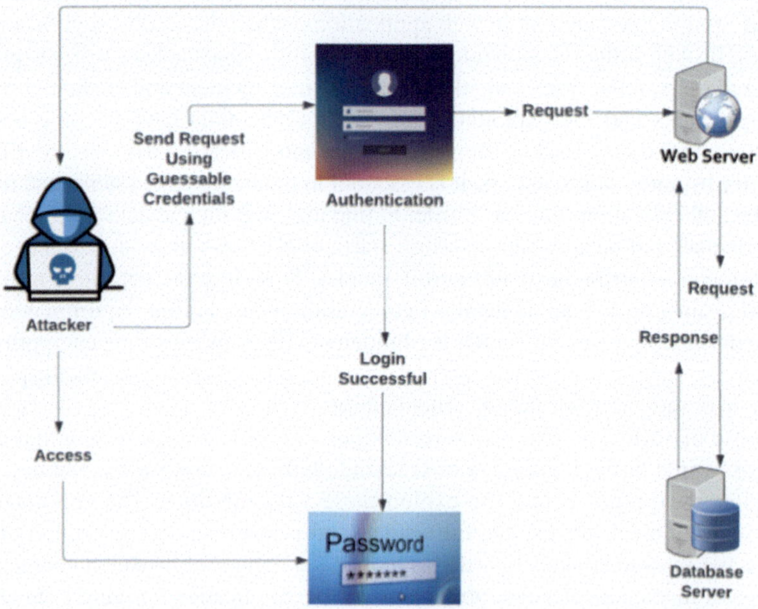

Fig. 2 Exploitation of generally broken authentication and session management [8]

of their remote accessibility, cross-platform interoperability, and quick creation. To protect websites, web applications, and web services, it underlines the significance of web application security. It focuses primarily on the authentication and session management weaknesses that are frequently not implemented effectively, giving attackers access to passwords, keys, session tokens, and other vulnerabilities that they can use to assume user identities. It highlights the effects of several attack types on web application security, including brute force assaults, session hijacking, replay attacks, and session fixation attacks. They have also recommended a Python algorithm to identify and fix broken authentication and session management flaws in web applications. The method comprises steps to check for various vulnerabilities, including implementing anti-cross-site scripting safeguards, correct input validation, session timeout settings, authentication settings, and the availability of security configuration settings. The algorithm's goal is to produce a report with a list of the web application's source code's vulnerabilities that were found. In conclusion, the chapter underlines how critical it is to fix issues with session management and broken authentication in online applications in order to assure their security. It emphasizes the necessity for security measures to be implemented properly from the start of the development process and for website/application owners to thoroughly check their programs for vulnerabilities before making them available to the public. The suggested method acts as a tool to help developers lessen vulnerabilities and boost the security of web applications.

Attacks on web applications that exploit session Management vulnerabilities are covered in [13]. Such assaults, which are the second most common according to the OWASP list, are highlighted as being of serious concern. This paper gives an overview of the different kinds, instances, and defenses against this vulnerability. Multifactor authentication (MFA) devices are also recommended for deployment and configuration as an additional security layer. To secure the security of the application, it underlines the value of keeping up with security reports like OWASP.

The importance of website security and the potential consequences of web application vulnerabilities are discussed in [14]. This research paper aims to provide developers with a place to start when addressing common vulnerabilities in their applications. It focuses on the top ten OWASP security flaws and aims to design and build a secure web application in accordance with OWASP standards, addressing security flaws like SQL injection, shoddy authentication, exposed sensitive data, shoddy access control, and XML external entities.

3 Results and Analysis

The findings from the review suggest that this broken authentication is one of the most common, vulnerable, and critical issues faced by web applications. In order to prevent unauthorized access and safeguard sensitive data, it also emphasizes the significance of putting strong authentication and session management mechanisms into place. In order to confirm these mechanisms' efficacy and find any potential vulnerabilities, some studies also advise routine testing and auditing.

The table (Table 1) contrasts various solutions for dealing with compromised authentication while highlighting their drawbacks. It is obvious that while many different approaches have been suggested, none of them completely address the security risks connected to weak authentication and session management. This highlights the issue's complexity and enduring nature.

Unreliable authentication must be addressed using a multifaceted approach. Strong password policies must be implemented, two-factor authentication must be enabled, user activity must be monitored and audited, and the most recent security patches and updates must be applied to software and systems. The risks associated with inadequate authentication cannot be entirely eliminated, but they can be greatly reduced. Research shows that insufficient session management and weak authentication can lead to significant data breaches. Attackers take advantage of session management and credential weaknesses by employing techniques like phishing attacks and the use of stolen credentials. To prevent such attacks, secure session management techniques must be used, and vulnerabilities must be regularly tested.

Using complex passwords, hiding session IDs in URLs, encrypting data with powerful algorithms, deleting session IDs after inactivity, and displaying generic error messages are some security precautions that should be taken to increase the security of session cookies and login credentials. To recognize and prevent weak authentication attacks, it's also important to test and comprehend the authentication schema.

The research analysis shows the ongoing challenges of addressing session management and compromised authentication flaws in web applications. Although strategies and suggestions have been made, more study, testing, and improved security protocols are still required.

4 Conclusion

In conclusion, broken authentication poses a serious risk to web services, opening the door to unwanted access to user accounts and sensitive data. It is a common issue caused by defective, misconfigured, or easily guessable authentication processes. Security experts are extremely concerned about this vulnerability because it has been linked to several data breaches. Attackers may be able to acquire access by abusing weaknesses in session management and credentials. These vulnerabilities cannot be found by automatic scanners, hence manual testing is required to find and fix them. The dangers associated with compromised authentication must be minimized through preventive measures such as strong password constraints, multifactor authentication, and routine monitoring of the authentication process. In today's digital environment, it is crucial for enterprises to make investments in thorough security processes and maintain their knowledge of new dangers.

References

1. Broken Authentication: What is it and how to prevent it. https://www.authgear.com/post/broken-authentication-what-is-it-and-how-to-prevent-it. Last accessed 12 May 2023
2. Internet and social media users in the world. https://www.statista.com/statistics/617136/digital-population-worldwide/. Last accessed 11 May 2023
3. Cyber. Cyber security. https://www.jigsawacademy.com/blogs/cyber-security. Last accessed 12 May 2023
4. Nugroho DCS, Hidayat MF, Soewito B (2022) Concurrent implementation of WAF and hardening broken authentication to secure web application. J Syntax Lit 7(8)
5. Nagpal NB, Nagpal B (2014) Preventive measures for securing web applications using broken authentication and session management attacks: a study. In: International conference on advances in computer engineering and applications (ICACEA) vol 2014
6. Nadar VM, Chatterjee M, Jacob L (2018) A defensive approach for CSRF and broken authentication and session management attack. In: Ambient communications and computer systems: RACCCS 2017. Springer, Singapore, pp 577–588. https://doi.org/10.1007/978-981-10-7386-1_49
7. Hanan A, Khawla A, Ruyuf A. Broken authentication and session management vulnerabilities. ISC Int'l J Inf Secur 13:11–19
8. Hassan MM, Nipa SS, Akter M, Haque R, Deepa FN, Rahman M, Siddiqui MA, Sharif MH (2018) Broken authentication and session management vulnerability: a case study of web application. Int J Simul Syst Sci Technol 19(2):1–11
9. Sharma R, Sheth R (2017) Secure ASP.NET web application by discovering broken authentication and session management vulnerabilities. Orient J Comput Sci Technol 10(2):359–363
10. Takamatsu Y, Kosuga Y, Kono K (2012) Automated detection of session management vulnerabilities in web applications. In: 2012 tenth annual international conference on privacy, security and trust. IEEE, pp 112–119
11. Joshi C, Singh UK (2016) Performance evaluation of web application security scanners for more effective defense. Int J Sci Res Publ (IJSRP) 6(6):660–667
12. Haque MF, Miah MBA, Al Masud F (2017) Enhancement of web security against external attack. Eur Sci J (ESJ) 13(15):228
13. Alahmad M, Alkandari A, Alawadhi N (2022) Survey of broken authentication and session management of web application vulnerability attack. J Eng Sci Technol 17(2):0874–0882
14. Lala SK, Kumar A, TS (2021) Secure web development using OWASP guidelines. In: 2021 5th international conference on intelligent computing and control systems (ICICCS). IEEE, pp 323–332. https://doi.org/10.1109/iciccs51141.2021.9432179

Skin Lesion Segmentation Using Deep Learning

Kaustav Sarkar, Varun Haralalka, Vishal Shaw, Rishav Raj Singh, Rayirth Jaiswal, and Tridip Pramanick

Abstract Dermoscopic image analysis has gained significant importance for dermatological applications due to its potential in eliminating observer bias. Segmenting lesion areas automatically in these images is crucial for expedited disease diagnosis. In this study, an innovative deep learning-based approach for precise lesion delineation has been proposed. Leveraging a deep convolutional neural network, where input images are included after appropriate preprocessing. Pedro Hispano Hospital (PH2) 2013 dataset has been employed, comprising dermoscopic images and lesion binary masks as ground truth for evaluation. The approach uses the SegNet architecture, which achieves remarkable accuracy of 99.14%, surpassing benchmark standards (DC 96.94, JI 99.11).This research offers a promising solution for automated lesion segmentation in dermoscopic images, advancing the field of medical image analysis.

Keywords CNN · Skin lesion segmentation · Deep learning · Dermoscopic images · Image augmentation · Neural networks

1 Introduction

A deep CNN architecture in this research has been provided to extract the lesion region from skin images. Segmenting skin lesions represents a pivotal phase within the realm of skin diagnostics process because it enhances both manual and computer-aided diagnostics by directing medical staff attention to certain areas of the skin. Deep neural networks are effective for the typical computer vision problem of image segmentation, which divides a digital image into many parts. For diagnosing patients, tracking treatment progress, and evaluating clinical trials for novel therapies, automatic and precise segmentation of lesions is crucial. From the detection of skin lesions to the identification of brain lesions, lesion segmentation is essential for many

K. Sarkar (✉) · V. Haralalka · V. Shaw · R. R. Singh · R. Jaiswal · T. Pramanick
Department of Computer Application and Science, Institute of Engineering and Management, Kolkata, India
e-mail: kaustav.sarkar@iem.edu.in

S. Dutta et al. (eds.), *Cyber Intelligence and Information Retrieval*, Lecture Notes in Networks and Systems 1025, https://doi.org/10.1007/978-981-97-3594-5_6

activities related to medical diagnosis. A critical stage in the diagnosis of melanoma involves the demarcation of the lesion area from the adjoining normal skin. This study introduces a technique that leverages the potency of deep learning methodologies to achieve precise extraction of the lesion region. The input image is primed through preprocessing to mitigate the impact of extraneous noise and artifacts. Subsequently, the image is channelled through a deep convolutional neural network (CNN). This CNN adeptly amalgamates both local and overarching contextual cues within the image, ultimately issuing a pixel-wise label assignment. This endeavor culminates in the creation of a segmentation mask, which delineates the contours of the lesion region.

2 Review of Existing Studies

Frequently employed segmentation methodologies encompass various image analysis techniques, with the thresholding approach being a prominent example [1–3], dynamic boundary model [4], Cluster analysis technique [5, 6], histogram statistical technique [7] and the technique of region growth [8–10], as well as techniques rooted in supervised learning, such as the support vector machine (SVM) paradigm [11] and deep learning. Primarily, these techniques heavily rely on the inherent attributes of the object itself, encompassing characteristics like color and texture [12] alternatively, acquiring knowledge from an annotated training dataset to execute the segmentation task, and for a comprehensive exploration of methodologies pertaining to the segmentation of pigmented skin lesions [13].

In the modern era, spurred by advancements in parallel processing through graphical processing units (GPUs) and the ready abundance of expansive training datasets like ImageNet, significant progress has been made [14], deep convolutional neural networks (DCNNs) have garnered significant interest for their applications in classification and recognition tasks. The journey commenced with the pioneering LeNet model, marking the inception of this transformative trajectory [15]. Subsequently, in contrast to the aforementioned, the AlexNet [16] and VGG-16 [17] models emerged. The core objective of this undertaking revolves around the identification of anomalies within dermoscopic images and the subsequent isolation of these anomalies from the adjacent skin area.

3 Methodology

3.1 Pre-Processing

Employing the PH2 dataset, encompassing a compilation of 200 dermoscopic images, forms the crux of this study. To foster optimal model performance, the dataset is partitioned, apportioning four-fifths for validation and one-fifth for testing. Consequently, this partitioning yields a training cohort comprising 180 images and a distinct test ensemble containing 20 images. To standardize the input, all images undergo resizing, transforming them into uniform 256 × 256 pixel dimensions.

Augmentation of images takes place within the training dataset, introducing transformations like horizontal flipping and random rotation. Consequently, the training dataset expands to encompass a total of 540 images. This augmented dataset is then partitioned into distinct training and validation sets, constituting the nourishment for the neural network's training process. Figure 1 shows instances of image augmentation involving rotations and flips of the masks were incorporated during the process. Figure 2 shows a generalized version of integration within the Convolutional Neural Network Framework.

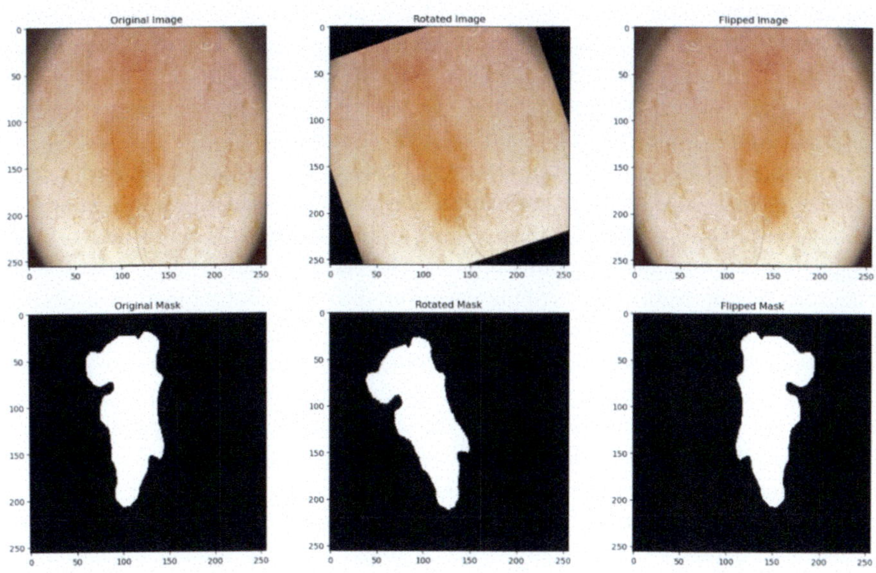

Fig. 1 Instances of image augmentation involving rotations and flips of the masks were incorporated during the process

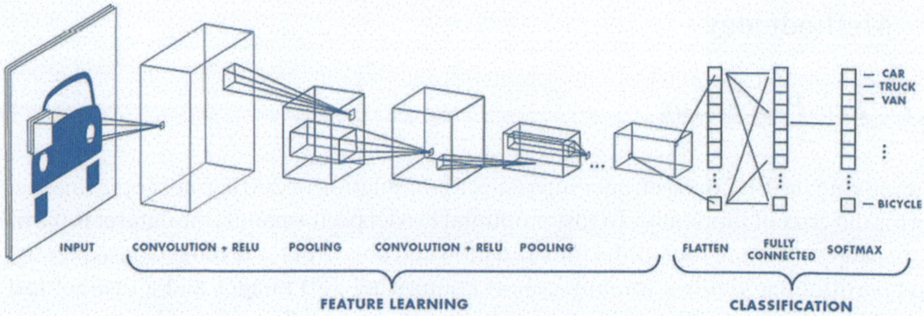

Fig. 2 A generalized version of integration within the Convolutional Neural Network Framework [12]

3.2 *Incorporation into the Convolutional Neural Network Paradigm*

Commencing the architectural construction of the neural network, the initial stride entails the creation of a convolutional layer, housing matrices of Feature Detectors. These matrices serve as adept filters that scan input images. These images are standardized in dimensions and are subjected to activation through the Rectified Linear Unit (ReLu) function. To perpetuate the refinement process, a strategic application of Max Pooling ensues. This pivotal maneuver not only engenders a reduction in spatial invariance but also serves to downsize the image, all while meticulously conserving the salient features pinpointed by the Feature Detectors.

Subsequent to this curatorial process, the feature map takes a metamorphic turn as it is flattened. This nuanced transformation morphs the matrix into a coherent columnar vector, prepped for the forthcoming journey through the neural network. At this juncture, the curtain rises upon our hidden layers, each harboring an assemblage of neurons. The activation duties of these neurons mirror the ReLu function, fostering non-linearity and increased model flexibility. Within this stratified architecture, the final layer stands poised to orchestrate the grand denouement, wielding the sigmoid activation function to issue forth the network's ultimate output. With this intricate maneuvering of layers and activations, the neural network is equipped to distill intricate patterns from input data and deliver refined predictions.

Consequently, the progression entails channeling the images into the neural network structure. Through the process of image augmentation, we ingeniously expand a moderately sized dataset, thereby magnifying its volume for the CNN to glean insights from. This dynamically augmented dataset, enriched in real-time, is then ingested by the CNN, embarking on a journey of enlightenment across epochs. At the culmination of each epoch, the CNN subjects itself to scrutiny via the lens of its loss function, a yardstick employed to gauge its accuracy in deciphering the underlying patterns. This introspection triggers a recalibration of the network's internal weights, priming it for even greater proficiency as the learning voyage persists. Concurrently, a designated test set stands as a sentinel to corroborate predictions, casting light upon the network's efficacy and precision in traversing the intricacies of the task at hand.

3.3 *Image Segmentation with the Proposed Method*

A shift is occurring in the realm of image processing, moving away from methodologies where both inputs and outputs consist of images. Instead, the trend is moving toward approaches where input remains static images, while the outputs encapsulate intrinsic attributes extracted from these images.

In this study, SegNet semantic segmentation model has been used which is an architecture comprised of encoders and decoders. A pixel wise classification layer follows it. An image is partitioned into its discrete regions or entities. The extent of this partitioning hinges upon the nature of the problem at hand, dictating the level of subdivision required. When an application's objects or regions of interest are identified, segmentation should come to an end. Due to the fact that one of the most challenging jobs in image processing is segmenting non-trivial pictures. One or two fundamental characteristics of intensity values, discontinuity, and similarity, are the foundation of segmentation algorithms. Implemented methodology hinges on segmenting an image into segments that share common attributes based on a predetermined set of criteria. Like, regions developing, combining, and dividing. The picture is divided or partitioned into a number of fragments. After partitioning the image into distinct segments, extraction of the essential components can be done from these segments for further image processing. Figure 3 shows the process flow diagram of the proposed system.

Initiating the process involves the application of image augmentation to the training dataset, a technique aimed at amplifying image variety and quantity within the training realm. Subsequently, this augmented dataset undergoes division into two subsets: the validation set and the training set. These delineated sets are then presented as input to the model.

The model's response yields predictions, notably encompassing both the mask representing the spatial layout and the segmented lesion extracted from the unaltered source image. To gauge the model's efficacy, this prediction output is subjected to testing against a designated test dataset. Evaluation metrics, carefully chosen, form the yardstick by which the model's accuracy is measured within this test scenario.

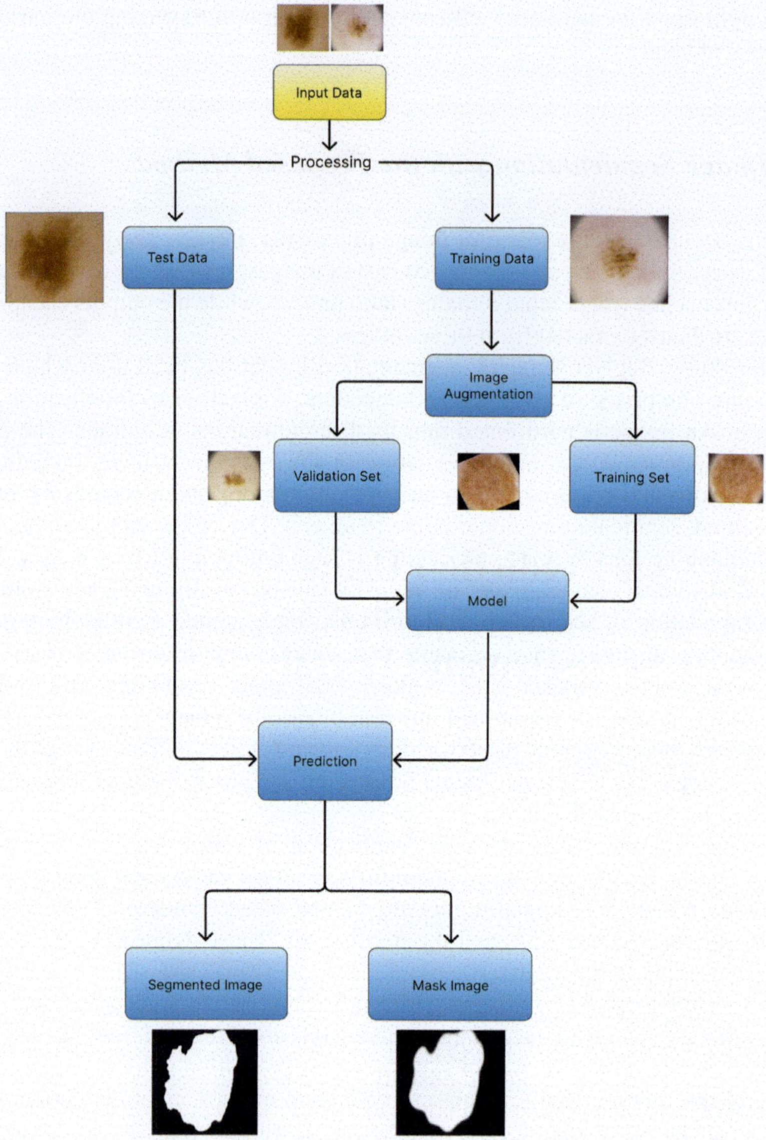

Fig. 3 Process flow diagram of the proposed system

3.4 Evaluation Metrics

Accuracy: Accuracy is quantified as the proportion of accurate predictions in relation to the complete count of predictions executed by the model.

$$Accuracy = \frac{\mathbf{True}_{positive} + \mathbf{True}_{negative}}{\mathbf{True}_{positive} + \mathbf{True}_{negative} + \mathbf{False}_{positive} + \mathbf{False}_{negative}}$$

Dice Coefficient: The Dice coefficient emerges as a widely employed measure for quantifying the likeness between a pair of data sets. These sets could manifest as binary images or delineated segments. The metric's formulation involves computing the proportion of the overlap shared by the two sets in relation to their collective size average.

$$Dice = \frac{2 \times \left(\mathbf{True}_{\mathbf{positive}}\right)}{\left(\mathbf{True}_{\mathbf{positive}} + \mathbf{False}_{\mathbf{positive}}\right) + \left(\mathbf{True}_{\mathbf{positive}} + \mathbf{False}_{\mathbf{negative}}\right)}$$

Jaccard Index (Intersection over Union): The assessment of resemblance between two segmentation masks—one delineating the ground truth and the other embodying the projected segmentation—entails the utilization of the Jaccard index.

$$Jaccard\ \mathbf{Index} = \frac{\mathbf{Intersection}}{\mathbf{Union}}$$

4 Results and Discussions

The comparison of the results with those emanating from specific alternate models, coupled with the findings from various Deep Learning architectures in the context of skin lesion segmentation, can be delineated in the below mentioned table (shown in Table 1). Figure 4 shows the outcomes are presented alongside the original images, their corresponding ground truth, and the predicted results, showcasing the segmented skin lesions.

Table 1 Outcomes from deep learning approaches for the segmentation of Skin Lesions

Ref	Year	Dataset	DC	JI	ACC
Peng Tang et al. [18]	2019	ISIC 2017	86.93	79.26	94.31
Xie et al. [19]	2019	PH2	91.9	85.7	94.9
Lei Bi et al. [20]	2019	PH2	92.10	85.90	95.30
Al-masni et al. [21]	2018	PH2	91.77	84.79	95.08
Peng Tang et al. [18]	2019	PH2	94.23	94.23	97.60
Proposed study	2023	PH2	**96.94**	**99.11**	**99.14**

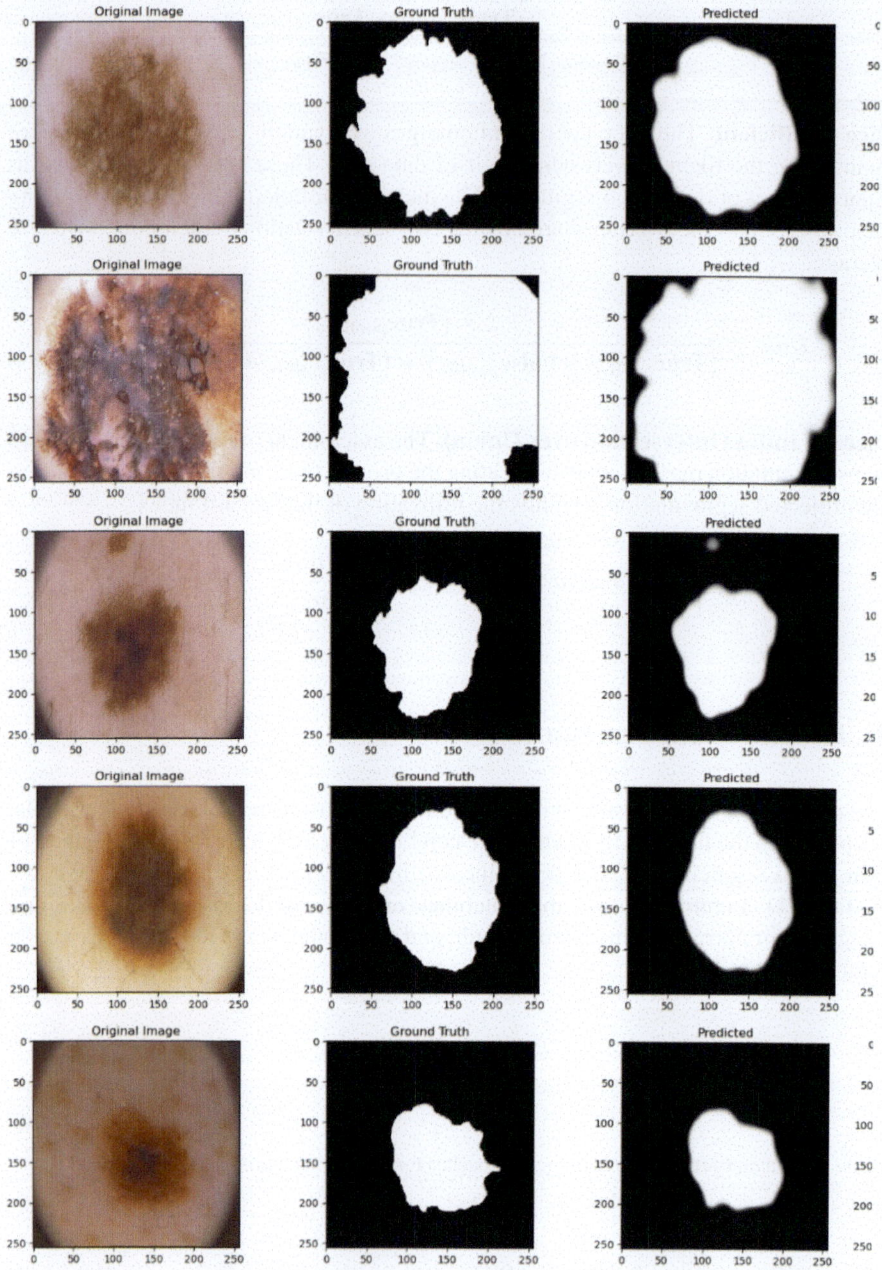

Fig. 4 The outcomes are presented alongside the original images, their corresponding ground truth, and the predicted results, showcasing the segmented skin lesions

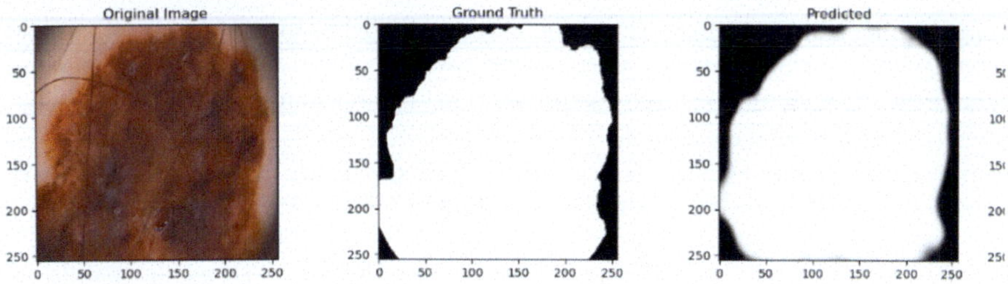

Fig. 4 (continued)

5 Conclusion

The results presented in this manuscript demonstrate the effectiveness of the innovative strategy for delineating skin lesions using convolutional neural networks, with the SegNet architecture at its core. The utilization of the PH2 dataset for both model training and assessment has yielded exceptional outcomes. Notably, the proposed model achieved an impressive accuracy rate of 99.15% in deciphering skin lesions from medical images. This remarkable achievement marks a significant advancement when compared to existing segmentation methodologies. It highlights the potential for deep learning approaches to redefine the field of medical image analysis.

The success of the proposed technique can be attributed to SegNet's ability to handle both localized intricacies and the overall context of the images. This dynamic interplay empowers the architecture to not only capture subtle details within the images but also maintain their spatial coherence, a crucial aspect in accurate medical image analysis. Additionally, the augmentation maneuvers, such as flips and rotations, integrated into the training process have enhanced the model's capacity to generalize and mitigate the risk of overfitting, making it a robust tool for real-world applications.

The implications of this study extend far beyond the confines of the research laboratory. In the realm of medical image scrutiny, especially in dermatology, the precision of the proposed skin lesion delimitation holds great promise. It could pave the way for improved diagnosis and treatment of cutaneous disorders, ultimately leading to enhanced patient outcomes. The potential impact on patient care is substantial, making this research highly relevant in the medical community.

Looking ahead, future investigations should consider deploying this approach across larger and more diverse datasets, exploring its applicability in various medical image analytics domains. The promising results obtained on the PH2 dataset suggest that this methodology may revolutionize the way we analyze medical images, opening doors to a wealth of opportunities for research and clinical applications. Moreover, the proposed approach might serve as a template for the development of similar models in other medical specialties, further expanding the frontiers of medical image analysis.

In summary, this research serves as a compelling testament to the efficacy of convolutional neural networks, particularly when coupled with the SegNet architecture, for the task of skin lesion segmentation. The outstanding 99.15% accuracy achieved on the PH2 dataset underscores the potential of this methodology to transform medical image analysis. It is our hope that this study will inspire and drive future explorations in this rapidly evolving field.

References

1. Peruch F, Bogo F, Bonazza M, Cappelleri V-M, Peserico E (2014) Simpler, faster, more accurate melanocytic lesion segmentation through MEDS. IEEE Trans Biomed Eng 61(2):557–565. https://doi.org/10.1109/TBME.2013.2283803
2. Emre Celebi M, Wen Q, Hwang S, Iyatomi H, Schaefer G (2013) Lesion border detection in dermoscopy images using ensembles of thresholding methods. Skin Res Technol 19(1):e252–8. https://doi.org/10.1111/j.1600-0846.2012.00636.x
3. Garnavi R, Aldeen M, Celebi ME, Varigos G, Finch S (2011) Border detection in dermoscopy images using hybrid thresholding on optimized color channels. Comput Med Imaging Graph 35(2):105–115. https://doi.org/10.1016/j.compmedimag.2010.08.001
4. Ma Z, Tavares JMRS (2016) A novel approach to segment skin lesions in dermoscopic images based on a deformable model. IEEE J Biomed Health Inform 20(2):615–623. https://doi.org/10.1109/JBHI.2015.2390032
5. Zhou H, Schaefer G, Sadka AH, Celebi ME (2009) Anisotropic Mean Shift Based Fuzzy C-Means Segmentation of Dermoscopy Images. IEEE J Sel Top Signal Process 3(1):26–34. https://doi.org/10.1109/JSTSP.2008.2010631
6. Ashour AS, Hawas AR, Guo Y, Wahba MA (2018) A novel optimized neutrosophic k-means using genetic algorithm for skin lesion detection in dermoscopy images. Signal Image Video Process 12(7):1311–1318. https://doi.org/10.1007/s11760-018-1284-y
7. Gómez DD, Butakoff C, Ersbøll BK, Stoecker W (2008) Independent histogram pursuit for segmentation of skin lesions. IEEE Trans Biomed Eng 55(1):157–161. https://doi.org/10.1109/TBME.2007.910651
8. Iyatomi H et al (2008) An improved Internet-based melanoma screening system with dermatologist-like tumor area extraction algorithm. Comput Med Imaging Graph 32(7):566–579. https://doi.org/10.1016/j.compmedimag.2008.06.005
9. Celebi ME et al (2008) Border detection in dermoscopy images using statistical region merging. Skin Res Technol 14(3):347–353. https://doi.org/10.1111/j.1600-0846.2008.00301.x
10. Emre Celebi M, Alp Aslandogan Y, Stoecker WV, Iyatomi H, Oka H, Chen X (2007) Unsupervised border detection in dermoscopy images. Skin Res Technol 13(4):454–462. https://doi.org/10.1111/j.1600-0846.2007.00251.x
11. Arakeri MP, Reddy GRM (2015) Computer-aided diagnosis system for tissue characterization of brain tumor on magnetic resonance images. Signal Image Video Process 9(2):409–425. https://doi.org/10.1007/s11760-013-0456-z
12. An N-Y, Pun C-M (2014) Color image segmentation using adaptive color quantization and multiresolution texture characterization. Signal Image Video Process 8(5):943–954. https://doi.org/10.1007/s11760-012-0340-2
13. Celebi ME, Wen Q, Iyatomi H, Shimizu K, Zhou H, Schaefer G, 4 A state-of-the-art survey on lesion border detection in dermoscopy images
14. Russakovsky O et al (2015) ImageNet large scale visual recognition challenge. Int J Comput Vis 115(3):211–252. https://doi.org/10.1007/s11263-015-0816-y
15. Lecun Y, Bottou L, Bengio Y, Haffner P (1998) Gradient-based learning applied to document recognition. Proc IEEE 86(11):2278–2324. https://doi.org/10.1109/5.726791

16. Krizhevsky A, Sutskever I, Hinton GE, ImageNet classification with deep convolutional neural networks. http://code.google.com/p/cuda-convnet/
17. Simonyan K, Zisserman A (2014) Very deep convolutional networks for large-scale image recognition, Sep 2014. http://arxiv.org/abs/1409.1556
18. Tang P et al (2019) Efficient skin lesion segmentation using separable-Unet with stochastic weight averaging. Comput Methods Programs Biomed 178:289–301. https://doi.org/10.1016/j.cmpb.2019.07.005
19. Xie F, Yang J, Liu J, Jiang Z, Zheng Y, Wang Y (2020) Skin lesion segmentation using high-resolution convolutional neural network. Comput Methods Programs Biomed 186:105241. https://doi.org/10.1016/j.cmpb.2019.105241
20. Bi L, Kim J, Ahn E, Kumar A, Feng D, Fulham M (2019) Step-wise integration of deep class-specific learning for dermoscopic image segmentation. Pattern Recognit 85:78–89. https://doi.org/10.1016/j.patcog.2018.08.001
21. Al-masni MA, Al-antari MA, Choi M-T, Han S-M, Kim T-S (2018) Skin lesion segmentation in dermoscopy images via deep full resolution convolutional networks. Comput Methods Programs Biomed 162:221–231. https://doi.org/10.1016/j.cmpb.2018.05.027

Information Retrieval

AI in Subconscious Language Learning for Error Remediation

Nayantara Mitra and Ayanta Banerjee

Abstract Artificial intelligence involves machines, particularly computer systems, simulating human intelligence processes. It enables machines not only to replicate but also enhance the capabilities of the human mind. Whether in the advancement of self-driving vehicles or the widespread use of intelligent assistants such as Siri and Alexa, AI is increasingly integrated into our daily lives. The Theory of Second Language Acquisition was developed by linguist and University of California professor Stephen Krashen. According to his approach, students should not focus on grammatical exercises if that is not how they choose to learn a language. He contends that whereas language acquisition is a subconscious process more in line with how children learn languages, language learning is an active, conscious process that imparts a language's norms. This approach is used by apps like LingQ and Rosetta Stone, which aim to teach languages through association and experience rather than through grammar rules. It is practically hard for a teacher to determine the best strategy for every student in a class of 25 pupils. Yet, the unique requirements of each learner can be addressed through the utilization of AI for language acquisition. AI-powered language learning platforms empower learners to progress at their own speed, reinforcing challenging concepts and emphasizing areas of difficulty. These platforms engage learners with tasks in which they excel, align with their interests, and consider factors such as cultural background. Additionally, the data generated provides teachers with insights into their students' cognitive processes, enabling them to anticipate future performance.

Keywords Artificial intelligence · Second language learning · Error correction · Remediation · Computer-assisted language learning (CALL) · Intelligent tutor (ITS) · Daedalus integrated writing environment (DIWE)

N. Mitra (✉) · A. Banerjee
Institute of Engineering & Management, Kolkata, India
e-mail: nayantara.mitra@iem.edu.in

A. Banerjee
e-mail: ayanita.banerjee@uem.edu.in

S. Dutta et al. (eds.), *Cyber Intelligence and Information Retrieval*, Lecture Notes in Networks and Systems 1025, https://doi.org/10.1007/978-981-97-3594-5_7

1 Introduction

An ongoing debate in second language acquisition (SLA) revolves around the question of whether to rectify the mistakes made by second language learners. Cognitive theorists assert that the crux of learning a second language lies in paying attention to [10] and noticing [11] the language features. This stands in contrast to nativists' [7] belief that language can be acquired without explicit attention. In alignment with the cognitive perspective, correction seems advantageous as it channelizes the learner's attention to specific language attributes, compelling them to observe and internalize systemic characteristics.

There is some evidence that methodically correcting common errors helps people learn second languages thus far [4]. This essay argues that error repair has real advantages that could eventually result in remediation [6] or the elimination of errors. Additionally, it asserts that artificial intelligence can be a highly helpful tool for remediating and correcting errors in second languages.

Around the world, L2 learners appear destined to make errors that immediately identify them as non-native speakers (NNS) when they are trying to learn a second language in addition to their mother tongue. Native speakers of the target language sometimes respond to the nature of their expression with tolerance and understanding, whereas other times, patience and understanding appear to wane. On the other hand, once the fundamental concepts and methods for using computer programs for language learning have been mastered, artificial intelligence (AI) can handle new issues. For instance, a computer program should contain the identical valid and improper forms to identify a student's inaccurate sentence. The system should also include a record of incorrect rules in this way, and so that the computer program can inform the pupil as to why the sentence contains problems.

Given the considerations mentioned above, this study aims to underscore the importance of technological applications and programs in addressing linguistic errors within the context of second language acquisition (SLA). The study outlines the key considerations that should be taken into account when incorporating computer applications into the realms of teaching and learning, language acquisition, and autonomous natural language processing.

2 Historic Framework of CALL

Computer-assisted language learning, or CALL, is a category of education technology created for using as a tool for learning. It simply refers to the use of computer programs for language learning and instruction. For many years, opportunities produced by the teacher in the classroom have been the only way to teach foreign languages, together with material gleaned from text books, tapes, and recordings, among other sources. Alternative methods of teaching languages, such as distant learning and self-directed learning online, were created in the 1990s with two goals in mind: first, to conserve resources, and second, to lessen the time and geographic restrictions associated with completing a foreign language course.

The integration of computers into language teaching and learning has indeed resulted in noteworthy enhancements in the pedagogy of second languages. This includes more efficient learning within a shorter time frame, increased retention of knowledge, and enhanced communicative proficiency [2, 14].

Commencing in the 1950s and implemented during the 1960 and 1970s, the initial phase of computer-assisted language learning (CALL) was grounded in the Behaviorist Theory of the time. This theory emphasized the repetition of language exercises [1, 2, 14]. At this stage, the concept revolved around the belief that a computer could function as a tutor, with arguments asserting: (1) computers were ideal for executing repetitive exercises as they could repeat the same linguistic material without fatigue or errors, while also providing immediate feedback; (2) presenting content individually to each student allowed computers to facilitate success by enabling learners to progress at their own pace.

In conclusion, the Behaviorist Theory put an emphasis on language skill development. Grammar or focus-on-form activities received the majority of its attention. Two significant developments in the 1970 and 1980s undercut Behaviorist CALL. The first reason is that behaviorist methods of teaching languages have been discredited in 5 theory and pedagogy.

Secondly, because the introduction of the microcomputer prepared the door for a new stage of CALL and a new variety of possibilities. The communicative theory of education, which became popular in the 1970 and 1980s, served as the foundation for the current phase of CALL [14]. The proponents of this approach contended that true communication was not given enough value by the exercise and practice programs of the preceding decade. Memorization was not appreciated as highly as creativity in self-expression. The negotiation of meaning had grown in strength and significance. In schooling, comprehension had grown in importance.

In other words, CALL's usefulness depends more on how it is applied into the concept of teaching and learning of languages than on the medium itself. This claim made by Garrett has a lot of merit because the computer's greatest promise for language learning comes in its capacity to offer a setting where each learner is encouraged to grow, develop, and improve their own language and communication skills in a foreign tongue. The development of their potential will undoubtedly have a considerable impact on how languages are taught and learned in this century, therefore which should be the focus of our efforts.

3 Literature Survey

From a variety of angles, AI is exciting to linguists and language instructors. Among other things, artificial intelligence (AI) and natural language processing (NLP) are being utilized to develop better-processed, more accurate representations of natural languages for gaining a better understanding of the mental processes that take place in human brains during verbal interaction. In computer linguistics, AI-powered technologies are also employed for the creation of computer languages, machine translation, and, among other things, improving human–machine interaction through speech recognition and speech synthesis.

Artificial intelligence (AI) can be used to simulate the actions of a teacher or a student in language learning and teaching activities. A machine must possess the "knowledge" of instructional techniques for mimicking the actions of a language teacher. A machine has to understand learning styles and tactics in order to replicate a learner. But in particular, both of these imitations demand linguistic expertise. Computational linguistics is a term used to describe the field that represents linguistic knowledge within AI [8].

Computational syntax has been used most commonly in computational linguistics to aid in the acquisition of second languages [8]. This branch of computational linguistics is in charge of automating the segmentation of sentences into elements of speech or natural language processing. Parsers are computer programs created to carry out this type of analysis [12]. Intelligent tutoring systems, or examples of intelligent computer-assisted language learning (ICALL), are being used in SLA that contains parsers (e.g., Hamburger et al., 1999; Holland et al., 1993). In order to remedy learner errors, many such systems have been created till date [5, 13].

Though the language produced by the student is meant to include non-standard elements, often categorized as errors, it is anticipated that the teacher's language adheres to correctness and conforms to the standard (James, 1998). Merging these two linguistic models poses a challenging task for a parser. The parser needs to be meticulous and unforgiving of errors to uphold the accepted and accurate language. The tool examined in here is the Intelligent Tutor, a software application crafted to recognize and correct prevalent mistakes committed by mature learners of the English language. Following specific mal-rules or bug rules [5], the program is capable of identifying a grammar. In essence, this implies that the program understands the student's interlanguage and includes a built-in mechanism for recognizing incorrect grammar. As indicated in [9], parsers are expected to be able to recognize all appropriate language expressions. An argument could be made that even native speakers, to differing extents, achieve this task, providing rationale for the presence of parsers.

Given that humans themselves occasionally fall short, it is reasonable not to expect a computer to unfailingly excel in such situations. Accepting the performance of a parser that efficiently processes specific language elements is fitting, especially as pedagogical demands often target isolated language features rather than the entirety of language at all times [3]. This becomes particularly relevant when there is a shortage of human tutors to address all students' needs. Parsers contribute to the authenticity of the learning experience when employed alongside accurate simulations of real-world scenarios [1]. This research suggests that students should validate their hypotheses regarding the second language (L2) through ongoing encounters with novel instances of language construction. Consequently, parsers prove valid and can generate significant learning benefits. As indicated by [1], the study affirms the positive impact of the Intelligent Tutor on learning outcomes.

There are many different suggestions about how to fix mistakes. Some integrationists, for instance, assert that this is the cognitive intrusion that has the least disruptive chance of producing the intended healing effect. Lyster and others counter that recasts do not force learners to self-correct despite the useful contrarian evidence they provide.

4 Errors and Correction in Second Language Learning

Many AI researchers were bullish about the possibility of a day when a computer might think like a human from the 1950s through the 1970s. Alan Turing and other experts in the area thought that over the following ten years, humans will be able to transcend the constraints of computing power. During this time, the first CAI programs that presented information to learners linearly were created. These early initiatives consisted of computerized flashcard systems that allowed learners to discern between correct and erroneous replies. Although these programs offered a variety of benefits, they also had drawbacks, such as: (a) The linear order of teaching encouraged a passive attitude to learning. (b) Students were unable to use natural language to come up with solutions. (b) Multiple-choice tests were required of learners, which reduced their drive, initiative, and inventiveness.

To get over these restrictions, CAI programs were created that provided the learner with information depending on their responses to various questions. These CAI programs, also known as "adaptive CAI" or "frame-oriented CAI," served as the basis for intelligent tutoring system (ITS). In order to determine the material and pedagogical approaches required for a given learning session, intelligent tutoring system actually performs experiments with the students.

The majority of ITS developers agreed that an intelligent tutoring system consists of four parts: the evaluation module, the teaching module, the expert module, and the student mode.

Expert Module—Technology is used to convey the whole course by CAI lecturers. The development of ITS was motivated by the need to teach a modeled curriculum. There is a presumption that because the system must function like an experienced human educator, it must have access to and be knowledgeable about all of the domain's knowledge.

Student Model—ITS needs to grasp the student's domain, learning patterns, and individual learning styles in order to perform better. Some ITS just concentrated on simulating the student's knowledge and did not take the learner's individual learning style into consideration. The development of a system that takes into account student preferences, learning styles, and knowledge is a significant challenge. When this methodology evaluates student knowledge through ongoing engagement or a specific "test-like" circumstance, it is successful.

Evaluation Module—Because an ITS is an integrated system of learning, it can occasionally be challenging to concentrate on just one component of it. The teaching module and the student model overlap significantly with the assessment module, which also assesses the knowledge and preferences of the learner. The assessment procedure might include conventional examinations, surveys, or ongoing evaluation of students based on their general contact with the ITS.

Tutorial Module—Through an ITS instructional module, fresh content is taught to pupils. The method used to access the instructional module is where there is the most variance among ITSs. Within one ITS, several learning methods may be employed. Some ITSs may engage in sophisticated dialogical engagement with the learner, while others are problem-driven. The lesson module can be influenced by the subject content and the creators' implicit or explicit educational philosophy.

5 Methodology

In the Intelligent Tutor's design, the Intelligent Tutor's capacity to identify and fix some learner faults is based on the errors' frequency, seriousness, and communicational relevance. Taking into account the needs of the target learners in particular as well as their familiarity with the previously described frameworks. One may say that this program uses a mixed approach to development.

The method is able to correct a few minor morphological flaws in addition to seven main structural faults. The program itself is structured similarly to the Daedalus integrated writing environment (DIWE) program as an instructional interaction between the learner and the machine. The key difference between these two programs lies in the fact that DIWE does not utilize artificial intelligence; instead, it facilitates peer input and interaction among multiple learners. In contrast, the Intelligent Tutor employs a parser to deliver feedback.

Diverging from DIWE, this approach aims to assist students not only in the essay writing process but also in identifying and addressing learner flaws, although the latter is not its primary objective. The computer engages students by posing questions related to the topic and readings, with the requirement that each inquiry elicits a complete sentence from the student. This questioning approach is designed to target specific error types for rectification, as research on learner errors suggests its efficacy in language teaching. If the student's response is grammatically correct, she receives commendation along with the full parse tree of her sentence. In the case of a grammatically incorrect response, the algorithm notifies her and presents three options: try again, receive a suggestion, or obtain the solution. To illustrate, let us consider an example (Table 1).

One of the prompts created to induce the negative building error is, "Can the human immune system effectively fight malaria?" The user is then instructed to incorporate the ergative verb "fail" in their response to potentially commit the ergative construction mistake. Subsequently, the user likely provided a response and it is

Table 1 Some major error types recognized by the Intelligent Tutor

Error type	Example
Finite/non-finite verb	It will caused death of both mother and baby
Missing copula	Secondly, communities * affected
Distorted equivalent expressions of feelings/reactions/states	The disease had * dominant over human
Existential construction	There is a new problem occur
Challenging movement	More difficult to be realized...
Ergative construction	The immune system can be failed
Pseudo-passive	Malaria can find all over the world

confirmed that they did, saying, "The immune system can be failed." The initial part of the response highlights the error in the statement and presents three alternatives. If the learner opts for the hint, the computer guides them to change the phrase "be failed" to an infinitive. This choice is illustrated in Fig. 1.

Now the learner has a greater opportunity to revise her phrase. In case she succeeds, she receives a parse, as seen in Fig. 2. Figure 3 illustrates how the computer will answer with a grammatically sound statement if the solution is chosen as an option.

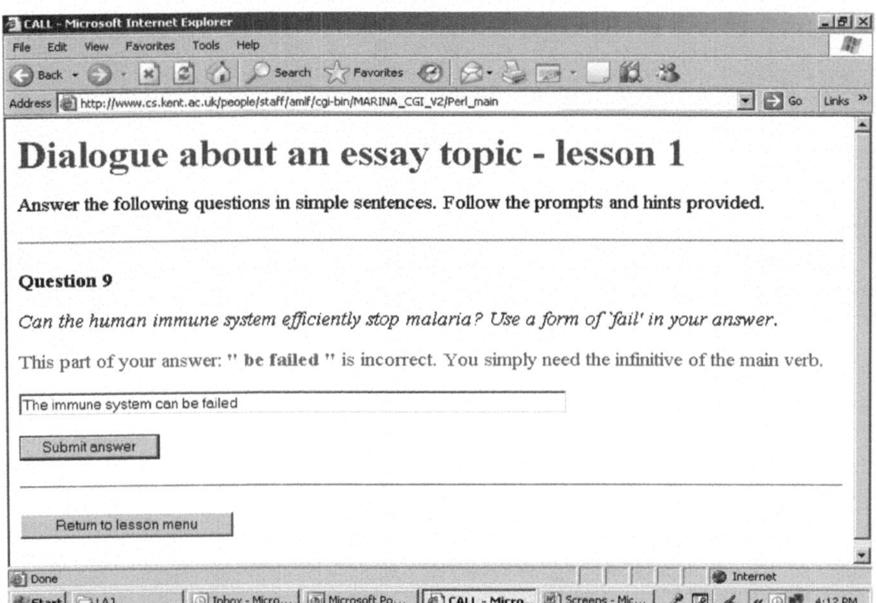

Fig. 1 System provides a hint

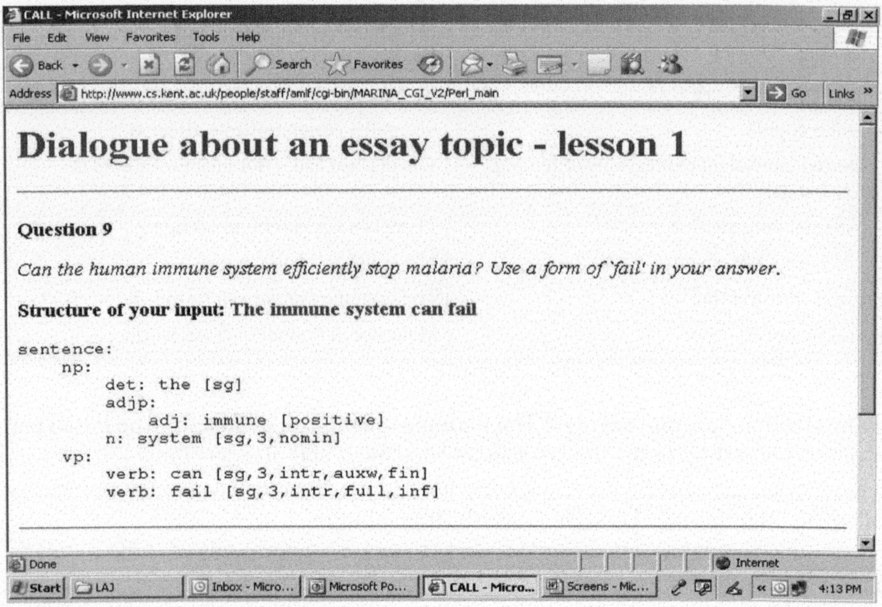

Fig. 2 Correct sentence rewarded with a parse

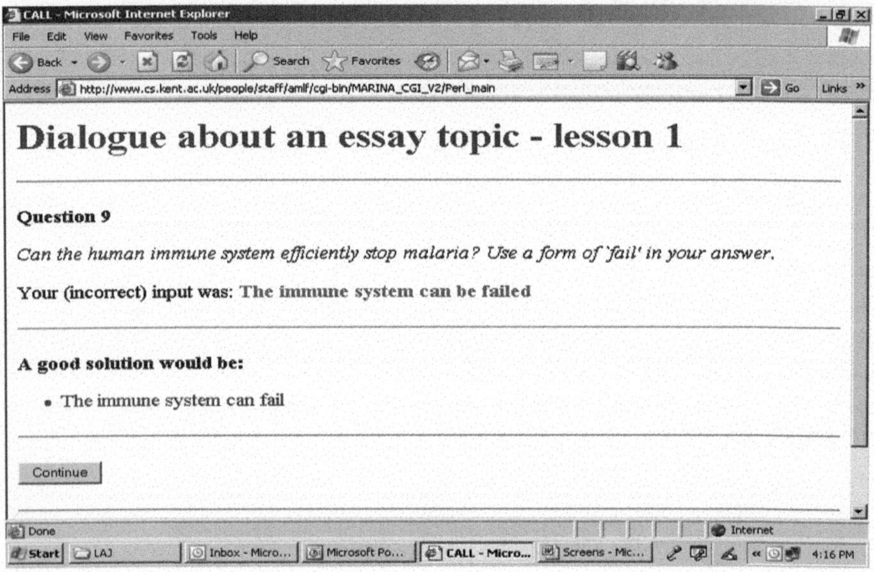

Fig. 3 Solution option

6 Study Design

Does interaction with the Intelligent Tutor, which involves systematic error rectification, influence the learning process? was the study's research question. The Intelligent Tutor (treatment)'s systematic mistake correction is hypothesized to have a huge impact on the learning results. Due to this, the study's design was limited to one group and one treatment method, as well as a pre- and post-test covering seven common structural defects and a few morphological ones.

In the experimental group, the Intelligent Tutor is utilized; however, there is no control group employing a different resource or receiving no treatment, as this study operates within a quasi-experimental setting. This is true in several ways, not the least of which is that the study's main goal is to ascertain whether and to what extent there is improvement after using the Intelligent Tutor. Second, the goal of this research was not to establish that the Intelligent Tutor was superior over a textbook, an instructor, or any other incompatible resource.

A grammar judgment exam is used as the pre-test [4]. The participants in this test must evaluate the grammaticality of statements in their literal sense. Their success on this exam is therefore interpreted as an indication of their familiarity with or proficiency with these structures [4]. The fact that the grammaticality judgment test will most likely disclose genuine competence issues rather than typos or insignificant performance faults is another justification for using it [4].

The pre-test consists of 12 questions formatted as a multiple-choice exam. Strikingly, each question closely resembled those posed by the Intelligent Tutor. Consequently, students were required to produce one or more precise paraphrases of the original language for each question. Consistent test results indicated that students in the test case consistently made four specific errors, including challenging movement, ambiguity in part of speech, confusion between finite and non-finite verbs, and errors in ergative construction. The short-response questions in the post-test provide students with the chance to create their own sentences. To prevent the influence of basic learning from the pre-test on the post-test results, the format is intentionally different from that of the pre-test. The degree of mistake identified in the sentences submitted by the students can be used to gage progress. Even though this tool focuses on expected errors, all mistakes made by students are logged, regardless of whether they were what was expected or not. The study's piloting has demonstrated that few unexpected mistakes really happen. The work with the Intelligent Tutor serves as the quasi-experimental task equivalent to the therapy process. To ensure variation, the individuals are drawn from three distinct target populations (Taiwan, Australia, and the UAE). There is just one therapy recommended; hence, there is no randomization or pair matching. Therefore, the subjects are college students or candidates between the ages of 19 and 21, with a TOEFL score of around 500.

After finishing the pre-test and before taking the post-test, the students had the chance to work with the Intelligent Tutor during block times. The pupils were given a number of literatures on the subject of malaria at the start of the session. The texts were changed to correspond to the pupils' TOEFL scores, which ranged from 500 to 550. While working alone, the students were free to ask their classmates or the teacher questions.

A total of 266 students served as participants in this study. Among them, 107 were located in Sydney, Australia (83 of who were in Sydney), 77 in Taichung, Taiwan, and 60 in the United Arab Emirates (UAE), specifically in the emirates of Sharjah and Dubai. The participating institutions included—Ling Tung College in Taichung, the American University of Sharjah (AUS), Zayed University (ZU), In Search Institute at the University of Technology Sydney, and lastly, the American University of Sharjah (ZU). The study spanned several months, conducted between 2004 and 2005. The subsequent section delves into the results.

7 Result

The post-test indicated an average 83% decrease in the mistake rate across the three student samples (Taiwan, Australia, and the Emirates) compared to the pre-test. Among the groups, Taiwanese students demonstrated the highest performance with a 94% reduction in mistakes; AUS students followed with an 85% error reduction rate, ZU students with a 79% error rectification rate, and international English learners in Australia with a 73% error reduction rate. Consequently, it can be concluded that the Intelligent Tutor likely contributed significantly to improved learning outcomes for a substantial portion of the students. This study suggests a broader implication that artificial intelligence can aid second language English learners in rectifying their L2 mistakes, marking a crucial initial step in understanding methods for teaching and learning a second language (Fig. 4).

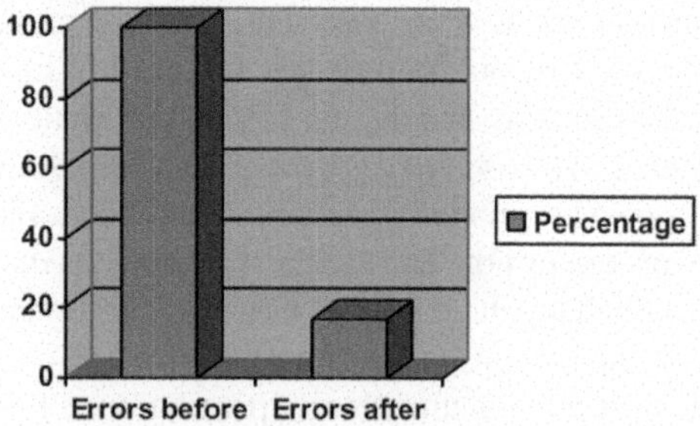

Fig. 4 Difference in error rates pre- and post-utilization of the Intelligent Tutor

8 Conclusion

One could contend that the outcomes appear remarkably impressive, perhaps even too much so. To gage the extent of their contribution to the results, it is essential to delve into potential issues within this study. A straightforward question arises: did participants merely sidestep mistakes in the open-ended post-test by opting for alternative grammatical structures? Nevertheless, the students encountered identical structures that presented difficulties in the first two sessions—the pre-test and the Intelligent Tutor—and exhibited substantial improvement this time. Another possibility is that learners memorized entire phrases while utilizing the Intelligent Tutor or the pre-test.

However, the majority of responses in the post-test were distinct and showed slight variations from the suitable phrases provided by both the pre-test and the Intelligent Tutor. Since the responses varied from student to student, it seemed unlikely that the pupils duplicated one another. Therefore, the outcomes seem to point to a sincere change in grammar.

It would be wonderful to carry out this study in the near future with an even bigger sample of pupils. It would be fascinating to observe, for example, if analytical students prefer to remedy their own mistakes rather than receiving hints, and so on. The Intelligent Tutor is a prototype that currently sustains a relatively restricted curriculum and is not suitable for extended usage. But the Intelligent Tutor is not a unique instance. As per research on intelligent computer-assisted language learning (CALL), the key obstacles hindering the widespread adoption of this technology in second language instruction include inconsistent funding for ICALL projects and the relatively modest size of the generated curriculum. The Intelligent Tutor is currently being developed, and inventive methods are being used to lessen the shortcomings of instructional techniques. The most significant enhancement is the addition of several characters to the Intelligent Tutor. With the help of Intelligent Tutor, students may take on roles that are either competitive, collaborative, or troublemaking. Some kids enjoy being in a setting where they may compete with other students, while others enjoy working with a fellow student who needs their assistance.

Here, a student may complete a work before learning the subjects more thoroughly by imparting them to the computerized students. Another scenario is a computerized student injecting false information that a human learner must weed out. A more varied learning environment for the student would be established by introducing the human student to various (computerized) learning partners inside the Intelligent Tutor in accordance with the student's preferred learning style. Language acquisition is made simpler by the maturation of computer science technology, particularly AI.

These technologies provide brand-new methods of learning that do not require a teacher. In this essay, the usage of ITSs as an illustration of ICALL technology was discussed in relation to language learning and teaching. The creation and use of Intelligent Tutors that are tailored to individual languages requires additional focus from the development team. By addressing these concerns, an Intelligent Tutor may become more effective and usable overall, which will enhance the learning and teaching environment.

References

1. Chapelle C (2001) Computer application in second language acquisition. Foundation for teaching, testing and research. Cambridge University Press, Cambridge
2. Chapelle C (2003) English language learning and technology. John Benjamins B.V., Philadelphia, PA
3. Ehsani F, Knodt E (1998) Speech technology in computer-aided language learning: Strengths and limitations of a new CALL paradigm. Lang Learn Technol 2(1):45. http://llt.msu.edu/vol2num1/article3/. Accessed 18 Oct 2003
4. Ellis R (1997) SLA Research and Language Teaching. Oxford University Press, Oxford
5. Heift T (2003) Multiple learner errors and meaningful feedback: a challenge for ICALL systems. CALICO J 20(3):553–548
6. James C (1998) Errors in language learning and use. Exploring error analysis. Longman, London
7. Krashen S (1987) Principles and practice in second language acquisition. PrenticeHall, London
8. O'Grady W, Dobrovolsky M, Arnoff M (1997) Contemporary linguistics. Bedford/St. Martin's, Boston
9. Salaberry M (1996) A theoretical foundation for the development of pedagogical tasks in computer mediated communication. CALICO J 14(1):5–36
10. Schmidt R (2001) Attention. In: Robinson P (ed) Cognition and second language instruction. Cambridge University Press, Cambridge
11. Swain M (1998) Focus on form through conscious reflection. In: Doughty C, Williams J (eds) Focus on form in classroom second language acquisition. Cambridge University Press, New York, pp 64–82
12. Swartz M, Yazdani M (eds) (1992) Intelligent tutoring systems for foreign language learning. Springer, New York
13. Tschichold C (1999) Intelligent grammar checking for CALL. ReCALL special publication, Language Processing in CALL, pp 5–11
14. Warschauer M (2000) The death of cyberspace and the rebirth of CALL. English Teachers' J 53:61–67

Amazon Sales Sentiment Prediction and Price Forecasting Using Facebook Prophet

Sudhanshu Vishwakarma, Deepanshi Garg, Tanupriya Choudhury, and Thipendra Pal Singh

Abstract In today's fast-evolving world, where people's sentiments have a massive influence on various marketing platforms, it is often overlooked how user reviews impact sales of the product/company. By taking every review as a parameter for inquiring about the company's value, it is seen that these public reviews can significantly impact sales. Getting accurate predictions can help E-commerce developers make correct decisions in favour of goods to increase their value. In this study, the model designed can predict the public's sentiments and the sales of the product/company over a specified period. The time-series model is based on Facebook Prophet Library, an open-source library, to forecast the historical data of the brand's sales. Also, the SARIMA model is used for sentiment prediction to fetch desired results.

Keywords Sentiment analysis and forecasting · Time-series prediction · Facebook Prophet Library

---

S. Vishwakarma · D. Garg
CSE Department, University of Petroleum and Energy Studies (UPES), Dehradun, Uttarakhand 248007, India

T. Choudhury
CSE Department, Symbiosis Institute of Technology, Symbiosis International (Deemed University), Pune, Maharashtra 412115, India

T. Choudhury (✉)
CSE Department, Graphic Era Deemed to Be University, Dehradun, Uttarakhand 248002, India
e-mail: tanupriyachoudhury.cse@geu.ac.in; tanupriya.choudhury@sitpune.edu.in; tanupriya1986@gmail.com

T. P. Singh
School of Computer Science Engineering and Technology, Bennett University, Greater Noida, UttarPradesh 201306, India

S. Dutta et al. (eds.), *Cyber Intelligence and Information Retrieval*, Lecture Notes in Networks and Systems 1025, https://doi.org/10.1007/978-981-97-3594-5_8

1 Introduction

With the development of machine learning and artificial intelligence encapsulated with technology enhancement, forecasting is a crucial aspect of marketing. Sales forecasting is determining the future trends in the sales of a respective product/brand based on historical data. The trends seen throughout the year can be predicted by the efficient use of an open-source library designed for making forecasts for univariate time-series datasets. Virtually every business decision and process are based on a forecast. An organization uses its sales information to forecast what its income extent might be for the upcoming 12 months. Any statistics related to the time, depending on time-associated subjects, may be termed time-series statistics.

By efficiently using an open-source library developed by Facebook's core Data Science team, the Facebook Prophet Library, time-series analysis for sales can be implemented. The Facebook Prophet Library is developed to carry out forecast-related tasks for time-series data with minimal dataset requirements. This library can perform all the fine-tuning parts within itself whilst the data is fed to the model. The problem of eradicating any seasonality and non-stationarity can vanish using the Facebook Prophet Library. It is simple to apply and designed to routinely discover an appropriate set of hyperparameters for the model to make skilful forecasts for data with trends and seasonal structure by default.

Seasonal Sentiment Review data can refer to the type of data whose trend is followed over time, irrespective of the year. In such kind of data, trends, non-stationarity, and seasonality are primarily based on a daily, weekly, every year basis. These data points are adversely affected whilst any above parameters are tuned. In this study, the model analyzed this seasonal data using Seasonal Autoregressive Integrated Moving Average (SARIMA), a machine learning model to predict sentiments over time.

Furthermore, considering the users' views and sentiments, the study can assist the company in enhancing its sales by knowing the target audience's appraisal. In addition, companies can also gain access to user sentiment forecasts, i.e. knowing what the user's view would be in the coming season. This is possible by considering the historical customer reviews of that product/brand. Recent trends in social media platforms, such as Instagram, Twitter, and Facebook, have extended the worldwide sharing of public statistics and sentiments on all domain names with fast communication. Social media platforms get public attention faster than traditional newspapers and news channels. Moreover, reviews and comments are substantial in today's world. They are considered as a form of customer perspective and feedback. The Internet is widely used by everyone today. People give their opinions regarding a brand on the web through reviews/comments. Only classifying reviews as positive or negative is futile. The hidden factors in the sentiments of these reviews/comments are to be analyzed. The current research focuses on sentiment analysis of time-series data of annual sales of an e-commerce platform. To achieve this, the Facebook Prophet Library is implemented for forecasting purposes. It is used to estimate the hike in the sales of a respective online E-commerce platform by observing past trends over

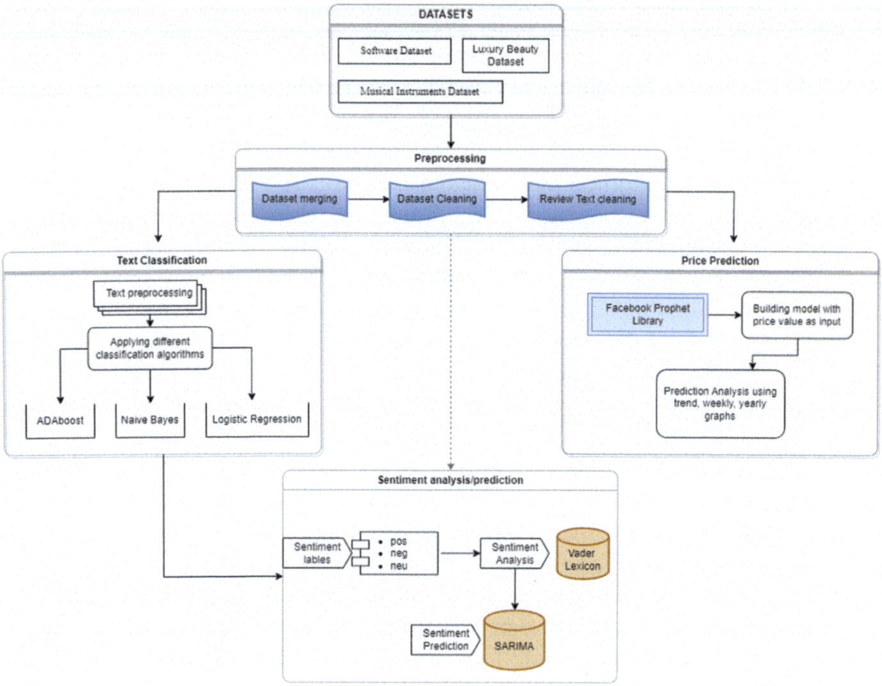

Fig. 1 Graphical abstract

the year. This data would likely help us predict sales and the sentiment analysis. Furthermore, thorough sentiment analysis is performed on the Amazon dataset, a collection of public reviews of a specific category.

A well-known NLTK tool, Vader Lexicon, is used for sentiment analysis. It is a rule-based sentiment analyzer. Classification algorithms such as logistic regression, Naive Bayes, and AdaBoost classify sentiments. Additionally, the SARIMA model, which is hyper-tuned for seasonal review data, is implemented to predict the public's sentiments for a given period. With these results, the brand could work on making necessary changes to the advertisement to positively impact the public for better sales. As mentioned, our study's proposed workflow and contents in Fig. 1 cumulate both the time-series and sentiment analyses with respective predictions.

2 Related Works

In ML-based prediction and sentiment analysis, many studies have been conducted to successfully apply machine learning and artificial intelligence techniques. In this literature survey, various studies on time-series forecasting and the different methods employed by researchers have been thoroughly explored.

Mohan et al. [1] have successfully predicted the impact of the third wave of COVID-19 in India using various statistical ML models. The authors have utilized the hybrid autoregressive integrated moving average (ARIMA) and Prophet model to forecast daily and cumulative confirmed cases. The proposed model forecasts succeeding daily confirmed and cumulative confirmed cases. This study helps the management plan appropriate public health measures for the upcoming waves of COVID-19. Moving on, Khayyat et al. [2] proposed a time-series-based model using the Facebook Prophet for a similar prediction of COVID-19 outbreak as discussed in [1]. By applying Python to their proposed model, the authors have found that the proposed model has a high ability to forecast death cases of the COVID-19 dataset. This study has guided the way through the application of the Facebook Prophet Library in various fields of time-series analysis.

In addition to the previous approach of applying the prophet library, Jha and Pande [3] proposed the implementation of the same library for forecasting supermarket sales. The research has examined the FB Prophet additive and ARIMA model for forecasting purposes. To refine the forecasting, Chou and Ramachandran [4] have successfully implemented prediction using sentimental information. The authors have sufficiently gained accurate results by minimizing the difference between the actual and the predicted values. An analogy can be drawn between this study [4] and the proposed study in this research. Similarly, Liapis et al. [5] leveraged sentiment analysis usage in multivariate forecasting. In addition, the survey consists of two case studies exploring the performance of various multivariate time-series forecasting schemes utilizing sentiment analysis. Another brilliant work by Gonsalves [6] highlights the comparison of various forecasting techniques, concluding that Holt's Winter Method performs better than others. By using the deep learning approaches, Maqsood et al. [7] have developed a model for sentiment-based efficient stock exchange forecasting. With the use of 11.42 million tweets, implementation has been done. The results show that the performance improves using the sentiment for these events. With adequate ARIMA models, Khan et al. [8] compared different ARIMA models and found that ARIMA (1, 1, 33) showed the most accurate results. This is done for accurate time-series stock forecasting, specifically using Netflix stock historical data for five years.

Similarly, Jagdale et al. [9] researched how sentiment analysis can be performed on specific product reviews. Using machine learning techniques, Naive Bayes got an accuracy 'of 98.17%', and the support vector machine got an accuracy 'of 93.54%' for camera reviews. Sharma et al. [10] proposed a method to forecast stock prices with the help of sentiment analysis and decomposable time-series model along with multivariate-linear regression. Additionally, Pathak and Shetty [11] mention that machine learning approaches highlight the use of sentiment analysis in predicting the stock market in India. This study guided in providing insights about the prediction analysis.

Navratil and Kolkova [12] developed a model using the Prophet library for its application in the business economy. However, there is no such comparison with other forecasting models regarding efficiency and accuracy. Moving on, Catal et al. [13] explore sales forecasting in the fashion industry. The study has implemented regression methods and time-series analysis to predict sales. Regression methods are found to perform better than the time-series analysis approaches. Valencia et al. [14], by deriving insights into market behaviour, made precise predictions on stock market behaviour using sentiment analysis.

In this literature survey, various studies are explored, which reflect the proposed work of sales prediction along with elaborated sentiment analysis with its forecasting for the betterment of future sales. In this survey, the works of some renowned authors who have worked on the same in the past have been thoroughly explored. This ushered us in gaining some insights into our work.

3 Methodology

In this section, the dataset used in the model, different algorithms used, and the proposed workflow of the model are stated. The datasets from different sales sectors, including the Luxury Beauty Sector, the Musical Instrument Sector, and Software Sector, are considered in this study.

3.1 Dataset Used

Amazon Sales dataset was collected by the PhD Scholars of UCSD who were guided by Julian McAuley (Professor) at the Computer Science Department @ UCSD. This dataset is used here for the forecasting and analysis. This dataset contains review text, price, and date of the different products from a specific sector. This dataset is divided into multiple sectors of products, namely musical instruments, software, electrical, books, home appliances, fashion, automotive, gift cards, video games, digital music, etc. For this project, the following datasets have been used—software, musical instruments, and fashion datasets.

Sample dataset contains:

- Reviewer ID: A2SWNJ2A3GNN3B.
- Amazon Standard Identification Number(ASIN): "0023113714"
- Reviewer name: John MacCarthy.
- Helpful: [3, 3].

- Review text: The music is sometimes hard to read because we think that the book was published for singing more than playing. Still it can help us learn guitar for the music festival in the next month. A worthy investment, however!
- Overall: 5.0.
- Summary: Guitar Hymns.
- Unix Review Time: 12225777800.
- Review time: 19 February 2007.

Number of entries for each dataset:

- Software dataset—(499831 rows, ten columns).
- Musical instruments dataset—(1731703 rows, ten columns).
- Luxury beauty dataset—(600230 rows, ten columns).

Timespan: May 1996–July 2014.

3.2 Algorithms Used

– **Sentiment Analysis**

As for sentiment analysis, VADER LEXICON is used, which is a rule-based sentiment analysis approach that relies on a set of predefined rules to determine the sentiment of a text and provides text back with its sentiment polarity, which is a numerical score that represents a degree of positive and negative sentiments conveyed by the phrase or word.

Algorithm 1. Sentiment Analysis

- **Input**: User Product/ Company Review
- **Output**: Sentiment Polarity Value

1. Function calculate sentiment scores(data)
2. Sentiment scores = empty array
3. t= 0
4. for i, row in tqdm(data.arrows(), total=len(data)):
5. text = row['reviewText']
6. : sentiment.score[t] =
7. SentimentIntensityAnalyzer().polarity.scores(text)
8. : t = t + 1
9. : **return** sentiment.scores=0

– **Text Classification**

Logistic regression and Naive Bayes approaches are implemented to classify sentiments of the users. This would split the dataset into test and train datasets and then use it to train a model which would be used for prediction. Model evaluation is done based on the confusion matrix, i.e. calculation of accuracy, precision, and score.

Algorithm 2. Text Classification

- **Input**: Review Text, sentiment, Polarity Values
- **Output**: Classified Text Reviews into Pos/Neg Sentiments

function classify sentiment(pdf):
X = df['reviewText']
Y = df['Sentiment']
Y = LabelEncoder().fittransform(Y)
Xtrain, Xtest, Ytrain, Ytest = traintestsplit(X, Y, test-size=0.3, rando state=42)
vectoriser = TfidfVectorizer (parameters)
Xtrain = vectorizer.fit transform(Xtrain)
Xtest = vectorizer.transform(Xtest)
model1 = classificationmodel().fit(Xtrain, Ytrain)
ypred1 = model1.predict(Xtest)
return ypred1 =0

After vectorization is done, training this data with different classification models like logistic regression, Naive Bayes, and AdaBoost is implemented.

– **Sentiment Prediction**

The Seasonal Autoregressive Integrated Moving Average (SARIMA) model is a time-series model used to predict future values of a time-series variable based on its past values. SARIMA models are beneficial for modelling data with seasonal patterns and trends, common in many time-series datasets, including sentiment analysis.

Algorithm 3. Sentiment Prediction

- **Input**: Review Polarity, Review Date
- **Output**: Sentiment Predictions for Future

function SARIMA1(pdf):
model 2 = SARIMAX(sentiment,parameters).fit()
forecast1 = model2.forecast(len(test['sentiment']) + for-caste.period)
return forecast1 =0

– **Price Prediction**

Facebook Prophet Library is used for time-series analysis in this project. Facebook Prophet is an open-source library developed by Facebook for time-series forecasting. It is based on an additive model that includes trend, seasonality, and holiday effects and can be customized to include additional regressors.

Algorithm 4. Price Prediction

- **Input**: Historical Product Price and Date
- **Output**: Price Prediction Values

```
function forecastwithprophet(data):
model = Prophet()
data.columns = [' ds', 'y']
model. fit(data)
future = model.makefuturedataframe(periods=3650)
lastrows = future.tail()
forecast = model.predict(future)
lastforecastrows = forecast[['ds', 'yhat']].tail()
return lastforecastrows =0
```

3.3 *Proposed Workflow*

(1) **Data Collection and Preprocessing**: The first step starts with the data collection and its effective preprocessing to make the data more eligible for further implementation. Historical stock market data and related news articles or social media posts were collected here. This data was then preprocessed by cleaning and formatting it for further analysis. This preprocessing involved removing duplicate values to handle the missing values along with standardization of the format of the data available.

(2) **Sentiment Analysis**: Next, sentimental analysis was performed on the previously collected news articles and the stock market data. This involved the effective utilization of the nltk Vader lexicon for the purpose of sentimental analysis on the dataset preprocessed.

(3) **Sentiment Prediction**: Based on sentimental analysis, the public's sentiment was predicted along with the price module for the company's upcoming timeline.

(4) **Time-Series Analysis**: As part of the next step forward, the famous Prophet library by Facebook was utilized to perform the time-series analysis on the historical stock market data preprocessed earlier.

(5) **Visualization and Reporting**: The results of this extensive analysis were visualized using various graphs, charts, and tables. The reports and summaries were also generated based on the analysis to better understand and effectively communicate with the stakeholders.

(6) **Optimization**: The analysis also needed to be refined along with the algorithms, which were improved to increase its accuracy and performance. This process involved fine-tuning the hyperparameters of the machine learning algorithms along with the exploration of new methods of analysis.

This workflow is just a general overview; the specific steps and algorithms used may vary depending on the needs of the project and the data available.

4 Evaluation and Results

The model built successfully analyzed the different sectors of product datasets. Different analyses were done using different algorithms and all the necessary results are given below for the musical instruments dataset. Figure 2 describes the Cumulative Distribution Function. On the classification of sentiments using the logistic regression classification model, an accuracy of 88% with the musical instrument product dataset of Amazon sales is achieved, as shown in Fig. 3. The graph shows predictions made by the model for the next ten years based on historical price data, as in Fig. 4. From Fig. 5, the upcoming trend for user sentiments over a period is deduced. The graph in Fig. 6 describes the seasonal decompose of sentiments and the dataset's recent trends, seasonality, and residual. Figure 7 describes the recent trends in price in the musical instruments sector.

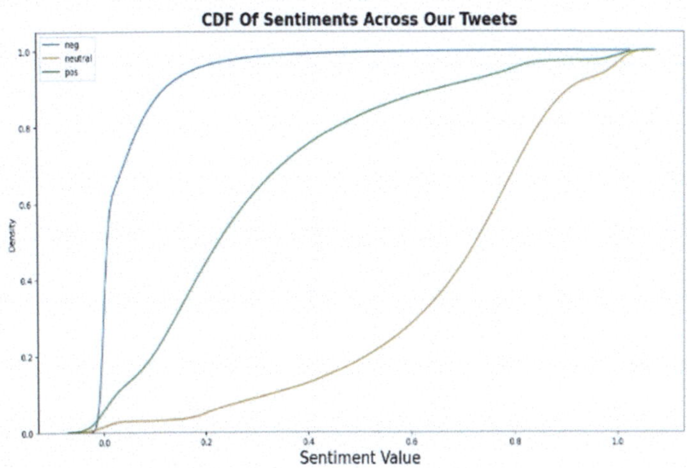

Fig. 2 Cumulative distribution function

	precision	recall	f1-score	support
neg	0.74	0.65	0.69	43331
neu	0.44	0.09	0.15	25188
pos	0.90	0.98	0.94	299879
accuracy			0.88	368398
macro avg	0.70	0.57	0.59	368398
weighted avg	0.85	0.88	0.86	368398

Fig. 3 Logistic regression confusion matrix

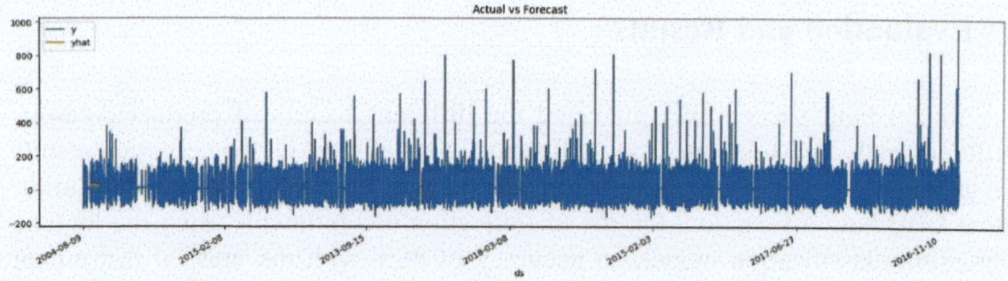

Fig. 4 Price prediction using Facebook Prophet

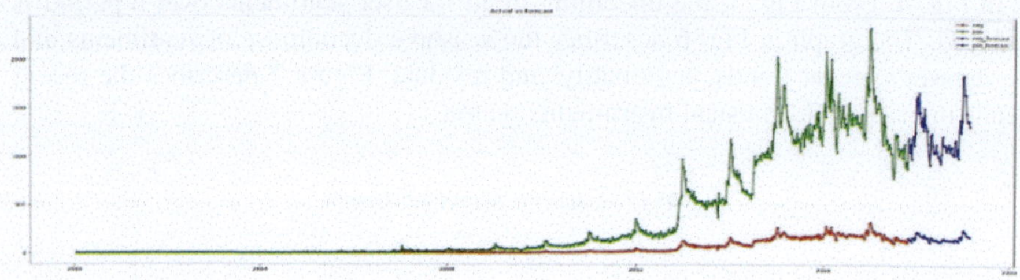

Fig. 5 SARIMA model for sentiment forecast

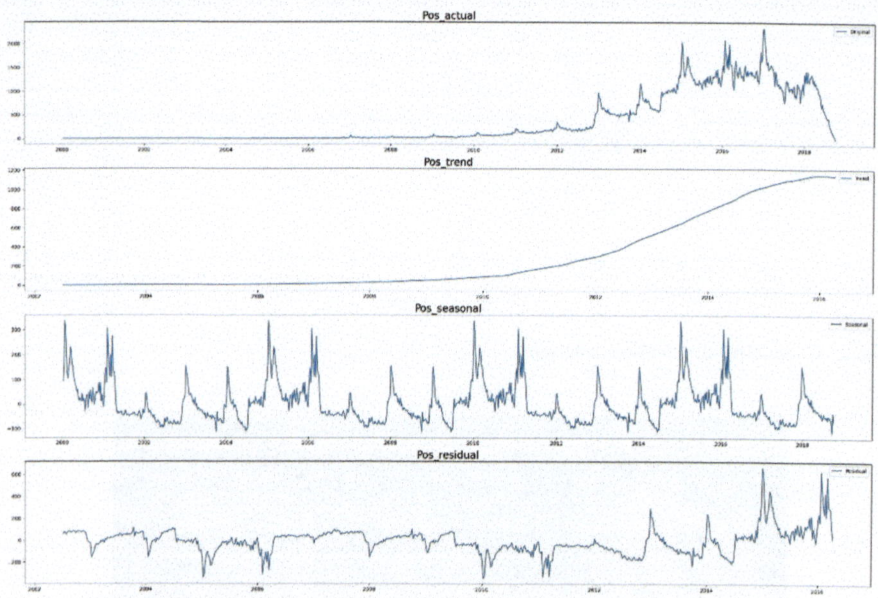

Fig. 6 Seasonal decompose of sentiments

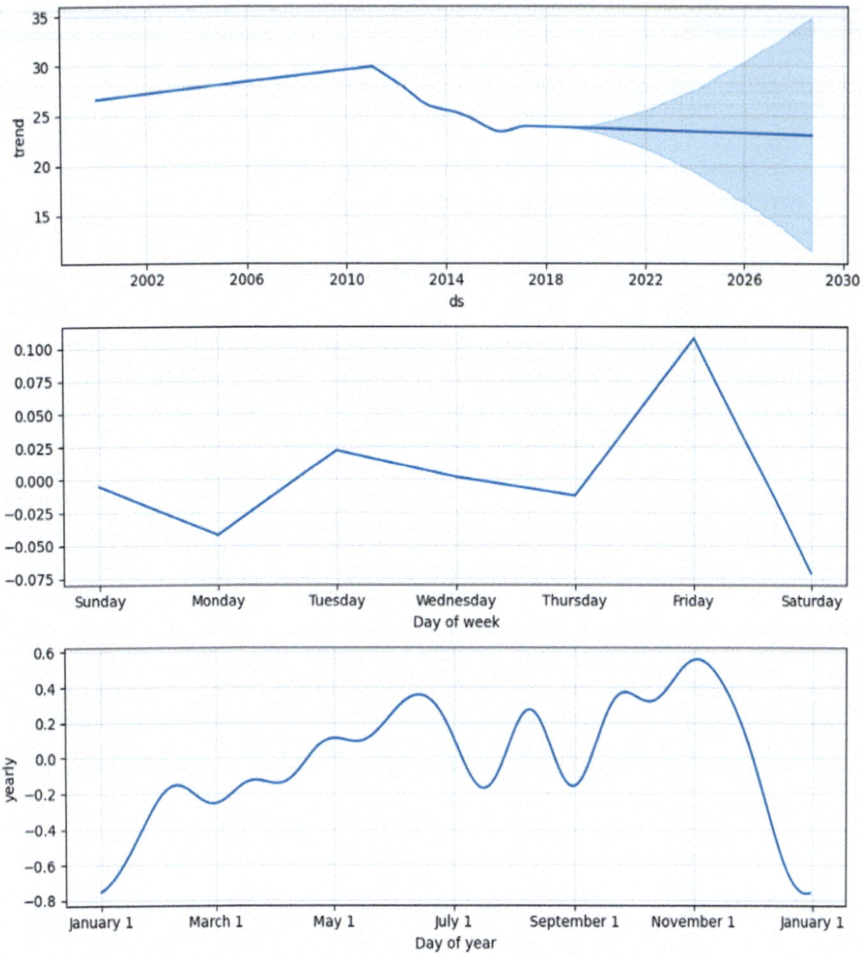

Fig. 7 Trends of price in the musical instruments sector

5 Comparative Study

In the comparative study, a derived comparison by the Dickey–Fuller test for checking the stationarity of the time series is highlighted. Below are the p values and RMSE values of a dataset concerning their positive and negative sentiments.

A. *Musical Instruments Dataset*

- Positive sentiment—0.5568440663406161.
- Negative sentiment—0.5657465401055206.

B. *RMSE values*

- Pos.—44.242488.
- Neg.—49.086680.

6 Conclusion and Future Work

In this study, the usability of Facebook Prophet Library for the time-series analysis on price prediction for ten years of respective datasets is proposed. Additionally, sentiment analysis on customer reviews of the products is performed. On analyzing the sentiments of the customers, future sentiments are also predicted. This study explores the predictive power of customer reviews. Sentiment analysis is an essential domain of interest and impact as these meaningful insights by the audience provided us with information about how the customers perceive the products. Hence, the proposed study uses the Vader Lexicon for sentiment analysis and the SARIMA model for sentiment prediction. Furthermore, prediction and analysis are plotted by identifying our data's patterns, trends, and seasonality. In terms of future work, this study's predictive power of reviews in predicting customer behaviour can be used to utilize and derive insights about a company's business.

Acknowledgements We owe our gratitude to the exceptional guidance of Prof. (Dr.) Thipendra P. Singh. Your guidance and expertise have been instrumental in shaping the direction and quality of this project. We also deeply appreciate our research mentor, Dr. Tanupriya Choudhury, whose unwavering commitment to excellence and valuable insights guided us through this research journey.

References

1. Mohan S, Solanki AK, Taluja HK (2022) '&' Singh, A Predicting the impact of the third wave of COVID-19 in India using hybrid statistical machine learning models: a time series forecasting and sentiment analysis approach. Comput Biol Med 144:105354
2. Khayyat M, Laabidi K, Almalki N, Al-Zahrani M (2021) Time series facebook prophet model and python for COVID-19 outbreak prediction. Comput Mater Continua 3781–3793
3. Jha BK, Pande S (2021) Time series forecasting model for supermarket sales using FB-prophet. In: 2021 5th international conference on computing methodologies and communication (ICCMC), Apr 2021. IEEE, pp 547–554
4. Chou HC, Ramachandran KM (2021) Combining time series and sentiment analysis for stock market forecasting
5. Liapis CM, Karanikola A, Kotsiantis S (2021) A multi-method survey on the use of sentiment analysis in multivariate financial time series forecasting. Entropy 1603, 23.12
6. Gonsalves O (2020) A study of sentiment analysis and sales prediction: tourism domain
7. Maqsood H, Mehmood I, Maqsood M, Yasir M, Afzal S, Aadil F, Muhammad K (2020) A local and global event sentiment based efficient stock exchange forecasting using deep learning. Int J Inf Manag 50:432–451
8. Khan S, Alghulaiakh H (2020) ARIMA model for accurate time series stocks forecasting. Int J Adv Comput Sci Appl 11.7

9. Jagdale RS, Shirsat VS, Deshmukh SN (2019) Sentiment analysis on product reviews using machine learning techniques. In: Cognitive informatics and soft computing: proceeding of CISC 2017. Springer Singapore, pp 639–647
10. Sharma V, Khemnar R, Kumari R, Mohan BR (2019) Time series with sentiment analysis for stock price prediction. In: 2019 2nd international conference on intelligent communication and computational techniques (ICCT), Sept 2019. IEEE, pp 178–181
11. Pathak A, Shetty NP (2019) Indian stock market prediction using machine learning and sentiment analysis. In: Computational intelligence in data mining: proceedings of the international conference on CIDM 2017. Springer Singapore, pp 595–603
12. Navratil M, Kolkova A (2019) Decomposition and forecasting time series in the business economy using prophet forecasting model. Central European Bus Rev 8(4):26
13. Catal C, Kaan ECE, Arslan B, Akbulut A (2019) Bench-marking of regression algorithms and time series analysis techniques for sales forecasting. Balkan J Electr Comput Eng 7(1):20–26
14. Valencia F, Go´mez-Espinosa A, Valde´s-Aguirre B (2019) Price movement prediction of cryptocurrencies using sentiment analysis and machine learning. Entropy 21(6):589

House Price Prediction Using Random Forest Regression Considering the Nearest In-Demand Location

Manjima Saha, Shubhamita Saha, and Shreya Majumder

Abstract Machine learning has become more important over the past few years in a variety of fields, including image detection, spam recognition, speech commands, product suggestions, and medical diagnosis. Today's sophisticated machine learning algorithms are assisting us in enhancing security warnings, enhancing public safety, and advancing medical research. Moreover, machine learning technologies offer safer automotive systems and improved customer service. In this study, we concentrate on applying machine learning techniques to forecast future home prices. We examined and investigated different prediction systems in order to choose the optimal one. Therefore, we selected random forest regression as our model because of its versatility and probabilistic model selection process. Our findings demonstrate the effectiveness of our method for solving the issue and its capacity to generate forecasts that are on par with those from other models for predicting housing costs. The establishment of home value indexes, which in turn promotes real estate policies and plans, is another contribution made by this study. To develop a housing cost prediction model, we used machine learning techniques as a research strategy. To assess the performance of different machine learning algorithm models, including gradient boosting boost, lasso regression, and neural random forest, we tested and compared them. According to our research, the random forest regression method predicts housing costs more accurately than other models on a constant basis. Overall, our study advises making more educated decisions about a home's valuation with the use of the lasso regression method for sellers or real estate brokers. We can create dependable and accurate housing cost prediction models with the aid of machine learning, which will ultimately help the real estate market as a whole.

M. Saha (✉) · S. Saha · S. Majumder
Institute of Engineering and Management, Kolkata, India
e-mail: manjima.saha@iemcal.com

S. Dutta et al. (eds.), *Cyber Intelligence and Information Retrieval*, Lecture Notes in Networks and Systems 1025, https://doi.org/10.1007/978-981-97-3594-5_9

1 Introduction

Machine learning is a branch of artificial intelligence (AI) that focuses on using technology and algorithms to extract information that is usable from data, especially in the case of enormous data where manual processing is impractical. In contrast to mathematical approaches, machine learning makes use of algorithms that give computers the ability to learn to solve issues algorithmically. Supervised and unsupervised machine learning fall into two broad types. Unsupervised learning focuses on finding connections and hidden patterns within data, while supervised learning includes training a program on a specified set of data in order to predict fresh data [8].

Today's real-world applications employ a range of machine learning algorithms, each performing differently depending on the situation. This thesis compares how well artificial neural networks and regression algorithms perform at forecasting values from a given dataset. Many other papers used several other algorithms in order to predict the accuracy, and we have taken the dataset from Kaggle kc house data [7]. Since the prediction of house prices relies on a wide range of criteria that can vary in price depending on location and other circumstances, house prices will specifically be used as a benchmark for comparison. We added new variables to the local dataset in order to examine the relationship between these variables and the sale price. We came to a conclusion that our model is more efficient to predict the price of the house compared to any other model.

2 Methodology

2.1 Data Preprocessing

Data preprocessing includes data collection, data formation, and data modification. We have collected the exact dataset that they have used in their paper [7] from Kaggle. We added some new attribute in the dataset kc house data which consists of 23 attribute and 21614 columns. This dataset shown in Fig. 1 is a collection of the data information of all the houses there and surrounding king county, and it is all other cities. After collecting and modifying first thing we checked for any missing or null data values. Then that clean data was uploaded for next procedure [4]. That dataset is given following in Fig. 1.

	id	date	price	bedrooms	bathrooms	sqft_living	sqft_lot	floors	waterfront	view	...	grade	sqft_above	sqft_basement	yr_built	yr_renovated	zipcode	lat
0	7.129301e+09	20141013T000000	221900.0	3	1.00	1180	5650	1.0	0	0	...	7	1180.0	0	1955	0	98178	47.5112
1	6.414100e+09	20141209T000000	538000.0	3	2.25	2570	7242	2.0	0	0	...	7	2170.0	400	1951	1991	98125	47.7210
2	5.631500e+09	20150225T000000	180000.0	2	1.00	770	10000	1.0	0	0	...	6	770.0	0	1933	0	98028	47.7379
3	2.487201e+09	20141209T000000	604000.0	4	3.00	1960	5000	1.0	0	0	...	7	1050.0	910	1965	0	98136	47.5208
4	1.954401e+09	20150218T000000	510000.0	3	2.00	1680	8080	1.0	0	0	...	8	1680.0	0	1987	0	98074	47.6168

5 rows × 21 columns

Fig. 1 Name of the attributes present in the dataset

2.2 Experimental Data Analysis

Data analyzing is a very essential step to perform. Analytically the dataset that we have collected and cleaned that was not enough to predict best result as we see in other papers [5–7]. Some machine learning that they used was getting more accurate result than other algorithms [10]. So, we add some new attribute in the dataset and tried to make it efficient enough to predict best result using other algorithms. We consider a nearby city Bellevue and its longitude and latitude 47.6101°N, 122.2015°W. We tried to experiment with adding new attribute and do subtraction operation with the entire element in the longitude and latitude column with the location of that nearest state which is mentioned above. Doing this experiment, we got best results accordingly.

2.3 Model Selection

Next most important step is selecting suitable models. In this case regression algorithms are most suitable that we used for prediction [9]. We select a few machine learning algorithms like linear regression algorithm, lasso regression algorithm, random forest regression algorithm, and gradient boosting algorithm [1, 3]. Figure 2 shows the flow chart of Data mining process.

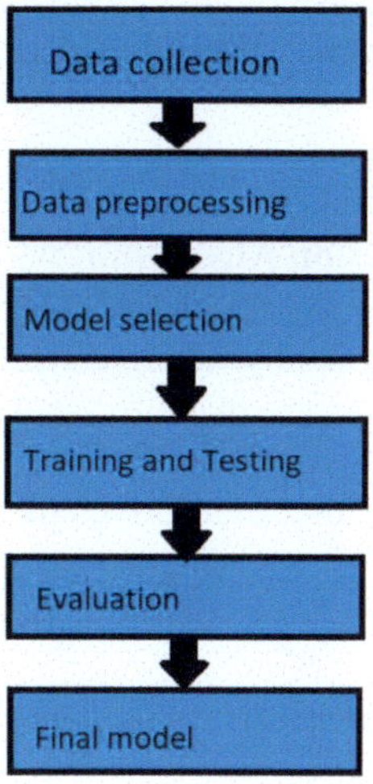

Fig. 2 Flowchart

2.3.1 Lasso Regression Algorithm

Lasso was originally formulated for linear regression models. This simple case reveals a substantial amount about the estimator. These include its relationship to ridge regression and best subset selection and the connections between lasso coefficient estimates and so-called soft thresholding. It also reveals that the coefficient estimates do not need to be unique if covariates are collinear [2]. Though originally defined for linear regression, lasso regularization is easily extended to other statistical models including generalized linear models, generalized estimating equations, and proportional hazards models.

2.3.2 Linear Regression

The algorithm that we have selected is linear regression, where the value for the dependent variable is calculated using multiple independent variables. The value of variable which is to be predicted depends on its strength of relationship with the other independent variables. This factor is called correlation.

$$A = ny + c \tag{1}$$

[A is dependent variable, y is independent variable, and n is intersection factor.]

In linear regression, the relationships are modeled using linear predictor functions whose unknown model parameters are estimated from the data. Such models are called linear models

2.3.3 Random Forest Regression Algorithm

The random forest algorithm works by taking different decision trees together and combining them. Next predictions are made for each, and averaging is done on the basis of these predictions. The prediction is made on the basis of most accurate results. The following flowchart Fig. 2 describes the working of the random forest algorithm.

2.3.4 Gradient Boosting Algorithm

Gradient boosting works by multiple residual models together. The error value and the y value are taken as input at each start of residual model. Then after calculating the prediction, its error value is fed into next residual model. This process goes on until the error becomes minimum.

2.4 *Training and Testing of Dataset*

The dataset was divided into s and t model to train and testing. The model was tested with t train and stored in t predict. Both t train and t predict were compared.

3 Result

Different model gives different results; among that random forest model gives the most accurate result.

The scatter plot shows how much one variable is affected by another and also the relationship between them. We have generated the graph as following.

We have displayed the relationship between the attributes using scatter plot in Fig. 3. It helps us to find the relationship between both the axes. In scatter plot, X axis denotes prices and Y axis denotes predicted prices. The graph shows that the dots distilled around the value 200000 on the axis x, and 200000 on the axis using plt.show() function we have generated the graph in Fig. 4. We can also see that the spread is wider on the lower middle part.

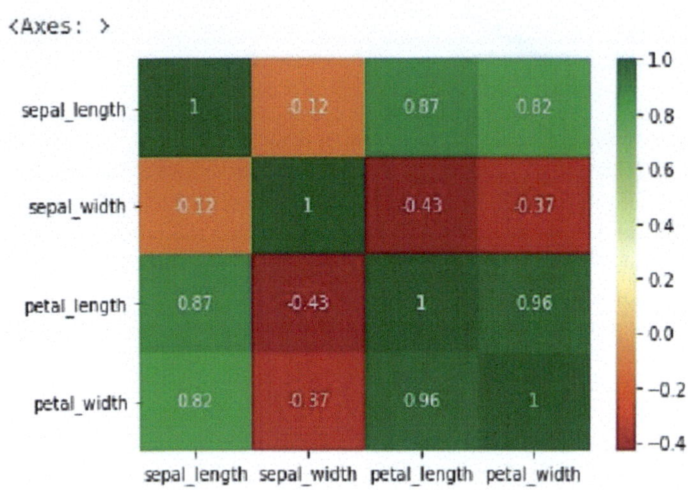

Fig. 3 Heatmap

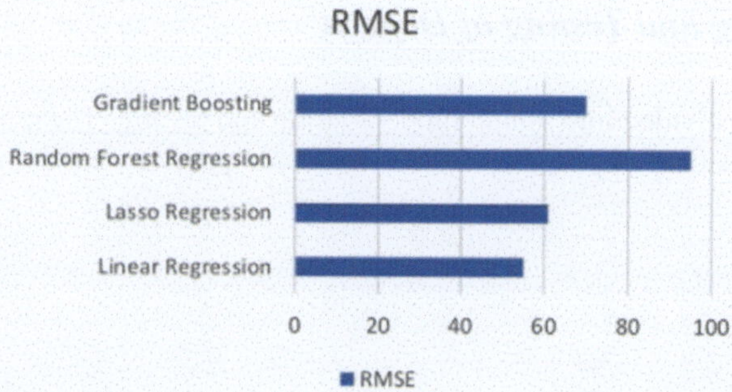

Fig. 4 Comparison graph

Model	R2 adjust matrix
Linear regression	0.55
Lasso regression	0.61
Random forest regression	0.95
Gradient boosting	0.70

Here using random forest we got the most accurate result.

A heatmap (shown in Fig. 3) is a visualization that shows a relationship between two dimensional matrices. In heatmap, i.e., in Fig. 3 each value of a matrix is represented by different color patterns. We have also added the labels to represent each matrix. After predicting the code we have executed the heatmap, and we have got the following output.

Here is a bar graph to emphasis our result that we have found previously (shown in Fig. 4).

4 Discussion and Conclusion

Several research papers on the same topic were studied which adopted several methodologies to predict housing prices. In the paper [7] kc house data named dataset was used which was found from Kaggle. New attributes to predict better result than that paper [7] have been added. The city that we consider is one of the most important places in king county US or any nearest area. Because it is the best desirable place for living, here houses are 138% more expensive than national average. More nearer to this place will be more expensive. We got contrasting result from different model. In random forest regression algorithm we got the most accurate result 0.95% accuracy.

References

1. Abdullahi A, Usman H, Ibrahim I (2018) Determining house price for mass appraisal using multiple regression analysis modeling in Kaduna North, Nigeria. ATBU J Environ Technol 11(1):26–40
2. Basysyar FM, Dwilestari G (2022) House price prediction using exploratory data analysis and machine learning with feature selection. Acadlore Trans Mach Learn 1(1):11–21
3. Engle RF, Lilien DM, Watson M (1985) A dynamic model of housing price determination. J Econometr 28(3):307–326
4. Park B, Bae JK (2015) Using machine learning algorithms for housing price prediction: The case of fairfax county, virginia housing data. Expert Syst Appl 42(6):2928–2934
5. Ravikumar AS (2017) Real estate price prediction using machine learning. PhD thesis, Dublin, National College of Ireland
6. Rawool AG, Rogye DV, Rane SG, Vinayk A (2021) House price prediction using machine learning. Int J Res Appl Sci Eng Technol 9:686–692
7. Satish GN, Raghavendran CV, Sugnana Rao MD, Sriniva-sulu C (2019) House price prediction using machine learning. J Innov Technol Explor Eng 8(9):717–722
8. Varma A, Sarma A, Doshi S, Nair R (2018) House price prediction using machine learning and neural networks. In: 2018 second international conference on inventive communication and computational technologies (ICICCT). IEEE, pp 1936–1939
9. Wang P-Y, Chen C-T, Su J-W, Wang T-Y, Huang S-H (2021) Deep learning model for house price prediction using heterogeneous data analysis along with joint self-attention mechanism. IEEE Access 9:55244–55259
10. Zulkifley NH, Rahman SA, Ubaidul-lah NH, Ibrahim I (2020) House price prediction using a machine learning model: a survey of literature. Int J Mod Edu Comput Sci 12(6)

Enhancing Legal Document Understanding Through Text Summarization: A Study on NLP and Wavelet Tree Techniques

Utkarsh Dixit, Sonam Gupta, Arun Kumar Yadav, and Divakar Yadav

Abstract The summarization of text resulted in faster access to the information inside the documents, reducing the time required to comprehend them. The application of text summarization to legal documents proved beneficial for the legal profession to gain a better understanding of the cases. In this paper, we discussed various text summarization techniques specifically designed for summarizing legal documents, including the use of NLP and Wavelet trees. Additionally, this paper included the presentation of our previous research work's results on text summarization.

Keywords NLP · Wavelet tree · Summarization

1 Introduction

The digital era has brought an unparalleled data flood, resulting in an exponential increase of online information with each passing day. The legal realm has not been immune to this boom, with a substantial increase in the amount of legal papers [1]. Due to the exponential increase in data, text summarization has emerged as a critical challenge in today's information era. This activity is divided into two parts: finding the most important sections of text and writing logical summaries that try to compress massive volumes of information into succinct and comprehensible forms [2]. The

U. Dixit · S. Gupta (✉)
Ajay Kumar Garg Engineering College, Ghaziabad, India
e-mail: guptasonam6@gmail.com

U. Dixit
e-mail: utkarsh2010016m@akgec.ac.in

A. K. Yadav
National Institute of Technology, Hamirpur, HP, India
e-mail: ayadav@nith.ac.in

D. Yadav
IGNOU, New Delhi, India
e-mail: dsy99@rediffmail.com

S. Dutta et al. (eds.), *Cyber Intelligence and Information Retrieval*, Lecture Notes in Networks and Systems 1025, https://doi.org/10.1007/978-981-97-3594-5_10

exponential development of legal papers offers an insurmountable barrier, making manual reading and comprehension practically impossible. Automatic summarizing appears as a critical method for extracting key material to address this issue. Unlike human summarizing, which uses an unlimited vocabulary and does not impose artificial linguistic limits, automatic summary focuses on extracting relevant sentences from the original text. These collected sentences provide the basis for summaries, which may then be enhanced using post-mining methods. Text summarization, a broad topic of NLP, combines techniques from NLP and machine learning. Among these are part-of-speech (POS) tagging, named entity recognition (NER), language modelling, and a variety of other NLP methods. These techniques and approaches are critical in the creation of successful text summarizing systems, which strive to compress enormous amounts of data into succinct and logical summaries [3].

Text summary has great promise for improving the efficiency and productivity of legal practitioners in India. In India, the legal landscape is comprised of a vast corpus of complicated documents, such as court judgements, case briefs, contracts, and legal opinions. These documents are frequently long and dense with legal language, making it difficult for attorneys, judges, and legal researchers to extract essential information. Legal professionals can enhance information retrieval and workflow by using text summarizing techniques. Automatic summarizing algorithms may provide short and clear summaries of large legal documents, emphasizing significant facts, major arguments, and the final judgement. This may save a lot of time and effort, allowing lawyers to focus on more important activities like case analysis, strategy building, and client counselling. Text summary can also help with legal research and case preparation. Researchers may swiftly browse through summarized versions of many case precedents and legal opinions, spot important trends, and form meaningful conclusions. Such qualities may be extremely beneficial in developing compelling arguments and boosting legal strategy. Furthermore, legal papers frequently include very sensitive and secret information, demanding stringent data protection and security precautions for any summarizing system utilized in this context. Compliance with legal and ethical norms is critical to preserving the integrity and secrecy of legal procedures. Text summarizing techniques have the potential to revolutionize the way legal practitioners manage information, improve decision-making, and optimize their workflow in the Indian legal realm. By automating the summarizing process, attorneys and judges may more effectively explore huge numbers of legal papers, creating a more accessible and effective legal system for the benefit of all parties.

This paper is structured in such a way that the first section provides an overview of text summarization, the second section discusses how a legal document differs from a normal document, the third section highlights what techniques are available for text summarization, the fourth section deals with the use of NLP in summarization, the fifth section is about text summarization using wavelet trees, the sixth section provides an inside look at the related work done in summarization, the seventh section is about the previous research result, and the last section is about the conclusion for the paper followed by the references.

2 Decoding Legal Documents: Recognizing Significant Differences from Regular Texts

Examining specific qualities and aspects within the text can help distinguish between a legal document and a regular document. Here are some important considerations to consider:

Title and Heading: Legal papers frequently have special titles and headers that reflect their legal nature. In legal papers, phrases such as "Contract," "Affidavit," "Complaint," "Judgement," "Agreement," "Terms and Conditions," and so on are widespread and can serve as indications.

Legal vocabulary and Formal Language: Legal papers often employ specialized legal vocabulary and formal language. These phrases may not appear in conventional papers, and their use can indicate the text's legal character.

Signatures and Notarization: To be legally legitimate, many legal documents require signatures and notarization. The existence of authority signatures, seals, or stamps might suggest that the document is of a legal character.

Legal Authorities: To support their claims or establish legal rights, legal papers frequently mention laws, statutes, rules, or judicial precedents. Such allusions are uncommon in regular texts.

Party Information: Legal papers frequently include information on the parties involved, such as names, addresses, and legal entities. These particulars may be irrelevant in standard texts.

Date and Jurisdiction: Specific dates and the jurisdiction under which they are controlled are frequently included in legal documents. This material contributes to the legal environment.

Legal Clauses and Sections: Specific clauses, sections, and paragraphs defining the rights, duties, and parameters of the agreement or legal subject at hand are common in legal documents. These structural features are uncommon in standard papers.

Legal Formatting: Depending on the jurisdiction's regulations and court standards, legal documents typically have certain formatting requirements, such as margins, line spacing, and font sizes.

Seals and Logos of Legal Authorities: Official legal papers may contain seals or logos of courts, government agencies, or notaries in some situations.

It should be noted that not all legal papers will have all of these qualities, and even regular documents may contain legal phrases or signatures if they are linked to legal concerns. As a result, a detailed examination of the document's content and structure is required to appropriately assess its legal character.

3 Different Types of Summary

To categories summarized texts, many classification criteria have been developed in the field of text summarizing. A consistent categorization system, however, has yet to be created. Several research publications have investigated various criteria for this aim, each providing useful insights into the categorization process. This section aims to offer a complete review of the various text summarizing approaches. Grasp these various techniques allows us to have a better grasp of the multidimensional nature of summarizing methods and their implications for producing accurate and effective summaries. We will get significant insights into the developments achieved in the field of text summarization by exploring the many methodologies used, laying the framework for a closer study of their strengths, shortcomings, and prospective applications [4].

3.1 Approaches

Text summarizing systems are classified as extractive or abstractive based on how they analyse texts and generate summaries.

- **Extractive summaries** are created by reusing sections of the original text, such as words and phrases. The most important information is copied to the final summary, which is commonly found in the opening sentence of each paragraph, special names, italic or bold words. Extractive summaries, on the other hand, may suffer from inconsistencies, a lack of balance, and a lack of continuity.
  Example: Surrey University's Summ-It applet is one such system that use an extractive technique to creating summaries [5].
- **Abstractive summarization** is an interpretation of the original material. By rewriting the information in a shorter version, the algorithm analyses the primary text and delivers its comprehension in a human-understandable manner. To communicate the substance of the text, wordy notions are replaced with concise terms.
  Example: SUMMARIST is an example of a text summarizing system that employs an abstractive approach. It has modules for subject interpretation and summary production, so it can generate abstract summaries [6].

3.2 Classification Based on Details

Summarized texts may be divided into two groups based on the degree of detail they include, resulting in two types of summarization systems: indicative and informative summarizing systems.

- **Indicative Summarization**: An indicator summarization presents the user with the key concept of a text. These summaries are often quite brief, ranging from 5 to 10% of the length of the original text. They are intended to encourage readers to read the entire primary document.
- **Informative Summarization**: Informative summaries provide succinct yet thorough information from the original text and can be used in place of the main document. These summaries are more thorough than suggestive summaries and typically range from 20 to 30% of the length of the main text.

3.3 Content

Another distinction is between general and query-based summaries, which are based on the significance of the material in the original text.

- **Generic summaries**, also known as extractive summaries, are created in the absence of a specific query or user input. These summaries are intended to offer a quick summary of the key points and important information included in the original material. They are not dependent on any specific user context or information requirements. Generic summaries are often used in news stories, research papers, and document summary systems to provide a brief overview of a document's content.
- **Query-based summaries**, also known as abstractive summaries, are created in response to specific user requests or information demands. These summaries take the user's input, context, or inquiry into consideration and strive to produce a meaningful and short summary customized to the user's needs. Query-based summaries, as opposed to generic summaries, entail natural language creation, in which the system analyses the query and provides a summary that is not a literal extraction of the source text but rather an interpretation of the relevant information in response to the question.

4 NLP Techniques for Text Summarization

Text summarization relies heavily on NLP (Natural Language Processing) methods. These strategies aid in the comprehension and processing of language, the extraction of vital information, and the generation of succinct and coherent summaries. Some of the most important NLP approaches for text summarization are:

- Tokenization is the process of dividing a text into individual words or tokens. This procedure is required for subsequent text analysis and processing.
- Tagging of Parts of Speech (POS): POS tagging adds a grammatical label to each word in the text, such as noun, verb, adjective, and so on. POS tagging aids comprehension of the text's syntactic structure.

- Named Entity Recognition (NER): NER recognizes and categorizes named entities such as people's names, organizations' names, places, dates, and so on. NER is vital for detecting key items in the text that may be significant for the summary.
- Stopwords are frequent words (e.g., "the," "and," "is") that do not add anything to the sense of the text. Stopwords are removed to decrease noise and focus on more important material.
- Text Ranking Algorithms evaluate the relevance of sentences or paragraphs based on a variety of parameters, including word frequency, word location, and semantic similarity. The ranking process aids in the selection of the most relevant and useful material for the summary.
- Word embeddings represent words in a dense vector space and capture their semantic links. These embeddings aid in comprehending the context and meaning of words, resulting in improved summarization.
- Text Compression: Text compression techniques shorten the summary while maintaining important information. Techniques such as sentence fusion and deletion aid in the creation of short summaries.
- Language modelling is utilized in abstractive summarizing, where the machine creates summaries in a more human-like manner by predicting the next word based on context.
- Topic Modelling (e.g., Latent Dirichlet Allocation - LDA): Topic modelling techniques, such as LDA, aid in the identification of prominent themes in a collection of documents, guiding the summarizing process by concentrating on essential subjects.

5 Wavelet Tree for Text Summarization

The Wavelet Tree is a powerful data structure used for compressing documents over a network. It excels in storing strings in a compressed form and extends the capabilities of rank and select operations to support arbitrary alphabets. This tree-like structure efficiently divides the data into smaller parts, enabling easier compression.

In the compression process, documents are tagged or labelled in a manner that ensures efficient searching and retrieval. The Wavelet Tree represents a succinct data structure that recursively partitions a string into two parts until homogeneous data remains.

During pattern matching, the tree data structure proves highly relevant, particularly for recursively partitioning the alphabet into pairs of subsets. Each node's bitvector efficiently determines whether a symbol of the string belongs to one subset or the other [7].

With the Wavelet Tree's capabilities, processing and compressing documents become more efficient and effective, making it an invaluable tool for handling data over networks. The phrase "Wavelet Tree" comes from an analogy with the wavelet transform, which is utilized for signals. The wavelet transform, in this comparison, iteratively breaks down a signal into its low-frequency and high-frequency components [8].

6 Literature Review

In this paper Prashed et al. [9] proposed WBTC_WT, an efficient technique for pattern matching inside compressed texts. For rapid searching without decompression, it employs word-based tagged code (WBTC) and wavelet tree. On large datasets, the technique surpasses previous approaches, assuring accurate matches and overcoming false match concerns.

Author of the paper Khetan et al. [10] present a method for compressed parameterized word matching, which is relevant in a variety of disciplines such as graph isomorphism, plagiarism detection, and picture processing. The suggested method compresses data using Word-Based Tagged Code (WBTC) and searches efficiently using a Wavelet tree, reducing space and execution time. Notably, the approach reduces the need for additional verification steps, which results in a shorter total execution time.

Biswas et al. [11] offer in their work a unique technique dubbed DCESumm for effective summary of long legal texts. Unsupervised clustering-based score enhancement is used in conjunction with supervised sentence-level relevance prediction. Experiment findings on the BillSum and FIRE datasets show considerable gains over existing state-of-the-art approaches, assisting legal practitioners in extracting critical information from legal documents.

In the article Solanki et al. [12] present T2SAM, an Abstract Text Summarization (ATS) model based on a Transformer with Self-Attention Mechanism, in their study. The self-attention process resolves textual coreference, resulting in better understanding. The suggested model was trained using the Inshorts News dataset and the DUC-2004 shared tasks dataset. T2SAM beats previous baseline models in ROUGE metrics, with a training loss reduction from 10.3058 to 1.8220 (at 30 epochs) and a model accuracy of 48.50% F1-Score on both the Inshorts and DUC-2004 news datasets.

The article was discussed Gupta et al. [13] conduct a comprehensive study of legal document summarization in the paper, analysing several machine learning models and evaluation criteria. The study compares summarization quality with and without deep reinforcement learning. The rational augmented model [RAG] with deep reinforcement learning surpasses previous models, achieving 79.8% accuracy, 70.5% precision, 90.75% recall, and 75.09% F-measure. Furthermore, the long short-term memory network model without deep reinforcement learning performs well, with 0.93 recall and 0.98 F-measure.

In their study Rathi et al. [14] undertake a thorough assessment of text summarizing methodologies, including Extractive, Abstractive, and Hybrid methods. Despite great progress, computerized summaries still fall short of those written by humans. The study is a great resource for ATS approaches, methodologies, techniques, datasets, assessment methods, and future research goals in the domain of NLP.

In their article, Vaissnave et al. [15] present an automated optimum DL-based legal text summarizing (ODL-LTS) technique. The ODL-LTS approach measures similarity using TF-IDF and Similarity based on Rouge-L scores (SROUGE), and it combines glowworm swarm optimization (GSO) with a bidirectional gated recurrent neural network (BiGRNN) model for summary creation. On their own dataset, the model achieves an accuracy of 35.54%, a recall of 51.60%, and an F-score of 39.06%. The ODL-LTS technique is unique in that it does not rely on handmade features or domain-specific expertise, making it extensible to a wide range of domains.

Hanne et al. [16] focus on abstractive text summarization in German utilizing the German BERT multilingual model and the English BART monolingual model, taking into account translation possibilities. The experiment used the German Wikipedia article dataset to compare the multilingual model's performance to machine-translated text summaries from monolingual English models. The ROUGE-1 measure was used to assess the quality of the text summary.

In their work Deroy et al. [17] investigate the usefulness of combining several summarizing algorithms for court case judgement papers. The study shows that ensemble summaries outperform individual base algorithms in terms of ROUGE and METEOR scores by integrating multiple ensembling strategies. The Indian Supreme Court datasets are utilized, one with extractive gold standard summaries and the other with abstractive gold standard summaries. The study sheds light on how to improve legal case judgement summary by intelligent ensembling.

In this article Dalal et al. [18] evaluate past techniques to automated summarization and compare their output on six distinct legal papers. They offer an abstractive summarization and paraphrasing approach based on the LexRank algorithm and the PEGASUS transformer. The LexRank model's summary outperforms other models, attaining a higher ROUGE-F1-score. The final summary efficiently maintains significant information and paraphrases legal papers into simpler language, making them more accessible to both attorneys and the general public.

Bauer et al. [19] handle legal extractive summarization in the research using a dataset of 430 K US court decisions with annotated important portions. The MemSum model, which is based on reinforcement learning, outperforms transformer-based methods in automated summary quality criteria. MemSum successfully captures significant elements in long court decisions, according to expert human review. The authors open-source their models to the public as a step towards democratizing law, making US court rulings more available to a larger audience.

In the work Bagewadikar et al. [20] provide a complete review of text summarizing strategies, including extractive and abstractive methods as well as deep learning-based models. The study investigates several assessment criteria and text summarizing applications. The inclusion of external information sources and the strengthening of language generating skills are suggested as future research objectives. This publication is a significant resource for text summarizing and its applications scholars and practitioners.

In the paper, Gupta et al. [21] explore text classification for legal documents. They examine machine learning and deep learning algorithms, finding that CNN achieves 90% accuracy in the machine learning category, while CRF performs better with 97% accuracy in the deep learning category. Text categorization proves effective in organizing complex legal papers and extracting relevant information efficiently.

7 Results

We had done some research work in the field of text summarization, in which we had come across some research questions that had helped us get a better understanding of the text summarization domain. Our main focus had been on the use of machine learning and deep learning, with the application of wavelet trees, as they compressed the data in a much more fluent way, helping the retrieval of information in a better way.

Here are some of the research question:

Q1: Why is there greater emphasis on extractive summarizing and less emphasis on abstractive summarization over the legal document?

Ans: While using abstraction summarization in the field of legal document summarization, it tended to generate the content in a shorter version. The algorithm had analyses the primary text and delivered its comprehension in a human-understandable manner, which had resulted in the loss of information that had not been manageable in the legal document.

Q2: How do we find out if our text summarization results are performing better or not?

Ans: For analysing the performance of the text summarization, ROUGE Score had been used, and some of the evaluation metrics had also been used, such as Precision, Recall, F1-Score, and accuracy.

Q3: Which model performs the best for text summarization?

Ans: There were most of the models that performed better in terms of generating the summary, and all of them had their own unique strengths and weaknesses. For the extractive summarization, there had been the use of TextRank and LexRank models, and for the abstractive summarization, there had been the use of neural networks such as RNN or Transformer-based architectures like BERT, GPT, etc. They had been good at understanding the meaning of the text and had been able to generate more human-like summaries.

Q4: What are the current challenges and limitations of DL techniques for text summarization, and how can they be addressed?

Ans: Text summarizing approaches had confronted issues such as providing high-quality and coherent summaries, dealing with extensive input texts, generating different summaries, domain-specific summarization, and a lack of interpretability.

We had implemented two wavelet trees (single % parallel), which had been deployed over the Indian high court dataset, and their performance had been analysed over two processors (I7 & I5). Result of that implementation is mentioned below in Tables 1 and 2.

In the Bombay High Court compression, the time taken to generate a single wavelet tree on an I7 processor was 54 s, while it took 60 s on an I5 processor. In contrast, the time taken to generate a parallel wavelet tree on an I7 processor had been 96 s, while it had taken 89 s on an I5 processor. For the Uttarakhand High Court, the time taken to generate a single wavelet tree had been the same for both the I7 and I5 processors, at 1 s. The time taken to generate a parallel wavelet tree had also been the same for both processors, at 1 s.

Table 1 Use of wavelet tree over Bombay High Court

Parameters	I5 Processor	I7 Processor
Enter the site url	https://bombayhighcourt.nic.in/index.php	
Enter the site name	bombayhighcourt.nic.in	
Enter the url count	5	
Time taken in fetching urls	41,409	30,594
Time taken in reading pdfs	38,880	36,508
Time taken in lemmatization	200,204	124,753
Time taken in inverse index generation	34 s	6 s
Time taken in generating single wavelet tree	60 s	54 s
Time taken in generating parallel wavelet tree	89 s	96 s

Table 2 Use of wavelet tree over Bombay High Court

Parameters	I5 Processor	I7 Processor
Enter the site url	https://highcourtofuttarakhand.gov.in/	
Enter the site name	highcourtofuttarakhand.gov.in	
Enter the url count	7	
Time taken in fetching urls	800	754
Time taken in reading pdfs	3	4
Time taken in lemmatization	98	101
Time taken in inverse index generation	5	4
Time taken in generating single wavelet tree	1	1
Time taken in generating parallel wavelet tree	1	1

8 Conclusion

In this paper, we discussed text summarization techniques specifically tailored for the summarization of legal documents in the context of Indian legal content. Various summarization techniques and the utilization of NLP for text summarization were explored and analysed. The use of wavelet trees for summarization purposes was also detailed, including an explanation of how they work. Additionally, we presented research questions and the results obtained during our work in the summarization domain.

Acknowledgements A Council of Science and Technology, Lucknow, Uttar Pradesh, supported this research via Project Sanction letter number CST/D-3330.

References

1. Turtle H (1995) Text retrieval in the legal world. Artif Intell Law 3:5–54
2. Brunn M, Chali Y, Pinchak CJ (2001) Text summarization using lexical chains, workshop on text summarization in conjunction with the ACM SIGIR conference, New Orleans, Louisiana
3. Kumar R, Raghuveer K (2012) Legal document summarization using latent dirichlet allocation. Int J Comput Sci Telecommun 3(7):8–23
4. Gholamrezazadeh S, Salehi MA, Gholamzadeh B (2009) A comprehensive survey on text summarization systems. In: 2009 2nd international conference on computer science and its applications. IEEE, pp 1–6
5. https://web.science.mq.edu.au/~swan/summarization/projects_full.html
6. Hovy E, Lin CY (1999) Automated text summarization in SUMMARIST. MIT Press, pp 81–94
7. Grossi R, Gupta A, Vitter JS (2003) High-order entropy-compressed text indexes
8. Grossi R, Vitter JS, Xu B (2011) Wavelet trees: from theory to practice. In: 2011 first international conference on data compression, communications and processing. IEEE, pp 210–221
9. Mishra SP, Prasad R, Singh G (2018) Fast pattern matching in compressed text using wavelet tree. IETE J Res 64(1):87–99
10. Khetan R, Agarwal S, Prasad R (2016) An efficient approach towards compressed parameterized word matching using wavelet tree. J Inf Optim Sci 37(2):285–301
11. Jain D, Borah MD, Biswas A (2023) A sentence is known by the company it keeps: improving legal document summarization using deep clustering. Artif Intell Law 1–36
12. Kumar S, Solanki A (2023) An abstractive text summarization technique using transformer model with self-attention mechanism. Neural Comput Appl 1–20
13. Shukla B, Gupta S, Yadav AK, Yadav D (2022) Text summarization of legal documents using reinforcement learning: a study. In: Intelligent sustainable systems: proceedings of ICISS 2022. Springer Nature Singapore, Singapore, pp 403–414
14. Rathi K, Raj S, Mohan S, Singh YV (2022) A review of state-of-the-art automatic text summarisation. Int J Creat Res Thoughts
15. Vaissnave V, Deepalakshmi P (2023) Modeling of automated glowworm swarm optimization based deep learning model for legal text summarization. Multimed Tools Appl 82(11):17175–17194
16. Alcantara THM, Krütli D, Ravada R, Hanne T (2023) Multilingual text summarization for German texts using transformer models. Information 14(6):303

17. Deroy A, Ghosh K, Ghosh S (2023) Ensemble methods for improving extractive summarization of legal case judgements. Artif Intell Law 1–59
18. Dalal S, Singhal A, Lall B (2023) LexRank and PEGASUS transformer for summarization of legal documents. In: Machine intelligence techniques for data analysis and signal processing: proceedings of the 4th international conference MISP 2022, vol 1. Springer Nature Singapore, Singapore, pp 569–577
19. Bauer E, Stammbach D, Gu N, Ash E (2023) Legal extractive summarization of US Court Opinions. arXiv:2305.08428
20. Kumar A, Kumar S, Dev RK, Bagewadikar V, Kumar A, A comprehensive study of text summarization: techniques, applications and future directions
21. Gupta S, Yadav A, Yadav D, Dixit U (2022) Analysis of automatic text classification of legal documents. Available at SSRN 4288439

Neural Network Participation to Enhance Hydrogen-Biofuel CI Engine Performance and Combat Emissions

Atanu Roy, Ahmed J. Obaid, Sabyasachi Pramanik, Kalyan Mitra, and Manashi Chakraborty

Abstract To efficiently run, combat emissions, and predictive maintenance of the compression ignition engine (CIE), artificial intelligence participation plays a vital function. In this study, a one-cylinder CIE was outfitted with blended fuel (hydrogen, biofuel, and water) to combat emissions and enhance engine running efficiency. Simulink was used to collect data, which was then preprocessed and analyzed to predict emission characteristics of CIEs using deep reinforcement learning (DRL) and artificial neural networks (ANN). This paper presents a hybrid model. In this study, mean square error (MSE), R^2, and accuracy are evaluated to show how well the prediction model worked to improve CIEs' running characteristics and fuel types. The proposed method is promising to enhance engine performance and combat emissions. Additionally, the model was found to have a low MSE, indicating that it can make accurate predictions for engine running characteristics and fuel types.

Keywords Hydrogen · Biofuel · CIE · Emissions · BSFC · DRL-ANN

A. Roy · M. Chakraborty (✉)
The Neotia University (TNU), D.H. Road, 24 Pgns (S), Sarisa, WB 743368, India
e-mail: manashi.chakraborty@tnu.in

A. Roy
e-mail: atanuroy@tnu.in

A. J. Obaid
University of Kufa, Kufa, Iraq
e-mail: ahmedj.aljanaby@uokufa.edu.iq

S. Pramanik (✉)
Haldia Institute of Technology (HIT), Hati Beria, Haldia, WB 721657, India
e-mail: sabyalnt@gmail.com

K. Mitra
Indian Maritime University - Kolkata Campus, P - 19, Taratalla Road, Kolkata 700088, India

S. Dutta et al. (eds.), *Cyber Intelligence and Information Retrieval*, Lecture Notes in Networks and Systems 1025, https://doi.org/10.1007/978-981-97-3594-5_11

1 Introduction

The growing energy sectors endanger both human health and the environment by emitting toxic exhaust into the atmosphere by burning fossil fuel resulting in pollution. To minimize the pollution effect, various strategies were adopted by the industries. Alternative fuel, technology enhancement, and stringent policy implementation are the primary tools to deal with the situation. While considering alternative fuel, biofuel is a renewable energy source made from organic materials like plant oils and animal fats and shows considerable emission reduction while burning [1]. Also, hydrogen (H_2) fuel is a clean fuel that produces zero emissions when burned [2]. They can reduce emissions and improve air quality as alternatives to conventional fossil fuels.

Several researchers have attempted to develop a "dual-fuel engine" that runs on H_2 and vegetable oil, biodiesel, or diesel as pilot fuel [3]. However, it takes time and money to research to analyze performance and engine emission behavior for dual-fueled CIE based on H_2 under various operating conditions. Computer modeling can be used to accomplish this task. Artificial neural network (ANN) may be used to forecast performance and emissions more accurately [4].

This study's novel neural network (NN) architecture is meticulously curated for engine monitoring. It reveals how NN can be optimized for predicting engine performance and emissions with high accuracy.

An overview of the literature on neural networks for CIE in the literature review section organizes the paper. The background study discusses the machine learning (ML) approach for emission analysis. The proposed methodology explains it in detail. In conclusion, the report summarizes the survey with critical findings after examining the presented model results and discussion of findings.

2 Literature Review

An inclusive review of the literature on ML practices for CIE investigation is exhibited in this section. A concise overview of the research gaps and critical findings identified in the reviewed literature is presented in Table 1. CIE characteristics are predicted and analyzed based on methodologies, techniques, and evaluation metrics developed in this field.

Through numerical analysis, Liu et al. [5] analyzed the fuel-burning behavior and engine emission of a dual-fuel natural gas/diesel engine under high load. Fuel injection nozzles increased air utilization and combustion heat release, improving engine efficiency and reducing soot emissions. As an analysis of the study, it was identified there was a limit to engine efficiency and that 11 nozzles were optimal. The engine efficiency reached 47.6% by optimizing the natural gas substitution ratio and commencement of injection timing, while nitrous oxide (NOx) and soot exhaust remained lesser than Euro VI emission limits. Kashyap et al. [6], in this study, honge

Table 1 Research gaps analysis from literature review

Fuel type	Tool used	Methodology	Outcome	Gap evaluation	Ref.
PG + HOME	CCD-RSM	Predicting engine operating parameters	P-value < 0.5; $R^2 > 0.97$, Mean error max. 4.53%	A higher misclassification rate reduces the efficiency	[6]
PM + DEE	ANN, BTE, and NO	Prediction of emission features	R^2 93 to 95%	Data constraints	[7]
Diesel	MLP-MOACO	Prediction of emission	R^2 0.99978	Nonlinear mapping and learning week	[8]
Diesel + Natural Gas	LSTM + NSGA-II	PCCI combustion mode, IMO	RMSE <1.3; MAPE <0.15%	Features cause data dimensionality issues.	[9]

oil methyl ester (HOME) and producer gas (PG) were utilized as primary fuels and pilot fuels, respectively, to optimize engine operating parameters. Experimentally analyzing performance and emissions characteristics, an RSM model was developed, and the model was used to optimize operating parameters. With an overall desirability value of 0.96, 61.10 % engine load, 18 compressions, and 23.30° before the top dead center (TDC) were determined as the optimal operating parameters.

Magesh et al. [7] examined whether pumpkin and maize (PM) could be used to produce biodiesel by transesterification and diethyl ether (DEE) addition. Performance, combustion, and emission characteristics were evaluated using pumpkin-maize biodiesel and different amounts of DEE under various load conditions. As compared to diesel fuel, adding 5 ml of DEE significantly improved brake thermal efficiency by 31.91%. In addition, the DEE blend reduced brake-specific fuel consumption by 9.519%. As a result, NO emissions at peak loads were reduced by 34.91%, smoke opacity improved by 27.24%, and HC emissions increased by 32.43%.

Jose et al. [8], an optimization model for engine emissions in a four-stroke, spark ignition (SI) gasoline engine was proposed using a multilayer perceptron (MLP) model coupled with an ANN. Three objective functions were used during optimization using the multi-objective ant colony optimization (MOACO) algorithm. From the Pareto front obtained by MOACO, the VIKOR decision method was applied to select the final compromise solution. A correlation coefficient of 0.99978 was achieved for estimating NOx emissions in gasoline engines using the MLP-MOACO model.

In Ma et al. [9] a predictive regression model, extended short-term memory networks (LSTM) with non-dominated sorting genetic algorithms (NSGA-II), for calibrating parameters and outturn responses in a dual-fuel engine was proposed. Across all working situations, the dual-fuel engine reduced NOx emissions by 77%. In addition, the optimized dual-fuel engine consumed the same amount of fuel as the original diesel engine. With a four-condition test cycle, the optimized dual-fuel engine lesser NOx emissions by 26.6% compared to Tier-III legal necessities. Model train and test results have root mean squared error (RMSE) of less than 1.3 and mean absolute percentage error (MAPE) less than 0.15%.

Considering this study's literature review, a data analytics model was proposed that provides a promising solution for predicting CIE characteristics. The objectives of the research are

1. To ensure that engine data used for training prediction models is accurate and high quality.
2. Advanced ML and deep learning (DL) algorithms have been developed to improve the accuracy and efficiency of engine characteristic predictions.
3. To predict engine characteristics, it is necessary to explore feature extraction techniques to determine the most relevant and informative features.
4. An ensemble learning study was conducted to improve the performance of multiple prediction models by analyzing them together.

3 Background Study

Hariharan et al. [10] proposed predicting the relationship between engine inputs and output responses, including load, lemongrass oil, and H_2. As an add-on, optimized engine input parameters based on response surface methodology (RSM) to reduce emissions and increase thermal efficiency also has been proposed. Implementing a more efficient premixed combustion phase releases more energy, which allows for this enhancement. A combination of RSM and ANN improved engine performance and emission features within 5% of reality. With ANN [11–13] models, complex interactions between engine reactions and input factors can be captured, allowing accurate forecasts. This has significant implications for improving engine performance and reducing emissions.

As illustrated in Fig.1, the emission prediction experiment utilizes the modified one-cylinder diesel engine in dual-fuel mode. As part of the compression stroke, lemongrass oil (LGO) is introduced 23° before the top dead center (TDC) in the combustion chamber. The intake valve opens 4.5° before TDC and closes 35.5° after bottom dead center (BDC), and the exhaust valve opens 35.5° before BDC and closes 4.5° after TDC. H_2 gas is introduced at the intake manifold along with air throughout the suction stroke to ensure safe and efficient operation. With a H_2 gas flow meter, the flow rate of H_2 is measured in liters per minute (LPM). A pressure regulator and flame arrester are used to control the H_2 supply and prevent back flames.

To test the engine shaft, an eddy current electrical dynamometer is used. Multiple sensors and analytical tools are used to gather data, ranging from a stopwatch and burette to a manometer, air stagnation chamber, and orifice meter. Diesel fuel consumption is monitored with a stopwatch and burette, while airflow proportion is assessed using a manometer, air stagnation chamber, and orifice meter.

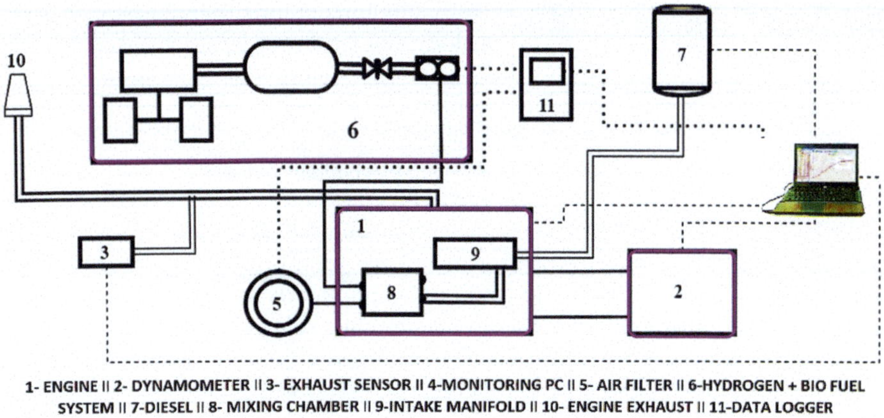

Fig. 1 Schematic view of experimental setup

A charge amplifier amplifies data from a piezoelectric pressure sensor on the engine head, which converts a data-gathering system to a digital signal. The heat release rate pattern is measured using an AVL "Indicom mobile 2012" data-gathering system and combustion analyzer connected to a crank angle encoder with 0.1° resolution. The AVL 444n analyzer detects and measures gases such as hydrocarbon (HC), carbon monoxide (CO), and carbon dioxide (CO_2). Non-dispersive infrared gas analyzers are used to measure CO, CO_2, and HC, while electrochemical sensors are used for NOx emissions.

4 Proposed Methodology

The proposed work includes a set of processes, such as preparing data, extracting features, and forecasting data.

Ranges of the fuel blending approach are modeled in Simulink using adequate experimental data. A standard preprocessing method is used to wipe up the data from artifacts and noise. The accuracy of predictions was enhanced by combining ResNet50 with chaotic vortex optimization (CVO) to pull out deep features from the data. Deep reinforcement learning and artificial neural networks (DRL-ANN) [14] are pooled to foresee CIE running performance and emissions characteristics. In this model, the DRL-ANN model was developed to capture complicated patterns to optimize prediction and improve the accuracy of the forecasts.

The model core involvement is illustrated in Fig. 2, where a hybrid deep learning technique and optimal feature extraction are used to build a data analytics model for predicting the characteristics of CIEs.

1. Experimental data are collected from a simulation (Simulink) with an engine running on H_2, LGO, and water mix to train and validate the prediction model.

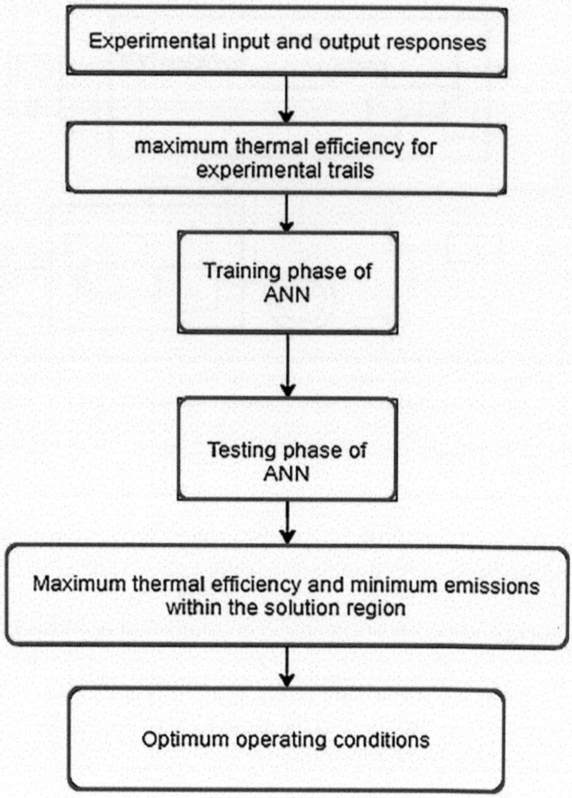

Fig. 2 System designs

2. A CVO approach enhances the model's predictive ability by extracting deep features from engine data.
3. A hybrid deep reinforcement learning (RL) model is presented to predict CIE running characteristics with ANN, DRL-ANN.

A deep learning tool ResNet50 was used to extract features for predicting the running characteristics of a cylinder CIE. By exploring solution spaces efficiently, CVO avoids local optimums by exploring solution spaces efficiently inspired by natural chaos. To crack complex optimization problems, it starts at the midpoint of the search space, constantly updating, forming nesting circles as it progresses. DRL-ANN predicts and optimizes emissions and performance factors in CIEs. Based on inputs such as engine load, LGO, and H_2, it measures different operating conditions to calculate engine performance and emissions. Using DRL-ANN, the engine can sustain better control of its performance, reduce emissions, and improve fuel efficiency. System malfunctions can also be detected and diagnosed using DRL-ANN, reducing maintenance costs.

In Algorithm 1, the DRL-ANN model predicts CIE characteristics. To model the prediction, adjustment, and evaluation steps, resulting in an effective method of estimating emissions accurately the proposed algorithm was used.

Algorithm 1. DRL-ANN algorithm—CIE prediction

Initialization: The parameters of the policy function (θ) and the value function (ϕ) are initialized with a standard Gaussian distribution (N (0, 1)).

$$\theta, \phi \sim N(0,1) \quad (1)$$

The replay buffer (D) is set as an empty set.
The initial state (s_0) is obtained from the engine environment.

Policy Selection: A SoftMax policy utilized, where the probability of taking an action (a) in a state (s) is proportional to the exponent of its associated Q-value:

$$\pi\ (\mathrm{a} \mid \mathrm{s}'; \theta) = \exp\ (\mathrm{Q}(\mathrm{s}, \mathrm{a}; \theta))/\Sigma[\mathrm{a}']\exp\ (\mathrm{Q}(\mathrm{s}, \mathrm{a}'; \theta)) \quad (2)$$

This function yields a probability distribution over actions under state s, which are used to decide the action a_t at timestep t.

Loop for individually 'episode': For each timestep t in an episode:
Execute action a_t, receive reward r_t and observe the next state s_{t+1}.
Store the transition (s_t, a_t, r_t, s_{t+1}) into the replay buffer D:

$$D = D \cup \{(st, at, rt, s\_\{t+1\})\} \quad (3)$$

Prediction: The policy parameters θ updated and value parameters ϕ based on transitions sampled from the replay buffer D. Calculated the loss function L based on the Bellman equation and use stochastic gradient descent to adjust θ and ϕ. Define the target Q-value y as:

$$y = r + y^{*}max\_\{a'\}Q(s', a'; \phi\hat{}-) \quad (4)$$

The loss function L is defined as:

$$L(\theta, \phi) = E\_\{(s, a, r, s') \sim U(D)\}[(y - Q(s, a; \theta))\hat{}2] \quad (5)$$

And the gradients as:

$$\Delta\theta = -\eta * \int (y - Q(s, a; \theta))\nabla\_\theta Q(s, a; \theta)ds \quad (6)$$

$$\Delta\phi = -\eta * \int (y - Q(s, a; \phi))\nabla\_\phi Q(s, a; \phi)ds \quad (7)$$

The policy parameters θ and the value function parameters ϕ are then updated:

$$\theta = \theta + \Delta\theta \quad (8)$$

$$\phi = \phi + \Delta\phi \quad (9)$$

Evaluation: The performance of the policy π is evaluated by calculating the expected discounted reward over an episode:

$$\mathrm{R} = \mathrm{E}\,[\Sigma\_\{t = 0\}\hat{}Ty\hat{}t^{*}rt] \quad (10)$$

Here, the expectation is over the trajectories generated by the policy π

5 Results and Discussion

A DRL-ANN model is applied to reduce emissions and optimize engine performance by predicting the running time characteristics of engine parameters. Figs. 3, 4, and 5 show the bar chart comparison of varying NOx, opacity/smoke, and brake-specific fuel consumption (BSFC) while varying percentage of H_2 fuel injection blending with LGO and water concerning varying engine loads, whereas Figs. 6, 7, and 8 display the DRL-ANN model forecasting and comparison of NOx, opacity, and BSFC with the previous ANN experiment model. As shown in Fig. 3, when diesel fuel and LGO are compared, LGO exhibits generally higher NOx values at all load percentages. There is a difference between diesel and LGO NOx emissions of approximately 0.27 to 6.75%, with the most significant difference occurring at higher engine loads. Compared to LGO H10 W3 (blends of LGO, which include 10% H_2/3% water), LGO H10 emits approximately 0.35 to 14.86% more NOx.

A comparison of the opacity values between diesel fuel and LGO at varying engine loads is shown in Fig.4, indicating the degree to which opacity or smoke is produced during combustion for different fuel types. In general, LGO exhibited lower opacity values across all load percentages when compared to diesel fuel. Diesel and LGO have an opacity difference of approximately 0.38 to 6.15%, increasing as engine load increases. Comparing LGO H10 with LGO H10 W3, the percentage difference in opacity ranges from 0.45 to 11.47%.

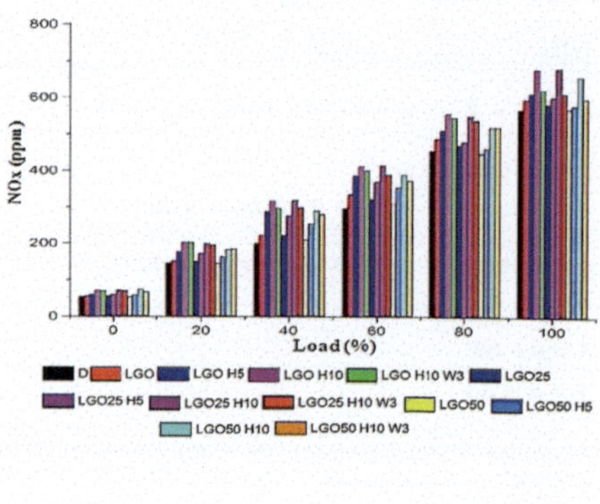

Fig. 3 NOx % emissions w.r.t loads. *Note* Color coding for bar graph

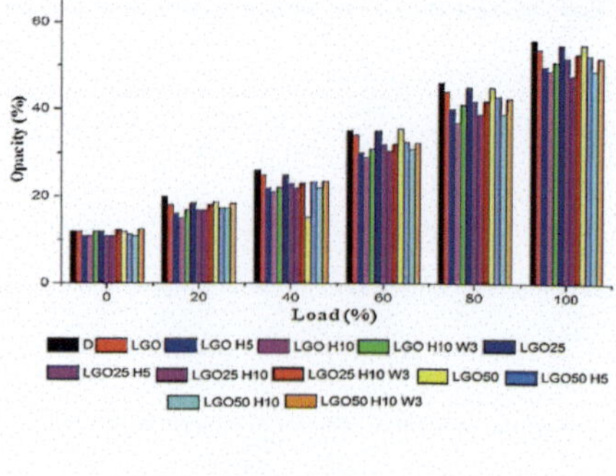

Fig. 4 Opacity (ppm) w.r.t varying loads. *Note* Color coding for bar graph

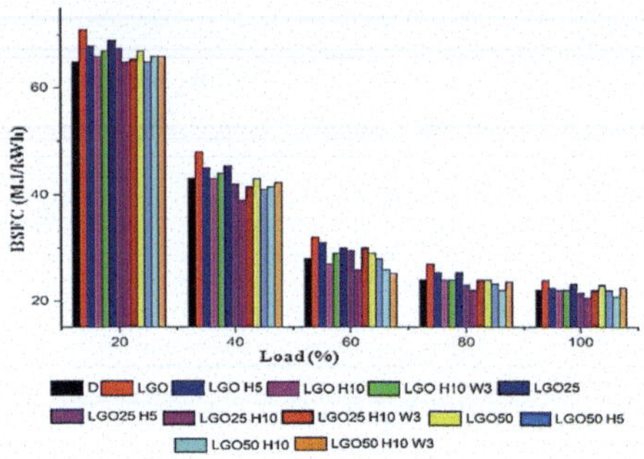

Fig. 5 BSFC w.r.t varying engine loads. *Note* Color coding for bar graph

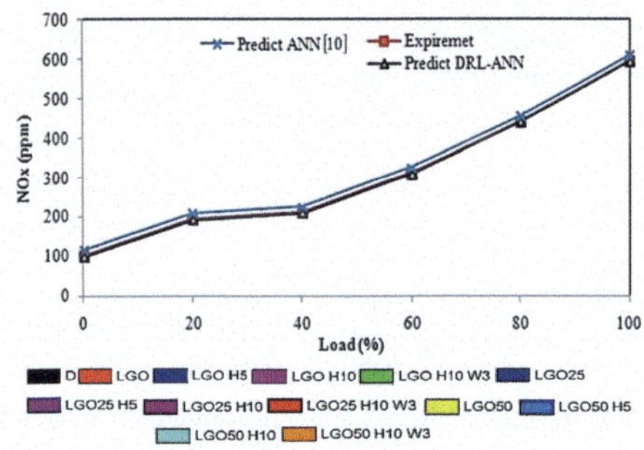

Fig. 6 Comparative study of NOx. *Note* Color coding for bar graph

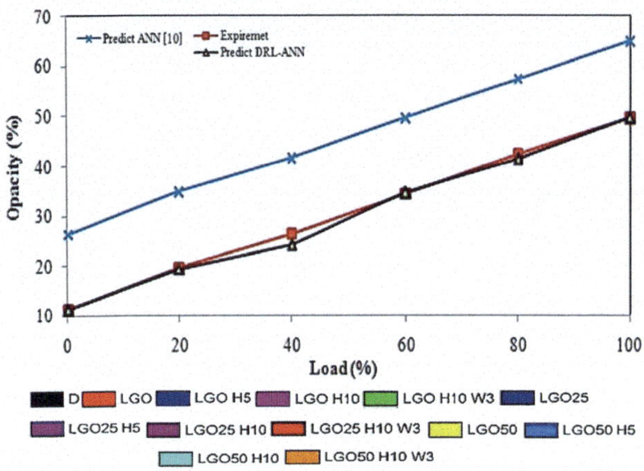

Fig. 7 Opacity emissions (ppm) vs Predicted value

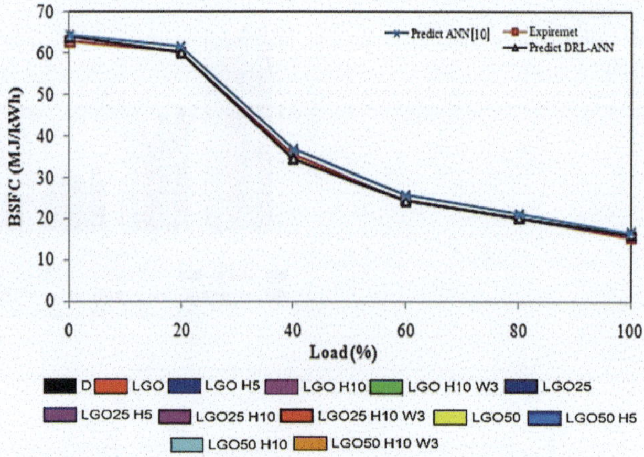

Fig. 8 Comparative study of BSFC. *Note* Color coding for bar graph

According to Fig. 5, LGO exhibits higher BSFC values across all load percentages when compared with diesel fuel. There is a 14.76 to 30.58% difference in BSFC between diesel and LGO, with the most considerable difference occurring at lower engine loads. LGO H10 exhibits approximately 0.89 to 23.21% less BSFC when compared to LGO H10 W3.

Figure 6 shows NOx emissions for different engine load percentages, along with experimental results, predicted values from the proposed DRL-ANN model, and predicted values from a previous ANN model. The DRL-ANN model provides reasonable predictions based on the observed NOx emissions. The DRL-ANN model effectively predicts NOx emissions with an error of approximately −0.90 to +0.55%. Similarly, comparing the experimental NOx emissions with the predicted values from the existing ANN model, a difference of roughly +14.86 to +5.4% exists between the observed predicted and experimental ANN values.

Opacity values for different engine load percentages are shown in Fig. 7, along with experimental results, predicted values from the proposed DRL-ANN model, and predicted values from an existing ANN model. The DRL-ANN model provides reasonably accurate predictions compared to experimental opacity values. There is a difference between train and test values of approximately 0 to 3.9%. The ANN model also provides forecasts with somewhat lower accuracy than the DRL-ANN model. There is a significant difference between predicted and observed ANN values of approximately 27.8 to 35.4%. Although the ANN model generally predicts higher values than the experimental values, it still follows a similar increase with load percentage.

Figure 8 shows the BSFC (MJ/kWh) for different percentages of engine load based on experimental results, values predicted by the proposed DRL-ANN model, and values predicted by an existing ANN model. Observe a reasonable degree of accuracy in the DRL-ANN model. BSFC predicted values at lower load percentages (0 and 20%) are slightly higher than the testing values, varying from 1.65 to 0.57%. As the

Table 2 Training and testing event assessment

	Training event			Testing event		
	Accuracy	R^2	MSE	Accuracy	R^2	MSE
NOx	0.91907	0.94387	0.00014	0.95131	0.938005	0.00062
Opacity	0.98165	0.98162	0.00097	0.96252	0.971454	0.00061
BSFC	0.95618	0.95082	0.00055	0.98052	0.94594	0.00059

load percentage increases, the predicted values align closely with the experimental values, with percentage differences ranging from approximately −1.59 to 1.08%. Also, by comparing observed and predicted BSFC values, we can see that the ANN model provides reasonably accurate predictions. Approximately −3.64 to 1.84% of the predicted ANN values differ from the experimental values.

5.1 Analysis of the Model Used to Estimate CI Engine Characteristics

Training and testing of the DRL-ANN model demonstrate its generalizability and sturdiness. Its reliability in measuring opacity/smoke emissions is one of its most appealing features. This precision is essential in emissions modeling, where inaccuracies can have serious environmental consequences. A reduced MSE [15, 16] for NOx, opacity, and BSFC indicates improved performance and demonstrates its exceptional prediction ability. In addition, the DRL-ANN model demonstrated remarkable accuracy. A model comparison given in Table 2 reveals DRL-ANN model performed better than other models at forecasting CI Engine emissions.

6 Conclusion

It is known that conventional diesel fuel releases several pollutants, resulting in greenhouse gas, acid deposition, and climate change and negatively affecting human health. Eco-friendly fuel and optimized engine running are crucial to mitigating these effects. This paper used an H_2, LGO, and water hybrid fuel mixture to operate a one-cylinder CI engine and predict its emissions and performance. Compared to earlier models, the DRL-ANN has improved accuracy. For NOx, opacity, and BSFC predictions, DRL-ANN has an MSE of 0.00062, 0.00061, and 0.00059, respectively, and R^2 97% for opacity, demonstrating model effectiveness for monitoring and reducing environmental pollution impacts by reducing, predicting, and classifying CI engine emissions accurately.

References

1. Priya PSD, Verma Y, Muhal RA, Goswami C, Singh T (2021) Biofuels: an alternative to conventional fuel and energy source. Materials today: proceedings, Sep 2021. https://doi.org/10.1016/j.matpr.2021.08.227
2. Chai WS, Bao Y, Jin P, Tang G, Zhou L (2021) A review on ammonia, ammonia-hydrogen and ammonia-methane fuels. Renew Sustain Energy Rev 147:111254. https://doi.org/10.1016/j.rser.2021.111254
3. Alias M, Hairuddin AA, Hassan MK, Anas K (2019) A review of hydrogen addition in an HCCI engine fueled with biofuels, Jan 2019. https://doi.org/10.1063/1.5085988
4. Bakay MS, Ağbulut Ü (2021) Electricity production-based forecasting of greenhouse gas emissions in Turkey with deep learning, support vector machine and artificial neural network algorithms. J Clean Prod 285:125324. https://doi.org/10.1016/j.jclepro.2020.125324
5. Liu X, Wang H, Zheng Z, Yao M (2021) Numerical investigation on the combustion and emission characteristics of a heavy-duty natural gas-diesel dual-fuel engine. Fuel 300:120998. https://doi.org/10.1016/j.fuel.2021.120998
6. Kashyap D, Das S, Kalita P (2021) Exploring the efficiency and pollutant emission of a dual fuel CI engine using biodiesel and producer gas: an optimization approach using response surface methodology, vol 773, pp 145633–145633, Jun 2021. https://doi.org/10.1016/j.scitotenv.2021.145633
7. Magesh N et al (2022) Experimental investigation and prediction of performance, combustion, and emission features of a diesel engine fuelled with pumpkin-maize biodiesel using different machine learning algorithms, vol 2022, pp 1–17, Oct 2022. https://doi.org/10.1155/2022/9505424
8. Martínez-Morales J, Quej-Cosgaya H, Lagunas-Jiménez J, Palacios-Hernández E, Morales-Saldaña J (2019) Design optimization of multilayer perceptron neural network by ant colony optimization applied to engine emissions data. Sci China Technol Sci 62(6):1055–1064. https://doi.org/10.1007/s11431-017-9235-y
9. Ma C, Song E, Yao C, Long Y, Ding S-L, Xu D (2021) Analysis of PPCI mode and multi-objective comprehensive optimization for a dual-fuel engine, vol 303, pp 121296–121296, Nov 2021. https://doi.org/10.1016/j.fuel.2021.121296
10. Hariharan N, Senthil V, Krishnamoorthi M, Karthic SV (2020) Application of artificial neural network and response surface methodology for predicting and optimizing dual-fuel CI engine characteristics using hydrogen and bio fuel with water injection. Fuel 270:117576. https://doi.org/10.1016/j.fuel.2020.117576
11. Praveenkumar S et al (2023) Prediction of patients' incurable diseases utilizing deep learning approach. In: Hassanien AE, Castillo O, Anand S, Jaiswal A (eds), Springer Nature Singapore, pp 33–44
12. Chellam VV, Veeraiah V, Khanna A, Sheikh T, Pramanik S, Dhabliya D (2023) A machine vision-based approach for tuberculosis identification in chest x-rays images of patients, pp 23–32, Jan 2023. https://doi.org/10.1007/978-981-99-3315-0_3

13. Ahamad S, Veeraiah V, Ramesh JVN, Rajadevi R, Reeja SR, Pramanik S, Gupta A (2023) Deep learning based cancer detection technique, thrust technologies' effect on image processing. IGI Global
14. Pramanik S, Obaid AJ, Niranjanamurthy M, Bandyopadhyay SK (2023) Applications of big data in clinical applications, Jan 2023. https://doi.org/10.1063/5.0119414
15. Pramanik S (2023) An adaptive image steganography approach depending on integer wavelet transform and genetic algorithm, Mar 2023. https://doi.org/10.1007/s11042-023-14505-y
16. Pramanik S, Singh RP, Ghosh R (2020) Application of bi-orthogonal wavelet transform and genetic algorithm in image steganography. Multimed Tools Appl 79(25–26):17463–17482. https://doi.org/10.1007/s11042-020-08676-1

Analyzing MACD Divergence: Challenges in Forex and Commodity Trading

Mahamudul Hasan, Md. Al-Amin, Izazul Hoq Imran, Tropa Das, and Nazib Al-Hoque

Abstract MACD stands for Moving Average Convergence, and it's a trading tool used by traders, analysts, and researchers in Forex and commodity trading to estimate the relationship between two moving averages of a security's price. It checks the currency pair price movements and briefly compares two moving averages, usually 12 and 26 periods (EMA 12 and EMA 26). We use here MACD divergence and simple and exponential moving averages. The bullish and bearish trends represent price increases and decreases, respectively, and the diverges indicate strong trends, whereas closed lines indicate weak or sideways trends. The MACD indicator effectiveness depends on assets and different time frames like EUR/USD and XAU/USD, which change with different time frames. It works effectively when used combined with other technical tools and current market news. The EUR/USD shows a 44% win rate in the 30-min time frame, and the XAU/USD shows a 45% win rate over 15- and 60-min time frames. Although MACD is a valuable asset in real-time forex and commodity trading, its productiveness is changeable, and traders should combine it with other technical tools, market news, and economic events for optimal results.

Keywords MACD divergence · Technical analysis indicator · EUR/USD and XAU/USD currency pair · Win rate · Loss rate · Net balance calculation

1 Introduction

The trading and investment process is always challenging due to the uncertain behaviors of the forex and commodity markets. The Moving Average Convergence Divergence indicator, over the years, is a necessary tool for traders, analysts, and researchers [1]. The Moving Average Convergence Divergence is a well-known and broadly used indicator that was developed by Gerald Appel in the late 1970s. It predicts the divergence form or not based on the current market data. It is popular to understand market movements and momentum, which helps traders, analysts, and

M. Hasan (✉) · Md. Al-Amin · I. H. Imran · T. Das · N. Al-Hoque
CSE, East West University, Aftabnagar, Dhaka 1212, Bangladesh
e-mail: Munna09bd@gmail.com

S. Dutta et al. (eds.), *Cyber Intelligence and Information Retrieval*, Lecture Notes in Networks and Systems 1025, https://doi.org/10.1007/978-981-97-3594-5_12

researchers to make their buy or sell decisions [3]. It is very critical to understand the actions of every asset in the global financial ecosystem. For example, the EUR/USD currency pair has a big effect on forex markets. On the other hand, XAU/USD is a secure and stable investment. However, this research problem is whether the MACD indicator predictions are accurate or not [6].

The purpose of this paper is to investigate the specifics of how well the MACD indicator works and attention to getting profit from the EUR/USD currency pair and XAU/USD. After the research, we find out the win, loss, and break-even rates for the EUR/USD currency pair and XAU/USD by trying them thoroughly using real-time data and the MACD Divergence Detection algorithm. In the further sections, we discuss in more detail, about the MACD Divergence Detection algorithm works and what its parts are, such as the short-period, long-period, MACD line, MACD histogram, and signal period (EMA 9). The win rate is 44%, the loss rate is 31%, and the break-even rate is 25% for the EUR/USD currency pair. The win rate is 41%, the loss rate is 33%, and the break-even rate is 26% for XAU/USD.

2 Literature Review

Literature involves finance, technical analysis, and trading. This research evaluates trade material and builds on previous studies [1]. Financial market literature exhibits complicated and dynamic Forex and Commodity Trading market activity. Examine how supply, demand, market efficiency, and external variables affect price volatility. Many research have studied how technical analysis indicators predict market movements. Moving averages, RSI, and MACD were evaluated. Authors measured these indicators' advantages and downsides [2]. The corpus includes day, swing, and long-term trading approaches. Much academic research has evaluated trading intervals and profitability. They discovered era-specific trading success. Authors in [3] disagree on investment returns in a scholarly active trading versus passive investing discussion. Risk and psychology are in trading. Some researchers concentrated on limiting negative outcomes, whereas Kahneman and Tversky (2015) explored decision-making biases. The risk management and trading psychology research covers all trading performance aspects. This literature review includes trade and financial market results from several regions. Our research found market efficiency, technical analysis, trading techniques, risk management, and psychology in trading. This study enhances prior studies on trading's dynamic environment [4].

3 Methodology

To accuracy and dependability, learning efforts demand a disciplined strategy. It relations unprocessed data to intriguing MACD indicator findings in the EUR/USD and XAU/USD markets. The analysis records transaction values, success, failure, and break-even rates in real-time EUR/USD and XAU/USD trading data. A data cleaning and preparation method provides correctness, coherence, and completeness. It analyzes MACD line crossovers and signal formation using EUR/USD and XAU/USD market data. Divergences and substantial differences are examined [5]. We have been done with simulated trades using MACD signals and historical price data to calculate profits, losses, expenses, and slippage to evaluate transaction efficiency. The comparative study compared the Moving Average Convergence Divergence indicator accuracy to real-world trading results including win, loss, and break-even. It explored MACD indicator performance in EUR/USD and XAU/USD over time, comparing trends, variances, and correlations across assets [6]. Here we employed statistical approaches to analyze data and support trends with quantitative data. Using the EUR/USD and XAU/USD Moving Average Convergence Divergence (MACD) signals, they examined ethics, data privacy, and compliance. The study informs traders, analysts, and academics [7].

3.1 Technical Analysis of MACD

Technical analysis anticipates financial asset value changes using prior price data and transaction activity. Patterns, trends, and market sentiment determine a company's intrinsic value. Historical events may declare Forex and Commodity Trading market price movements, including up, down, and stable [7, 8]. These tendencies are due to market psychology and collective behavior (Fig. 1). Scholarly studies employ candlestick charts to depict asset values, while support and resistance concepts are used to predict price levels at which a decrease in value would stop owing to increasing buying activity [9]. Trend lines show the course and development of a trend, whereas moving averages eliminate variations and offer a continuous trend line. Technical indicators predict financial market movements mathematically [10]. The Moving Average Convergence Divergence (MACD) momentum indicator analyzes trends, while the Relative Strength Momentum (RSM) indicator detects overbought or oversold market circumstances (Fig. 4). Billing bands show the N-period simple moving average and two standard deviation bands [11]. The stochastic oscillator compares a security's closing price and price range, whereas Fibonacci retirements indicate support and resistance levels based on important Fibonacci values (Fig. 2). Technical analysis of trailing indicators is difficult owing to their subjectivity and temporal latency (Fig. 3). Signal interpretation may lead to erroneous findings and financial loss. Diverse approaches and technological advances are needed to enhance technical analysis [12] (Fig. 2).

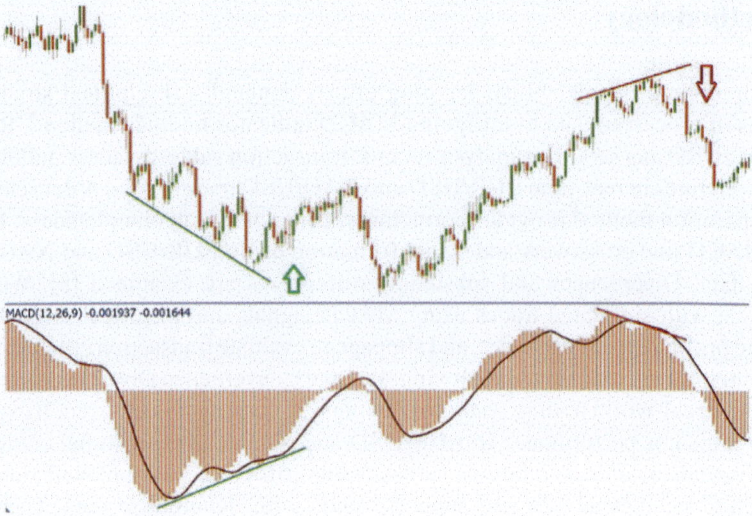

Fig. 1 Divergence indicator

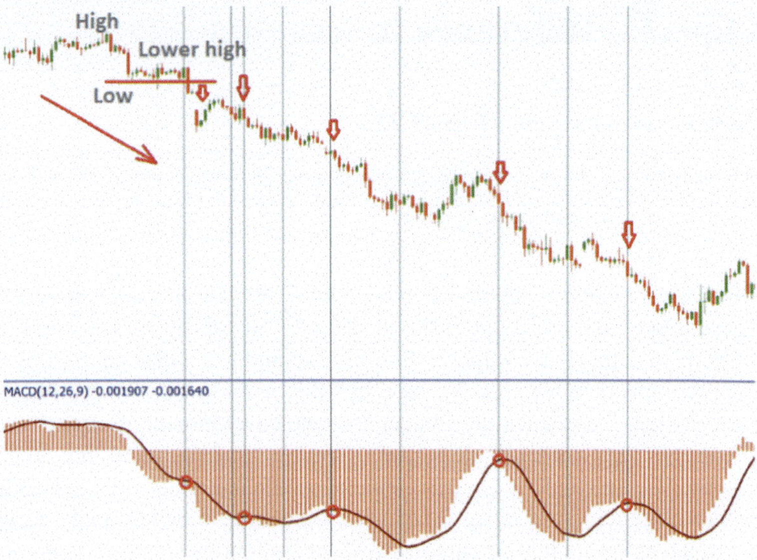

Fig. 2 Signal line crossing

Fig. 3 Divergence indicator

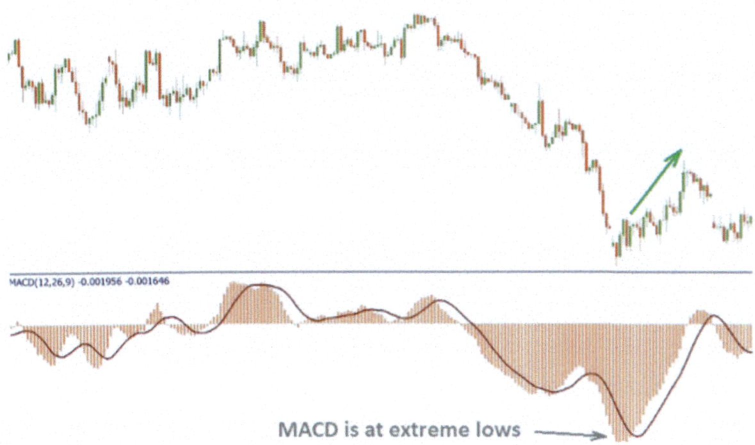

Fig. 4 Overbought oversold levels

3.2 Exponential Moving Average (EMA)

We use for MACD divergence to evaluate the trend's direction and momentum by monitoring the MACD and EMA. When the MACD line goes up to zero and the price is up on the EMA, then this is a possible buy signal, otherwise, a sell signal is provided when the MACD line crosses below zero and the price is below the EMA. The MACD line passes below the signal line, indicating a probable shift in momentum, the approach proposes leaving the trade [7] (Fig. 5).

Fig. 5 EMA indicator

$$\text{EMA-12} = (\text{Closing Price} - \text{Previous EMA-12}) \times \frac{2}{1+12} + \text{Previous EMA-12}, \tag{1}$$

$$\text{EMA-26} = (\text{Closing Price} - \text{Previous EMA-26}) \times \frac{2}{1+26} + \text{Previous EMA-26}, \tag{2}$$

$$\text{EMA-9} = (\text{Closing Price} - \text{Previous EMA-9}) \times \frac{2}{1+9} + \text{Previous EMA-9}. \tag{3}$$

3.3 MACD Calculation

The MACD is essential for computing exponential moving averages, and long-term time frames use a 100-day EMA, whereas short-term time frames use a 50-day EMA. For Signal Line, use a 9-day EMA. Subtract the 50-day and 100-day EMAs to get the MACD line. The MACD histogram shows the MACD Line and Signal Line divergence. The MACD Line is computed as

$$\text{MACD Line} = \text{EMA-12} - \text{EMA-26}. \tag{4}$$

The MACD histogram differentiates the MACD Line and the Signal Line:

$$\text{MACD Histogram} = \text{MACD Line} - \text{Signal Line}. \tag{5}$$

3.4 Net Balance Calculation

The Net Balance is calculated using:

$$\text{Net Balance} = (\text{total win rate} \times \text{profit per win}) + (\text{total loss rate} \times \text{loss per loss})\,. \tag{6}$$

Algorithm 1 MACD base trading algorithm

```
prev_swing_high ← 0, curr_swing_high ← 0, prev_swing_low ←
  0, curr_swing_low ← 0, ema_cross_over_signal ← none, trade ← none, EMA [0]
  = Current EMA and EMA [1] = Previous EMA
for i ← 0 to length(data) − 1 do
    if data[0].prev_swing_high > curr_swing_high then
        prev_swing_high ← curr_swing_high
        curr_swing_high ← data[0].prev_swing_high
    end if
    if data[0].prev_swing_low < curr_swing_low then
        prev_swing_low ← curr_swing_low
        curr_swing_low ← data[i].prev_swing_low
    end if
end for
if ema_50[0] > ema_200[0] and ema_50[i] < ema_200[i] then
    ema_cross_over_signal ← buy
else if ema_50[0] < ema_200[i] and ema_50[1] > ema_200[1] then
    ema_cross_over_signal ← sell
end if
for i ← 0 to length(data) − 1 do
    if data[0].prev_macd_swing_high > macd_curr_swing_high then
        macd_prev_swing_high ← macd_curr_swing_high
        macd_curr_swing_high ← data[0].prev_macd_swing_high
    end if
    if data[0].prev_macd_swing_low < curr_macd_swing_low then
        macd_prev_swing_low ← curr_macd_swing_low
        curr_macd_swing_low ← data[i].prev_macd_swing_low
    end if
end for
if Histogram[i] > 0 and Histogram[0] > Histogram[1] then
    macd_signal ← buy
else if Histogram[0] < 0 and Histogram[0] < Histogram[1] then
    macd_signal ← sell
end if
if ema_cross_over_signal = buy and macd_signal = buy then
    trade ← buy
else if ema_cross_over_signal = sell and macd_signal = sell then
    trade ← sell
else
    trade ← wait for better opportunity
end if
return trade
```

The algorithm is a Moving Average Convergence Divergence (MACD) trading method that generates buy or sell signals based on the interaction of the MACD with a predetermined set of swing highs and lows. It then iterates over market data, initializing numerous variables such as the previous and current swing highs and lows for both price and MACD values. When new extremes are discovered, the algorithm changes swing values. To establish the overall market trend, it looks for a crossover signal between two exponential moving averages (EMA), especially the 50-period EMA and the 100-period EMA. In parallel, it examines the MACD histogram values to determine momentum.

4 Result and Finding

The EUR/USD currency pair has a win rate of 44%, a loss rate of 3%, break-even rate of 25%. The positive results are within 60-min time frames the loss rate is 35%, the break-even is 25%, and the maximum win rate is achieved within 30-min time frames. The XAU/USD currency pair has a win rate of 41%, a loss rate of 33%, and a break-even of 26%. In 60-min time frames, 25% of loss rate, win rate 45%, and break-even 30% were achieved, with a significant correlation with a benchmark of 30% (Tables 1 and 2).

Table 1 Trading performance by time frame for EUR/USD

Time frame (min)	Win (%)	Loss (%)	Break-even (%)
1	40	35	30
5	35	40	25
15	45	35	25
30	50	30	25
60	40	35	25
Total	44	31	25

Table 2 Trading performance by time frame for XAU/USD

Time frame (min)	Win (%)	Loss (%)	Break-even (%)
1	35	45	20
5	40	35	25
15	45	25	30
30	40	35	25
60	45	25	30
Total	41	33	26

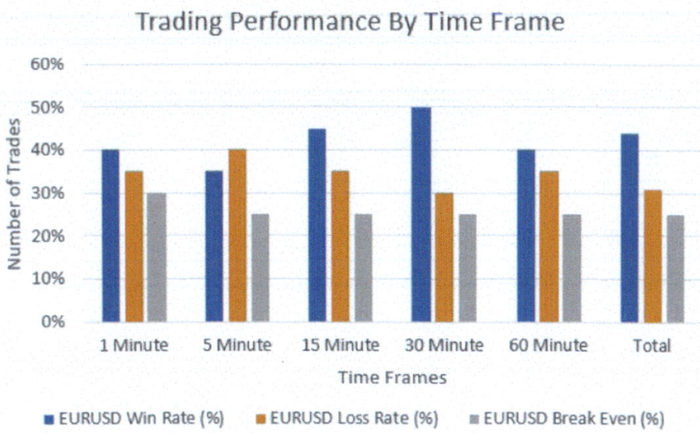

Fig. 6 EUR/USD trading performance

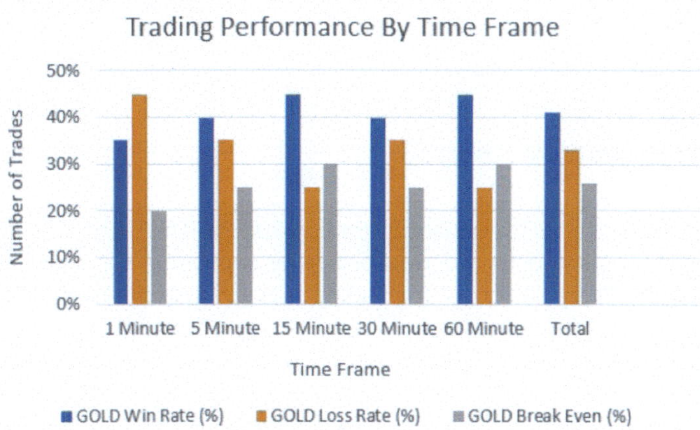

Fig. 7 XAU/USD trading performance

4.1 Experiment and Analysis

In 1-, 5-, 15-, 30-, and 60-min time frames, 40 wins, 35 losses, and 25 break-evens were achieved in 100 trades. In 30 min, 50 wins, 30 losses, and 25 break-evens were achieved. The 30-min time frame has the highest success rate, while 1- and 5-min timeframes have smaller positive net balances. 15- and 60-min time frames have higher win rates and lower loss rates, and break-even points vary between time intervals (Fig. 6).

Out of 100 trades, 35 wins, 45 losses, 20 break-even, 40 wins, 35 losses, 30 break-even, 45 wins, 25 losses, and 30 break-even in 1-, 5-, 15-, 30-, and 60-min time frames. Traders with high win rates in 15- and 60-min periods may find financial success. Smaller positive net positions indicate lower profit potential. The winning rate is slightly higher than the losing rate in 5-min periods (Fig. 7).

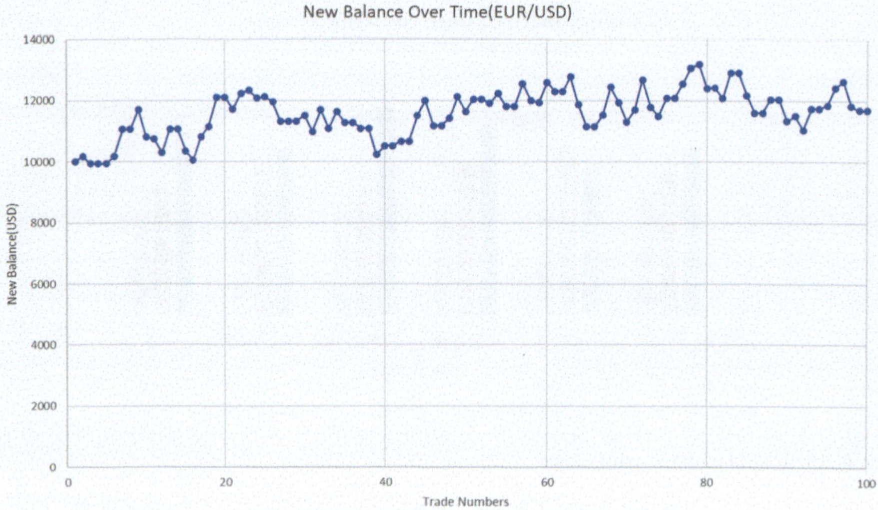

Fig. 8 Trading amount of EUR/USD

4.2 Trading Balance Analysis

The trade balance is crucial to evaluating a trading strategy. This research helps traders identify a strategy's strengths and weaknesses by comparing its performance. By understanding the likely outcomes of a trading strategy, traders may make educated decisions about risk management, position size, and profitability. The research analyzes trade balances for EUR/USD and XAU/USD datasets, computing net balances considering win, loss, and break-even rates. It evaluates profitability over different trading intervals (Fig. 8). The EUR/USD dataset shows success, failure, and neutral outcomes over 1- to 60-min intervals, providing early insight into the approach's performance. In a 1-min trading period, a net balance of 100 trades indicates marginal profit possibilities but is subject to market movements. In 5 min, a net balance of −0.05 indicates a little negative net balance, predicting losses over time. In 15 min, a net balance of 0.10 suggests profitability due to a higher win and low case rate value a balance of 0.20 indicates maximum profitability 7 potential due to a high win rate and low loss. In 60 balances, a net balance of 0.05 indicates fewer transactions and low-profit possibilities. The overall net balance for 100 deals is 44 winning transactions, 31 losing trades, and 25 break-even deals, with optimal net balances between 15 and 30 min (Fig. 9). The XAU/USD dataset measures success, failure, and equilibrium rates over temporal periods, evaluating technique efficacy by identifying successful transactions, unsuccessful deals, and trades without gains or losses. The net balance is −0.10, indicating losses over time. In 5-min intervals, a balanced net balance is observed with nearly equal win-loss rates, a low win rate, and a positive net balance, suggesting lucrative trading with a high victory rate. In 60-min inter-

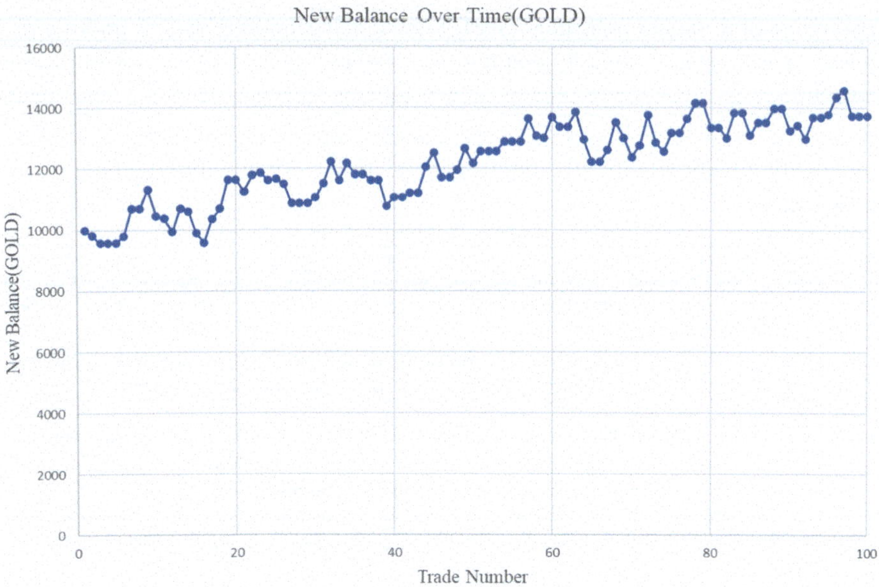

Fig. 9 Trading amount of XAU/USD

vals, a balanced net balance is found, with an 8-unit profit for 100 deals. Overall, the technique provides a near equilibrium between wins and losses for various trading periods.

5 Conclusion

In conclusion, the purpose of this paperwork finds out the effectiveness of the MACD divergence method for trading the EUR/USD, and XAU/USD currency pairs over different time frames. Here, we review trade data and apply the MACD calculations, then evaluate the MACD strategy performance in terms of win rate, loss rate, and break-even rate. The findings indicate that the MACD strategy identifies the short-term trade win rate as bigger than the long-term trade win rate, particularly within 15–60-min time frames. It highlights the importance of adjusting trading techniques, identifying which timeframes are more profitable, and the unique features of each currency pair. The MACD is a well-known technical indicator in the forex market, its variable outcomes across time frames show the importance of prudent risk management and additional strategy development. EUR/USD shows 44% win rate, 31% loss rate, and 25% break-evens, and XAU/USD shows 44% win rate, 33% loss rate, and 26% break-evens. This study offers the basis for a more comprehensive understanding of MACD Divergence in practice. Future research may benefit from evaluating a larger range of technical indicators, market news, and economic data to improve the MACD's prediction potentiality and traders make more profitable financial market decisions.

References

1. Chan KF, Gup BE (2017) International forex and commodity trading market efficiency and integration: a study of eighteen nations. J Bus Finance Account 21(6):785–804. https://doi.org/10.1111/1468-5957.00134
2. Menkhoff L, Sarno L, Schmeling M, Schrimpf A (2019) Carry trades and global foreign exchange volatility. J Finance 67(2):681–718. https://doi.org/10.1111/j.1540-6261.2012.01728.x
3. Sarno L, Taylor MP (2020) The economics of exchange rates. Cambridge University Press
4. Ito T, Hashimoto T (2016) Intra-day seasonality in activities of the foreign exchange markets: evidence from the electronic broking system. J Jpn Int Econ 39:17–33
5. Lo AW, Hasanhodzic J (2018) The statistical and economic role of jumps in continuous-time interest rate models. J Bus Econ Stat 28(2):226–238. https://doi.org/10.1111/j.1540-6321.2004.00632.x
6. Cartea A, Donnelly R, Jaimungal S (2016) Algorithmic trading with learning. J Finance 70(5):2227–2274. https://doi.org/10.1142/S021902491650028X
7. Hendershott T, Riordan R (2016) Algorithmic trading and information. J Finance 66(4):1055–1094
8. Tversky A, Kahneman D (2019) Judgment under uncertainty: heuristics and biases. Science 185(4157):1124–1131. https://www.science.org/doi/abs/10.1126/science.185.4157.1124
9. Kahneman D, Tversky A (2020) Prospect theory: an analysis of decision under risk. Econometrica 47(2):263–291. https://www.jstor.org/stable/1914185
10. Thaler RH (2015) Mental accounting and consumer choice. Market Sci 4(3):199–214. https://doi.org/10.1287/mksc.4.3.199
11. Kang BK (2017) Improving MACD technical analysis by optimizing parameters and modifying trading rules: evidence from the Japanese Nikkei 225 futures market. J Risk Financ Manag 14(1):37. https://www.mdpi.com/1911-8074/14/1/37
12. Jegadeesh N, Titman S (2018) Returns to buying winners and selling losers: implications for forex and commodity trading market efficiency. J Finance 48(1):65–91. Ohttps://doi.org/10.1111/j.1540-6261.1993.tb04702.x

GIS-Based Analysis of Water Integrity Differences Within the Madurai Region

Baranitharan Balakrishnan, Karthik Chandran, K. Kannabiran, Subrata Chowdhury, Tanupriya Choudhury, and Ketan Kotecha

Abstract This present study is mainly concerned with the variations of physicochemical parameters of groundwater samples from different areas of the Madurai region due to different land use and monsoon seasons. Groundwater samples were collected from 40 sample points in and around Madurai city. 14 parameters were considered for the study and physicochemical analysis was done for collected samples. Sampling was done for the monsoon season. Thematic mambos. The study was prepared by visual interpretation of IKONOS satellite imagery using ArcGIS to

B. Balakrishnan (✉)
Department of Agricultural Engineering, Kalasalingam Academy of Research and Education, Krishnankoil 626126, TN, India
e-mail: jack.barans@gmail.com

K. Chandran
Department of Robotics and Automation, Jyothi Engineering College, Chelakkara 679531, KL, India
e-mail: karthikmtech86@gmail.com

K. Kannabiran
Department of Horticulture, Kalasalingam Academy of Research and Education, Krishnankoil 626126, TN, India
e-mail: kannabiran@klu.ac.in

S. Chowdhury
Department of Computer Science and Engineering, SITAMS (A) Chittoor, Murukambattu, AP, India

T. Choudhury (✉)
CSE Department, Graphic Era Deemed to be University, Dehradun 248002, Uttarakhand, India
e-mail: tanupriyachoudhury.cse@geu.ac.in; tanupriya.choudhury@sitpune.edu.in; tanupriya1986@gmail.com

CSE Department, Symbiosis Institute of Technology, Symbiosis International (Deemed University), Pune 412115, Maharashtra, India

K. Kotecha
Symbiosis Center for Applied Artificial Intelligence, Symbiosis Institute of Technology, Symbiosis International (Deemed University), Pune 412115, India
e-mail: director@sitpune.edu.in

S. Dutta et al. (eds.), *Cyber Intelligence and Information Retrieval*, Lecture Notes in Networks and Systems 1025, https://doi.org/10.1007/978-981-97-3594-5_13

interpret the water quality index in graphical form. The relationship between Variation in parameters and land use change is interpreted using Statistical Package for the Social Science (SPSS) software.

Keywords Groundwater quality · Land use variation · Monsoon season · Water quality index

1 Introduction

Groundwater plays a crucial role in our environment in three major ways. Firstly, it provides the necessary base flow to keep most rivers flowing throughout the year, which helps sustain aquatic ecosystems. Almost all land-based activities, whether industrial, urban, or agricultural, have the potential to pollute groundwater. While obvious sources like industrial discharges, landfills, and chemical waste injection sites can be easily identified and regulated, the more challenging issue is diffuse sources of pollution such as agrochemical leaching, animal waste, and subsurface discharge from septic tanks and latrines. In a study conducted by [1], the Water Quality Index (WQI) was used as a reliable method to evaluate water quality. The resolve of the work was to measure the suitability of groundwater for domestic use in the Srungavarpu Kota village of Andhra Pradesh. Physicochemical properties were analyzed and the computed WQI was categorized to create zones indicating moderate to good quality. The study results can assist planners in managing and monitoring resources and determining the fitness of water for ingestion and domestic tenacities. Reference [2] conducted a study in which river trials were poised after 46 places during the season and 300 locations during the season. The samples were analyzed for their organic features. The researchers also evaluated the water's suitability for irrigation by considering factors. Reference [3] collected water samples from the Tunga River to investigate the influence of terrestrial use variations on its aquatic excellence. The authors used factor analysis to evaluate the river water quality and observed that agricultural pollutants with non-point sources from agricultural activities and urban waste along the river banks and beds are the major contributors to the degradation of the river's water quality. Reference [4] collected groundwater samples from 15 locations in different areas of the Aligarh district, which were used for irrigation, domestic, and other purposes. The samples were analyzed for alkalinity, Cl-, F-, SO4 2-, and other parameters. The study found that NO3-, F-, PO4-, and Cl- concentrations increased continuously during the study period. Additionally, the concentration of nutrients, heavy metals, and pesticides decreased with increasing well depth. Reference [5] aimed to characterize groundwater quality in 18 wells and created a spatial map to show the groundwater quality. They also divided the groundwater quality into five groups using cluster analysis, which can be used for resource management. The geochemical simulation showed that the formation of acidic and oxidizing groundwater. Reference [6] investigated the major ion concentrations, distribution patterns, and variability in groundwater chemistry in Jaipur City, Rajasthan (India). They collected 11 samples from different hand pumps and analyzed parameters such as pH, EC, total hardness, and various ions using standard methods [7]. Natural resource that supports to identification of the primary factors responsible for different hydrochemical facies (Benin Formation) aquifers was conducted on the hydrochemical data. A study was conducted on the hydrochemical data of the Benin Formation aquifers to identify the primary factors responsible for different hydrochemical facies. Kaiser Normalization was used to extract these factors. The anthropogenic activities in the area significantly dominated these processes, leading to changes in the groundwater chemistry. The method used to identify these factors showed high efficacy in characterizing the hydro-organic facies of the places [8]. The learning conducted in Erode District involved the collection of groundwater samples from 60 locations during the pre-monsoon. The drinking water in almost all the locations was highly contaminated during both seasons. WQI was found to be an effective and quick method for monitoring water quality. In [9] conducted a study in which they collected groundwater samples from various locations within a 4-km range around a sugar factory. During the monsoon period, nine different spots were selected and the physical characteristics of the collected groundwater samples were analyzed. Standard procedures were used to measure different parameters. The study examined the suitability of the groundwater samples for domestic use by comparing the analyzed results with the established standards. The study's major goal was to assess the quality of groundwater sources in the industrial region and compute an aquatic excellence score for every one of those. Reference [10] did a mathematical modeling study on 12 variables for Mumbai.

2 Materials and Methods

The drop in groundwater concentrations along with shifts in agricultural practices both led to a loss in river excellence, making it critical to track the purity of water through variable assessment. A thorough procedure is required for evaluating the variable, which involves choosing the research region and monitoring locations, establishing the suitable parameters to measure, collecting specimens, doing laboratory tests, and analyzing the findings. The evaluation of data may be done with particular programs such as Anova and geographic data sets such as ArcGIS. It is feasible to achieve a thorough understanding of the water's brilliance in a given location by implementing this procedure and identifying possible spots for enhancement. Madurai has a land area of 52 square kilometers and is located between 78,015′47.5E and 78,009′47.5 East equator and 9094′47.22 and 9090′00 Northern degree. After analyzing the land use map of Madurai, several distinct areas were identified, including residential areas with underground sewerage facilities such as Annanagar, K.K.nagar, Kalavasal, and S.S colony, as well as residential areas without underground sewerage facilities like Iyyerbungalow and Pudur (Karpaga Nagar). Additionally, there was a commercial area located around the Meenakshi Amman temple, an industrial area consisting of Jaihindpuram and Sellur, and an agricultural area in Karuppayurani.

To collect water samples, five sample points were chosen from each of these areas, for a total of 40 samples collected in and around Madurai city. The findings had been portrayed in Table 1.

2.1 Total Dissolved Solids

The taste of drinking water is greatly influenced by its Total Dissolved Solids (TDS) content. Generally, water with TDS levels below 600 mg/L is considered to be pleasant tasting, while levels above 1200 mg/L are usually deemed unpalatable. The ideal aesthetic objective for TDS is around 500 mg/L to maintain good taste and prevent excessive scaling. It's important to note that low levels of TDS can enhance the taste of drinking water. Moreover, high TDS levels can lead to the corrosion of metal equipment and accessories, despite having balanced water chemistry. It can also cause skin and eye irritation, even if the water's pH is correct and chloramines are absent.

2.2 Turbidity

According to the Interim Primary Drinking Water Regulations set by the EPA, the turbidity level of drinking water should ideally be below 1 unit. However, under specific circumstances, it can be up to 5 units.

2.3 Nitrates

Usually, nitrate concentration in groundwater is low, but it can become dangerously high due to leaching or runoff from agricultural lands and contamination from animal or human waste. High levels of nitrate and nitrite can lead to the formation of methemoglobinemia, which is commonly known as "blue-baby syndrome". To minimize water contamination, proper management of nitrogen sources such as fertilizers and manures is essential for maintaining safe drinking water supplies.

2.4 Hardness

The presence of very hard water is undesirable for many household applications because it can leave a mineral deposit on the interior of pipes, boilers, and tanks. It

Table 1 Sampling points and location

S. no	Location	Notations
I	AnnaNagar & K.K.Nagar	
1	Near Bus stand	AN1
2	Kamaraj Street, Munthiri Thoppu	AN2
3	Near Sundaram Park	AN3
4	Near Arch	AN4
5	Devar Nagar	AN5
II	Kalavasal & S.S.Colony	
6	Ram Nagar	KV1
7	PRC Shed Area	KV2
8	Ponmeni	KV3
9	Near Anantha Metals	KV4
10	Near Germanus Hotel	KV5
III	Iyyerbungalow	
11	Vishwanatha Das Nagar	IB1
12	Bama Nagar	IB2
13	Pamban Nagar	IB3
14	EB Stop	IB4
15	UcchaParambu Medu	IB5
IV	Pudur (Karpaga Nagar)	
16	7th Street	PU1
17	6th Street	PU2
18	5th Street	PU3
19	4th Street	PU4
20	3rd Street	PU5
V	Around Meenakshi Temple	
21	Jewellery Street	MT1
22	Amman Sannathi Street	MT2
23	Near Pothys	MT3
24	North Masi Street	MT4
25	BSNL Office	MT5
VI	Karuppayurani	
26	MasthanPatti	KA1
27	KabirNagar School	KA2
28	Muthumari Nagar	KA3
29	Near CheckPost	KA4
30	Near PC Perungayam	KA5
VII	Jaihindpuram	
31	Near TVS Sundaram School	JP1

(continued)

Table 1 (continued)

S. no	Location	Notations
32	JaihindPuram Market	JP2
33	West Side of Channel	JP3
34	East Side of Channel	JP4
35	Behind TN WPT	JP5
VII	Seller	
36	Near Jain Vidyalaya	SL1
37	Swarajpuram Main Road	SL2
38	Seller Road	SL3
39	Pandalgudi Main Road	SL4
40	Jamburopuram Market	SL5

is possible to soften hard water at a relatively low cost, but it is not always necessary to remove all of the minerals that cause water hardness. Water that is too soft, on the other hand, may corrode metal surfaces. Despite this, very soft water is preferable for washing clothes, dishes, and bathing.

2.5 *Alkalinity*

Water's alkalinity can result from the presence of various ions, such as hydroxides, carbonates, and bicarbonates. These ions can be found in water sources naturally or may be added during water treatment processes, such as lime soda ash softening. Among these ions, bicarbonates are the most prevalent source of alkalinity. In most cases, having moderate levels of alkalinity is desirable as it counteracts the corrosive effects of acidity in water supplies. However, excessive amounts of alkalinity can lead to several issues. Water with a high level of alkalinity can have a disagreeable taste, often described as "soda-like". To remove troublesome amounts of alkalinity, reverse osmosis treatment can be employed, which eliminates not only excess alkalinity but also other total dissolved solids.

2.6 *Chlorides*

The geological composition of an area's bedrock largely determines the natural concentration of salts present. Regions with arid or semi-arid climates, where evaporation surpasses precipitation, are expected to have high salt concentrations. The presence of chloride in water increases its electrical conductivity, leading to higher corrosivity.

2.7 Sodium

Naturally, water contains sodium and chloride due to erosion or intrusion of saltwater. Human activities such as road salting in residential, commercial, and industrial areas can cause an influx of sodium into both surface and groundwater sources. While sodium is crucial for maintaining good health, excessive intake can lead to hypertension or high blood pressure. Hence, balancing the sodium intake is necessary for a healthy life.

2.8 Land Use Change and Groundwater Quality

Rapid changes in land use patterns, driven by growing pollution, increased demand for food, fodder, fuel, and industrial activities, have been observed in developing countries changes in agricultural practices, and an increase in non-sewered communities. These land use changes are likely to have an impact on groundwater quality. For example, developed land use settings have been observed to have increased concentrations of major ions, changes in oxidation–reduction conditions, and increased concentrations of minor elements compared to undeveloped settings. Factors such as age and density of septic systems in areas without municipal sewers and cropping practices in agricultural areas also affect groundwater quality.

2.9 Sodium Adsorption Ratio

Sodium is a crucial element that can impact and displace other cations, as indicated by its Sodium Adsorption Ratio (SAR). When irrigation water has high levels of sodium, it can have negative effects, such as adsorption of sodium onto soil cation exchange sites. This process can cause soil aggregates to disintegrate, block soil pores, and make it difficult for water to flow. SAR is a useful measure for evaluating whether groundwater is appropriate for use in irrigation.

2.9.1 Geographic Information System (GIS)

By using GIS, we can investigate, interpret, and exhibit data in numerous ways, revealing correlations, patterns, and trends via maps, charts, and reports. Maps are commonly generated with GIS, as they are an effective tool for sharing the outcomes of GIS analyses with others.

2.9.2 GIS in Groundwater Assessment

The contamination of drinking water is a significant challenge facing environmentalists in rapidly developing nations. The contamination of drinking water is a significant challenge facing environmentalists in rapidly developing nations. The exposure of a growing number of individuals to contaminated water has raised numerous issues involve not only addressing the contaminated water but also preventing similar occurrences in the future. Decision support systems, such as GIS, can be helpful in making prompt decisions by providing graphical representations that are easy for policymakers to comprehend. Mapping with GIS can aid in efficient decision-making and facilitate the prevention of further contamination.

3 Objective

Including rapid urbanization, an increase in the discharge of wastewater onto land (up to 460,000 m^3/day), greater use of nitrogen fertilizers, and improper disposal of industrial waste from industries the water level decreases. As Madurai undergoes rapid industrialization and significant land use changes, its groundwater quality has been widely affected. The fluctuation of various parameters throughout the year can significantly impact groundwater quality, making it crucial to monitor the quality of the groundwater to identify changes and take necessary remedial measures before utilizing the water for domestic purposes. Groundwater is an essential resource that significantly impacts human life, despite being invisible to the naked eye. Monitoring the fluctuation of parameters throughout the year is necessary to establish the interrelation between land use changes and groundwater quality. Identifying the primary parameters that affect groundwater quality can lead to implementing suitable counteractive measures. The ideas of this learning are to compare the groundwater quality of industrial, commercial, residential, and agricultural areas in Madurai, establish correlations between land use types and variations in groundwater quality parameters, use SPSS software for Factor Analysis to obtain statistical relationships among the water quality parameters, and analyze the results and present them in the form of a thematic map using ArcGIS.

4 Objective of the Work

Including rapid urbanization, an increase in the discharge of wastewater onto land (up to 460,000 m^3/day), greater use of nitrogen fertilizers, and improper disposal of industrial waste from industries the water level decreases. As Madurai undergoes rapid industrialization and significant land use changes, its groundwater quality has been widely affected. The fluctuation of various parameters throughout the year can significantly impact groundwater quality, making it crucial to monitor the quality of the groundwater to identify changes and take necessary remedial measures before utilizing the water for domestic purposes. Groundwater is an essential resource that significantly impacts human life, despite being invisible to the naked eye. Monitoring the fluctuation of parameters throughout the year is necessary to establish the interrelation between land use changes and groundwater quality. Identifying the primary parameters that affect groundwater quality can lead to implementing suitable counteractive measures. The aim of this study is to compare groundwater quality and present the findings in the form of a thematic map using ArcGIS.

4.1 GIS as an Administrative Aiding Technology

A GIS is a database that enables the examination of integrated knowledge and the depiction of facts in a geographical setting. This renders GIS a useful instrument for performing subsurface analyses since it allows for the interpolation of observed readings for indicators of water quality from particular places as well as the linking of groundwater conditions with additional pertinent data such as development patterns and soil traits. In addition, GIS has extensive map-generation capacities, which are essential in communicating statistical analysis findings. Although GIS is being used across the globe to explain connections between the state of water and both environmental and human-related operations, its application in the present moment remains crucial.

4.2 Conceptually Map Creation

Land Control Stations were placed on the digital map using Google Earth to display them. Using our ERDAS Imagine 8.5 programs, ground-based Management Locations were then combined with the Madurai metropolitan border model. With origin Pro 8.5, graphs with bars representing the relevant water quality index have been generated for each region. ArcGIS was used to construct a packed bar graph visualization from the supplied bar graphs. Lastly, a map with a theme depicting the water condition in each location during various times of the year was created.

4.3 SPSS Program Evaluation

SPSS is a powerful and extensive data analysis and handling program. It can read information from a variety of file formats and generate tabular tables, graphs, and maps that depict dispersion and patterns, in addition to statistical data. Furthermore, SPSS can do extensive statistical analysis and a range of numerical examinations, such as t-tests and f-tests, to assess the degree of importance of the various water quality measures.

5 Results and Discussion

Madurai is a city witnessing significant expansion as well as shifts in how land is utilized, that are having a direct influence on the condition of groundwater. Water scarcity and pollution have grown as a result of growing population, urbanization, industry, and competitiveness for job creation. It is critical to monitor aquatic brilliance through the year since parameter variances fluctuate significantly throughout various periods of monsoon. Predetermined sampling locations were chosen, and specimens were taken in aluminum cans throughout the rainy season. The samples were stored in a laboratory and analyzed for various parameters. The results were then tabulated based on the land use of each sample point and the different monsoon seasons during which the samples were collected. To ensure accuracy, five sample points were selected from each area with a particular land use.

Changes in rainfall can affect the concentration of various water quality parameters as shown in Fig. 1, which can in turn impact the overall quality of the water. These parameters are particularly sensitive to fluctuations during the monsoon season. To best capture these variations in different areas and seasons, bar charts are the most effective way to represent the data. Figure 2 shows that water quality in Madurai is better during the monsoon season, with twice the normal rainfall received between October and December. However, the water quality index value during pre-monsoon indicates that the water is unsuitable for human consumption, with values mostly exceeding 100. The index value again increases during post-monsoon, indicating severe pollution of groundwater due to inorganic values as portrayed in Fig. 3. A thorough assessment of the inorganic parameters in the water was also done as shown in Fig. 4. Therefore, the area's groundwater needs treatment before consumption and protection from contamination. To improve the water quality of Madurai, remedial techniques such as rainwater harvesting, building ponds and lakes to store rainwater, and planting trees to prevent runoff could be implemented. By increasing groundwater recharge, its quality can be maintained year-round. According to recent findings, heavy rainfall during the monsoon season has led to a significant improvement in water quality across all areas. However, the water quality in Iyyerbungalow and Pudur is comparatively worse than in other regions, while the Sellur area has relatively better water quality. The water quality index indicates that higher values correspond to lower water quality. Specifically, monsoon season, the pH points in the Pudur area assortment between 6.3 and 7.0. In almost all areas, there is an increase in pH, with levels exceeding 8.0, especially in the Pudur area during the monsoon season. The pH level remains relatively consistent during the post-monsoon period, with no significant difference observed as shown in Fig. 5. The study found that the levels of nitrates are generally higher in residential areas without underground sewerage compared to those with underground sewerage. During the monsoon season, there is

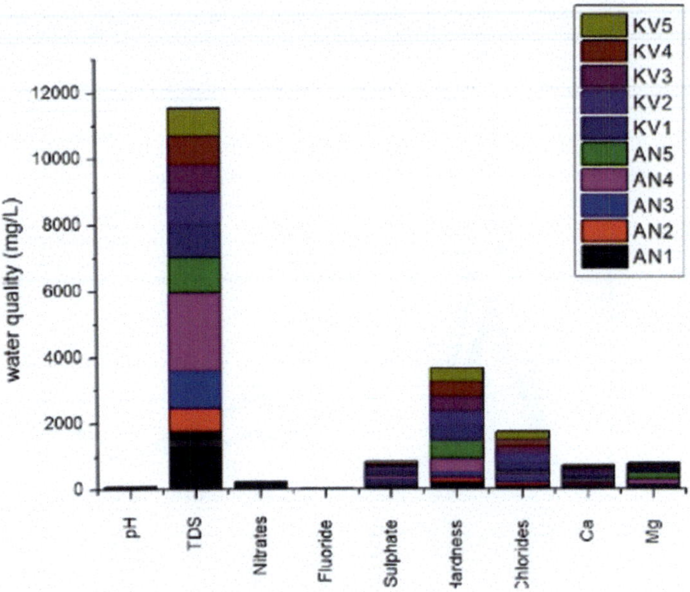

Fig. 1 Graph representing variation in water quality

a reduction in nitrate levels across all areas as shown in Fig. 6, which then increases during the post-monsoon season.

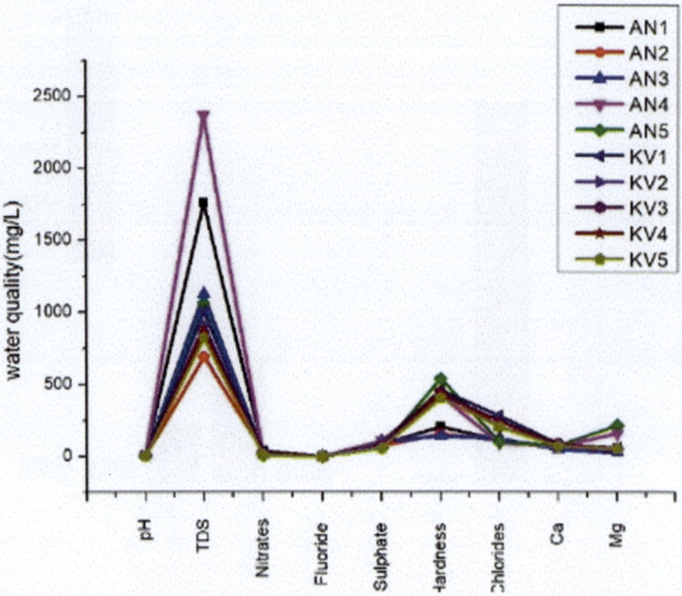

Fig. 2 Graph showing analysis for physiochemical values

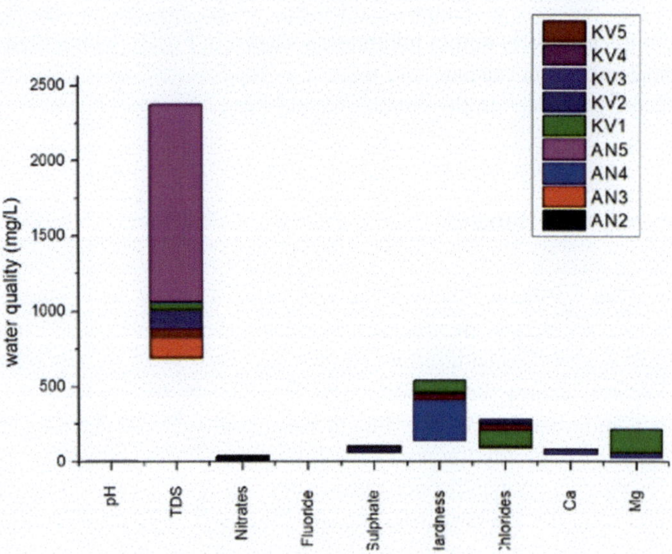

Fig. 3 Graph representing analysis for inorganic values

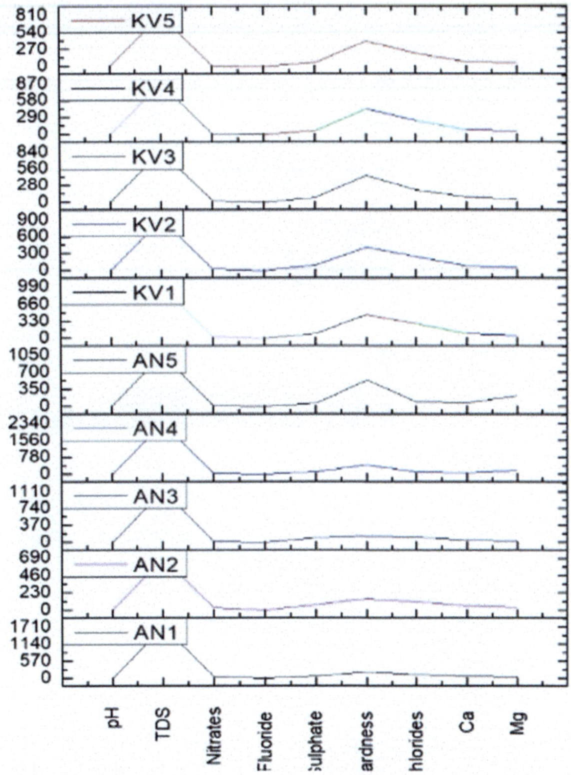

Fig. 4 Assessment of inorganic parameters

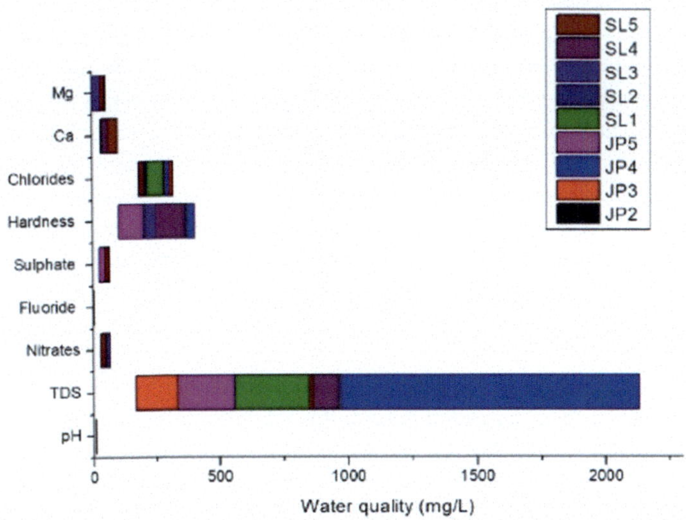

Fig. 5 Assessment of water quality values

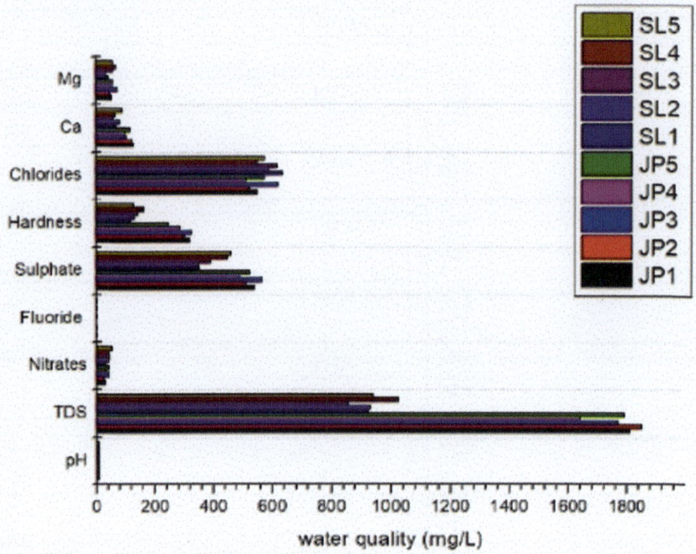

Fig. 6 Graph representing water quality values

6 Conclusion

Water is a dynamic natural reserve that is essential for anthropoid persistence, environmental stability, and economic development. However, excessive groundwater abstraction and poor waste disposal practices can lead to groundwater contamination. It is crucial to implement suitable management strategies that prioritize both development and environmental protection, as well as public health. Preventative measures are preferable to remediation techniques when it comes to groundwater contamination. Therefore, it is essential to regularly monitor various parameters that affect groundwater quality to prevent contamination. Groundwater pollution can have detrimental effects on human health, economic growth, and social progress. Due to its vulnerability to adsorption of domestic, industrial, and agricultural wastewater, groundwater quality needs to be monitored regularly. During the monsoon season, water quality is generally good due to excessive recharge. Therefore, implementing meticulous rainwater harvesting schemes and measures like laying pervious concrete pavements to increase the recharge area can help maintain good water quality. Planting more trees can prevent surface water runoff, and digging ponds and lakes can store rainwater for a more extended period. It is vital to understand that groundwater is not an infinite resource that can be utilized carelessly, even though it is available in abundant quantities. Issues such as waterlogging, salinity, agricultural toxins, and industrial effluents must be adequately addressed. There has been an increasing awareness among the general public regarding the need to conserve and develop groundwater in the last decade. Water usage needs to be integrated effectively with water regeneration, as was practiced in many traditional technologies.

References

1. Swarna Latha P (1998) Water quality assessment by using an index at Village Level—A case study. Poll Res 26(4):619–622
2. Sadashivaiah C (2006) Hydrochemical analysis and evaluation of groundwater quality in Tumkur Taluk, Karnataka State, India. Int J Environ Res Public Health 5:158–164
3. Yogendra K, Puttaiah ET (2007) Water quality assessment of Tunga River using factor analysis. Nat Environ Pollut Technol 6(3):393–397
4. Bansal OP (2008) Groundwater quality of Aligarh District of Uttarpradesh, India, A 11 year study. Poll Res 27(4):721–724
5. Liu C-W, Jang C-S (2008) Characterization of groundwater Quali-ty in Kinmen Island using multivariate analysis and geochemical modelling. Hydrol Process J 22:376–383
6. Sharma S, Singh Chandel CP (2008) Studies on physico-chemical parameters of groundwater of sambher block of Jaipur District Poll Res 27(4):671–677
7. Amadi AN, Olasehinde PI (2010) Characterization of groundwater chemistry in the coastal plain-sand aquifer of Owerri using factor analysis. Int J Phys Sci 5(8):1306–1314
8. Kavitha R, Elangovan K (2010) Ground water quality characteristics at Erode district, Tamilnadu, India. Int J Environ Sci 1(2)
9. Ratnakant C (2010) Water quality assessment of groundwater resources around Sugar Factory of Jamkhandi Town, Bagalkot District, Karnataka State. Int J Environ Sci 2(1)
10. Kumar N (2010) Drinking Water quality management through correlation studies among various physicochemical parameters: a case study. Int J Environ Sci 1(2)

A Machine Learning Approach to Cardiovascular Disease Prevention in Smart Healthcare

Atul Garg, Nidhi Bansal Garg, Anurag Jain, Tanupriya Choudhury, and Ketan Kotecha

Abstract Cardiovascular diseases (CVDs) remain a prominent cause of death globally, but the incorporation of machine learning techniques into smart healthcare systems has demonstrated encouraging outcomes in detecting and preventing heart disease. Smart healthcare harnesses emerging technologies like wearable devices, electronic health records, and real-time data streaming to continuously monitor an individual's health indicators. By employing machine learning algorithms such as deep learning, support vector machines, and random forests, these extensive and diverse datasets can be analyzed to identify patterns, risk factors, and predictive models associated with CVDs. Additionally, this research emphasizes the significant role of predictive analytics in personalized medicine. Machine learning models can utilize patient-specific information to evaluate the likelihood of developing CVDs, enabling targeted interventions and customized treatment plans. Integrating patient preferences with clinical guidelines enhances decision-making processes that ultimately lead to enhanced patient outcomes.

A. Garg · N. B. Garg
Chitkara University Institute of Engineering and Technology, Chitkara University, Rajpura, Punjab 140401, India
e-mail: atul.garg@chitkara.edu.in

N. B. Garg
e-mail: garg.nidhi@chitkara.edu.in

A. Jain (✉)
School of Computer Sciences, University of Petroleum and Energy Studies (UPES), Dehradun, Uttarakhand 248007, India
e-mail: dr.anuragjain14@gmail.com

T. Choudhury
CSE Department, Symbiosis International (Deemed University), Symbiosis Institute of Technology, Lavale Campus, Pune, Maharashtra 412115, India
e-mail: tanupriya.choudhury@sitpune.edu.in

K. Kotecha
Symbiosis Centre for Applied Artificial Intelligence, Symbiosis International (Deemed University), Symbiosis Institute of Technology, Pune, Maharashtra 412115, India
e-mail: director@sitpune.edu.in

S. Dutta et al. (eds.), *Cyber Intelligence and Information Retrieval*, Lecture Notes in Networks and Systems 1025, https://doi.org/10.1007/978-981-97-3594-5_14

Keywords Health · Healthy lives · Diseases · Heart disease · Machine learning

1 Introduction

The importance of good health surpasses wealth, status, and other measures. Being in good health enables individuals to engage in various fulfilling activities. The utilization of the Internet, World Wide Web, and smartphones has significantly impacted human intelligence and cognition. Urban planning, manufacturing, and healthcare have transformed due to Industry 5.0. This revolution has given rise to advanced infrastructures equipped with state-of-the-art technologies and systems that leverage modern advancements such as the Internet of Things (IoT), Artificial Intelligence (AI), and big data analytics to enhance efficiency, productivity, and sustainability.

Consequently, urban connectivity, workplace automation, and access to healthcare have improved [1, 2]. Despite a decrease in physical activity and output from humans themselves, groundbreaking discoveries have multiplied. With the shift toward automated or digital jobs replacing manual labor comes sedentary work environments characterized by technological integration and increased reliance on cognitive tasks—resulting in health concerns arising from this paradigm shift. It is important to remember that contaminants like pesticides enter our bodies through the digestive system [3, 4].

Machine learning techniques are vital in various sectors like education, agriculture, health, sharing, etc. [5]. In contemporary times, myocardial infarctions, also known as heart attacks, have emerged as the predominant etiological factor leading to mortality. In healthcare, a paramount objective revolves around the timely identification of diseases during their nascent stages, thereby impeding their advancements [6, 7]. Recent advancements in the field of machine learning (ML) have presented significant opportunities for the creation of prediction models aimed at mitigating the occurrence of heart attacks. Machine learning algorithms can potentially analyze extensive collections of patient data, enabling the identification of previously undetected illnesses, the identification of individuals at elevated risk levels, and the recommendation of suitable treatment plans [8, 9]. It is crucial to highlight that healthcare systems play a significant role in providing information, while machine learning algorithms demonstrate notable efficacy when presented with extensive datasets. Both facts are deserving of attention. Machine learning (ML) can extract substantial insights from diverse data types, encompassing medical records and diagnostic tests, lifestyle data, and genetic data [10]. Machine learning models, including decision trees, random forests, and deep neural networks, have demonstrated superior capabilities in identifying intricate correlations between potential risk factors and the incidence of heart attacks, in contrast to conventional statistical techniques like correlation and regression analysis [11].

In machine learning, the concept of intervention is important alongside the commonly discussed predictive modeling. The utilization of machine learning algorithms in conjunction with an individual's risk profile has the potential to yield personalized treatment strategies. In certain circumstances, medical professionals may advise implementing alterations to one's lifestyle, prescribing pharmaceutical interventions, or, in some cases, resorting to invasive procedures to mitigate risk [12]. Based on the empirical evidence at hand, it is evident that the act of personalizing interventions has a direct correlation with the enhancement of their efficacy. Customized therapies have the potential to yield a more focused and individualized approach to treatment for every patient [7]. The customized approach considers the individual's comprehensive medical history, daily routines and choices, personal preferences, and socioeconomic circumstances. When one considers the prevailing circumstances, it becomes apparent that interventions can be customized to more effectively align with the unique needs and objectives of each individual. The enhancement of healthcare therapies is achieved through the implementation of personalized approaches. Furthermore, empirical evidence has demonstrated that it can foster patient well-being and active engagement in their healthcare trajectory [13].

Machine learning has great potential for heart attack prevention, but it faces various hurdles. Healthcare data are delicate and must be protected. To achieve real-world accuracy and reliability, ML models must be carefully validated and tested. ML models' "black box" nature makes it difficult to grasp their decisions, requiring openness and explainability. This article explores the promising use of machine learning as a preventative measure for heart attacks.

2 Literature Review

The primary aim of this research endeavor is to systematically classify and discern the indispensable attributes requisite for implementing diverse machine learning methodologies in the realm of cardiac disease prognosis.

Machine learning, a field of study within artificial intelligence, possesses the remarkable capability to predict forthcoming outcomes through the identification and assessment of intricate patterns inherent within vast quantities of data. The utilization of input is employed to accomplish this task. Machine learning algorithms, when trained on historical patient data, possess the capability to approximate the probability of a myocardial infarction occurring within a specific period [14]. It is conceivable that the potential for enhanced precision in diagnosis and treatment planning exists. The utilization of more intricate models facilitates the provision of comprehensive risk profiles, encompassing various factors such as age, gender, medical and familial history, cholesterol levels, blood pressure, and lifestyle choices [15]. This invaluable tool enables medical providers to strategically focus their preventive efforts on individuals who possess a higher susceptibility to adverse health outcomes [2]. In their work [11], they out the notion of applying different models accessible within the WEKA program, notably K Star, J48, SMO, Bayes Net, and multilayer perceptron, for the aim of forecasting heart disease. Similar work done by renowned authors is compared in Table 1.

Table 1 Similar work done in the literature

Reference	Model/classifier/ algorithms	Parameter
[14]	ElasticNet	Hemoglobin, leukocyte volume, high-density lipoprotein-to-total-cholesterol ratio, and age
[16]	SVM, KNN	Sound of heart, aortic stenosis (AS), mitral valve prolapse (MVP), mitral stenosis (MS), mitral regurgitation (MR)
[17]	2D CNN	Artifact, sound of hearth, murmur
[18]	CRNN	Sound of hearth, normal (N), mitral regurgitation (MR), aortic stenosis (AS), mitral valve prolapse (MVP), mitral stenosis (MS)
[19]	SVM, KNN, and random forest	Normal and abnormal heart sound
[20]	Logistic regression, random forest, neural networks	Smoking, sex, race, and age

Utilizing machine learning techniques facilitates the analysis of real-time data, thereby facilitating the creation and advancement of early warning systems. Wearable devices and health monitoring applications can consistently gather data about heart rate, activity levels, and other vital signs. Machine learning algorithms can perform real-time analysis of the provided data, thereby enabling the prompt detection of any irregularities or indications of potential heart-related problems. This timely identification facilitates the immediate notification of patients and medical practitioners, allowing them to take swift action and implement timely interventions [21].

3 Methodology

In literature, machine learning techniques are helpful to reduce the risk of heart attacks. This study introduces an automated method based on machine learning to predict patients based on their previous health history. In the present study, the dataset is the foundational resource for developing automated forecasting models to classify approaches to cardiovascular disease (CVD) prevention in smart healthcare. The dataset under consideration pertains to CVD, encompassing a comprehensive collection of information related to various aspects of patients. This dataset is a valuable resource for researchers and educators alike, facilitating the exploration and analysis of dataset evaluation provided by Kaggle [22], which consists of data from 918 patients. Attributes of the dataset are depicted in Table 2. Figure 1 depicts the visualization of attributes.

Table 2 Attributes in the dataset

Sr. no	Attribute	Abbreviation	Description
1.	Age	Patient's age	Numeric value
2.	Sex	Patient's age	Male or female
3.	ChestPainType	Type of pain	TA: Typical angina, ATA: Atypical angina, NAP: Non-anginal pain, ASY: Asymptomatic
4.	RestingBP	Blood pressure	Numeric value
5.	Cholesterol	Serum cholesterol	Numeric value
6.	FastingBS	Blood sugar (fasting)	Yes or no
7.	RestingECG	Electrocardiogram results	Normal, ST: abnormal, LVH: left ventricular hypertrophy
8.	MaxHR	Maximum heart rate	Numeric value
9.	ExerciseAngina	Exercise-induced angina	Yes or no
10.	Oldpeak	Old peak	Depression in measured
11.	ST_Slope	The slope of the peak exercise	Upsloping (up), flat, down sloping (down)
12.	heart disease	Heart disease	Yes or no

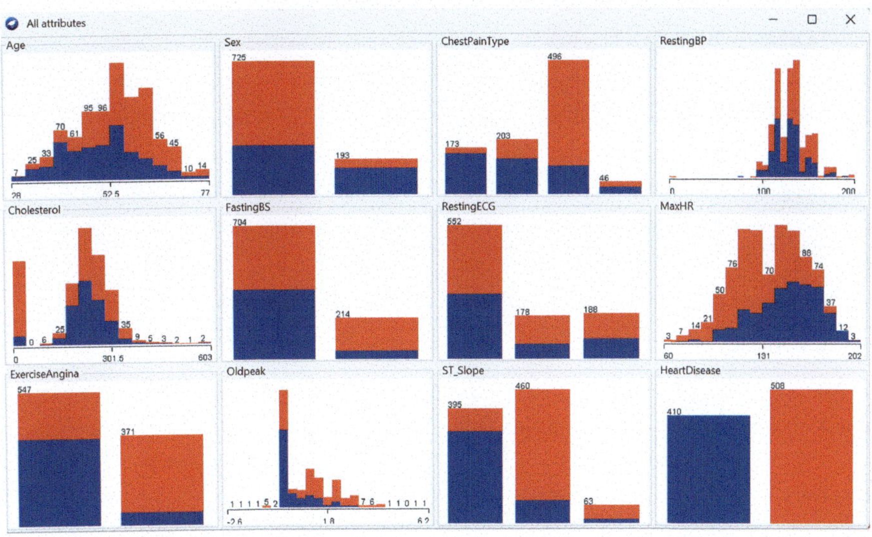

Fig. 1 Visualization of attributes

4 Results and Discussions

In this work, ChestPainType (type of chest pain) and FastingBS (blood sugar report) parameters are considered to predict CVD early. Weka 3.8.6 is used, and three employed models, including the Naïve Bayes, J48, and random forest classifications, are used with ten folds. Machine learning criteria have developed the models to ensure the accuracy of user data. Various metrics, such as precision, F-measure, recall, and accuracy, are employed to evaluate the effectiveness of machine learning (ML) methodologies. Tables 3 and 4 depict the confusion metrics received on the selected parameters, respectively:

Precision, recall, and F-measure results were measured based on ChestPainType (type of chest pain) and FastingBS (blood sugar report) attributes using Weka 3.8.6. The measured results are depicted in Tables 5 and 6, Figs. 2 and 3. The weighted average of different sub-parameters using different parameters and AI classifications is shown in Figs. 4 and 5.

In the parameter of chest pain, the class category is subdivided into four distinct categories: ATA, NAP, ASY, and TA. Conversely, when examining the attribute of blood sugar, the classification is simplified to a binary system consisting of only two classes: yes and no. The comparative results of the two parameters, along with their corresponding classes, have been depicted in Figs. 2, 3, 4, and 5. The experimental findings indicate that the performance of the random forest classifier surpasses that of other classifiers [23, 24].

Table 3 Confusion matrix with parameter type of chest pain using different classifiers

	Naïve Bayes				J48				Random forest			
Class category	ATA	NAP	ASY	TA	ATA	NAP	ASY	TA	ATA	NAP	ASY	TA
ATA	36	07	05	00	26	10	11	01	24	12	12	00
NAP	33	6	26	01	20	17	28	01	22	13	31	00
ASY	29	06	114	00	16	11	119	03	14	06	128	01
TA	04	01	05	02	04	02	06	00	03	01	06	02

Table 4 Confusion matrix with parameter fasting blood sugar (sugar: yes, no sugar: no)

	Naïve Bayes		J48		Random forest	
Class category	No	Yes	No	Yes	No	Yes
No	170	55	210	15	212	13
Yes	20	30	32	18	36	14

Table 5 Measured results with parameter type of chest pain using different classifiers

Classification	Class category	Precision	Recall	F-measure
Naive Bayes	ATA	0.353	0.750	0.480
	NAP	0.300	0.091	0.140
	ASY	0.760	0.765	0.763
	TA	0.667	0.167	0.267
Weighted avg.		0.574	0.575	0.542
J48	ATA	0.394	0.542	0.456
	NAP	0.425	0.258	0.321
	ASY	0.726	0.799	0.760
	TA	0.000	0.000	0.000
Weighted avg.		0.564	0.589	0.569
Random forest	ATA	0.381	0.500	0.432
	NAP	0.406	0.197	0.265
	ASY	0.723	0.859	0.785
	TA	0.667	0.167	0.267
Weighted avg.		0.585	0.607	0.576

Table 6 Measured results after opting for the parameter fasting blood sugar (FastingBS)

Classification	Class category	Precision	Recall	F-measure
Naive Bayes	No	0.895	0.756	0.819
	Yes	0.353	0.600	0.444
Weighted avg.		0.796	0.727	0.751
J48	No	0.868	0.933	0.899
	Yes	0.545	0.360	0.434
Weighted avg.		0.809	0.829	0.815
Random forest	No	0.855	0.942	0.896
	Yes	0.519	0.280	0.364
Weighted avg.		0.794	0.822	0.800

5 Conclusions

The application of machine learning techniques in mitigating the risk of cardiovascular disease is a groundbreaking advancement in medical practice. Machine learning can reduce heart attacks by utilizing data-driven insights, developing prediction models, and providing personalized treatment approaches. Effective communication and collaboration between data scientists, medical practitioners, and policymakers are crucial to fully harnessing the potential of machine learning in preventing heart disease. This investigation assesses chest pain and blood sugar levels as key

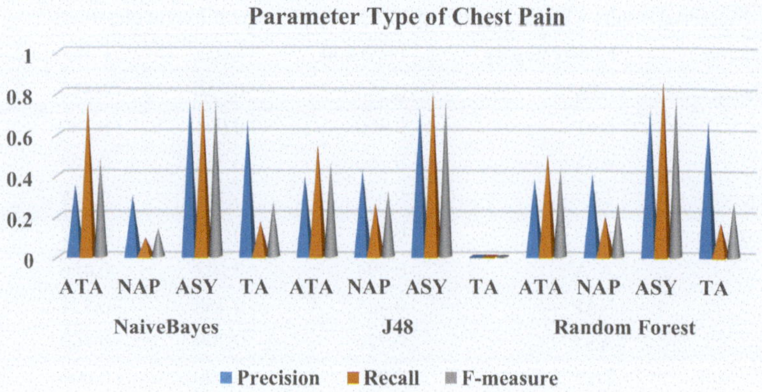

Fig. 2 Measured results with parameter type of chest pain using different classifiers

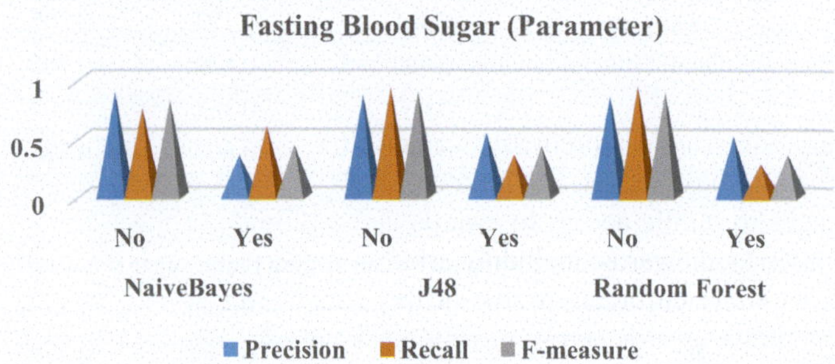

Fig. 3 Measured results with parameter fasting blood sugar using different classifiers

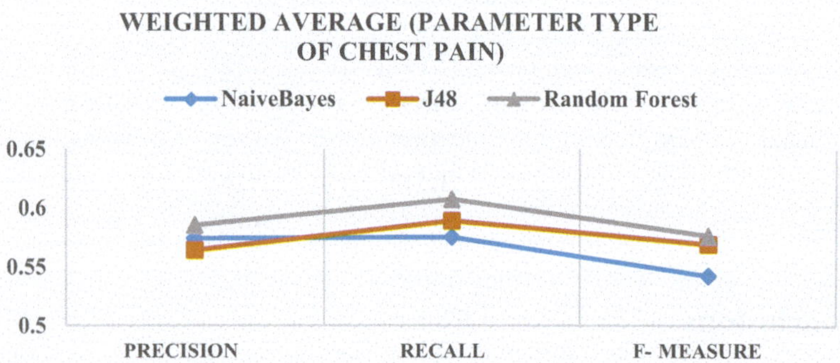

Fig. 4 Weighted average results with parameter type of chest pain using different classifiers

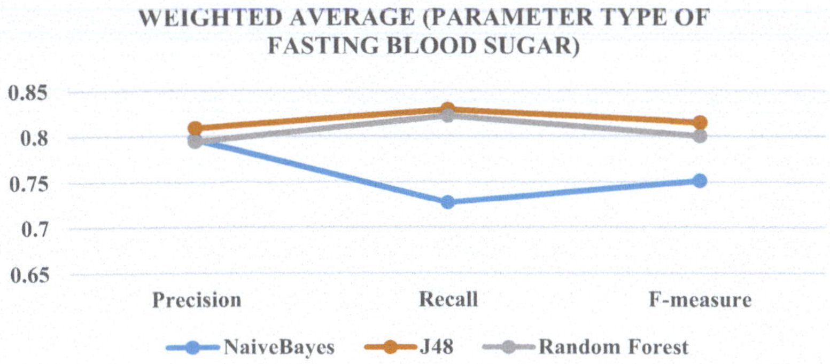

Fig. 5 Weighted average results with parameter fasting blood sugar using different classifiers

variables. Machine learning algorithms can be integrated with routine patient monitoring procedures to provide early awareness for timely medical intervention. This proactive approach has been successful in reducing mortality rates associated with cardiovascular diseases.

References

1. Lee S, Chu Y, Ryu J, Park YJ, Yang S, Koh SB (2022) Artificial intelligence for detection of cardiovascular-related diseases from wearable devices: a systematic review and meta-analysis. Yonsei Med J 63:S93-107
2. Talukdar J, Singh TP (2023) Early prediction of cardiovascular disease using artificial neural network. Paladyn 14(1)
3. Bhattacharya A, Roy B, Sur SN, Mallik S, Dasgupta S (2022) Internet of things and data mining for modern engineering and healthcare applications. Internet Things Data Min. Mod. Eng. Healthc Appl 1–278
4. Ahmed I (2022) A study of heart disease diagnosis using machine learning and data mining. Electron theses, Proj Diss 1591
5. Garg A, Lilhore UK, Ghosh P, Prasad D, Simaiya S (2021) Machine learning-based model for prediction of student's performance in higher education 162–168
6. Tiwari S, Jain A, Sharma AK, Mohamad AK (2021) Phonocardiogram signal based multi-class cardiac diagnostic decision support system. IEEE Access 9:110710–110722
7. Ramesh TR, Lilhore UK, Poongodi M, Simaiya S, Kaur A, Hamdi M (2022) Predictive analysis of heart diseases with machine learning approaches. Malaysian J Comput Sci 2022(Special Issue 1):132–48
8. Xie S, Yu Z, Lv Z (2021) Multi-disease prediction based on deep learning: a survey. C - Comput Model Eng Sci 127(3)
9. Anuradha R (2022) An assessment on cardiovascular disease prediction and diagnosis using machine learning algorithms. Asian J Converg Technol 8(1):56
10. Johnson KW, Torres Soto J, Glicksberg BS, Shameer K, Miotto R, Ali M et al (2018) Artificial intelligence in cardiology. J Am Coll Cardiol 71(23):2668–2679
11. Yan Y, Zhang JW, Zang GY, Pu J (2019) The primary use of artificial intelligence in cardiovascular diseases: What kind of potential role does artificial intelligence play in future medicine? J Geriatr Cardiol 16(8):585–591

12. Das A, Mallick C, Dutta S (2020) Deep learning-based automated feature engineering for rice leaf disease prediction. In: Das A, Nayak J, Naik B, Dutta S, Pelusi D (eds) Computational intelligence in pattern recognition. Advances in intelligent systems and computing, vol 1120. Springer, Singapore. https://doi.org/10.1007/978-981-15-2449-3_11
13. Kondoro A, Tongora EF (2022) Smart healthcare system for cardiovascular patients using machine learning. Prim Heal Care [Internet] 12(4):1–5. https://www.iomcworld.org/open-access/smart-healthcare-system-for-cardiovascular-patients-using-machine-learning.pdf
14. Sánchez-Cabo F, Rossello X, Fuster V, Benito F, Manzano JP, Silla JC et al (2020) Machine learning improves cardiovascular risk definition for young asymptomatic individuals. J Am Coll Cardiol 76(14):1674–1685
15. Russak AJ, Chaudhry F, De Freitas JK, Baron G, Chaudhry FF, Bienstock S et al (2020) Machine learning in cardiology—ensuring clinical impact lives up to the hype. J Cardiovasc Pharmacol Ther 25(5):379–390
16. Upretee P, Yüksel ME (2019) Accurate classification of heart sounds for disease diagnosis by a single time-varying spectral feature: preliminary results. 2019 Sci Meet Electr Biomed Eng Comput Sci EBBT 2019:1–4
17. Banerjee M, Majhi S (2020) Multi-class heart sounds classification using 2D-convolutional neural network. In: Proceedings of the 2020 international conference computer and communications security ICCCS 2020
18. Shuvo SB, Ali SN, Swapnil SI, Al-Rakhami MS, Gumaei A (2021) CardioXNet: a novel lightweight deep learning framework for cardiovascular disease classification using heart sound recordings. IEEE Access 9:36955–36967
19. Narváez P, Gutierrez S, Percybrooks WS (2020) Automatic segmentation and classification of heart sounds using modified empirical wavelet transform and power features. Appl Sci 10(14)
20. Weng SF, Reps J, Kai J, Garibaldi JM, Qureshi N (2017) Can machine-learning improve cardiovascular risk prediction using routine clinical data? PLoS ONE 12(4):1–14
21. Ghosh P, Lilhore UK, Simaiya S, Garg A, Prasad D, Kumar A (2022) Prediction of the risk of heart attack using machine learning techniques. Lect Notes Electr Eng 907:613–621
22. Fedesoriano, "Heart failure prediction dataset," Kaggle. https://www.kaggle.com/datasets/fedesoriano/heart-failure-prediction. Accessed 20 Aug 2023
23. Rani P et al (2023) PCA-DNN: a novel deep neural network oriented system for breast can-cer classification. EAI Endorsed Trans Pervasive Health Technol 9:1–18. https://doi.org/10.4108/eetpht.9.3533
24. Sachdeva RK et al (2023) Empowering hepatitis diagnosis using RFE feature selection. In: 2023 7th international symposium on multidisciplinary studies and innovative technologies (ISMSIT), Ankara, Turkiye, pp 1–5. https://doi.org/10.1109/ISMSIT58785.2023.10304999

Examination of Sustainability 4.0 with Special Reference to International Legal Doctrines

Maheshwar Narayan Joshi, Anil Kumar Dixit, Satyam Sharma, Sagar Saxena, and Gagandeep Kaur

Abstract Enabling Sustainability 4.0 is the efficient implementation of cutting-edge innovations that include the Internet of Things (IoT), artificial intelligence (AI), machine learning (ML), augmented reality, big data (BD), additive production (AD), and other contemporary innovations. Due to the efficient use of natural resources and energy with less waste, these technologies facilitate significantly cheaper services. Producers are always searching for ways to reduce the operational costs related to their manufacturing procedures. Implementing Sustainability 4.0 technologies could help manufacturers optimize their value chain's production and related processes. These technologies will aid manufacturers in selecting the optimal facilities and personnel, reducing operational costs, boosting productivity and resource utilization, and identifying gaps that can be resolved. Sustainability is predicated on the efficient use and reuse of resources throughout the entire product life cycle, including resources, procedures, machinery, and skills. Sustainable production is the creation of green products through cost-effective processes that minimize adverse environmental impacts while conserving power and natural assets. This paper describes Sustainability 4.0 and its essential requirements. Finally, we selected some of the international environmental sustainability Doctrines or standards that act as guidelines or Doctrines for the nations to follow to achieve SDGs. Thus, the use of environmentally friendly practices, cutting-edge technologies, and environmental legal Doctrines will aid in resolving environmental, social, and economic development issues.

Keywords AI · Sustainability 4.0 · ML environment sustainability · UN · Egocentrism · Polluter pays · Environment

M. N. Joshi (✉) · A. K. Dixit · S. Sharma · S. Saxena
Law College Dehradun, Uttaranchal University Dehradun, Dehradun, India
e-mail: maheshwarbhai@gmail.com

G. Kaur
School of Law, University of Petroleum & Energy Studies, Dehradun, India
e-mail: gkaur@ddn.upes.ac.in

S. Dutta et al. (eds.), *Cyber Intelligence and Information Retrieval*, Lecture Notes in Networks and Systems 1025, https://doi.org/10.1007/978-981-97-3594-5_15

1 Introduction

AI is now an integral part of modern technologically advanced societies' daily lives. AI is currently pervasive in industries and businesses alike. It can transform how we identify, acquire knowledge, interact with one another, and function. It possesses enormous social and financial prospects. As the world continues to enter the period of sustainable growth, in which the 17 SDGs are defining the global agenda for development, AI is significantly establishing a new frontier in the disciplines of company operations, business strategies, and government regulation. The intelligence of robots and machinery with DL capabilities is already resolving intellectual issues typically associated with intelligence among Homo sapiens. Stride by stride, AI is speedily penetrating various fields of expertise by superseding humankind with greater predictive ability, higher effectiveness, and superior outcomes. However, humans will be replaced progressively in the future as we grow more immersed in the mechanisms of intelligence and as intelligent machines become an integral part of our bodies and lives. Authors postulated that the world has entered an era of sustainable growth, which means all nations are required to work together to contribute to dealing with the toughest challenges of relentless extreme impoverishment, marginalization, financial inequality, poor governance or administrations, and deterioration of the environment. Using the four cornerstones of economic growth, social advancement, preservation of the environment, and effective governance, he suggested a structure to examine environmentally friendly development. Every single one of these four elements is a distinct yet complementing pillar, but they are all crucial to the world's environmentally conscious growth. Thus, AI is such a unique, volatile, and quickly evolving thing that it positively affects the mission of promoting the SDGs.

Environment-related problems are the negative consequences of human actions on the planet. These include contamination, population growth, failed management, variations in the climate, increasing temperatures, and the GHG effect. The goal of various individual, corporate, and government-level environmental protection programs is to establish a harmonious relationship between mankind and Mother Nature. Climate variations, water contamination, rising temperatures, isothermal layer depletion as a result of the increasing mean temperature, pollution in the atmosphere, forest loss, industrialization, etc., are some of the pressing environmental concerns that require immediate consideration. This demonstrates the significance of the planet and its ecosystem, as well as the necessity for minimizing contamination increases, as "it is easier to discharge anything into the water, air, and land, but complex to take it out again" [1]. The continued development across various forms of society, industrial development, and urban growth has necessitated the promotion of environmentally friendly practices to ensure that people enjoy longer, more nutritious, and more satisfying lives. Numerous human actions result in the discharge of harmful substances or components known as pollutants into our natural environments, thereby jeopardizing people's well-being, happiness, and existence. Residing a more fulfilling and environmentally friendly existence is contingent on adhering to the Doctrines of management of the environment, where contamination mitigation and progressive methods of treatment are required to discourse SD challenges; hence, the purpose of this study is to use efficient control strategies and advanced methods for addressing environmental damage incidents [2]. In the words of Damania et al., as quoted by the UN, certain actions conducted by people in pursuit of material satisfaction or prosperity are detrimental to the well-being of the planet. Therefore, it is essential to carry out laws and cleaner technologies to halt the pattern and potentially create a more sustainable future. This is referred to be the ecological Kuznets curve, which illustrates why economic growth is the most effective means to improve the environment [3]. The manufacturing and production industries are currently confronted with the universal challenge of environmental sustainability. The Agenda 2030 for SDG, which was agreed upon by all UN member states in 2015, relies on the four pillars described above. Concerning these four cornerstones, the efficient execution of the tenets of environmentally friendly production must take into account the volatile market conditions caused by local instability or even exceptional incidents such as the corrosive COVID-19 pandemic, as well as the required speed of response at the company's and operational scales [4]. Nowadays corporations and non-profit corporations expend an immense amount of energy, resources, and time developing new strategies to combat the detrimental impacts of previous generations. The technology of Sustainability 4.0 enables collaboration among production lines, company procedures, and teams irrespective of time, place, area, connection, or any other factor. In an intelligent manufacturing facility, it is speedier to increase or decrease output. This ends up resulting in increased revenue for a manufacturing facility. Cloud computing and big data will make sure that the IoT and Industrial Internet of Things (IIoT) machines create extremely efficient production and connect with the consumer experience. Cloud storage is incredibly effective and full of features, as well as adaptable, current, and secure. Cloud computing is also an effective tool for linking products to businesses across nations, and it can manage IoT-generated big data [5]. Implementing ethically conscience and ecologically sound acts must necessitate production with positive effects on the world's wellness on an enormous scale. Sustainability that concentrates on resolving company issues can have considerable commercial and environmental effects. Utilizing sustainable growth as a guiding Doctrine rather than an addendum would revolutionize how businesses operate. AI and ML can determine appropriate remedies for environmental problems such as minimizing pollution, maximizing resource utilization, neutralizing carbon emissions, and freshwater [6].

A sustainable perspective generates a positive feedback cycle that will increase the number of people and businesses. A broader shift in views would likely result from the unraveling of rules and existing laws that encourage equitable development. Through wireless connectivity and IoT capabilities, Sustainability 4.0 links technologies, detectors, and other equipment to the individuals accountable for observing the steps for both productivity and effectiveness. The transparency of these innovations provides operators with the exhaustive data required to make suitable choices. It enhances communication, allowing users in every step of the manufacturing chain to collect big data and professional proficiency to aid the creation of and identify important areas for advancement and transformation [7]. Constantly communicating, creating new ideas, and establishing ecologically advantageous, economical, and person-cantered practices are required to promote an environmentally friendly mind-set. Due to the severity of the imminent climate crisis, we must take action now and accomplish everything we can if we wish to protect the planet and achieve a sustainable future. Sustainability 4.0 incorporates innovations in cutting-edge technology, advancements in robotics and AI, advanced detectors, cloud computing, the IoT, and numerous other applications. Decades of advancements in manufacturing techniques and processes have enabled corporations to increase profitability, worker productivity, and effectiveness [8]. Thus, at present, the literature review on Sustainability 4.0 and sustainable environment focuses on this review study predominantly on the scientific aspects of AI-ML technologies that are related to climate variation mitigation under ethical and legal constraints and consequences. In addition, we propose investigating the future of the environmentally friendly development of AI in the wake of climate variation. In summary, Sustainability 4.0 can revolutionize our response to environmental degradation using sustainable environmental techniques by providing innovative resources and knowledge to help others achieve a future that is more environmentally friendly. Thus, this review study offers an organized and exhaustive study of the relevant interdisciplinary academic literature to clarify existing information and comprehend the connection between Sustainability 4.0 and climate variation. In the following pages, the authors aspire to accomplish the following objectives will be discussed:

A. How do several cutting-edge technologies AI, ML, AR, BG, IoT, IIoT, etc., use the Sustainability 4.0 standard to contribute to the preservation of the natural world?
B. What are the legal Doctrines that are acting as guidelines for the nations curbing the issues related to environmental sustainability?

The paper consists of seven main sections. The first part begins with a short background on Sustainability 4.0 and environmental sustainability. Section 2 states the benefits of Sustainability 4.0 for Earth especially for humankind in the future. Section 3 investigates the use of AI against climate variation. Section 4 describes the environment and Sustainability 4.0. Section 5 states various applications of advanced industrial models and technologies contributing to environmental sustainability. Section 6 provides recommendations from the author's side. Lastly, Section 7 concludes with the conclusion.

2 Substantial Sustainability 4.0 Advantages

Producers have effectively executed their strategies for Sustainability 4.0 breakthroughs, putting them ahead of the competition. They can enter the next phase of production and distribution based on data, data-optimized, modular automation that provides complete oversight over the supply and flow of materials. Sustainability 4.0's benefits consist of better competitiveness and efficiency, enhanced adaptability and durability, and higher profit margins. Additionally, Sustainability 4.0 would improve customer experience. Intelligent manufacturing technology is intriguing and enthralling, but the benefits of Sustainability 4.0 should remain the focal point of every discussion. This includes technologies that improve automation, machine-to-machine interaction, generic production, and taking decisions. Sustainability 4.0 technologies allow producers to create superior, products that are more productive. In other words, the capital will be less expensive and dependable, while also being able to generate more revenue efficiently [9]. These advances enable the democratization of data, the provision of expanded perspectives, and the incorporation of Sustainability 4.0 and its applications. The vision of Sustainability 4.0 will link together devices and establish a connection outside the production plant's boundaries. Metadata has become a cutting-edge asset for several companies. Massive amounts of information gathered from detectors and devices are extremely important and well-organized. At present, there is a drive in industries to enhance ingenuity rather than supplement it. The most significant action is spending on developing skills and culture change. Training the workforce in statistics and cutting-edge technology would prepare them for a world in flux and prime them for further education while keeping them pertinent [10]. In addition, it is essential to take advantage of these breakthroughs throughout every step of the value chain and inter-corporation distribution systems. It would involve an effective application of AI and ML based on real-time data accumulated throughout the supply network, and it would provide intelligent conclusions for making better choices. All of this is infeasible without a robust network of collaborators, such as entrepreneurs and technological enterprises, that will develop accessible and affordable innovations that will make this transformation a reality. Additionally, academic institutions will conduct studies and experiments to improve technologies. Sustainability 4.0-enabled manufacturing can observe, analyze, and identify raw materials and process work through multiple production steps for effective and precise outbound transportation. Using a combination of automation and detectors, producers will have total control over the information and content flow of their facilities. This level of regulation is essential for stability and efficacy in the age of technology, where customer requirements and working conditions change swiftly [11]. Thus, the idea of Sustainability 4.0 has recently become commonplace and was explicitly referred to in a 2011 German government proposal as a plan for industrialization in that nation. To this vision, Sustainability 4.0 is an integral part of a global economy that has become integrated and interlinked as a result of the ICT transformation. This technological shift in Sustainability 4.0 is facilitated by the IoT and IOS, which electronically link the industry's internal and external supply chain networks. All of this makes the sector "smart," and cyber-physical systems (CPS) support it. The CPS addressed in this paper can be described as linked software as well as hardware elements that produce huge-scale results [12]. A comprehensive analysis of the papers on Sustainability 4.0 focuses on the preservation of energy and its obstacles and issues. Consequently, Sustainability 4.0 is an important element of CPS and IoT. Numerous net-related Sustainability 4.0 concepts, which include the

Internet Operating System (IOS), the Internet of Everything (IOE), and the Internet of People (IOP), have been examined in many studies. In addition to IOE and CPS, Sustainability 4.0 integrates machinery that includes big data, cloud computing, additive production, and blockchain technology [13].

Through detectors alongside additional machines, CPSs link the mechanical rudiments of the sector, such as machinery and other rudiments, to a computerized layer that makes up systems for information. This technology facilitates the transfer of data, carries out processes, and connects humans and machines. IoT can be described as the internet's association with the real world that is furnished with detectors and actuators [14]. Using this link, every signal from the real-world manufacturing chain can be observed and recorded as "big data" for later utilization in operations such as the development of fresh value, notably within the context of changing users' appetite for novel services and fresh employment models. The number of the innovations utilized during the Sustainability 4.0 revolution is "cloud computing" which can be described as resources for computing and services of visualizing, that is an amalgam of creativity and the implementation of cutting-edge innovations such as Sustainability 4.0 which has a significant and advantageous effect on the day-to-day activities of corporations to minimize expenses and boost productivity [15]. The amalgamation of all extant technologies with the physical elements of manufacture and all associated entities, suppliers, clients, users, and workers produces numerous benefits for Sustainability 4.0, including computerization and digital transformation. This facilitates choice-making because comprehensive information is readily available, thereby limiting administrative dominance.

3 Ecology and Sustainability 4.0

Environmental sustainability in Sustainability 4.0 is a key topic covered in many academic works. Studies disagree on whether Sustainability 4.0 affects ecological sustainability [16, 17]. This study focuses on environmental sustainability, so we discussed findings that are important to this topic. Other research suggests that entrepreneurs and emerging industries have harmonized their Sustainability 4.0 and ecologically sustainable production practices. Researchers showed the benefits of using technology in manufacturing to protect the environment, but they also showed its immaturity and limitations [18, 19]. On the other hand, Sustainability 4.0 presents huge opportunities to focus industrial value on environmental sustainability. In another study, researchers surveyed German and Chinese enterprises to assess Sustainability 4.0's ecological impact on energy conservation and supplies. According to Burritt and Christ et al., Sustainability 4.0 improves ecological sustainability by handling external events more precisely, efficiently, and quickly [20, 21]. In another study, the researchers proposed a framework based on the Triple Bottom Line (TBL) to account for the challenges and opportunities of implementing Sustainability 4.0. They found that there is a strong and positive link between Sustainability 4.0's environmental benefits and its use, so businesses are more likely to adopt it. Tim et al. have proposed in other studies a plan to encourage the most efficient and ecologically sound use of natural resources by encouraging circular economy standards in organizations using a Sustainability 4.0 strategy to waste reuse and reprocessing [22].

Another researcher examined the connection between the fortification of the environment and the method of security parameters within Sustainability 4.0 by examining the presence of a prevalent keyword within prior literature. They discovered that there is more research on ecological safeguarding and Sustainability 4.0 than procedure security. In other research work, he stated that value formation positively aids sustainable growth through the development of new progressions, a view of the life cycle, and novel company structures [23]. In another research study [24], the researchers acknowledged 18 obstacles that Sustainability 4.0 may confront when enhancing the production chain's viability. Finally, we examined prior research relating Sustainability 4.0 to ecological resilience. The research on the topic has employed a variety of methodologies, including in-depth investigations by Ford and Despeisse et al. [18, 19]; conversations with professionals or field specialists; analysis of content and data collection; acceptance of resource effectiveness indices for ecological viability evaluation; indices of effectiveness for substances, water, power, and the amount waste [19]; ecological expenses; and ecological effect diminution [25]. In deduction, after investigating all of these studies, the difficulties and possibilities related to the execution of Sustainability 4.0 remain unreliable and the advances linked with this sector of sustainable development have not been sufficiently researched because they are still relatively new. Strategies for the future do not yet address how to incorporate effective utilization of finite resources, raw materials, data, ethical consumption, and power with environmentally friendly objectives. To minimize pollution to the environment and attain long-term viability the 4Rs reduce, reuse, recycle, and replace can be implemented. Sustainability 4.0 and ecology will achieve efficiency and green innovation as a result. This is a novel discovery, despite the work of multiple writers.

4 Sustainability 4.0 Contributing to Environmental Sustainability

Figure 1 demonstrates several essential aspects of the Sustainability 4.0 ideology for creating an environmentally friendly world as a whole. The dominant Sustainability 4.0 aspects of sustainability are power, physical substances, raw materials, contaminated materials and diffusion, among others. These scopes are further subcategorized as: the utility of land, water, air and minerals, and reusable useful in case of raw materials; inexhaustible, decline percentage, application or usage in case of power section; remnant rate, the supplemental technique, and intelligent manufacturing and volume in case of physical substance; and perilous level, isothermal dwindling, and global warming or climate change effect in case of contaminated materials' diffusion. The appropriate implementation of these dimensions provides

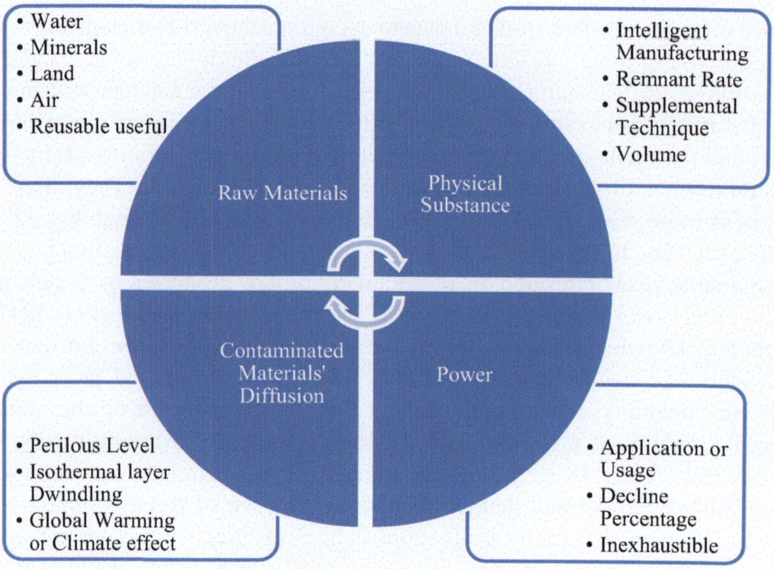

Fig. 1 Sustainability 4.0 ideology for environment

an additional advantage for environmentally conscious growth [26]. A prosperous society uses its people, environment, and economy to meet current needs and assure future survival. Sustainability 4.0 enables manufacturers and production units to use intelligent equipment and other technology. We must evaluate the role and value of employees. Additive manufacturing has advanced for years. Technologies have been around for a while, but applications were often prototypes. Over the past few decades, innovation and application have spread beyond design and engineering into smaller and medium-sized production. As the IIoT gains acceptance in global production [27], manufacturers expect many innovations and solutions to be free. Implementing Sustainability 4.0 technologies reduces waste, time, and power and creates an efficient manufacturing system. Plastic packaging is vital in the medicine and pharmaceutical industries. Automation and adaptable technology have helped plastic producers reduce water use and rapid wastewater clean-up, minimizing polymers' impact on water systems. Imitating reality with computers helps improve industry control. Immersive technologies simplify resource management, including equipment and labor. Enhanced reality aids business inventory management. A worker will use a smartphone to find a part in a vast warehouse. AR will also be the fastest way to convey an aspect from another industry. It could improve worker productivity and simplify difficult procedures. In a sustainable society, many personal goals are met. It is a safe and efficient place for varied cultures and viewpoints, where each industry makes its own decisions' bench [28].

5 International Environmental Legal Doctrines

5.1 *The Doctrine of Sustainable Development*

The Brundtland Report of 1987 was the initial document to address the concept of environmentally friendly development. It has been stated that the standards of sustainable growth refer to a type of advancement in which the necessary demands of the poor, in specific, are met without jeopardizing the potential of generations to come to fulfill their requirements. The preservation of the environment is the main objective of environmental sustainability in this respect. This Doctrine has gained worldwide and regional acceptance. Three components comprise sustainable growth. The three components are [29]:

(1) Integrational Equity—This implies that every person must leave at least some amount of wealth (in this case, resources from nature) for generations to come as they were able to acquire during their lives so that future generations can also enjoy the resources of nature.
(2) Sustainable use of natural resources endeavors for judicious, prudent, or sensible utilization of environmental resources.
(3) Environmental and socioeconomic integration—It is considered that sustainable growth cannot be accomplished without bringing together both the environment and economic growth. Consequently, natural and social growth must additionally be taken into account when establishing environmental responsibilities. In microeconomics, for instance, environmentally friendly growth would necessitate placing the expenses of harm to the environment on the state that inflicted the adverse effects.

5.2 *The Doctrine of Absolute Liability*

In the realm of environmental legal sense, the Doctrine of strict liability is enacted in a legal context. It is used in the context of environmental legal sense to evaluate the dangers of environmental legislation, allowing for the imposition of liability for illegal actions. In several states, neither the procedure nor the filing requirements for environmental petitions are clearly stated. In such situations, obtaining compensation for the damage is difficult. However, the implementation of strict accountability within the framework of the law on the environment makes it simpler to obtain compensation [30].

5.3 Polluter Pays Doctrine

The essence of this Doctrine is the belief that those who cause pollution are responsible for its administration and control so that it does not damage the environment and people. The Emitter Pay Doctrine was recognized in the 1992 Rio Declaration. This standard mainly applies to land, air, and water. We have all become aware of the impact that GHG has had on the environment. This concept applies to GHG. A carbon dioxide price may be used to implement the Doctrine. The carbon dioxide price is a modest fees that emitters of GHG pay in proportion to the possible cost of future climate variation, thereby compelling polluters to absorb the cost of emissions. Carbon pricing can be paid for in two ways [31]:

(1) Using a carbon tax (direct approach).
(2) In the form of a quota, a certain limit is established, and a fee must be paid if it is exceeded.

5.4 Ecocentrism Doctrine

Before addressing the notion of egocentrism, it is important to comprehend the traditional human perspective on the environment. Historically as well as from a religious standpoint, it has been noted that certain individuals or groups believe they have authority over Mother Nature. Consequently, they may pursue it in any manner they choose. However, egocentrism refers to a nature-focused philosophy. Here, it is mentioned that humans are the predominant consumers of the environment. They are responsible for preserving nature regardless of whether or not they receive any advantages from it. The mistreatment of the environment by humans is extremely hazardous because it can cause irreversible damage. Therefore, humanity should protect and preserve our planet [32].

6 Recommendation

Thus, after reviewing the literature, we authors would like to recommend the following two points:

A. Industry 5.0 is the next phase and an emerging pattern that has improved human–machine collaboration and interconnection. The change in cyber-physical systems characterized by Sustainability 4.0 has evolved into Industry 5.0, profoundly altering our way of life, employment, and linking practices. The next technological revolution must define human–machine interaction laws and characterize how we collaborate.

B. Sustainability in the environment will be promoted if Sustainability 4.0 is incorporated with the SDGs. However, sustainability results may suffer if Sustainability 4.0 and sustainable development are not well aligned. Beginnings in regards to the framework of industries can also affect future responses, as these responses can vary depending on the nation of origin and its stage of progress. To avoid an unfavorable scenario, environmentally favorable manufacturing business models can be developed.

7 Conclusion

For an environmentally sustainable society, industries are transforming into a comprehensive, completely sustainable model. The technological core of advanced technologies is intrinsically attached to digitalization. Sustainability 4.0 leads to quantification and reduces complexity and waste, while efficient implementation of Sustainability 4.0 innovations substantially improves information storage and computation capabilities. Businesses are becoming wiser and more efficient as a result of AI, ML, automated robotic processes, cloud computing, big data, intelligent sensors, and IoTs, among other technologies. Even with the best of intentions, the manufacturing industry employs Sustainability 4.0 technology to capture and store complex manufacturing operations to enhance its positive ecological impact. At every level, production methods and structures are becoming sustainable using cutting-edge technologies and all procedures becoming energy-efficient while maintaining the required quality. In zero-emission production, the production system necessitates the recycling and reprocessing of wastes within the production system.

To reduce the environmental impact of production procedures, Sustainability 4.0 is essential for enhancing the environmental sustainability of existing processes as well as creating novel greener practices. Further, international standards stated in various conventions and treaties act as guidelines, Doctrines, or standards for the nations to curb or mitigate the issues of negative effects of human actions or activities upon planet Earth. These Doctrines are strict and must be followed by every signatory nation so that future generations can enjoy these natural resources without any fear.

References

1. Wwap U (2012) World water assessment programme: the United Nations world water development report 4: managing water under uncertainty and risk
2. Rani M, Shanker U (2019) Advanced treatment technologies. In: Hussain C (ed) Handbook of environmental materials management. Springer, Cham. https://doi.org/10.1007/978-3-319-73645-7_33
3. Damania R, Desbureaux S, Rodella A-S, Russ J (2019) Quality unknown: the invisible water crisis. World Bank Publications

4. Colla V, Pietrosanti C, Malfa E, Peters K (2020) Environment 4.0: how digitalization and machine learning can improve the environmental footprint of the steel production processes. Matériaux Tech 108(5–6):507
5. Oláh J, Aburumman N, Popp J, Khan MA, Haddad H, Kitukutha N (2020) Impact of sustainability 4.0 on environmental sustainability. Sustainability 12(11):4674
6. Mathew D et al (2021) Artificial intelligence: hope for future or hype by intellectuals? In: 2021 9th international conference on reliability, infocom technologies and optimization (Trends and future directions) (ICRITO) (2021), pp 1–6
7. Braccini AM, Margherita EG (2018) Exploring organizational sustainability of sustainability 4.0 under the triple bottom line: the case of a manufacturing company. Sustainability 11(1):36
8. Sharma M, Kamble S, Mani V, Sehrawat R, Belhadi A, Sharma V (2021) Sustainability 4.0 adoption for sustainability in multi-tier manufacturing supply chain in emerging economies. J Clean Prod 281:125013
9. Jamwal A, Agrawal R, Sharma M, Kumar V, Kumar S (2021) Developing a sustainability framework for sustainability 4.0. Procedia CIRP 98:430–435
10. Vrchota J, Pech M, Rolinek L, Bednář J (2020) Sustainability outcomes of green processes about sustainability 4.0 in manufacturing: systematic review. Sustainability 12(15):5968
11. Margherita EG, Braccini AM (2020) Organizational impacts on the sustainability of sustainability 4.0: a systematic literature review from empirical case studies. In: Digital business transformation: organizing, managing and controlling in the information age. Springer International Publishing, pp 173–186
12. Ślusarczyk B (2018) Sustainability 4.0–are we ready? Pol J Manag Stud 17(1):232–248
13. Zawadzki P, Żywicki K (2016) Smart product design and production control for effective mass customization in the sustainability 4.0 concept. Manag Prod Eng Rev
14. Chan HC (2015) Internet of things business models. J Serv Sci Manag 8(04):552
15. Haug KC, Kretschmer T, Strobel T (2016) Cloud adaptiveness within industry sectors–measurement and observations. Telecommun Policy 40(4):291–306
16. Sachs JD (2015) The age of sustainable development. Columbia University Press
17. Smith JA (2011) Evaluating the contribution of interpretative phenomenological analysis: a reply to the commentaries and further development of criteria. Health Psychol Rev 5(1):55–61
18. Ford S, Despeisse M (2016) Additive manufacturing and sustainability: an exploratory study of the advantages and challenges. J Clean Prod 137:1573–1587
19. Jelonek M, Urbaniec M (2019) Development of sustainability competencies for the labor market: an exploratory qualitative study. Sustainability 11(20):5716
20. Burritt R, Christ K (2016) Sustainability 4.0 and environmental accounting: a new revolution? Asian J Sustain Soc Responsib 1:23–38
21. Müller E, Hopf H (2017) Competence center for the digital transformation in small and medium-sized enterprises. Procedia Manuf 11:1495–1500
22. Tim S, Michael O, Holger K (2018) Sustainability 4.0 as enabler for a sustainable development: a qualitative assessment of its ecological and social potential. Process Saf. Environ. Prot. 118:254–267
23. Junior JAG, Busso XM, Gobbo SCO, Carreão, H (2018) Making the links among environmental protection, process safety, and sustainability 4.0. Process Saf Environ Prot 11:372–382
24. Luthra S, Mangla SK (2018) Evaluating challenges to sustainability 4.0 initiatives for supply chain sustainability in emerging economies. Process Saf Environ Prot 117:168–179
25. Swain M, Blomqvist L, McNamara J, Ripple WJ (2018) Reducing the environmental impact of global diets. Sci Total Environ 610:1207–1209
26. Sajid S, Haleem A, Bahl S, Javaid M, Goyal T, Mittal M (2021) Data science applications for predictive maintenance and materials science in context to sustainability 4.0. Mater Today: Proc 45:4898–4905
27. Manavalan E, Jayakrishna K (2019) A review of Internet of Things (IoT) embedded sustainable supply chain for sustainability 4.0 requirements. Comput Ind Eng 127:925–953
28. Ashima R, Haleem A, Bahl S, Javaid M, Mahla SK, Singh S (2021) Automation and manufacturing of smart materials in additive manufacturing technologies using internet of things towards the adoption of sustainability 4.0. Mater Today: Proc 45:5081–5088

29. Conroy MM, Berke PR (2004) What makes a good sustainable development plan? an analysis of factors that influence doctrines of sustainable development. Environ Plan A 36(8):1381–1396
30. Shavell S (2018) Strict liability versus negligence. In: Economics and liability for environmental problems. Routledge, pp 89–113
31. Luppi B, Parisi F, Rajagopalan S (2012) The rise and fall of the polluter-pays Doctrine in developing countries. Int Rev Law Econ 32(1):135–144
32. Preston BJ (2014) Internalizing ecocentrism in environmental law. In: Wild law-in practice. Routledge, pp 75–94

A Study into the Application of Ecological Drones for Wildlife Conservation

Satyam Sharma, Anil Kumar Dixit, Sagar Saxena, and Gagandeep Kaur

Abstract The preservation of wildlife has perpetually remained a paramount issue on a worldwide magnitude. In light of the advent of Industry 4.0, also known as the Fourth Industrial Revolution, there has been a notable upswing in the application of sophisticated, intelligent technology to safeguard and promote the welfare of our precious biological wildlife. Integrating drones within the realm of wildlife conservation has been recognized as a groundbreaking instrument, fundamentally reshaping conventional methodologies for the surveillance and preservation of biodiversity. Through the utilization of advanced technological advancements, the realm of wildlife conservation finds itself on the cusp of a transformative era. This pivotal moment presents us with auspicious prospects to protect imperiled species, rehabilitate fragile ecosystems, and address the escalating challenges posed by the encroachment of human activities on wildlife habitats. The paper meticulously scrutinizes techniques utilized in the preservation of wildlife. Ultimately, the paper puts forth recommendations and underscores the inherent limitations found within diverse wildlife conservation models. It presents a range of recommendations to promote the optimal and sustainable conservation of our invaluable wildlife legacy.

Keywords Wildlife · Conservation · Technology · Drones · Ecosystems

S. Sharma · A. K. Dixit · S. Saxena (✉)
Law College Dehradun, Uttaranchal University Dehradun, Dehradun, India
e-mail: adv.sagarsaxena@gmail.com

S. Sharma
e-mail: satyamkabira@gmail.com

A. K. Dixit
e-mail: anilkdixit44@gmail.com

G. Kaur
School of Law, University of Petroleum & Energy Studies, Dehradun, India
e-mail: gkaur@ddn.upes.ac.in

S. Dutta et al. (eds.), *Cyber Intelligence and Information Retrieval*, Lecture Notes in Networks and Systems 1025, https://doi.org/10.1007/978-981-97-3594-5_16

1 Introduction

The current actions undertaken by the human populace have undeniably presented a substantial peril to the existence of wildlife and the preservation of biodiversity, thereby jeopardizing humanity's overall welfare and impeding sustainable development. In instances where natural habitats undergo degradation, resulting in the loss of biodiversity, it is imperative to acknowledge that the consequential compromise of crucial ecosystem services extends to the human population. Regrettably, it is the impoverished and the most susceptible members of society, namely women and children, who bear the brunt of these adverse effects. Wild animals and forests provide play a vital role in maintaining biodiversity. However, wildlife is being lost at an alarming rate all across the world. The loss, fragmentation, and degradation of forests due to human influences have resulted in a dramatic drop in and homogeneity of biodiversity [1]. Problems like unlawful poaching, forest loss, climate change, and other similar phenomena pose serious threats to wildlife. Protecting animal populations is crucial to keeping the ecosystem in check. Over 80% of global trade is carried out on the ocean, and its health is vital to the lives of more than three billion people. There has been a 26% rise in ocean activity since pre-industrial times, and another 125% increase is predicted for this century consequently posing a severe threat to ocean sustainability [2]. Similarly, terrestrial ecosystems are essential to human survival in every way. However, changes and deterioration of terrestrial ecosystems are occurring at an unprecedented rate due to human-influenced factors. The Sustainable Development Goals (SDGs) represent a comprehensive compilation of aims that duly acknowledge the intricate interrelations between individuals and the terrestrial sphere they inhabit, as well as the inherent reliance of human welfare upon a healthy and sustainable ecology. The articulated aim of Sustainable Development Goal 14 is to diligently preserve and judiciously harness the vast expanses of the oceans, seas, and marine resources to foster sustainable development. Conversely, Sustainable Development Goal 15 sets forth the explicit objective of safeguarding, reinstating, and fostering the sustainable utilization of terrestrial ecosystems, while simultaneously exercising prudent stewardship over forests, combating the encroachment of desertification, and decisively arresting and reversing the deleterious effects of land degradation and the erosion of biodiversity [3]. The advent of artificial intelligence (AI) and its accompanying digital technologies has ushered in a new era wherein the utilization of AI-powered systems to address worldwide predicaments in wildlife conservation has become feasible. In light of the aforementioned, it is evident that artificial intelligence (AI) possesses immense potential to expedite the attainment of the Sustainable Development Goals [4]. Given the advent of state-of-the-art technology, it has come to pass that ecologists and conservationists, in their pursuit of enhanced efficacy, have embraced the utilization thereof for more effectively monitoring and surveilling various species and ecosystems. In light of the formidable environmental challenges confronting the global community, it becomes increasingly imperative to duly harness the prospective capacity of technology in facilitating the preservation and protection of natural resources. This necessitates the unimpeded realization of innovation, interdisciplinary inquiry, and international collaboration, thereby ensuring the optimal utilization of technological advancements for conservation [5]. The exploration of intelligent machines (AI) is being propelled by a convergence of exigencies within a diverse array of endeavors for the preservation of biodiversity [6]. The objective of this review paper is to highlight the importance of emerging technology intervention in the effective protection and preservation of wildlife. The researcher aims to enlighten the issues and challenges concerning wildlife conservation using smart technology like drones.

2 Overview of Technology in Wildlife Conservation

The concept of wildlife conservation encompasses the systematic and deliberate efforts undertaken to safeguard the integrity and well-being of various plant and animal species, as well as the preservation and maintenance of their respective habitats. In the context of global ecological systems, it is widely acknowledged that wildlife plays a crucial role in maintaining equilibrium and ensuring the stability of natural processes. The primary objective of wildlife conservation endeavors is to diligently safeguard the continued existence and well-being of various species inhabiting our natural ecosystems. Furthermore, it is imperative to impart knowledge and awareness to individuals regarding the imperative need to coexist harmoniously with these diverse species, thereby promoting sustainable living practices [7]. The fundamental underpinning of robust and fruitful ecosystems lies in the realm of biological diversity. The advantages humans derive from a flourishing natural milieu are of considerable magnitude. The global community is currently grappling with an array of unparalleled challenges that pose significant risks to the preservation and conservation of wildlife. The diminution of wildlife's natural habitat resulting from agricultural activities, mineral extraction endeavors, and the establishment of novel urban infrastructures has engendered a substantial reduction in available ecological territory. The persistence of illicit commerce in wildlife presents a tangible and substantial threat to the preservation of biodiversity, the integrity of ecosystems, and the well-being of human populations. This peril is exacerbated by the fact that various novel diseases, originating from animal-derived commodities, whether sourced from domestic or untamed species, have emerged and proliferated [8]. The emergence of artificial intelligence (AI) and its concomitant digital technologies has inaugurated a novel epoch in which the employment of AI-driven systems to tackle global challenges in the realm of wildlife conservation has become practicable. Upon careful examination of the aforementioned statements, it becomes apparent that artificial intelligence (AI) harbors significant capacity to accelerate the achievement of the Sustainable Development Goals [9]. In light of the emergence of cutting-edge technology, it has become evident that ecological researchers and environmentalists, in their endeavor to achieve heightened efficiency, have adopted its application for the objective of better and more proficiently observing and overseeing diverse species and ecosystems. Given the substantial and pressing environmental obstacles that confront the international community, it is becoming increasingly crucial to effectively leverage the potential capabilities of technology to facilitate the conservation and safeguarding of natural resources. The aforementioned proposition underscores the imperative nature of unfettered actualization of novelty, cross-disciplinary investigation, and global cooperation, thereby guaranteeing the utmost exploitation of advancements in technology progressions with the aim of preservation [5]. Traditionally, the management and preservation of animal species have been predicated upon the acquisition of empirical data through the diligent efforts of human field operatives. These individuals engage in the meticulous task of quantifying animal populations, scrutinizing their behavioral patterns, and patrolling designated areas of ecological significance. The aforementioned endeavors necessitate a substantial investment of time, significant allocation of labor, and considerable financial resources [10]. The coexistence of human activity within natural environments introduces potential hazards to indigenous fauna and their respective ecosystems. The inherent constraints imposed by the physical and cognitive limitations of the human species inevitably restrict the capacity to observe multiple individual animals concurrently, the time resolution and intricacy of data collection, and the scope of physical territory which may be efficiently observed [11]. The present discourse acknowledges the existence of a prominent lineage of human-centered data gathering, which serves as the fundamental underpinning for a substantial portion of our comprehension regarding the geographical distribution of species, their behavioral patterns, and the underlying causes for their interrelationships. However, contemporary obstacles encountered within the realm of wildlife ecology and conservation have brought to the fore the inherent constraints associated with these conventional methodologies. There exists a pressing need within the realm of conservation to address a multitude of requirements in a manner that is both cost-effective and adaptable. This necessitates the implementation of initiatives that encompass a diverse array of natural solutions, technological advancements, and innovative applications of existing technologies [12]. Over the course of the past ten years, there has been a notable surge in the attention and exploration of drones, which are alternatively referred to as unmanned aerial systems (UAS) or UAVs. This heightened interest has been observed within various civilian and scientific domains, with drones being heralded as a novel and distinct epoch in the realm of remote sensing for environmental research [13]. Drones present an opportunity for the acquisition of data about natural phenomena in a manner that is characterized by a comparatively minimal level of risk and cost, thereby facilitating expeditious and methodical observation with a heightened level of spatial and temporal precision. In light of the aforementioned rationales, it is evident that unmanned aerial vehicles, colloquially referred to as drones, have emerged as a prominent phenomenon within the realm of wildlife research and management, as substantiated by scholarly discourse [14]. The burgeoning realm of conservation technology is currently witnessing a notable surge, wherein unmanned aerial vehicles, commonly referred to as drones, have emerged as a prominent and pioneering tool. Drones have been extensively employed within the military sector and have also found utility among prominent technology corporations for cartographic endeavors about road networks. Moreover, their utilization within conservation and ecology has been progressively expanding. Drones, in the context of contemporary times, have been observed to possess a multitude of applications within the realm of wildlife conservation and monitoring. This can be attributed to their diminished physical dimensions and the concomitant reduction in financial burden, rendering them more accessible to interested parties.

3 Drones in Wildlife

Unmanned aerial vehicles (UAVs), commonly referred to as drones, have garnered significant attention in the realm of animal conservation due to their potential as a promising technological tool for low-altitude image-based methods [15]. Drones, as agile systems, possess the inherent capability of swift deployment, potentially in response to immediate requirements, and at a relatively constrained expense, in principle. Therefore, it can be argued that they possess the optimal characteristics necessary for conducting local population monitoring. The employment of lower altitude flights, specifically, has the potential to offer oblique viewpoints that, to some extent, alleviate the issue of occlusion caused by vegetation. The diminished expenses and mitigated operational hazards associated with unmanned aerial vehicles (UAVs) serve to bolster their burgeoning status as a progressively feasible substitute for manned aircraft operating at low altitudes. The efficacy of unmanned aerial vehicles, colloquially known as drones, can be attributed, in part, to their remarkable adaptability in accommodating a diverse range of sensors and apparatus. The determination of the optimal combination of aerial platform and payload is contingent upon the delineation of the scope of the application. Drones, in their manifold manifestations characterized by diverse configurations and dimensions, are commonly employed to capture visual media, notably encompassing both dynamic video footage and static photographic imagery. The utilization of consumer-grade unmanned aerial vehicles (UAVs) in conjunction with lightweight cameras and multispectral sensors has facilitated the provision of professional mapping solutions at a significantly reduced expense compared to conventional photogrammetric methodologies employed in the past. According to scholarly research, it has been observed that drones of moderate size have the capability to be outfitted with thermal vision cameras of a compact nature, hyperspectral sensors, and laser scanning technologies like LiDAR. These advancements hold significant promise in the fields of wildlife ecology [16]. The utilization of drones presents a cost-effective and comparatively safer alternative to employing manned aircraft for research purposes. Moreover, drones possess the capability to survey expansive and otherwise unreachable regions that are ill-suited for conventional ground-based surveys. It is worth noting that drone systems possess the potential to make a valuable contribution to the realm of law enforcement by discreetly identifying and pinpointing human activities. This is achieved primarily through the utilization of night vision and thermal cameras, which enhance the efficacy and efficiency of such endeavors [17]. Furthermore, it is worth noting that drone flight paths possess the inherent capability to be meticulously and autonomously programmed, thereby affording them the capacity to effectively traverse and survey areas that may otherwise be deemed inaccessible due to challenging topographical conditions or dense vegetation. The significance of accessing areas that are deemed inaccessible within environments characterized by seasonal environmental fluctuations cannot be overstated. The findings derived from wildlife surveys utilizing unmanned aerial vehicles (UAVs) in environments characterized by a relatively uniform composition are demonstrated to be commensurate with those obtained through the deployment of manned aircraft. This correlation has been substantiated through empirical evidence about avian species and arctic mammals, as expounded upon in reference [18]. Acquiring data from privately-owned satellites entails a substantial financial burden and frequently proves unviable. However, the utilization of drones offers researchers the opportunity to procure data of superior quality at a reduced expense, while also enabling them to focus their data collection efforts on their precise area of interest. Unmanned aerial vehicles (UAVs), commonly known as drones, have been employed in the monitoring and surveillance of wildfires, thereby contributing to the overall efforts in disaster mitigation. Drones have the potential to address concerns about the safety of manned aerial surveys, as they eliminate the need for human pilots and can be deployed in remote areas that would otherwise be inaccessible [19]. The utilization of manned aircraft has conventionally served as a supplementary method to ground-based wildlife surveys. However, the limited allocation of resources to numerous protected areas hinders the broader implementation of this approach. Moreover, it is imperative to acknowledge that throughout history, there has been a notable prevalence of aerial accidents resulting in fatalities [20]. Drones have emerged as a viable and practical solution to circumvent the aforementioned difficulties encountered in small-scale operations, while simultaneously serving as a valuable supplement to contemporary wildlife conservation initiatives. The utilization of remotely sensed capabilities inherent in unmanned aerial vehicles presents an alternative method for monitoring that is characterized by reduced invasiveness, absence of hazards, repetitive data collection, and enhanced reliability. This technique facilitates the acquisition of information about species abundance and distribution, as well as the documentation of wildlife habits, life-history patterns, and health positions [21]. In the context of a collision event, it is plausible to assert that the comparatively diminutive dimensions of said entities are prone to engender a reduced magnitude of harm, thereby mitigating the peril of bodily harm for the personnel responsible for their operation. It is posited that rotary-wing unmanned aerial vehicles (UAVs) may possess a higher degree of peril compared to their fixed-wing counterparts, primarily due to the vertical descent trajectory that the former is prone to in the event of mechanical malfunction. Conversely, fixed-wing UAVs can execute a controlled glide descent, thereby facilitating a safer and more controlled landing procedure [22]. The prioritization of measures for ensuring the successful administration of terrestrial and marine protected areas is predominantly centered on the implementation of effective management and surveillance systems to address and mitigate unlawful activities. The aforementioned conservation measures are implemented to preserve the overall soundness and authenticity of imperiled species and ecosystems, despite the adverse impacts imposed by human activities.

However, it is imperative to acknowledge that these measures are currently beset by notable inadequacies and shortcomings in their practical implementation [23]. The task of enforcement poses considerable difficulties within expansive protected areas wherein emblematic species teeter on the brink of extinction as a result of illicit hunting, fishing, encroachment, or habitat degradation. Drones, as a manifestation of technological progress, serve as a supplementary measure to address the inadequacy of personnel and resources in the context of anti-poaching efforts [24].

4 Issues and Challenges

The understanding of drones by wildlife is intrinsically linked to various factors, including but not limited to the specific habitat in which the wildlife resides, the particular species under consideration, and the precise characteristics of the drones themselves, as well as how they are operated. It has been observed that the presence of voluminous and sonorous drones has the potential to elicit a primal behavioral reaction in fauna, commonly known as the fight-or-flight response. This reaction is particularly pronounced when drones intrude upon wildlife in an obvious and expeditious manner, thereby exacerbating the aforementioned response. According to empirical investigations, avian species are disproportionately affected by the utilization of drones. Furthermore, it has been observed that collective animal species, such as herds, exhibit more pronounced reactions when confronted with the presence of approaching unmanned aerial vehicles [25]. According to a recent empirical investigation, it has been observed that engaging in aerial activities at reduced altitudes can exert a substantial influence on the collective and individual conduct of mammals. However, the extent of disruption experienced by wildlife due to the utilization of unmanned aerial vehicles is contingent upon various factors, including the specific species under scrutiny, the equipment employed, and the specific attributes of the drone flight, such as the velocity and altitude of the approach. It is important to note that this realm of study is characterized by rapid evolution, and advancements aimed at mitigating noise disturbances will probably emerge in due course [26]. The prevailing body of scholarly literature about the utilization of conservation drones has been found to exhibit a notable dearth of acknowledging the inherent social implications that are intricately associated with their deployment. Further investigation into the societal ramifications of drone utilization is imperative. To ensure compliance with established protocols and ethical standards, it is imperative that research endeavors involving the utilization of unmanned aerial vehicles, commonly referred to as drones, adhere to the requisite clearance procedures mandated by the institutional research board. Furthermore, it is incumbent upon the ethics committee to meticulously assess and evaluate the prospective advantages that may be derived from the deployment of drones in the course of said study. To ensure compliance with the requisite legal framework, the drone operator must complete a comprehensive training program specifically tailored to drone piloting. Furthermore, it is incumbent upon the operator to adhere to the established civil aviation regulations governing the operation of unmanned aerial vehicles. Notwithstanding the considerable potential it holds, the utilization rate of unmanned aerial systems (UASs) in the context of remote sensing technologies, specifically for conservation and management objectives, such as within protected areas, continues to exhibit a notable deficiency. The restricted scope of its applicability can be attributed to the diminished acceptance of the methodologies, given the absence of established protocols or universally applicable suggestions at present [27]. Furthermore, when juxtaposed with manned aerial surveys, it is noteworthy that the prevailing drone technologies frequently necessitate the examination of considerably smaller survey areas. This limitation arises from the restricted flight time and speed inherent in drone operations. Consequently, the utilization of drones for the preservation of wildlife endeavors on expansive terrains becomes less appealing.

Presently, it is evident that the financial outlay associated with investing in cutting-edge technology remains considerably elevated in comparison with the costs incurred for ground surveys. Moreover, it is imperative to underscore the indispensability of comprehensive training programs aimed at equipping personnel with the requisite proficiencies to effectively operate drones. The implementation of long-range unmanned aerial vehicle (UAV) operations in numerous jurisdictions is impeded by prevailing legal restrictions and the persistence of intricate permit procedures. Furthermore, the imposition of stringent regulations about the collection and storage of data could create a sense of reluctance toward the widespread implementation of such systems on a significantly grander magnitude. Furthermore, it is imperative to ensure jurisdictional and ethical authorization about ownership of data, safety, and application to mitigate the risk of data misuse. However, it is worth noting that there is currently a dearth of comprehensive recommendations elucidating these specific aspects in the context of data derived from drone surveys. Accordingly, it is arguable that the current state of affairs suggests that drones may not serve as a comprehensive substitute for satellite images or manned aerial vehicles in the realm of data collection. However, it is plausible to assert that drones could potentially serve as a supplementary tool to augment the capabilities of these existing methods.

The use of drones is subject to a multitude of legal frameworks and regulations that differ across nations and territories. These regulations and laws are particularly stringent in areas designated as protected, where the preservation of wildlife habitats takes precedence. The intricate web of legal frameworks poses a formidable challenge, impeding the noble cause of preservation and potentially engendering unforeseen legal ramifications.

5 Recommendations

After examining the study or research of distinguished scholars or researchers:

A. Unmanned aerial vehicles can quickly cover huge areas, reducing human presence in environmentally sensitive habitats and wildlife disruptions. Drone pilots should keep a safe distance from animals to avoid stressing and disrupting their natural lifestyles. Conservation requires habitat identification, delineation, and monitoring. In areas with limited terrestrial access, drones should be used to minimize impacts on native flora and animals. Drones can also help protect endangered species' breeding areas. This proactive strategy helps save these rare and delicate species, which are endangered. Use cutting-edge drone technologies to fight poaching with local governments and organizations. This method preserves traditional values while bolstering conservation efforts.
B. Drones' unique perspective can draw attention and raise environmental awareness. Drone footage of protected areas, their wildlife, and conservation efforts can inspire public support for biodiversity conservation. Involving the public in such efforts can increase financial resources and volunteer participation in environmental preservation efforts. Comprehensive instructional programs that promote safe drone use, with a focus on protecting ecological systems and wildlife during drone operations, are recommended.

6 Conclusion

In the end, the utilization of drones in the realm of wildlife conservation has undeniably emerged as a transformative and remarkably efficient instrument in the preservation and protection of our invaluable ecological legacy. The utilization of unmanned aerial vehicles, commonly known as drones, has brought about a transformative impact on how we observe, safeguard, and oversee the well-being of various species and their respective ecosystems. The myriad benefits they provide have greatly bolstered the capacities of environmental advocates, scientists, and wildlife stewards in their endeavors to safeguard imperiled species and fragile ecosystems. The utilization of this technology has facilitated expedited and precise assessments of wildlife, thereby furnishing invaluable information concerning the fluctuations in population, spatial dispersion, and ecological conduct. Furthermore, it is important to acknowledge the significant role that drones have played in addressing and mitigating the detrimental impact of wildlife crimes, including but not limited to illegal hunting and illicit forest extractions. By harnessing the power of real-time surveillance abilities, we can effectively detect and track activities that may raise concerns, thereby enabling agencies to react quickly to emerging threats. The implementation of this preventive strategy has effectively dissuaded unlawful activities, thereby making a significant contribution toward safeguarding the well-being of endangered species. Notwithstanding the notable advantages, the utilization of drones in the realm of protecting wildlife also gives rise to a set of challenges and ethical deliberations. It is of utmost importance to thoroughly address concerns about confidentiality, information safety, and possible disruptions to wildlife. Ensuring a harmonious coexistence between the benefits of drone technology and the imperative to mitigate adverse effects on ecosystems and wildlife is of utmost importance. In summary, drones are emerging as a formidable instrument in the domain of wildlife conservation, providing unmatched capacities in the surveillance, safeguarding, and administration of ecological diversity. Through responsible and prudent utilization of drones, we have the remarkable opportunity to foster an everlasting and beneficial impact on the invaluable wildlife and delicate ecosystems of our planet. By embracing this technology, we can pave the way for an environmentally friendly tomorrow that will benefit future generations to follow.

References

1. MacDicken KG (2015) Global forest resources assessment 2015: what, why, and how? For Ecol Manage 352:3–8
2. Wu SR, Shirkey G, Celik I, Shao C, Chen J (2022) A review on the adoption of AI, BC, and IoT in sustainability research. Sustainability 14(13):7851
3. Carter SD (2022) Synergies between COVID-19, supply chain, sustainable development, and climate change: lessons from UNCTAD. In: International conference on current business issues in African countries. Cham: Springer International Publishing, pp 3–7
4. Theodorou A, Nieves JC, Dignum V (2022) Good AI for good: how AI strategies of the Nordic countries address the sustainable development goals. arXiv:2210.09010
5. Lahoz-Monfort JJ, Chadès I, Davies A, Fegraus E, Game E, Guillera-Arroita G, Joppa L (2019) A call for international leadership and coordination to realize the potential of conservation technology. Bioscience 69(10):823–832
6. Norouzzadeh MS, Nguyen A, Kosmala M, Swanson A, Palmer MS, Packer C, Clune J (2018) Automatically identifying, counting, and describing wild animals in camera-trap images with deep learning. Proc Natl Acad Sci 115(25):E5716–E5725
7. Tuia D, Kellenberger B, Beery S, Costelloe BR, Zuffi S, Risse B, Berger-Wolf T (2022) Perspectives in machine learning for wildlife conservation. Nat Commun 13(1):792
8. Arti, K (2020) Property status of feral animals in India. Eur J Mol Clin Med 7(4)
9. Sætra HS (2021) AI in context and the sustainable development goals: factoring in the unsustainability of the sociotechnical system. Sustainability 13(4):1738
10. Witmer GW (2005) Wildlife population monitoring: some practical considerations. Wildl Res 32(3):259–263
11. Kays R, Crofoot MC, Jetz W, Wikelski M (2015) Terrestrial animal tracking as an eye on life and planet. Science 348(6240):aaa2478
12. Pimm SL, Alibhai S, Bergl R, Dehgan A, Giri C, Jewell Z, Loarie S (2015) Emerging technologies to conserve biodiversity. Trends Ecol Evol 30(11):685–696
13. Whitehead K, Hugenholtz CH (2014) Remote sensing of the environment with small unmanned aircraft systems (UASs), part 1: a review of progress and challenges. J Unmanned Veh Syst 2(3):69–85
14. Rodríguez A, Negro JJ, Mulero M, Rodríguez C, Hernández-Pliego J, Bustamante J (2012) The eye in the sky: combined use of unmanned aerial systems and GPS data loggers for ecological research and conservation of small birds. PLoS ONE 7(12):e50336
15. Linchant J, Lisein J, Semeki J, Lejeune P, Vermeulen C (2015) Are unmanned aircraft systems (UASs) the future of wildlife monitoring? a review of accomplishments and challenges. Mammal Rev 45(4):239–252
16. Näsi R, Honkavaara E, Lyytikäinen-Saarenmaa P, Blomqvist M, Litkey P, Hakala T, Holopainen M (2015) Using UAV-based photogrammetry and hyperspectral imaging for mapping bark beetle damage at tree level. Remote Sens 7(11):15467–15493

17. Bondi E, Fang F, Hamilton M, Kar D, Dmello D, Choi J, Nevatia R (2018) Spot poachers in action: augmenting conservation drones with automatic detection in near real time. In: Proceedings of the AAAI conference on artificial intelligence, vol 32, no 1
18. Chabot D, Francis CM (2016) Computer-automated bird detection and counts in high-resolution aerial images: a review. J Field Ornithol 87(4):343–359
19. IV GPJ, Pearlstine LG, Percival HF (2006) An assessment of small unmanned aerial vehicles for wildlife research. Wildl Soc Bull 34(3):750–758
20. Jiménez López J, Mulero-Pázmány M (2019) Drones for conservation in protected areas: present and future. Drones 3(1):10
21. Alibhai SK, Gu J, Jewell ZC, Morgan J, Liu D, Jiang G (2023) 'I know the tiger by his paw': a non-invasive footprint identification technique for monitoring individual Amur tigers (Panthera tigris altaica) in snow. Eco Inform 73:101947
22. Sandbrook C (2015) The social implications of using drones for biodiversity conservation. Ambio 44(Suppl 4):636–647
23. Struhsaker TT, Struhsaker PJ, Siex KS (2005) Conserving Africa's rain forests: problems in protected areas and possible solutions. Biol Cons 123(1):45–54
24. Mulero-Pázmány M, Stolper R, Van Essen LD, Negro JJ, Sassen T (2014) Remotely piloted aircraft systems as a rhinoceros anti-poaching tool in Africa. PLoS ONE 9(1):e83873
25. Mohan M, Richardson G, Gopan G, Aghai MM, Bajaj S, Galgamuwa GP, Cardil A (2021) UAV-supported forest regeneration: current trends, challenges and implications. Remote Sens 13(13):2596
26. Schroeder NM, Panebianco A, Gonzalez Musso R, Carmanchahi P (2020) An experimental approach to evaluate the potential of drones in terrestrial mammal research: a gregarious ungulate as a study model. R Soc Open Sci 7(1):191482
27. Baxter PW, Hamilton G (2018) Learning to fly: integrating spatial ecology with unmanned aerial vehicle surveys. Ecosphere 9(4):e02194

The Transformative Role of Artificial Intelligence in the Legal Profession and International Arbitration

Jaffar Alkhayer, Gagandeep Kaur, and Chander Mohan Gupta

Abstract The profound technological revolution driven by artificial intelligence (AI) is impacting all aspects of business and personal life, including dispute resolution. AI has the potential to transform dispute resolution by reducing time and costs, enhancing predictability, and discouraging unmeritorious claims. However, concerns about decision-making, data access, and transparency persist. This article examines the implications of AI in the legal profession and international arbitration, focusing on its core characteristics, evolution through machine learning and data mining, and its possible impact on the legal profession and international arbitration. The paper addresses challenges facing AI in international arbitration, focusing on the limitations and implications of AI in three critical aspects: research and outcome prediction, the decision-making process, and policymaking. The paper concludes that while AI is rational and unbiased, understanding its predictions and lack of clear explanations raise concerns. The risks of perpetuating biases and the absence of emotional intelligence make the full replacement of humans doubtful. AI is more likely to assist arbitrators and lawyers, providing valuable insights, but human input remains crucial for accuracy and addressing legal complexities.

Keywords Artificial intelligence · Arbitration · Legal profession · Researching · Prediction · Decision · Policy

J. Alkhayer · C. M. Gupta (✉)
Faculty of Legal Sciences, Shoolini University, Solan, Himachal Pradesh, India
e-mail: Cg30aug@yahoo.com

G. Kaur
School of Law, University of Petroleum & Energy Studies, Dehradun, Uttarakhand, India
e-mail: gkaur@ddn.upes.ac.in

S. Dutta et al. (eds.), *Cyber Intelligence and Information Retrieval*, Lecture Notes in Networks and Systems 1025, https://doi.org/10.1007/978-981-97-3594-5_17

1 Introduction

The world is in the middle of a profound technological revolution, comparable to the industrial revolution, impacting every aspect of business and personal life, including dispute resolution. Artificial intelligence (AI) stands at the core of this revolution, influencing business practices, transactions, and dispute resolution methods [1]. AI has the potential to reduce time and costs in resolving disputes, enhance predictability, and discourage unmeritorious claims, but concerns about decision-making, data access, and transparency persist [2]. The definition of AI varies widely, but two prominent ones stand out. According to the Oxford Dictionary, AI involves creating computer systems capable of performing tasks that typically require human intelligence, such as visual perception, speech recognition, decision-making, and language translation [3]. John McCarthy, a renowned computer scientist, defines AI as the process of making machines behave intelligently, akin to human behavior [4]. Both definitions draw parallels with human intelligence and highlight AI's superiority in tasks like visual perception and speech recognition, where computers outperform humans. Frequently, voice recognition on smartphones is utilized to dictate quick emails and messages while on the move. Despite aspiring to enhance language skills beyond translation programs, it is acknowledged that computers excel in comprehending numerous languages compared to humans [5].

Artificial intelligence (AI) refers to the process of combining large datasets with advanced data processing systems and intelligent algorithms. This enables software to learn from patterns in the data, making it "intelligent." AI encompasses various subjects such as machine learning, deep learning, and natural language processing. It empowers software to analyze extensive amounts of data, learn from it, and provide answers, predictions, and complex reasoning without human intervention [6].

It is essential to clarify the distinctions within AI. Specialized literature differentiates between artificial intelligence (AI) and artificial general intelligence "AGI" [7]. AGI also known as "Strong AI" can achieve any goal, including learning new things, while weak AI cannot perform beyond its designated tasks; hence, it is limited to that specific task, for example, it cannot write a novel if it is programmed for playing chess. Weak AI mimics human thinking without possessing it, whereas strong AI possesses general intelligence and can learn, understand, and adapt across a wide range of tasks, aims to replicate human-level intelligence and problem-solving abilities, and can transfer knowledge from one domain to another [8].

In the context of the legal profession and arbitration, the question arises whether computers can outperform human decision-makers. The Oxford Dictionary includes "decision-making" as a comparison task for AI. In arbitration, AI is instrumental in analyzing both micro-data (related to specific disputes) and macro-data (pertaining to the resolution process and outcomes). Although AI's application to arbitral awards is still in its early stages, it is anticipated to bring about a revolution in dispute resolution by predicting trends and outcomes. To implement AI effectively in this regard, access to relevant arbitral awards is crucial [1].

The article delves into the role of artificial intelligence in the legal profession, specifically exploring the implications of machine learning and data mining. It focuses on the impact of these technologies on the legal profession. The paper examines AI's transformative role in international arbitration and its implications and limitations in three critical aspects related to international arbitration: research and outcome prediction, the decision-making process, and legislation enactment, agreements drafting, and policymaking.

2 The Role of Artificial Intelligence in the Legal Profession

In order to grasp the significance of artificial intelligence in the legal profession, one must first comprehend the fundamental traits of AI and its evolution through machine learning and data mining. This understanding is crucial to assess the potential influence of AI on the legal profession.

2.1 Understanding Machine Learning and Data Mining Implications

A significant number of lawyers lack a practical understanding of artificial intelligence [9], and attorneys who are well-versed in AI are scarce, akin to vegan butchers. [10]. However, it is essential for lawyers to understand the fundamental characteristics of AI to assess its potential impact on the legal profession and legal thought. AI is defined as making a machine behave intelligently, akin to human behaviors [11]. According to the Oxford Dictionary, "AI involves the theory and development of computer systems capable of performing tasks that typically require human intelligence, such as visual perception, speech recognition, decision-making, and translation between languages" [5]. Intelligence itself is challenging to define and has sparked debates among experts [12]. Essentially, intelligence is the ability to learn, understand, and make reasoned judgments or opinions, distinguishing humans from non-intelligent or less intelligent life forms [13]. In the early stages of AI research, there were attempts to imitate human cognitive processes by developing algorithms that could understand and replicate language acquisition. However, these efforts had limited success, especially when dealing with complex activities [14]. Present-day AI systems, referred to as expert systems or rule-based programs, rely on sets of rules expressed as "if–then" statements to reach logical conclusions [15]. However, these programs have their limitations, as constructing the knowledge base with manual rules is a time-consuming process and may not adequately address complex and dynamic situations [16].

The rapid progress in artificial intelligence (AI) research was accompanied by a significant increase in data, often referred to as the "data quake" [15]. This surge in data became possible due to the improved processing speed of computers, as described by Moore's Law, which doubles every twelve to eighteen months, and the lower costs of data storage, known as Kryder's Law, which also double at a comparable rate. This data explosion led to the emergence of "Big Data" and caused a shift in AI development, moving the focus away from constructing complex algorithms and toward the utilization of machine learning techniques [17].

Machine learning is a branch of AI that involves computer systems learning from experiences and improving their performance over time. Unlike traditional approaches, machine learning models like neural networks do not require explicit programming for tasks like language translation. Instead, they learn from vast amounts of available data and improve through immediate online feedback. Machine learning is based on inferring hidden patterns from observed data [15]. Rather than coding algorithms directly, computers derive them from large samples and powerful computation [14]. This approach is particularly useful for tasks that are challenging to define through explicit instructions. For instance, the software can effectively identify spam emails by recognizing patterns in labeled data [18].

Data mining is akin to mining valuable materials from tons of soil, but in the context of AI, it involves computers sifting through extensive datasets to discover hidden patterns [15]. These patterns are then used to develop models for future predictions, such as categorizing emails as spam or non-spam [17]. Furthermore, AI can analyze transaction data to identify patterns indicative of fraudulent activities [19], and AI and data mining are used to analyze vast amounts of climate data to model and predict weather patterns and climate changes [20].

Pattern recognition in machine learning relies on statistical and probability calculations. Machine learning AI utilizes probabilistic theories to achieve successful outcomes [15]. One popular type of probabilistic theory used in machine learning is Bayesian networks. Bayesian networks are a type of probabilistic graphical model that represents the relationships between variables as a directed acyclic graph [21]. There are other types of probabilistic graphical models used in machine learning, such as Markov random fields and conditional random fields. These models have various applications in computer vision, speech recognition, and social network analysis, among others [22].

Briefly, learning programs adjust model parameters based on knowledge extracted from sample data, using simple and straightforward algorithms instead of complex formulas. Machine learning in AI comes in various forms based on human involvement. Supervised learning requires human input to train the algorithm with labeled data and feedback, while unsupervised learning involves minimal human intervention and identifies co-occurring patterns for future occurrences [23].

2.2 *The Impact of Machine Learning and Data Mining on the Legal Profession*

Overall, the development of AI, particularly through machine learning and data mining, has significantly enhanced its capabilities and applications in various fields, including the legal environment [17].

AI models can produce "intelligent" outcomes similar to human cognitive processes, like understanding and classifying emails as spam, but they achieve this through probabilistic models instead of relying on human-like intelligence [24]. These statistical or probabilistic approaches are examined in detail regarding their implications for legal decision-making.

Statistical or probabilistic models in the context of AI refer to the use of mathematical techniques that involve probabilities and statistics to make predictions, classifications, or decisions. These models are widely used in various AI applications, including natural language processing, image recognition, and data analysis [25]. When applied to legal decision-making, these models play a crucial role in assisting lawyers and judges in making informed and objective decisions. In legal decision-making, there are often complex and uncertain factors involved, such as interpreting laws, assessing evidence, and predicting the outcomes of cases. Human judges and lawyers may rely on their experience and intuition, but they can also be influenced by cognitive biases or emotions, leading to potentially inconsistent or subjective decisions. Statistical or probabilistic models, on the other hand, are designed to handle uncertainty and make decisions based on probabilities and data [26]. They analyze large amounts of data, learn patterns, and estimate the likelihood of different outcomes, reducing the impact of biases. For example, AI models can predict case success based on historical data, helping lawyers and clients assess risks [27]. They also aid in contract analysis, saving time and identifying risks [28]. AI-powered legal research tools improve access to relevant legal information, enhancing the accuracy of legal arguments and decisions [27]. However, it is crucial to ensure these models are well-designed, trained on diverse data, and regularly updated to maintain fairness and effectiveness. The advancement of AI technology has the potential to revolutionize the legal profession by enabling more informed and efficient decision-making processes [26].

3 The Transformative Role of AI in International Arbitration

Currently, AI is extensively used in arbitration to analyze vast digital micro-data held by parties and their counsel. This helps identify relevant information and presents it more effectively, addressing the cost and time challenges brought by digitization [2]. Looking ahead, AI's potential in reducing arbitral uncertainty is remarkable. It can analyze arbitral awards, provide legal reasoning, and offer insights into past company performance, arbitrator decisions, and damages approached. These predictive capabilities will lower uncertainty in the dispute resolution process, discourage unmeritorious claims, and expedite settlements, positively impacting the economy and society [5]. AI also promises to transform case preparation in international arbitration. Parties can better select arbitrators, present more persuasive arguments, save time and costs on research and data analysis, and develop more realistic budgets [1]. However, accessing the necessary systems, data, and processing power is crucial for maximizing these benefits.

The policy implications for the future of the arbitral playing field are contingent on affordable and widespread AI availability. If accessible, AI can empower more arbitral actors with valuable insights on success chances, optimal strategies, arbitrator selection, and reduced risks at a lower cost. However, equitable access to reliable datasets and systems is crucial; otherwise, it could exacerbate disparities. Important questions arise: how will AI development impact participation? Who will have access to AI systems and data to mitigate litigation risk? Will small and large players in different countries benefit equally, or will it favor big corporations and law firms? Open or closed access to algorithms and macro-data (like awards) also affects data protection and privacy concerns [1]. AI can revolutionize the imbalance of resources between parties, but success hinges on affordable access to both AI technology and diverse dispute data. Extracting insights automatically through text analytics can overcome some obstacles, but challenges remain due to limited access to relevant arbitral awards and decentralized data points [5].

AI faces challenges in using awards to predict results in international arbitration due to limited award publication practices [29]. While some arbitration institutions publish redacted versions of selected awards, many commercial arbitral awards are not published. Investor-state arbitration, maritime arbitration, and sports arbitration often see unredacted awards published. However, the lack of comprehensive data on arbitral awards hinders AI's ability to make accurate predictions. Some institutions, like ICC and ICDR, publish summaries or redacted versions, but key details such as arbitrator, party, and counsel names are often excluded. The inconsistent publication practices among different arbitration centers create obstacles to AI analysis [29].

Arbitral institutions are increasingly publishing studies on the time and cost of proceedings under their rules. For instance, the AAA conducted a study comparing arbitration timelines to U.S. federal courts, highlighting the cost implications of longer court resolutions [30]. Additionally, organizations have developed databases to provide arbitration-related information. Notable examples include Arbitrator Intelligence [31], which collects surveys on arbitrators and aims to make arbitral awards available, and Dispute Resolution Data [32], which gathers arbitration data from major international institutions. The Global Arbitration Review Arbitrator Research Tool (GAR ART) [33] offers information on individual arbitrators, including their procedural preferences and counsel associations. While these data sources enhance understanding, the lack of access to full award reasoning and key identities limits their usefulness for AI analysis. Data protection laws, such as the GDPR, must be considered when granting access to awards. Redaction of personal data may be necessary but including full award reasoning and key identities (with appropriate permissions) would greatly enhance AI's predictive capabilities. However, concerns over confidentiality, costs, issue conflicts, and unintentional release of confidential information pose challenges to public access to awards and arbitrator names [29]. The current dataset of underacted awards, including arbitrator, party, and counsel identification, is limited to publicly enforced awards or those made public by parties. Obtaining and accessing these awards across jurisdictions and arbitral institutions is cumbersome and expensive. Decentralization of available awards and various platforms poses additional hurdles for AI implementation in arbitration, potentially reducing its efficacy and increasing costs. Predicting future arbitrator behavior is complicated by three-person tribunals, where attributing outcomes solely to the chair may overlook the influence of other members on the decision and reasoning. This challenge needs to be addressed by algorithms. The lack of an easily accessible dataset and decentralized decision-making may increase upfront and ongoing costs for AI use in predicting arbitration outcomes. While AI can offer a more complete analysis when full arbitral awards are available, it may face slower adoption and higher costs initially [1]. Furthermore, as international arbitration is widely utilized for complex trans-border disputes, there is a strong impetus to enhance AI's effectiveness in this domain. This may lead to increased demand for more accessible arbitral awards and a trend toward greater transparency [34]. The future of artificial intelligence in international arbitration is inevitable, given the significant sums involved and the substantial benefits it offers. However, as a community, it is critical to consider the beneficiaries, costs, and broader impacts not immediately apparent.

4 Implications and Limitations of AI in the International Arbitration

The integration of AI in international arbitration has profound implications and limitations, impacting three critical aspects: research and outcome prediction, the decision-making process, and legislation enactment, agreements drafting, and policymaking.

4.1 Research and Outcome Prediction

Several AI-powered research tools have emerged, offering vast datasets and powerful search engines. By using these tools, lawyers and practitioners of the legal profession can quickly access relevant case law, statutory provisions, and scholarly articles. Some of these tools even integrate deep learning capabilities, continuously improving their performance by learning from feedback. While these tools can reduce research time, human input remains crucial and cannot be underestimated [5]. According to the "Will a Robot Take My Job?" online test by McKinsey, AI could potentially handle 23% of a lawyer's tasks, but the overall job security for lawyers is considered safer than many other professions [35]. Tasks such as researching relevant legal materials and preparing legal documents are among those that the test suggests AI could handle.

However, expertise is still required to utilize research tools effectively, formulate appropriate queries, and analyze and interpret the findings. Additionally, big data treatment has become crucial in dispute resolution, including arbitration, involving the use of specialized programs to organize and analyze large datasets beyond the capabilities of traditional databases. These techniques are employed in e-discovery processes, such as predictive coding, which automate portions of document review by employing keyword search, filtering, and sampling to reduce the manual review of irrelevant documents. Lady Justice Gloster, former judge of the Court of Appeal of England and Wales and chair of the UK-wide Disclosure Working Group, highlighted the significant role of predictive coding in commercial court litigation in the UK. This application is likely to extend to arbitral proceedings utilizing e-discovery [5].

Thirdly, AI is poised to be applied in outcome prediction. Clients often inquire about the chances of success in a dispute, and computer programs now offer assistance in providing estimates based on predictive technology. Sir Geoffrey Vos, Chancellor of the English High Court, acknowledged the development and application of predictive technology in forecasting dispute outcomes, particularly in the USA and Europe [36]. Recent studies support the efficacy of computer programs in predicting dispute outcomes. For example, a 2016 study analyzed cases from the European Court of Human Rights [37], developing a dataset and employing a computer program to assess word frequency, sequences, and clusters, assigning weights to these factors in relation to the outcome determination.

In simple terms, the program analyzed word relationships, sequences, and clusters to predict case outcomes. In the study involving European Court of Human Rights cases, the program achieved 79% accuracy when predicting the outcome of remaining cases. Another study on US Supreme Court decisions [38], covering a broader range of areas, attained a 70% success rate. While commercial software tools exist to predict case outcomes in various fields, no specific technology is designed exclusively for international arbitration. Confidentiality issues pose a challenge as international arbitration awards are not as accessible as decisions from other courts. Limited data availability hinders effective outcome prediction, though investor-state arbitration, with more published awards, faces this obstacle to a lesser extent than international commercial arbitration [5].

4.2 The Decision-Making Process

The question arises whether AI can effectively replace human decision-makers. Despite the potential difficulties arising from confidentiality and limited datasets in international arbitration, the complex and unique nature of such cases makes it unlikely for AI to replace arbitrators in the near future [5, 35]. However, considering the possibility, the debate extends to philosophical aspects. Could AI programs be superior decision-makers compared to humans? Two arguments in favor of AI's superiority emerge. Firstly, AI programs are rational and not influenced by irrational factors, as demonstrated by a study on parole cases where the timing of the decision was affected by judges' hunger levels [39]. Secondly, AI programs are considered unbiased, unlike humans who are susceptible to unconscious cognitive biases, such as the anchoring effect studied by Daniel Kahneman. Daniel Kahneman, along with Amos Tversky and their students [40], conducted experiments on the anchor effect. In one experiment, participants were asked about Mahatma Gandhi's age when he died, and the average response was around 75. In a second experiment, participants were divided into subgroups and asked different anchor questions. Those given an anchor of 144 estimated Gandhi's age as higher than 75, while those given an anchor of 35 estimated it as lower. This illustrates the cognitive bias of the anchor effect, where the initial value given (the anchor) influences subsequent estimations of an unknown quantity.

When considering the implications of decision-making with AI, the potential influence of cognitive biases on judges or arbitrators is concerning. For example, anchoring biases could impact the outcome of damages claims based on the initial number suggested. AI programs, however, may not be subject to the same biases as humans [41]. While some argue that AI could lead to better decision-making due to rationality and lack of biases, there are counterarguments. Initially, concerns arose about programmers defining algorithms and influencing outcomes, but strong AI, which learns to learn, might be less subject to such programming influence. In studies on the European Court of Human Rights and US Supreme Court decisions, the programs drew inferences independently without prior algorithm definitions.

Hence a second and potentially stronger counterargument against AI decision-makers is brought. The challenge lies in understanding how and why a program makes its predictions. For example, a program that predicts European Court of Human Rights decisions with 80% accuracy may analyze previous decisions and certain word occurrences, but the exact inferences are unclear. Computer scientists face difficulties when trying to explain AI results [42]. The program's high accuracy lacked clear explanations, raising concerns about AI's inability to provide understandable reasons for its decisions. The ability to provide reasons is essential for decision-making in courts and tribunals, allowing parties to understand outcomes, adapt behavior, and ensure consistent rulings. While AI may make accurate predictions, its inability to offer comprehensible reasons casts doubt on its potential to replace human decision-makers. Instead, AI may assist human decision-makers, who remain crucial in providing meaningful decisions and explanations [5].

4.3 Legislation Enactment, Agreements Drafting, and Policymaking Process

Exploring the potential advantages of AI in legislation enactment, agreement drafting, and the policymaking process involves assessing the benefits that AI can provide in these domains. In a noteworthy 2016 study, researchers aimed to have a computer program draft an international investment agreement. The study suggested that policymakers worldwide could utilize AI programs to assist in crafting such agreements, aiming to create a more equitable playing field and address power asymmetries among states [43]. The study used a dataset of 1,682 English-language bilateral investment agreements, training the program to identify best practice provisions. The results of the study had two main aspects. Firstly, the program excelled at replicating individual provisions of international investment agreements, making it difficult to distinguish them from "real" provisions. However, on the other hand, the program struggled with drafting complete agreements. The computer-generated international investment agreements contained repetitive clauses, lacked essential provisions, and had an unsatisfactory order of provisions. This outcome highlighted the current limitations of AI in handling complex, comprehensive drafting tasks [5]. Nonetheless, this study also showcased the potential of AI as a valuable tool in the legislative and policymaking processes. By assisting in identifying best practices and streamlining certain aspects of agreements, AI has the capacity to enhance the efficiency and effectiveness of policy development. However, it is essential to recognize the need for human oversight and expertise to ensure coherence, comprehensiveness, and strategic vision in crafting legislative instruments and agreements [44].

AI has the potential to assist and enhance the policymaking process, but it cannot fully replace human policymakers due to several limitations. These include the risk of perpetuating biases, while it is true that AI lacks emotional and cognitive biases that humans may have, it can still perpetuate biases present in the data it is trained on. AI systems learn from historical data, and if that data contains biased information or reflects societal prejudices, AI will replicate those biases in its decision-making. The future of policymaking is likely to involve a collaborative approach where AI supports policymakers in data analysis while humans retain final responsibility for critical judgments and policy choices [45].

As AI technology continues to evolve, policymakers must carefully weigh the benefits and limitations of its application in these crucial domains. Utilizing AI as a complementary tool, alongside human expertise, can lead to better-informed, better-rounded policy decisions and agreements that effectively address complex global challenges.

5 Conclusion

AI is being utilized for outcome prediction, providing clients with estimates based on predictive technology. Recent studies validate the effectiveness of computer programs in predicting dispute outcomes. However, challenges persist, such as limited data availability and confidentiality issues in international arbitration. While AI programs are deemed rational and unbiased, understanding how they make predictions and decisions, along with their inability to provide clear explanations, raises questions. Additionally, AI faces risks of perpetuating biases, lacks the capacity to engage with the public and be held accountable, and may struggle to predict unforeseen consequences due to the absence of emotional intelligence. All these factors cast doubt on AI's potential to fully replace humans in tasks related to research and outcome prediction, the decision-making process, and policymaking. Instead, AI is more likely to assist arbitrators, lawyers, and human decision and policymakers by providing valuable insights and support in complex processes. Despite its advancements, human input remains invaluable in ensuring accuracy and addressing unique legal complexities.

References

1. Paisley K, Sussman E (2018) Artificial intelligence challenges and opportunities for international arbitration. New York Disput Resolut Lawyer 11(1):35–40. https://nysba.org/app/uploads/2020/03/DisputeResolutionLawyerSpring18.pdf#page=35
2. Li B, Sourdin T, Burke T (2019) Just, quick and cheap?: civil dispute resolution and technology. Macquarie Law J 19:17–38. https://search.informit.org/doi/10.3316/agispt.20200929037458
3. Jarek K, Mazurek G (2019) Marketing and artificial intelligence. Cent Eur Bus Rev 8(2):46–55. https://doi.org/10.18267/j.cebr.213
4. Moore PV (2018) Jerry kaplan artificial intelligence: what everyone needs to know. Organ Stud 40(3):466–470. https://doi.org/10.1177/0170840618792173
5. Scherer M (2019) International arbitration 3.0—how artificial intelligence will change dispute resolution. Austrian Yearb Int Arbitr 503. https://ssrn.com/abstract=3377234
6. Ashley KD (2017) Artificial intelligence and legal analytics: new tools for law practice in the digital age. Cambridge University
7. Baggaley K (2017) There are two kinds of AI, and the difference is important Most of today's AI is designed to solve specific problems. R Soc Open Sci. https://www.popsci.com/narrow-and-general-ai/
8. Ben-Ari D, Frish Y, Lazovski A, Eldan U, Greenbaum D (2017) Artificial intelligence in the practice of law: an analysis and proof of concept experiment. Richmond J Law Technol 23(2):1–36. https://www.kiip.re.kr/webzine/1707/resource/file/library04.pdf
9. Queen Mary School of International Arbitration Survey, The Evolution of International Arbitration (2018). https://arbitration.qmul.ac.uk/research/2018/
10. Lauritsen M (2018) Toward a phenomenology of machine-assisted legal work. J Robot Artif Intell Law 1(2):67–83. https://static1.squarespace.com/static/571acb59e707ebff3074f461/t/5a7e4abde2c48393c5a56afe/january-february2018/01edintro.pdf
11. Pelau C, Dabija D-C, Ene I (2021) What makes an AI device human-like? the role of interaction quality, empathy and perceived psychological anthropomorphic characteristics in the acceptance of artificial intelligence in the service industry. Comput Human Behav 122:106855. https://doi.org/10.1016/j.chb.2021.106855

12. Legg S, Hutter M (2007) A collection of definitions of intelligence. Front Artif Intell Appl 157(IDSIA-07-07):17–24. https://doi.org/10.48550/arXiv.0706.3639
13. Tegmark M (2017) Life 3.0: being human in the age of artificial intelligence. Knopf
14. Surden H (2014) Machine learning and the law. Washingt Law Rev 89(1):87–115. https://digitalcommons.law.uw.edu/cgi/viewcontent.cgi?article=4799&context=wlr
15. Alpaydin E (2016) Machine learning: the new AI. MIT Press
16. Domingos P (2012) A few useful things to know about machine learning. Commun ACM 55(10):78–87. https://doi.org/10.1145/2347736.2347755
17. Malhoutra A, Ahmad F (2022) Artificial intelligence and international arbitration. Novos Estud Jurídicos 27(2)SE-Artigos:258–281. https://doi.org/10.14210/nej.v27n2.p258-281
18. Ahmed N, Amin R, Aldabbas H, Koundal D, Alouffi B, Shah T (2022) Machine learning techniques for spam detection in email and IoT platforms: analysis and research challenges. Secur Commun Netw 2022:1862888. https://doi.org/10.1155/2022/1862888
19. Lokanan ME (2023) Predicting mobile money transaction fraud using machine learning algorithms. Appl AI Lett 4(2):1–15. https://doi.org/10.1002/ail2.85
20. Chen L et al (2023) Artificial intelligence-based solutions for climate change: a review. Environ Chem Lett. https://doi.org/10.1007/s10311-023-01617-y
21. Briganti G, Scutari M, McNally RJ (2022) A tutorial on bayesian networks for psychopathology researchers. Psychol Methods. https://doi.org/10.1037/met0000479
22. Li Y et al (2022) A comprehensive review of markov random field and conditional random field approaches in pathology image analysis. Arch Comput Methods Eng 29(1):609–639. https://doi.org/10.1007/s11831-021-09591-w
23. Flach P (2012) Machine learning: the art and science of algorithms that make sense of data. Cambridge University Press. URL: https://www.programmer-books.com/wp-content/uploads/2018/08/Machine-Learning.pdf
24. Kasparov G (2017) Deep thinking: where machine intelligence ends and human creativity begins. Hachette UK, John Murray
25. Ghahramani Z (2015) Probabilistic machine learning and artificial intelligence. Nature 521(7553):452–459. https://doi.org/10.1038/nature14541
26. Sourdin T (2018) Judge v robot?: artificial intelligence and judicial decision-making. Univ N S W Law J 41(4):1114–1133. https://search.informit.org/doi/10.3316/informit.040979608613368
27. Alarie B, Niblett A, Yoon AH (2018) How artificial intelligence will affect the practice of law. Univ Tor Law J 68(supplement 1):106–124. https://doi.org/10.3138/utlj.2017-0052
28. Hacker P, Krestel R, Grundmann S, Naumann F (2020) Explainable AI under contract and tort law: legal incentives and technical challenges. Artif Intell Law 28(4):415–439. https://doi.org/10.1007/s10506-020-09260-6
29. NYCBAR (2014) Publication of international arbitration awards and decisions. Report by the committee on international commercial disputes of the association of the bar of the city of New York. https://www2.nycbar.org/pdf/report/uploads/20072645-PublicationofInternationalArbitrationAwardsandDecisions.pdf
30. Weinstein R (2017) Arbitration offers efficiency and economic benefits compared to court proceedings. New York Disput Resolut Lawyer 10(2):36–39. https://nysba.org/app/uploads/2020/03/DisputeResolutionLawyerFall17.pdf#page=36
31. Rogers CA (2018) Arbitrator intelligence: from intuition to data in arbitrator appointments. In: Sussman E, Kahn S, Kaster LA (eds) New York dispute resolution lawyer, vol 11, no 1, NYSBA New York Dispute Resolution Lawyer, pp 41–44
32. Canada B, Slate D, Slate B (2018) A data-driven exploration of arbitration as a settlement tool: does reality match perception? New York Disput Resolut Lawyer 11(1):45–48. https://nysba.org/app/uploads/2020/03/DisputeResolutionLawyerSpring18.pdf#page=35
33. Samuels D (2018) The unusual suspects—easier to find with GAR's ART. New York Disput Resolut Lawyer 11(1):49–52. https://nysba.org/app/uploads/2020/03/DisputeResolutionLawyerSpring18.pdf#page=35

34. Queen Mary School of International Arbitration Survey: Choices in International Arbitration (2010). https://arbitration.qmul.ac.uk/media/arbitration/docs/2010_InternationalArbitrationSurveyReport.pdf
35. Johnson D (2017) Find out if a robot will take your job. https://time.com/4742543/robots-jobs-machines-work/
36. Vos G (2017) Presentation to a legal business seminar in Frankfurt the future for the UK's jurisdiction and English law after Brexit. http://9bri.com/wp-content/uploads/2017/12/Sir-Geoffrey-Vos-Frankfurt-presentation-nov2017.pdf
37. Aletras N, Tsarapatsanis D, Preoţiuc-Pietro D, Lampos V (2016) Predicting judicial decisions of the European court of human rights: a natural language processing perspective. Peer J Comput Sci 1–19. https://doi.org/10.7717/peerj-cs.93
38. Moorhead R (2017) Lawyers learning about prediction. Lawyer Watch. https://lawyerwatch.wordpress.com/2017/01/19/lawyers-learning-about-prediction/
39. Benforado A (2017) Can science save justice? Judicature 101(2):24–33. https://judicialstudies.duke.edu/sites/default/files/centers/judicialstudies/judicature/may2017_benforado.pdf
40. Kahneman D (2013) Thinking, Fast and Slow 119 et seq
41. Knight W (2017) Forget killer robots—bias is the real AI danger. MIT Technol Rev. https://www.technologyreview.com/2017/10/03/241956/forget-killer-robotsbias-is-the-real-ai-danger/
42. Kuang C (2017) Can A.I. be taught to explain itself? New York Times. https://www.nytimes.com/2017/ll/21/magazine/can-ai-be-taught-toexplain-%0Aitself.html
43. Alschner W, Skougarevskiy D (2016) Mapping the universe of international investment agreements. J Int Econ Law 19(3):561–588. https://doi.org/10.1093/jiel/jgw056
44. Stix C (2021) Actionable principles for artificial intelligence policy: three pathways. Sci Eng Ethics 27(1):15. https://doi.org/10.1007/s11948-020-00277-3
45. Valle-Cruz D, Criado JI, Sandoval-Almazán R, Ruvalcaba-Gomez EA (2020) Assessing the public policy-cycle framework in the age of artificial intelligence: from agenda-setting to policy evaluation. Gov Inf Q 37(4):101509. https://doi.org/10.1016/j.giq.2020.101509

Impact of Health Information Systems on Healthcare Service Quality in Supporting Non-communicable Disease Management in Kosovo

Alba Ymerhalili, Debabrata Samanta, and Suplab Podder

Abstract The service quality in healthcare sector seeks constant improvement for ensuring the first relief from disease and taking precautionary measures. Health information system brings the revolutionary changes in the service quality in healthcare sector that may reach to all the remote areas of Kosovo with required services for general public. The research motive indicates the connectivity of emerging technology in healthcare services through information systems that facilities for effective management of hospital service quality in Kosovo. Research activities start with the identification of health information systems input and output model, collection of data through primary and secondary sources, and analysis of data using the statistical tools and techniques. The research outcomes were summarized as the challenges of non-communicable disease management with the support of existing systems and technology. The significant findings include the positive impact of health information systems on healthcare service quality in supporting non-communicable disease management in Kosovo. The present study indicates the effectiveness of health information systems for better delivery of healthcare services to connect the general public through the efficient and economic healthcare facilities for all categories of people. The principal outcomes are well connected to the future research initiatives in the area of health informatics, data analytics in the healthcare sector.

Keywords Service quality · Health information system · Health informatics · Healthcare service · Non-communicable disease · Data analytics

A. Ymerhalili · D. Samanta (✉)
Rochester Institute of Technology – RIT Global Campus, N.N., Germia Campus, 10000 Prishtina, Europe, Kosovo
e-mail: debabrata.samanta369@gmail.com

S. Podder
Postdoctoral Researcher, Rochester Institute of Technology, Prishtina, Europe, Kosovo

S. Dutta et al. (eds.), *Cyber Intelligence and Information Retrieval*, Lecture Notes in Networks and Systems 1025, https://doi.org/10.1007/978-981-97-3594-5_18

1 Introduction

The rapidly increasing prevalence of non-communicable diseases has become a grave global concern. Kosovo, a country recognized for its young population, is foreseen to face difficulties managing and combatting NCDs. Addressing and promoting the risk factors requires an inclusive analysis and understanding of the current healthcare infrastructure and the effectiveness of implementing health information systems in supporting non-communicable disease management [1]. Researching the challenges and opportunities of HIS and NCDs and the correlation between them will add valuable insights for policymakers, researchers, and healthcare professionals to develop efficient interventions and strategies [2]. Kosovo has not been able to integrate and functionalize the health information system. Without proper data collection, storage systems, and adequate sharing mechanisms, it is not possible to effectively manage non-communicable diseases. Thus, this research aims to bring to light solutions for improving information exchange and data integration in Kosovo [4]. The main principle of efficient and effective healthcare decision-making is based on reliable, timely, and trustworthy information. These principles come to hand especially when designing programs and implementing new policies in preventing and promoting the management of non-communicable diseases [3]. The main purpose behind databases is to collect and interpret data so that medical professionals are able to choose the best course of action for their patients. The health information system serves as an enabling mechanism for identifying health factors that lead to non-communicable diseases.

2 Components of Health Information Systems (HIS)

Health information system is the combination of various innovative activities that can facilitate the free flow of information among the stakeholders and better connectivity of service delivery with effective management of healthcare services [5]. Various components of health information systems can be described with the following model (Fig. 1):

Electronic Health Records (EHRs): EHRs are digital repositories that store and manage patient health information, including medical histories, diagnoses, medications, treatment plans, and lab results. They provide a comprehensive view of a patient's health.

Practice Management Systems (PMS): PMS systems focus on administrative and financial aspects of healthcare practices [7]. They include features for appointment scheduling, billing, insurance claims processing, and revenue cycle management.

Health Information Exchange (HIE): HIE systems enable the secure exchange of patient data across different healthcare organizations and systems. They support interoperability and coordinated care among providers.

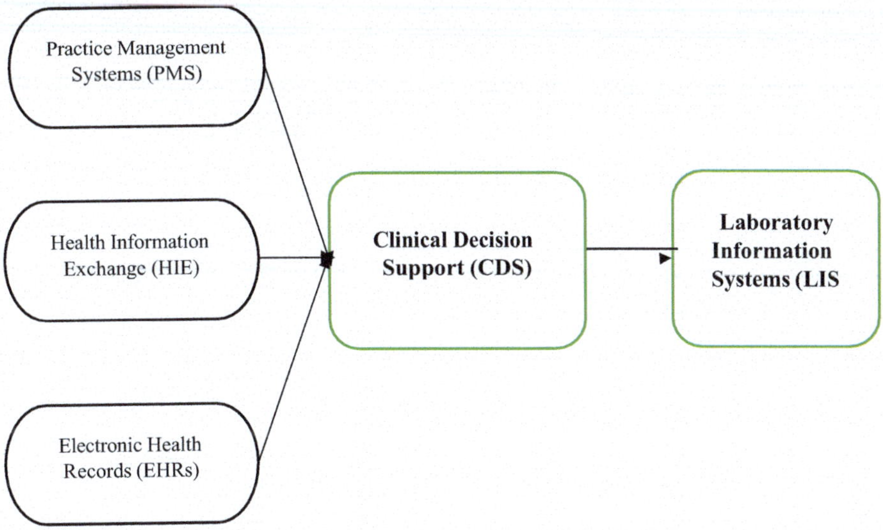

Fig. 1 Components of health information systems

Clinical Decision Support (CDS): CDS systems provide healthcare professionals with real-time clinical guidance and recommendations based on evidence-based guidelines and patient data [6]. They help improve the quality of care and reduce errors.

Laboratory Information Systems (LIS): LIS systems manage laboratory test orders, results, and data. They ensure the accuracy and reliability of lab results and support communication between laboratories and healthcare providers.

2.1 Correlations Between Input and Output

Patient registration is the first requirement to ensure that all individuals are entered into the database. Patient registration will ensure accurate demographic data in order to ensure efficient and effective care coordination (Fig. 2).

Data collection will ensure competent institutions accurate health data such as patient's history, information on patients' lifestyle, cholesterol levels, blood pressure, and blood sugar. This will enable competent institutions to create individualized assessments and care plans. Conduction of examinations will enable health professionals to accurately determine non-communicable diseases which contribute to empirical treatment decisions [8, 9]. Disease classification will be used to narrow and target interventions based on prevalence and disease. The initial data collected is updated in order for the patient to receive accurate and quality care [14]. All patients with non-communicable diseases are scheduled for follow-up appointments

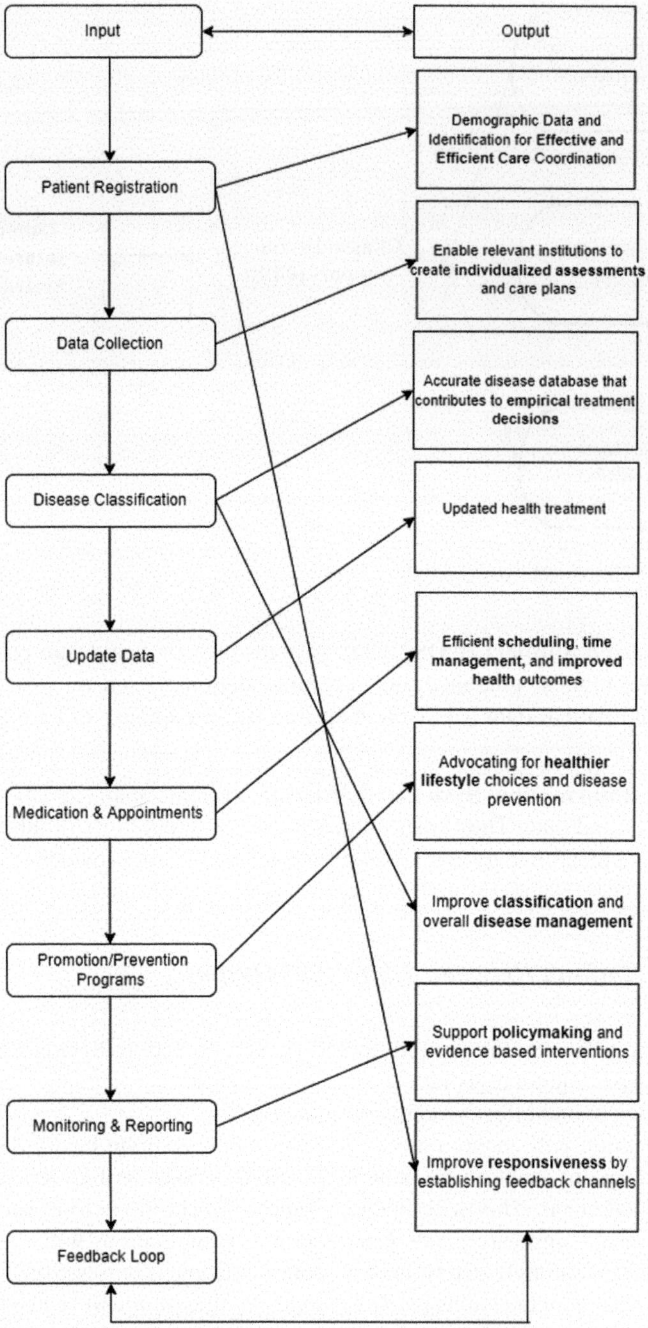

Fig. 2 Health information systems factors (input) and outcomes (output)

to ensure progress and that the medication prescribed is working [11]. With the data generated from HIS, health and policy-making institutions can easily establish targeted programs to advocate for healthier lifestyle choices and disease prevention [10]. Patients will be continuously monitored and tracked through HIS. Additionally, monitoring and surveillance will support policy-making and evidence-based interventions [12]. The feedback loop serves as a learning mechanism for health institutions to enhance data collection and analysis to improve responsiveness [13].

3 Related Efforts Toward Research

The relevant research articles were the inspiration and motivation factors to take the initiative of present research study and few articles are addressed with the major findings:

Authors in [15] described the utilization and integration of big data in the healthcare industry, with an emphasis on management, analysis, and potential. Considering this, healthcare institutions need to adopt contemporary technologies in order to enhance public health. Dash et al. (2021) indicate that the emerging technology in healthcare sectors ensure seamless data collection and interpretation that have highlighted the importance of managing and analyzing healthcare sources, such as patient/hospital reports and IoT technologies.

Authors [16] analyzed that the importance of understanding applications related to healthcare administration and the mechanisms behind them. The modern healthcare applications were established in order to categorize big data based on illness warnings and improve access to healthcare services.

Authors in [17] have shown that there is a regressive trend in unstructured data. As such, the paper highlights key components of structuring big data, like proper collection, management, interpretation, monitoring, and visualization.

International accreditation ensures the better service quality in healthcare sector that develop the relationship of customers and service. The research outcomes indicate the positive impact of latest technology for better services in hospitals.

4 Objectives of the Study

4.1 The Key Objectives Were Identified After Reviewing the Previous Research Outcomes and Analyzing the Statement of Problems Which Are Given Bellow

(i) To identify the components of health information systems (HIS) to support for effective healthcare services.

(ii) To understand the correlation between input and output for patient registration in the health information systems (HIS).
(iii) To analyze the impact of health information systems on healthcare service quality in supporting non-communicable disease management in Kosovo.
(iv) To examine the level of impact of health information systems on healthcare service quality in supporting non-communicable disease management in Kosovo.

4.2 Research Hypothesis

H0 = The impact of health information systems on healthcare service quality is not accountable in supporting non-communicable disease management.

H1 = The impact of health information systems on healthcare service quality is highly accountable in supporting non-communicable disease management.

5 Result and Discussion

5.1 Description Related to the Components of Health Information Systems (HIS) to Support for Effective Healthcare Services

Research study summarizes the expert opinion regarding the most common components of HIS that presents through the following table (Table 1):

The expert's opinion reflects the highest frequency of distribution of 28% toward health information exchange, 24% clinical decision support, 18% electronic health records, 16% practice management system, and 14% toward laboratory information systems. Figure 4 presents the impact of health information systems on healthcare service quality in supporting non-communicable disease management in Kosovo.

Table 1 Presents the expert opinion regarding the common components of health information systems

Sl. No	Components of health information systems (HIS)	Frequency of respondents (%)
1	Electronic Health Records (EHRs)	18
2	Practice Management Systems (PMS)	16
3	Health Information Exchange (HIE)	28
4	Clinical Decision Support (CDS)	24
5	Laboratory Information Systems (LIS)	14
Total		100

Figure 3 presents the expert opinion regarding the common components of health information systems.

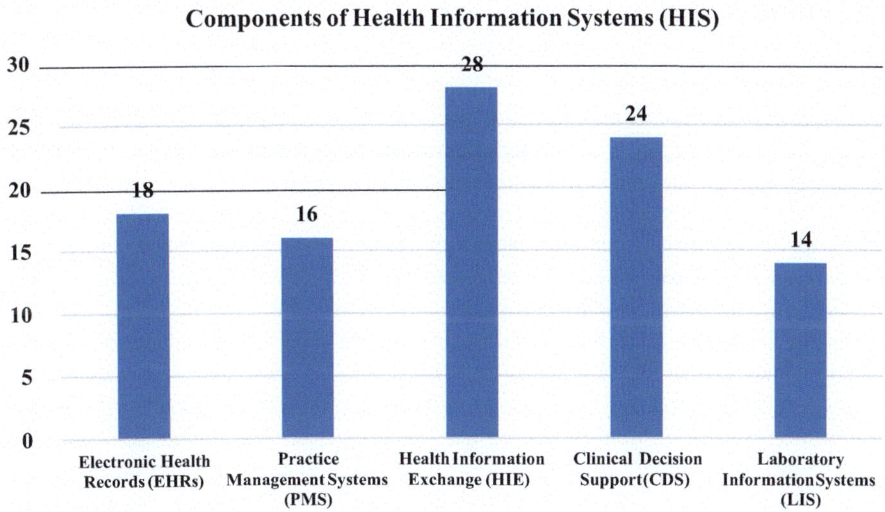

Fig. 3 Presents the expert opinion regarding the common components of health information systems

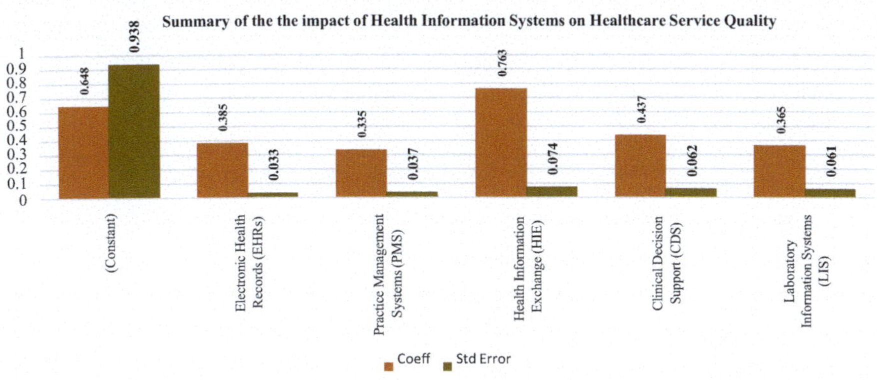

Fig. 4 Impact of health information systems on healthcare service quality in supporting non-communicable disease management in Kosovo

5.2 *The Brief Analysis Related to the Impact of Health Information Systems on Healthcare Service Quality in Supporting Non-communicable Disease Management in Kosovo*

The rapid growth of social networks, information technologies, cloud computing, and methodologies has led to an extensive increase in raw data among different disciplines. The study pinpoints the vitality of extensive research in enhancing predictions and pattern extraction. This is where machine learning plays a key role in uncovering patterns and predicting future values on existing data. The healthcare sector is one of the enterprises that benefit the most from big data applications since electronic health records play a vital role in determining services and patient health outcomes. The workflow of analyzing big data comprises four stages: the first one being sources, such as public health records, clinical or omics), followed by data warehouses and analytics (insightful and foresight) and enhanced results. These four stages combined provide healthcare professionals and institutions with credible insight into health conditions, resulting in quality patient care and outcome. The integration of electronic health records and omics presents a true potential in the revolution of health care, resulting in quality patient care, and empirical decision-making. Nevertheless, there are many challenges, especially when it comes to the distribution and allocation of healthcare data across administration, government, and hospitals.

There is a momentous level of impression of health information systems on healthcare service quality is highly accountable in supporting non-communicable disease management. Figure 5 presents the summary of the ANOVA with F-ratio and Sig.p. Table 2 shows impact of health information systems on healthcare service quality in supporting non-communicable disease management in Kosovo.

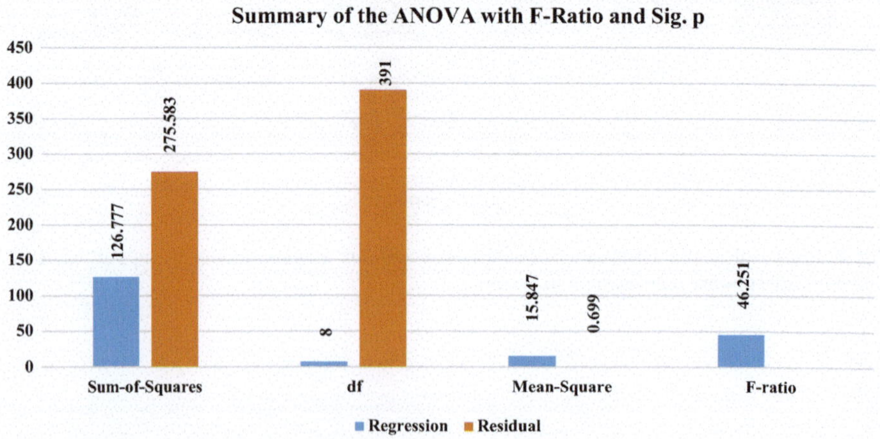

Fig. 5 Summary of the ANOVA with F-ratio and Sig.p

Table 2 Impact of health information systems on healthcare service quality in supporting non-communicable disease management in Kosovo

Summary of multiple regression analysis					
Factors influencing blended education system	Coeff	Std Error	Std Coeff	t	Sig
(Constant)	0.648	0.938		0.69	0.013
Electronic Health Records (EHRs)	0.385	0.033	0.768	8.023	0.001
Practice Management Systems (PMS)	0.335	0.037	0.501	6.124	0.002
Health Information Exchange (HIE)	0.763	0.074	0.775	10.342	0.011
Clinical Decision Support (CDS)	0.437	0.062	0.523	3.041	0.002
Laboratory Information Systems (LIS)	0.365	0.061	0.485	5.944	0.012
Significant at 0.05 level					
ANOVA					
Source	Sum-of- Squares	df	Mean square	F-ratio	Sig
Regression	126.777	8	15.847	22.671	0.014
Residual	275.583	391	0.699		

6 Summary of Findings

Summary of findings include the expert opinion and the analysis through statistical tools regarding the healthcare service quality in Kosovo. Following the summary of findings:

(i) The expert's opinion reflects the highest frequency of distribution of 28% toward health information exchange, 24% clinical decision support, 18% electronic health records, 16% practice management system, and 14% toward laboratory information systems.

(ii) Reminiscing on the vitality of the implementation of the health information system, the latest results from the Kosovo National Audit Office have shown that Kosovo is facing many difficulties

when it comes to the implementation phase. The failure to properly implement the system has been a result of several factors, such as irregularities in the management process, limited technological infrastructure, incomplete data on non-communicable disease information, limited awareness, delays in the coordination and establishment of inter-institutional monitoring, lack of human capacities, and insufficient training.

(iii) The implementation of health information systems is the smart decision for healthcare service sectors to improve the quality of treatment that ensure the better support system for non-communicable disease management in Kosovo.

(iv) There is a significant level of impact of health information systems on healthcare service quality is highly accountable in supporting non-communicable disease management.

7 Conclusion and Future Scope of the Study

In order to functionalize and enhance the effectiveness of the health information system, Kosovo should consider strategies that have been implemented by other developed countries [18]. The strategies presented will enable a seamless flow of information that will improve interoperability, resulting in an optimization of data management [19]. The opportunities for improvement rely on governance and policy initiatives, technological innovations and advancements, knowledge sharing and collaborative partnerships, training and capacity building, and data standardization [20]. The need to collect, properly use, and allocate big volumes of data has gathered a substantial interest in the field of healthcare, especially in the last two decades. This research paper aims to thoroughly explore the link between data, and the storage and analysis of such data among public and health sectors [21]. Nowadays, it is vital for both private and public sectors to extract and know how to utilize data. The analysis and interpretation of the data enable health institutions to predict and intervene, two factors that have a great impact on patient health outcomes.

References

1. Ahmed S, Tarique KM, Arif I (2017) Service quality, patient satisfaction and loyalty in the healthcare sector. Int J Health Care Qual Assur 30(5):477–488
2. Anell A (2006) The monopolistic integrated model and health care reform: the Swedish experience. Health Policy 37(1):19–33. https://doi.org/10.1016/0168-8510(96)00821-4
3. Crow R, Gage H, Hampson S, Hart J, Kimber A, Storey L et al (2012) The measurement of satisfaction with healthcare: implications for practice from a systematic review of the literature. Health Technol Assess 6(32):1–244. https://doi.org/10.3310/hta6320
4. Dandis AO, Jarrad AA, Joudeh JMM, Mukattash IL, Hassouneh AG (2022) The effect of multidimensional service quality on word of mouth in university on-campus healthcare centers. TQM J 34(4):701–727. https://doi.org/10.1108/TQM-12-2020-0295
5. Dwight-Johnson M, Lagomasino IT, Aisenberg E, Hay J (2014) Using conjoint analysis to assess depression treatment preferences among low-income Latinos. Psychiatr Serv 55(8):934–936. https://doi.org/10.1176/appi.ps.55.8.934
6. Devadasan N, Criel B, Van Damme W, Lefevre P, Manoharan S, Van der Stuyft P (2011) Community health insurance schemes & patient satisfaction-evidence from India. Indian J Med Res 133(1):40
7. Hart M (2015) Improving out-patient clinic waiting times: methodological and substantive issues. Int J Health Care Qual Assur 8(6):14–22. https://doi.org/10.1108/095268695100988 1316
8. Hardavella G, Aamli-Gaagnat A, Frille A, Saad N, Niculescu A, Powell P (2017) Top tips to deal with challenging situations: doctor–patient interactions. Breathe 13(2):129–135

9. Lee MA, Yom YH (2017) A comparative study of patients' and nurses' perceptions of the quality of nursing services, satisfaction and intent to revisit the hospital: a questionnaire survey. Int J Nurs Stud 44(4):545–55. https://doi.org/10.1016/j.ijnurstu.2006.03.006
10. Miller L, May D (2006) Patient choice in the NHS: how critical are facilities services in influencing patient choice? Facilities 24(9/10):354–364. https://doi.org/10.1108/02632770610677637
11. Merle V, Germain JM, Tavolacci MP, Brocard C, Chefson C, Cyvoct C et al (2009) Influence of infection control report cards on patients' choice of hospital: pilot survey. J Hosp Infect 71(3):263–268. https://doi.org/10.1016/j.jhin.2008.11.025
12. Mosadeghrad AM (2014) Patient choice of a hospital: implications for health policy and management. Int J Health Care Qual Assur 27(2):152–164. https://doi.org/10.1108/IJHCQA-11-2012-0119
13. Panchapakesan P, Sai LP, Rajendran C (2015) Customer satisfaction in Indian hospitals: moderators and mediators. Qual Manag J 22(1):10–29
14. Podder SK (2019) Information and communication technology in sustainable reporting towards paradigm shift in business. Int J Emerg Technol Innovat Res 240–250
15. Podder SK, Samanta D, Gurunath R (2020) Impact of business analytics for smart education system and management functions, Advances in computing and intelligent systems. Algorithms for intelligent systems. Springer, Singapore. https://doi.org/10.1007/978-981-15-0222-4_45
16. Roh CY, Lee KH, Fottler MD (2018) Determinants of hospital choice of rural hospital patients: the impact of networks, service scopes, and market competition. J Med Syst 32(4):343–53. https://doi.org/10.1007/s10916-008-9139-7
17. Rahmqvist M (2017) Patient satisfaction in relation to age, health status and other background factors: a model for comparisons of care units. Int J Qual Health Care 13(5):385–390. https://doi.org/10.1093/intqhc/13.5.385
18. Podder SK (2019) Information and communication technology in sustainable reporting towards paradigm shift in business. J Emerg Technol Innov Res 6(4):240–250
19. Podder SK, Karuppiah M, Thomas B, Samanta D (2022) Research initiative on sustainable education system: model of balancing green computing and ICT in quality education. In: 2022 interdisciplinary research in technology and management (IRTM). Kolkata, India, pp 1–5.https://doi.org/10.1109/IRTM54583.2022.9791758
20. Podder SK, Arun BK (2018) Comparison of effectiveness of employee engagement through permanent employees or outsourced employees. Int J Acad Res Dev 3(2):1406–1408, ISSN: 2455-4197
21. Samanta D, Podder SK (2019) Level of green computing-based management practices for digital revolution India. Int J Eng Adv Technol (IJEAT) 8(2):123–131

Contemporary Challenges in Metaverse: A Techno-Legal Analysis

Partho Banerjee and Gagandeep Kaur

Abstract In recent years, the concept of the metaverse has gained significant attention due to advancements in related technologies. The metaverse is envisioned as a realistic 3D virtual environment that offers direct and physical interactions, potentially breaking down societal barriers such as race, gender, and physical disabilities. It is a virtual community where avatars serve as the user's alter ego and interact with each other. The metaverse has been touted as a technology that could replace the internet in the future and become the next generation of humanity. Unlike the two-dimensional internet, the metaverse offers a three-dimensional virtual world where users can experience activities in a more immersive way, such as watching a film or a cricket match in 360° mode. Companies such as Facebook, Google, and Microsoft are investing in the metaverse, which is expected to be the next platform and medium for gaming, education, social networking, and more. Blockchain technology, cryptocurrency, and fast 5G internet are expected to play vital roles in the development of the metaverse. While the metaverse has enormous potential, it is still in its early stages of development. This field gained renewed interest with Facebook's rebranding as "Meta." This paper explores the metaverse and its nature, discussing its 3D-based virtual reality environment, the presence of avatars, and the merging of physical and virtual worlds. The paper therefore also highlights the legal challenges of the metaverse in India, including intellectual property rights, data privacy, governance, online harassment, financial regulations, jurisdictional issues, and consumer protection. It emphasizes the need for collaboration between technology and legal experts and suggests recommendations for addressing these challenges. The metaverse represents a significant opportunity for innovation and transformation across industries; however it requires careful consideration of the legal implications to ensure its responsible and inclusive development.

P. Banerjee (✉) · G. Kaur
Department of Law, University of Petroleum and Energy Studies, Dehradun, India
e-mail: partho.banerjee6@gmail.com; 500070272@stu.upes.ac.in

G. Kaur
e-mail: gkaur@ddn.upes.ac.in

S. Dutta et al. (eds.), *Cyber Intelligence and Information Retrieval*, Lecture Notes in Networks and Systems 1025, https://doi.org/10.1007/978-981-97-3594-5_19

Keywords Metaverse · Privacy · Artificial intelligence · Avatar · Blockchain technology · Legal regulation

1 Introduction

The COVID-19 pandemic has acted as a catalyst for innovation and transformation across various facets of our lives [1]. One such transformative concept that has gained significant attention is the metaverse. This term, initially coined by Neil Stephenson in his novel "Snow Crash," represents a 3D-based virtual reality where individuals engage in their daily activities and economic endeavors through digital avatars [2]. What makes the metaverse particularly intriguing is its potential to merge the physical and virtual realms seamlessly, transcending the boundaries of our current digital experiences. It is not beholden to the control of any single entity, making it a decentralized and open platform for exploration and interaction. In this context, our research seeks to address a central question: How can the metaverse be effectively developed, harnessed, and integrated into our society while mitigating the associated challenges and risks?

The proposed study makes several contributions to our understanding of the metaverse concept and its practical implications:

- **Comprehensive Examination**: We will conduct an in-depth analysis of the metaverse concept, tracing its origins, and examining the technological advancements that have made its development more feasible [3].
- **Identification of Benefits and Concerns**: Our research will elucidate the potential benefits of the metaverse, ranging from enhanced socialization and entertainment to transformative opportunities in commerce, work, and education. Simultaneously, we will explore and address the concerns related to privacy, security, and ownership, crucial for its successful implementation.
- **Assessment of Current State:** We will provide an up-to-date assessment of the metaverse's current state, highlighting platforms and technologies that offer virtual environments where users can interact through digital avatars [4].
- **Impact on Industries:** The study will shed light on how the metaverse has the potential to revolutionize various industries, including gaming, social media, education, and e-commerce [5], offering new avenues for interaction and innovation.
- **Legal Challenges and Mitigations**: We will delve into the challenges and risks associated with the metaverse, such as ensuring user safety, creating seamless environments, and addressing cyberbullying and potential mental health concerns. Additionally, we will propose mitigation strategies for these challenges.

By addressing these aspects, our research aims to provide a comprehensive understanding of the metaverse's potential and pitfalls, offering valuable insights for policymakers, businesses, and individuals as they navigate this emerging digital frontier [6]. As technology continues to evolve, the metaverse stands as a promising yet complex terrain, and our study endeavors to illuminate the path forward.

2 Review of Literature

This article provides an in-depth analysis of the legal challenges and policy implications of metaverse. It covers various issues such as intellectual property rights, data privacy, governance, and content management [7]. The authors emphasize the need for a new code system and propose possible ways to deal with the complex code of the metaverse. This paper highlights challenges and ethical considerations that arise in the metaverse. It explores the legal aspects of user rights, platform responsibilities, and resources [8]. The authors discuss the importance of effective governance mechanisms, including self-regulation, government intervention, and international cooperation, in dealing with legal issues related to the metaverse issue [9]. This conference paper focuses on virtual property rights in the metaverse. It provides a comprehensive legal review of ownership rights and challenges associated with real property. The authors explore the role of blockchain technology and smart contracts in secure and transparent virtual asset transactions [10]. In this article, the authors address important challenges related to data privacy and security in the metaverse. Topics such as user data collection, consent, retention, and creating security in virtual environments are covered. The paper also offers possible solutions including privacy enhancements and user-centric approaches to data management [11]. This research paper examines virtual worlds and their economies, discussing the emergence of virtual currencies and the concept of the metaverse [12]. Although not specifically focused on the metaverse, this paper introduces the concept of augmented reality and its position on the reality-virtuality continuum, which is relevant to understanding the development of the metaverse [13]. This paper explores the development of virtual reality in the context of the metaverse era, discussing its technological advancements, applications, and potential impacts [14]. While not focused solely on the metaverse, this paper provides insights into the functional building blocks of social media platforms, which are relevant to understanding the social aspects of the metaverse [15]. This book explores the concept of avatars and virtual worlds, providing a comprehensive introduction to virtual reality and its potential for building immersive environments in the metaverse [16]. This comprehensive book covers multimedia systems, including virtual reality and its applications, providing a foundational understanding of technologies relevant to the metaverse [17]. Written by the founder of Second Life, this book offers insights into the development and potential of virtual worlds, which are precursors to the concept of the metaverse [18].

It is observed that the concept of metaverse has not been discussed from both techno-legal perspectives. This paper provides a balanced approach in exploring both the technical as well as legal regulatory perspective of metaverse.

3 Metaverse and its Nature

The metaverse is a term used to describe a virtual cosmos or communal digital area in which individuals may connect, work, and socialize in three dimensions. It is a virtual reality that is immersive and linked, extending beyond the bounds of separate platforms and allowing for seamless interaction between the actual and virtual worlds [19]. Users may build and customize their digital avatars in the metaverse, explore virtual environments, participate in various activities, and interact with other users in real time.

The metaverse is a network of interconnected platforms, games, and virtual experiences that is not restricted to a particular virtual reality platform or application. By combining virtual reality, augmented reality, and the internet, it hopes to deliver a coherent and immersive experience [20]. Its nature encompasses the characteristics and fundamental aspects that define this virtual reality concept.

Firstly, the metaverse is a 3D-based virtual reality environment. It goes beyond traditional 2D interfaces and offers a more immersive and interactive experience [20]. Users navigate and explore virtual worlds, which can range from realistic simulations of physical environments to fantastical and imaginative realms [21].

The metaverse is a place where daily activities and economic interactions occur. Users can engage in a wide range of experiences, including socializing, attending virtual events, creating and selling virtual goods, participating in games and entertainment, collaborating on projects, and much more. It offers a multifaceted environment that merges aspects of work, leisure, communication, and creativity.

Another essential aspect of the metaverse is its role as an ecosystem that bridges the physical and virtual worlds. It creates a continuum between the two realms, allowing for seamless interactions and integration [22]. The metaverse is not limited to a single platform or company; it represents a new dimension of existence that transcends the control of any individual entity. Instead, it evolves through the collective contributions and activities of its users. Therefore, metaverse presents opportunities for socialization and community building. Users can connect with others, form relationships, join communities, and engage in shared experiences [23]. The metaverse fosters a sense of belonging and allows for the formation of virtual societies and cultures.

4 Legal Regulatory mechanism of metaverse in India

At present data protection system is governed by the Information Technology Act, 2000 as amended in 2008 and a set of guidelines called the Information Technology Rules, 2011, which fall under the Information Technology Act, 2000 [24]. The Information Technology Rules, 2011 in India offer a framework for facts protection and privacy (Fig. 1). These rules embody guidelines that entities and individuals coping with non-public data have to adhere to. One essential factor is the handling of touchy non-public data or facts (SPDI), which includes information associated with passwords, economic records, fitness conditions, and greater [25]. Consent is a key requirement underneath those rules, mandating that entities acquire knowledgeable and voluntary consent from people earlier than collecting and the use of their non-public information. Additionally, the policies emphasize the implementation of protection practices and strategies to protect personal facts from unauthorized get right of entry to or disclosure. They additionally deal with the switch of personal records outdoor of India, requiring entities to ensure similar stages of records safety or attain consent from people. Grievance redressal mechanisms and provisions for statistics retention and deletion are also outlined in these suggestions [26]. It is crucial for companies and individuals handling personal data to be aware of these recommendations and ensure compliance to guard individuals' privateness and data.

The **Information Technology (Intermediary Guidelines and Digital Media Ethics Code) Rules, 2021** had been delivered in India in February 2021. These policies' purpose to regulate the functioning of social media intermediaries and virtual media platforms to ensure responsibility, transparency, and safety of user rights. While those policies in the main recognition on content material moderation

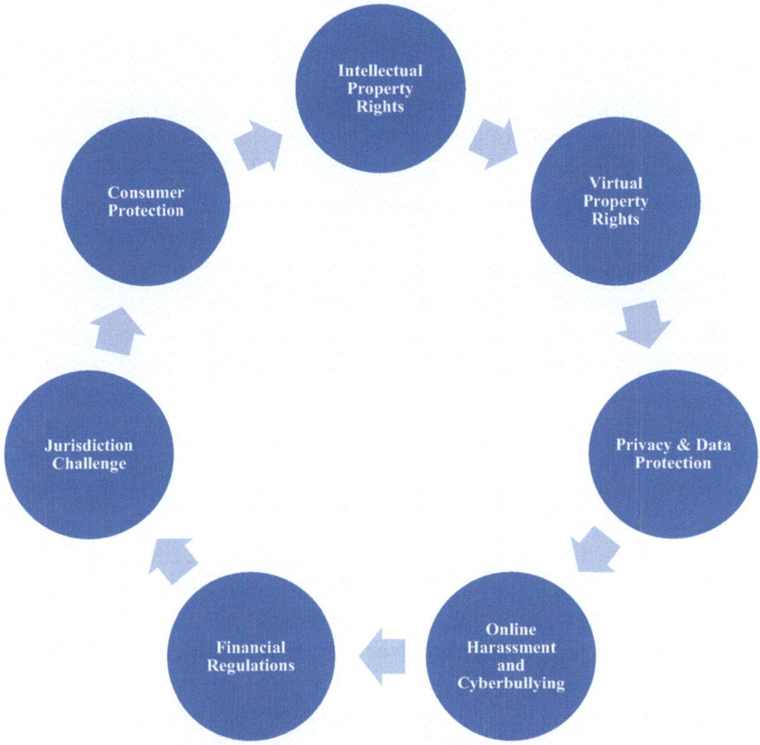

Fig. 1 A framework for facts protection and privacy

and intermediary legal responsibility, in addition they include certain provisions related to facts protection.

Under ***Section 69*** [27], the government can issue an order to any agency or authority for the purpose of national security or maintaining public order. This order may require intermediaries, service providers, or any person in control of a computer resource to assist in the interception, monitoring, or decryption of information.

The intermediaries or service providers who receive such orders are obligated to comply and provide assistance to the government [28]. Failure to comply with the order can result in penalties and imprisonment. While the legal panorama is evolving, here are some key legal challenges associated with the metaverse:

- ***Intellectual Property Rights***: Protecting high-value property in the metaverse may be complex. Issues may additionally stand-up concerning copyright [29], emblems, and patents associated with virtual property, avatars, virtual worlds, and digital creations. Determining ownership and implementing those rights in virtual surroundings poses big demanding situations.
- ***Virtual Property Rights***: The concept of digital belongings and ownership rights over virtual assets is a grey location [30]. Establishing clear felony frameworks for virtual property and addressing issues of ownership, transferability, and enforceability is crucial.
- ***Privacy and Data Protection***: The metaverse entails the gathering and processing of full-size quantities of user statistics [31]. Ensuring privacy and information protection for customers' private facts inside the metaverse is crucial. Laws and rules want to be advanced to address records privacy, consent, and protection in virtual environments.
- ***Online Harassment and Cyberbullying***: The metaverse may be prone to instances of online harassment, cyberbullying, and other kinds of harmful conduct. Developing felony frameworks to deal with and save such behavior inside the metaverse is important to create a safe and inclusive virtual space.
- ***Financial Regulations:*** The use of virtual currencies and in-game economies inside the metaverse increases questions on monetary rules [32]. Establishing pointers and rules for virtual currencies, inclusive of cryptocurrencies, to prevent fraud, cash laundering, and illicit activities is essential.
- ***Jurisdictional Challenges:*** The metaverse transcends geographical barriers, leading to demanding situations in figuring out jurisdiction and applicable legal guidelines in cross-border transactions and disputes [33]. Clarifying the prison framework and resolving conflicts in instances related to a couple of jurisdictions is vital
- ***Consumer Protection:*** Ensuring consumer protection within the metaverse is critical. Issues related to truthful practices, transparency, dispute resolution mechanisms, and safeguards against scams or fraudulent activities want to be addressed [34].

- ***Cybersecurity and Hacking:*** The interconnected nature of the metaverse exposes customers to cybersecurity risks and potential hacking threats. Robust cybersecurity measures and prison frameworks are essential to protect customers' non-public data and prevent cybercrime [35].

5 Conclusion

In a nutshell, the metaverse notion symbolizes a dynamic and transformative frontier that has caught the imagination of individuals, corporations, and industries alike. Our research has provided vital insights into its origins, prospective benefits, and plethora of problems. As we conclude our study, we recognize the significance of addressing both the metaverse's potential and dangers. Moving forward, various possibilities for future research in this topic exist:

- **Technical Advancements**: It is critical to continue exploring technological advancements. Future research can go deeper into metaverse technology development, increasing user experiences, and overcoming technological limits.
- **Privacy and Security**: It is critical to conduct research aimed at improving privacy and security in the metaverse. To safeguard users' personal data and protect them from potential risks, new solutions and procedures will be required.
- **Economic ramifications**: More research into the metaverse's economic ramifications, such as its impact on employment markets, revenue creation, and emerging business models, is required. Understanding the psychological and societal implications of long-term metaverse interactions will be critical. Critical insights will be gained via research into the effects on mental health and social behaviors.

Limitations

We acknowledge several limitations in our study, which can be categorized into practical and theoretical aspects:

Practical Limitations:

- **Analysis of Platforms**: The metaverse environment is wide, with multiple platforms and technologies. Our research may not have covered every platform, and specific subtleties of individual platforms may not have been thoroughly investigated.

Ethical Considerations:

While we examined privacy and security problems, a complete ethical analysis of the metaverse's implications was beyond the scope of this study. Future studies should explore deeper into ethical issues.

Theoretical Limitations:

- **Predictive Nature**: Based on available knowledge, our study primarily provides an appraisal of the metaverse's current status and potential. Predicting the metaverse's exact future trajectory remains speculative.
- **Generalization**: Because the metaverse is a diverse and growing concept, our findings may not be applicable to all metaverse implementations.
- **Interdisciplinary Nature**: The metaverse notion is interdisciplinary in nature, encompassing technology, psychology, economics, and society. Our research may not have adequately addressed the multidisciplinary complexity of this new topic.

Despite these limits, our research provides a framework for comprehending the intricacies, potentials, and legal challenges of the metaverse. We predict that as this digital frontier evolves, further studies will expand on these findings, leading to a more thorough knowledge of the metaverse's impact on society and individuals. The metaverse offers a unique potential as well as a slew of obstacles, and ongoing study will be critical in navigating this intriguing and complex virtual world.

References

1. Ravenscraft, The Metaverse Land: an Illusion. Wired. https://www.wired.com/story/metaverse-landillusion/
2. Tidy (2022) Billions spent in metaverse land grab in BBC news. https://www.bbc.com/news/technology-6454759
3. Ball L (2022) The metaverse: how it will revolutionize. Liverleft
4. Avital M, King JL, Beck R (2016) Rossi jumping on blockchain: lessons of the past and the future. In: Proceedings of 37th international conference on information systems held in Dublin, 11–14 Dec 2016, pp 1–6
5. Brenner SW (2020) The virtual weapon and international order. Yale University Press
6. Huang M, Liu D, Xiao B (2021) The rise of the metaverse: a survey on applications and implications. IEEE Access 9:91295–91307
7. PwC (2021) Navigating the metaverse: a market scan PwC
8. Damer B, de la Rosa A (2021) The origin and future of the metaverse. Computer 54(5):33–39
9. Narayanan A, Farooq B, Ren X (2022) A comprehensive survey of metaverse architectures: design, implementation, and challenges. IEEE Access 10:19062–19087
10. Reitman R (2021) The metaverse: what it is, where to find it, who will build it, and fortnite. Yale J Law Technol 23(1):1–65
11. Wozniak A, Wozniak K (2021) The metaverse: a paradigm shift in society. Eur J Sci Theol 17(4):231–241
12. Ahn J, Park J, Ahn M (2021) The metaverse: an overview and research directions. Int J Hum-Comput Interact 37(19):1809–1822
13. Gandal N, Halaburda H (2022) The economics of the metaverse. J Econ Perspect 36(4):87–110
14. Virtual World Society (2021) The metaverse: what it is and why it matters. Retrieved from https://virtualworldsociety.org/wp-content/uploads/2021/12/The-Metaverse-Report.pdf
15. Hessen A, Roehrich G (2021) The metaverse: a new virtual reality infrastructure with vast economic potential. SSRN Electron J
16. Bragg T, Griswold WG (2022) MetaVerse: an early architectural framework for a user-centric internet. arXiv:2201.03554

17. Hsu J, Lu S (2022) The future of entertainment: the metaverse as a new ecosystem. Int J Environ Res Public Health 19(2):439
18. Castronova E (2022) Virtual worlds and the metaverse: the impact of virtual worlds on real economies. J Virtual Worlds Res 15(1)
19. Vachhani S, Shah S, Agarwal R (2022) Metaverse and its potential in education. In: Proceedings of the 2022 international conference on advanced information technology, services and systems. ACM, pp 55–61
20. Gupta A, Chaturvedi R (2021) Legal challenges in the metaverse: an Indian perspective. J Intellect Prop Stud 4(2):113–126
21. Sharma K (2022) Metaverse and legal challenges in India: an analysis. Int J Innov Res Adv Stud 9(4):146–151
22. Law Street India (2021) The metaverse and legal issues in India. Retrieved from https://lawstreetindia.com/columns/legal-issues/metaverse-legal-issues-in-india/
23. Singh S (2021) Legal framework for the metaverse in India. NUJS Law Rev 221–246
24. Roy S, Pal S (2022) Regulatory and legal challenges of the metaverse in India. Int J Adv Res Comput Sci 13(4):1–5
25. Chakraborty A, Pandey N (2021) Metaverse and legal challenges in India. Indian J Law Technol 17(8):142–154
26. Madhvani K, Metaverse: legal and regulatory challenges in India. J Virtual Worlds Res 14(2)
27. Narsey R (2022) Legal and regulatory implications of the metaverse in India. J Intellect Prop Rights 27(1):40–47
28. Bhattacharya D (2022) Legal and regulatory challenges in the Indian Metaverse: an analysis. J Leg Stud 2(1):56–72
29. Mishra R (2021) Legal challenges of the metaverse in India: an overview. Int J Leg Stud Res 1(1):34–46
30. Thomas R, Bhattacharya K (2021) Legal and regulatory challenges of the metaverse in India. Indian J Law Technol 17(1):12–24
31. Barik RK (2022) Metaverse and legal issues in India: a study. J Cyber Law Inf Secur 3(1):85–99
32. Bansal V (2022) The metaverse: legal challenges and policy considerations in India. Indian J Intellect Prop Law 14(1):79–98
33. Chakraborty S, Mishra P (2021) Legal challenges in the metaverse: an Indian perspective. Indian J Law Technol 17(2):123–141
34. Singh D (2022) Regulating the metaverse: an Indian legal perspective. J Intellect Prop Rights Allied Sci 4(1):55–72
35. Mehta S (2021) Metaverse and legal challenges: an Indian outlook. Leg Asp Inf Technol 1(2):25–42

Artificial Intelligence in Disease Diagnostics: Rethinking Risk Factors for Cervical Cancer

Ambuj Kumar Agarwal, Naveen Tiwari, Soumi Dutta, Raj Gaurang Tiwari, and Naresh Kumar Trivedi

Abstract As machine learning evolves, many individuals and businesses utilize numerous algorithms to evaluate massive datasets and create actionable insights that aid in predicting behavior. And this type of technology is increasingly employed in the medical industry to forecast the early stages of certain severe diseases, such as cervical cancer. There has been a significant amount of research conducted on cervical cancer in recent years. Studies have focused on various topics such as risk factors for cervical cancer, early detection and screening, and the effectiveness of different treatment options. In this study, we conduct an in-depth comparison of the various machine learning methods, discussing their relative merits and shortcomings in terms of accuracy and overall performance. Staking which is an ensemble machine learning approach emerges as the best approach for cervical cancer classification.

Keywords Classification · Cervical cancer · Machine learning · Neural network · Stacking

---

A. K. Agarwal
Department of Computer Science and Engineering, School of Engineering and Technology, Sharda University, Greater Noida, India
e-mail: ambuj.agarwal@sharda.ac.in

N. Tiwari
School of Computing, Graphic Era Hill University, Bhimtal Campus, Uttrakhand, India

S. Dutta
Department of Computer Science, Sister Nivedita University, Kolkata, West Bengal, India

R. G. Tiwari (✉) · N. K. Trivedi
Chitkara University Institute of Engineering and Technology, Chitkara University, Punjab, India
e-mail: rajgaurang@chitkara.edu.in

N. K. Trivedi
e-mail: nareshk.trivedi@chitkara.edu.in

S. Dutta et al. (eds.), *Cyber Intelligence and Information Retrieval*, Lecture Notes in Networks and Systems 1025, https://doi.org/10.1007/978-981-97-3594-5_20

1 Introduction

One of the illnesses that are becoming more of a problem for women is cervical cancer. Among women, the rate of malignant tumors is greatest in developing countries and third highest in industrialized countries. Nearly a third (28.8%) of the world's cervical cancer cases are found in China each year, where 140,000 new cases are identified annually. In China, it is the most common malignancy of the female reproductive system [1]. When it first begins showing itself, it might be difficult to identify. Cervical cancer cannot be detected in any other way than via routine screenings. Magnetic resonance [2] and diffusion-weighted imaging [3] are ineffective in detecting the disease in its early stages. The human papillomavirus (HPV), having your first sexual encounter at a young age, smoking, and so on are all factors that increase your chance of developing cervical cancer. The human papillomavirus (HPV) is the single greatest cause of cervical cancer [4]. Researchers have shown that in comparison to nonsmokers, those who are infected with HPV and also smoke have a threefold increased risk of developing cervical cancer [5]. These days, early illness detection makes use of a wide range of detection methods, some of which are based on machine learning.

Vapnik [6] developed SVM, now one of the most used classification methods, based on the idea of using a partition hyperplane to split a sample space into subspaces corresponding to different classes. For its intended purposes, it has been successfully applied in a wide variety of fields, including image recognition, text classification, data forecasting, etc. [7].

Decision trees are widely used in many fields for data classification, including business, health care, and education. Samples are primarily categorized according to a division of characteristics, which is its defining feature. Many decision tree-based ensemble methods, such as Random Forests and XGBoost, employ decision trees as their core classifier, reaping the benefits of both their superior accuracy and generalizability. XGBoost is a powerful and efficient approach for designing classifiers.

When it comes to labeling large datasets with many attributes, the Random Forest method is the gold standard. It employs a tree-structured method of analysis for making categorizations and predicting outcomes. This research compares the risk features of the dataset from the UCI repository using three machine learning-based approaches (XGBoost, SVM, and Random Forest) [8].

Five distinct parts make up the framework of this study. Methodologies employed in data analysis are outlined in Section III. In the IV section, we assess the features and quality of the cervical cancer dataset. The results and effective strategies are also discussed in this section. In this final portion, the study reaches the conclusion.

2 Related Works

Studies have been conducted to determine what variables, such as human papillomavirus (HPV) infection, smoking, and a compromised immune system, increase a person's chance of developing cervical cancer. The early identification and screening of cervical cancer, using tools like the Pap test and HPV testing, are another field of study [9, 10].

Surgery, radiation therapy, and chemotherapy are only some of the treatment options for cervical cancer, and their efficacy has been the subject of many studies. Research on immunotherapy and targeted therapy for the treatment of cervical cancer has also been performed.

In addition, there has been an increasing focus on the development of new technologies, such as liquid-based cytology, to improve cervical cancer screening and to develop new biomarkers to detect cervical cancer early [11, 12].

A classifier for cervical cancer was coupled with the revised segmentation using machine learning methods; Random Forests were shown to be the most effective [14]. To handle the plethora of images that are produced by the recovered objects, unsupervised learning and robust refinement approaches are used, such as AdaBoost detectors [15], SVM supports [16], and Gaussian mixture models [17]. It was developed and implemented a new method of cell segmentation based on the Markov random field that makes use of superpixels to distinguish between cells that do not overlap. To identify cervical cells, the authors use the multifilter SVM using the settings described in [19]. Artificial neural networks (ANNs) were developed, evaluated, and found to classify cervical cells with 78% accuracy [20]. An unsupervised method [18, 21] was used to evaluate the quality of the medical evidence for each subtype of cervical cancer. When compared to other simple classification models, PSO with KNN membership values fared better [22]. Cervical cancer cells are segmented and classified using a Gabor-based technique based on their form and texture. The percentage of correctly identified cancerous cells rose to 89% [23]. One of the recommended model's reference components, the least-square support vector machine (LSSVM), was used to categorize the CNN features, with even better results. Comparatively, RBF-based support vector machines (SVMs) outperformed logistic regression and Random Forest methods [25]. Accuracy based on features was predicted to be between 90 and 95%.

Overall, research on cervical cancer is ongoing and continues to focus on improving early detection, developing new treatment options, and increasing understanding of the disease.

3 Matcrial and Method

3.1 Dataset

Cervical cancer dataset [13] has been used in this study.

3.2 Model Diagram

The proposed method is shown in a model diagram (Fig. 1). The picture illustrates how the research for the model developed.

3.3 Brief Description of Algorithms Used

3.3.1 k-Nearest Neighbors (k-NNs)

In the realm of machine learning, it is a well-liked approach for solving classification and regression issues. In order to make predictions, the k-NN algorithm looks for the k-nearest training instances in the feature space. The k-NN algorithm uses a distance metric, such as Euclidean distance, to measure the similarity between the input data and the training examples. Since no assumptions about the data distribution are made by the k-NN technique, it falls under the category of non-parametric algorithms. It is simple to implement and does not require a lot of computational resources, making it a popular choice for many applications.

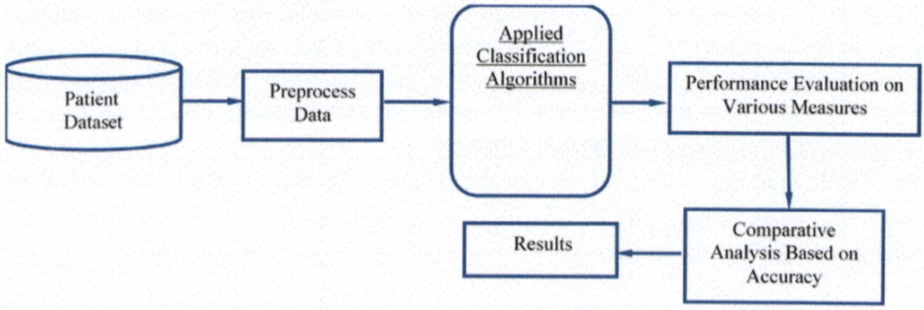

Fig. 1 Proposed model diagram

3.3.2 Neural Networks

Machine learning algorithms based on the human brain's structure and function are known as neural networks. Known as artificial neurons, its structure consists of stacked clusters of nodes that work together to analyze and send data. Before reaching the output layer, the input is altered and processed by numerous layers of these artificial neurons. The weights of the connections connecting the neurons are adjusted during training to bring the gap between the expected and actual outputs closer together. Neural networks are used extensively in many disciplines for things like voice recognition, language processing, and picture recognition. They are considered powerful algorithms, as they can learn to model highly nonlinear and complex relationships between inputs and outputs [24, 26].

3.3.3 Stochastic Gradient Descent (SGD)

It is an optimization algorithm used for training machine learning models, particularly for linear models such as linear regression and neural networks. The algorithm starts with a random set of weights and iteratively updates the weights by adjusting them in the direction of the negative gradient of the loss function. In each iteration, the algorithm uses a small set of training examples, called a batch or mini-batch, to calculate the gradient, instead of using the whole dataset. This makes the algorithm computationally efficient and able to handle large datasets. The algorithm is sensitive to the learning rate, which controls the step size of the update, and the choice of a suitable learning rate is critical for the performance of the algorithm. SGD is a widely used optimization algorithm as it is simple to implement and can be used to train large datasets on distributed computing architectures.

3.3.4 Stacking

Combining numerous models into one is called "stacking," and it is a common machine learning strategy for boosting model accuracy. With stacking, you take the predictions from many base models and feed them into a single meta-model as features. The final prediction is made by the meta-model using the characteristics given by the basic models as input.

There are two main types of stacking: homogeneous and heterogeneous. In homogeneous stacking, all the base models are of the same type, while in heterogeneous stacking, the base models are of different types. Stacking's primary value is that it may boost model performance by using the best features of many models while mitigating the worst. Additionally, stacking might enhance the model's interpretability by revealing which base models are most influential in producing the final forecast inside the meta-model. The performance of machine learning models in a wide range of applications, including image classification, natural language processing, and time series forecasting, may be improved using stacking, a strong ensemble approach.

3.4 K-fold Cross-Validation

It is a common way to gauge a model's efficacy in the field of machine learning. The procedure entails partitioning the data into k "folds" of equal size, with k–1 of the folds serving as training data and the remaining fold serving as test data. After collecting data from the k-test sets, the model's efficacy is determined by averaging the performance measure.

Since all the data are used for both training and testing, k-fold cross-validation yields a more accurate estimate of the model's performance. It may be useful for avoiding overfitting, which happens when a model is perfected on its training data but fails miserably when presented with novel data.

The size of the dataset is the primary factor in deciding what value of k to use. The model's performance can be estimated more accurately with a bigger k number, but this will need more time to train the model than with a lower k value. In machine learning, k-fold cross-validation is used for model comparisons, model evaluations, and hyperparameter optimization. You may apply this technique in both supervised and unsupervised and linear and nonlinear models to gain a more precise idea of how well they will perform [11].

3.5 Performance Evaluation Metrics

Predictions made by a machine learning model may be evaluated using these. The challenge at hand and the data at hand should guide your selection of a performance measure. There are a variety of indicators used to assess employee performance.

- Precision refers to how well the model fits the data used to make a forecast.
- Accuracy is the fraction of times a model correctly predicts a value out of a set of values. This is the fraction of cases when the model correctly predicted a good outcome (sensitivity).
- F1-score strikes a fine chord between accuracy and memorability.
- The area under the receiver operating characteristic curve (ROC-AUC) is a metric for evaluating a binary classifier's ability to separate positive and negative data.

4 Result and Discussion

This article investigates three different approaches to machine learning and makes use of ensemble learning in order to make a prediction about cervical cancer. The five characteristics that were used are outlined in the study's methodology and materials. A total of 1015 samples were collected after absenteeism and incidence statistics were examined independently. With 20-K-fold cross-validation, 20 different datasets were employed for testing and training.

Following are the five factors that conclude cervical cancer.

Dyskeratotic refers to cells or tissue that has abnormal, irregular, or degenerative keratinization. In other words, it is a term used to describe cells that have abnormal changes in the process of forming the protein keratin, which is a key component of the outer layer of skin and other structures in the body. Dyskeratotic changes can be seen in a variety of diseases and conditions and can be an indication of cell dysfunction or dysfunction of the tissue.

Koilocytotic refers to cells that have a characteristic appearance known as "koilocytes" when viewed under a microscope. Koilocytes are cervical cells that have a characteristic "halo" appearance, which is caused by changes in the cell's cytoplasm and nucleus. They are often seen in cervical infections caused by certain types of human papillomavirus (HPV) and are considered a sign of cervical dysplasia, which is a precancerous condition. Koilocytosis can also be caused by other factors such as irritation or trauma.

Metaplastic refers to a type of cell or tissue transformation in which one cell type is replaced by another. This process is also known as transdifferentiation. For example, metaplasia is the transformation of one type of epithelial cell into another type, such as squamous cells becoming glandular cells. Metaplasia occurs as a response to certain types of injury or chronic irritation and can be seen in different organs and tissues. In some cases, metaplasia can be a precursor to malignancy, but not always. It is important to note that metaplasia is a physiological process that can occur under normal circumstances as well as pathological conditions.

Parabasal cells are a type of cells that are found in the epithelium, which is the lining of certain organs and tissues in the body. These cells are located at the basal layer of the epithelium, which is the layer closest to the underlying tissue. They are typically larger and more irregular in shape than the more mature cells that are found in the upper layers of the epithelium. Parabasal cells can be seen in a variety of conditions, such as cervical dysplasia (precancerous changes in the cervix) and vaginal atrophy (thinning of the vaginal walls). They may also be seen in biopsy specimens taken from the cervix or vagina because of infection, inflammation, or other types of injury. High numbers of parabasal cells may indicate a need for further evaluation or treatment.

A superficial-intermediate cell is a type of cell found in the epidermis, the outermost layer of the skin. Superficial cells are in the uppermost layer of the epidermis and are responsible for forming the protective barrier of the skin. Intermediate cells are in the middle layers of the epidermis and play a role in the formation of new skin cells. Both superficial and intermediate cells are constantly shed and replaced by new cells produced in the lower layers of the epidermis [13].

Figures 2, 3, 4, 5 display confusion matrices for the k-nearest neighbor (k-NN), neural network, stochastic gradient descent, and stacking models, respectively. Here, dyskeratotic, koilocytotic, metaplastic, parabasal, and superficial-intermediate target classes have been abbreviated as D, K, M, P, and S, respectively.

Actual \ Predicted	D	K	M	P	S	∑
D	242	2	1	4	2	**251**
K	36	156	20	7	6	**225**
M	1	6	157	5	0	**169**
P	0	0	0	163	0	**163**
S	2	0	0	0	205	**207**
∑	**281**	**164**	**178**	**179**	**213**	**1015**

Fig. 2 k-nearest neighbor

Actual \ Predicted	D	K	M	P	S	∑
D	244	6	1	0	0	**251**
K	18	192	11	0	4	**225**
M	0	6	162	1	0	**169**
P	1	0	0	161	1	**163**
S	0	0	0	0	207	**207**
∑	**263**	**204**	**174**	**162**	**212**	**1015**

Fig. 3 Neural network

Actual \ Predicted	D	K	M	P	S	∑
D	242	7	1	1	0	**251**
K	15	183	15	5	7	**225**
M	0	7	159	3	0	**169**
P	0	0	0	163	0	**163**
S	0	0	0	0	207	**207**
∑	**257**	**197**	**175**	**172**	**214**	**1015**

Fig. 4 Stochastic gradient descent

Actual \ Predicted	D	K	M	P	S	∑
D	243	7	1	0	0	**251**
K	14	197	9	1	4	**225**
M	0	6	162	1	0	**169**
P	0	0	0	163	0	**163**
S	0	0	0	0	207	**207**
∑	**257**	**210**	**172**	**165**	**211**	**1015**

Fig. 5 Stack

Table 1 Comparison of different machine learning algorithms

Model	AUC	CA	F1	Precision	Recall
k-NN	0.988	0.909	0.905	0.913	0.909
Stack	0.994	0.957	0.957	0.957	0.957
SGD	0.962	0.939	0.938	0.939	0.939
Neural network	0.996	0.952	0.951	0.952	0.952

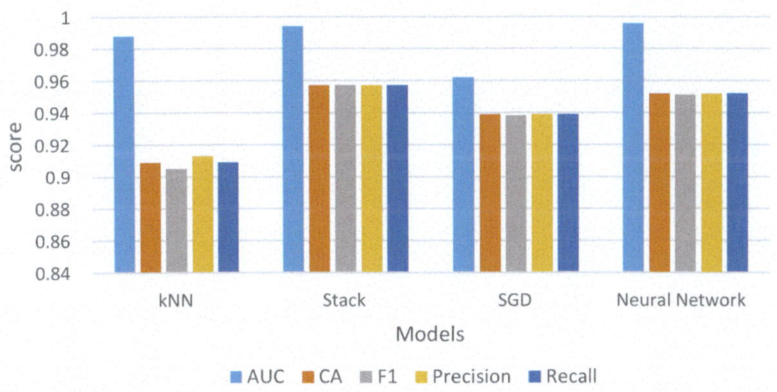

Fig. 6 Performance comparison

It is evident from Table 1's confusion matrices and Fig. 6 that k-NN and SGD perform the worst in trials (TN). The stacking model outperformed all other machine learning techniques with a maximum classification accuracy of 95.7%.

4.1 ROC for Performance Evaluation

The ROC is a curve that is generated from the FP ratio (x-axis) and the TP ratio (y-axis) (y-axis). When training data for both classes fluctuate over time, ROC becomes beneficial. If you want the most accurate classifier, your ROC area should be near to 1. According to Fig. 7, stacking surpasses all other methods for determining cervical cancer.

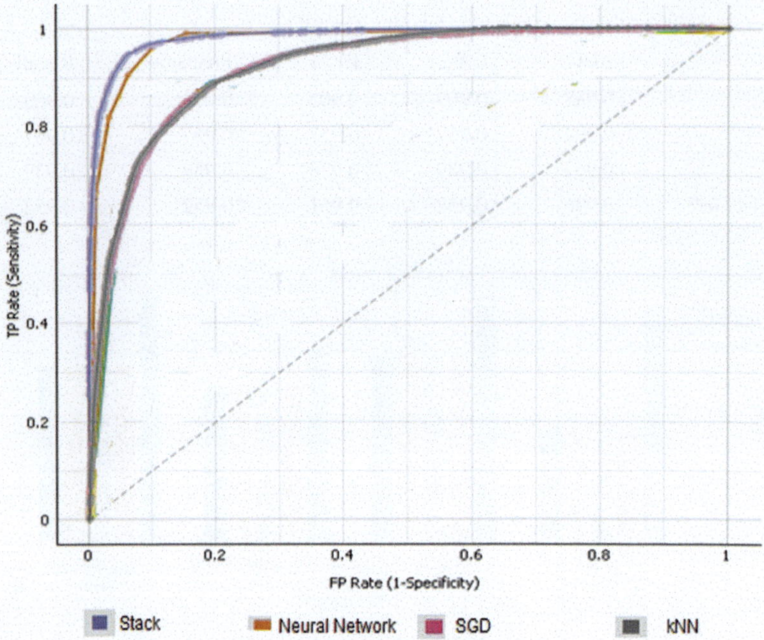

Fig. 7 ROC curve for different machine algorithms

5 Conclusion

Lifesaving and treatment decisions depend on an accurate cervical cancer diagnosis. This research shows that cervical cancer may be detected using machine learning methods. By the use of ensemble learning with 20-fold cross-validation, our neural network and SGD algorithms are able to achieve an accuracy of 95.7% with the stacking model, and our other classifiers also achieve a respectable degree of accuracy. This research used a classification approach to identify risk factors for cervical cancer. It is our goal to improve classification accuracy by extracting more features/characteristics from the present dataset in the future. For the greatest results, we will also employ a wider array of machine learning algorithms and deep learning approaches, including transfer and deep learning.

References

1. Parkin DM, Bray F, Ferlay J, Pisani P (2001) Estimating the world cancer burden: globocan 2000. Int J Cancer 94:153–156
2. Exner M et al (2016) 'Value of diffusion-weighted MRI in diagnosis of uter- ine cervical cancer: a prospective study evaluating the benefits of DWI compared to conventional MR sequences in a 3T environment.' Acta Radiol 57(7):869–877

3. Mcveigh PZ, Syed AM, Milosevic M, Fyles A, Haider MA (2008) Diffusion-weighted MRI in cervical cancer. Eur Radiol 18(5):1058–1064
4. Gadducci A, Barsotti C, Cosio S, Domenici L, Riccardo AG (2011) Smoking habit, immune suppression, oral contraceptive use, and hor- mone replacement therapy use and cervical carcinogenesis: a review of the literature. Gynecol Endocrinol 27(8):597–604
5. Luhn P, Walker J, Schiffman M, Zuna RE (2013) The role of co-factors in the progression from human papillomavirus infection to cervical cancer. Gynecol Oncol 128(2):265–270
6. Cortes C, Vapnik VN (1995) Support vector networks. Mach Learn 20(3):273–297
7. Sommer C (1998) Quantitative characterization, classification and reconstruction of oocyst shapes of Eimeria species from cattle. Parasitology 116:21–28
8. Agarwal AK, Kiran V, Jindal RK, Chaudhary D, Tiwari RG (2022) Optimized transfer learning for dog breed classification. Int J Intell Syst Appl Eng 10(1s):18–22
9. Lilhore UK, Poongodi M, Kaur A, Simaiya S, Algarni AD, Elmannai H, Vijayakumar V, Tunze GB, Hamdi M (2022) Hybrid model for detection of cervical cancer using causal analysis and machine learning techniques. Comput Math Methods Med
10. Koundal D, Gupta S, Singh S (2018) Computer aided thyroid nodule detection system using medical ultrasound images. Biomed Signal Process Control 40:117–130
11. Bhardwaj S, Jain S, Trivedi NK, Kumar A, Tiwari RG (2022) Intelligent heart disease prediction system using data mining modeling techniques. In: Soft computing: theories and applications: proceedings of SoCTA 2021. Springer Nature Singapore, Singapore, pp 881–891
12. Trivedi NK, Gautam V, Sharma H, Anand A, Agarwal S (2022) Diabetes Prediction using different machine learning techniques. In: 2nd international conference on advance computing and innovative technologies in engineering (ICACITE), Greater Noida, India, pp 2173–2177. https://doi.org/10.1109/ICACITE53722.2022.9823640
13. https://www.kaggle.com/datasets/prahladmehandiratta/cervical-cancer-largest-dataset-sip akmed
14. Meng Q (2020) Machine learning to predict local recurrence and distant metastasis of cervical cancer after definitive radiotherapy. Int J Radiat Oncol • Biol • Phys 108(3):e767
15. Shan J, Jiang R, Chen X et al (2020) Machine learning predicts lymph node metastasis in early-stage oral tongue squamous cell carcinoma. J Oral Maxillofac Surg 78(12):2208–2218
16. Saini SK, Bansal V, Kaur R, Juneja M (2020) ColpoNet for automated cervical cancer screening using colposcopy images. Mach Vis Appl 31(3):1–15
17. Sanyal P, Ganguli P, Barui S (2020) Performance characteristics of an artificial intelligence based on convolutional neural network for screening conventional Papanicolaou-stained cervical smears. Med J, Armed Forces India 76(4):418–424
18. Jany BR, Janas A, Krok F (2020) Automatic microscopic image analysis by moving window local Fourier transform and machine learning. Micron 130(102800)
19. Karunakaran V, Saritha VN, Joseph MM et al (2020) Diagnostic spectro-cytology revealing differential recognition of cervical cancer lesions by label-free surface enhanced Raman fingerprints and chemometrics. Biol Med 29:102276
20. Jia FAD, Zhengyi SB, Zhang TCC (2020) CNN-SVM network abstract. Neurocomputing
21. Jia AD, Li BZ, Zhang CC (2020) Detection of cervical cancer cells based on strong feature CNN-SVM network. Neurocomputing 411:112–127
22. Ren J, Zhang A, Wang X (2020) J Pre, Pharmacol Res, 104743
23. Ijaz MF, Attique M, Son Y (2020) Data-driven cervical cancer prediction model with outlier detection and over-sampling methods. Sensors 20(10):2809–2822
24. Ruiz EML, Niu T, Zerfaoui M et al (2020) A novel gene panel for prediction of lymph-node metastasis and recurrence in patients with thyroid cancer. Surgery 167(1):73–79
25. Stelzle D, Tanaka LF, Lee KK et al (2021) Estimates of the global burden of cervical cancer associated with HIV. Lancet Glob Health 9(2):e161–e169
26. Tiwari RG, Misra A, Khullar V, Agarwal AK, Gupta S, Srivastava AP (2021) Identifying microscopic augmented images using pre-trained deep convolutional neural networks. In: 2021 international conference on technological advancements and innovations (ICTAI). IEEE, pp 32–37

An Analytical Survey on Credit Scoring Mechanism in India

Ankan Bhowmik, Arkajyoti Basu, Arittra Basu, Anish Paul, and Atrri Roy

Abstract Credit score is an important indicator to lenders of your capability to repay loans. A consumer's creditworthiness is indicated by a number between 300 and 850 called a credit score. The advanced the score, the better a borrower looks to implicit lenders. It is predicated on data gathered from lenders and is consolidated in the Credit Information Report or CIR. This paper reviews to check one's report regularly, also helps you to identify any disagreement that may be lowered your score, and helps to recover the lower score to advanced situations score.

Keywords Credit score · Credit ratings · Net profit · Loan

1 Introduction

Credit scoring, this phrase refers to a statistical analysis carried out by lenders and financial organizations to assess the creditworthiness of clients, which can be any individual or proprietor. This process is used by lenders to determine whether to grant or deny credit. The client profile may be impacted by this score system when applying for financial goods like mortgages, auto loans, credit cards, and private loans. Popular credit scoring models include CRISIL, ICRA, CARE, SMERA, Fitch India, and Brickwork Ratings. Credit scoring is used by lenders in risk-based pricing, which bases loan terms, such as interest rates, on the likelihood that they will be repaid. While credit scores are used for individuals and small, family-owned enterprises, credit ratings are used for businesses and the government. The most well-known credit scoring model is the Credit Information Bureau (India) Limited (CIBIL), which has a license from the Reserve Bank of India. There are three further businesses with licenses from the RBI that are referred to as credit information companies. The range of a CIBIL credit score is 300–900. A good score is one that is greater than 700. This method is used by lenders and financial institutions to evaluate the creditworthiness

A. Bhowmik (✉) · A. Basu · A. Basu · A. Paul · A. Roy
Department of Computer Application and Science, Institute of Engineering and Management, Kolkata, India
e-mail: ankan.bhowmik@iem.edu.in

S. Dutta et al. (eds.), *Cyber Intelligence and Information Retrieval*, Lecture Notes in Networks and Systems 1025, https://doi.org/10.1007/978-981-97-3594-5_21

of an individual or a business. It involves assessing various factors such as credit history, payment behavior, debt-to-income ratio, and other financial data to assign a score that represents the likelihood of the borrower repaying their debt on time. This score helps lenders to determine whether or not to approve a loan or credit application as well as the terms and interest rate of the credit being offered. Credit scoring has become an integral part of the lending process as it enables lenders to make informed decisions based on data-driven analysis rather than relying solely on subjective opinions. A person's credit score determines their borrowing capacity for the rest of their lives. It takes time and several steps to build credit. There are not many studies, though, that describe how different variables affect credit ratings and how different variables affect interest rates. This study investigates how a person's credit score is impacted by their age, income, and debt-to-income ratio. It investigates how factors such as credit score, loan size, term, and loan type (consumer, real estate, and agricultural loans) affect the interest rate that a lender provides a borrower. Both consumers looking to raise their credit score and lenders seeking to understand what factors affect interest rates may find this study to be useful. Customers can concentrate on the key elements that affect their credit score using the findings. They can then focus on those areas to raise their score. Professional banks might also concentrate on the key factors affecting the interest rates they offer. This demonstrates to the loan industry how the customer interest rates are genuinely set. Also, the use of credit scoring in India has a very powerful economic impact by empowering the lenders to assess the creditworthiness of borrowers and make informed decisions on lending. This has led to increase the access to credit for individuals and businesses, which has in turn contributed to economic growth and development. Also, credit scoring has decreased default rates and minimizes the risk for lenders, which has led to more efficient and effective lending practices. In general, credit scoring has a very positive impact on the Indian economy by developing access to credit and promoting financial stability.

2 Literature Survey

In this work, the role of credit system in an economy is examined. In this paper, we have studied about the relationship between banking credit and growth in India. It helped us to learn about the credit, GDP, and the relationship between variables [1]. From Abby M. Kern's "Credit Score Analysis (2017)", we got an idea how to analyze credit score. This paper helped us on how to improve our credit score. It helped us to study about lenders who want to know what influences interest rates. It also helped us to know the goods credit score, quantum of the loan, term of the loan, and loan type have on the interest rate a client is offered by a lender [2]. From Dr. Saravanan Laxmanan and Dr. Sankaramuthukumar's "Awareness of Credit Score Mechanism in India: A Study with reference to Credit Information Bureau India Limited (CIBIL) (2017)", we also knew, lenders similar as banks and credit card companies use credit scores to determine who qualifies for a loan, at what interest rate, and what credit limits. Lenders also use credit scores to determine which guests are likely to bring in the most profit [3]. After this paper, to study the credit ratings, we can study Cristian Deritis's "Economically Conditioned Credit Scores: An Example from Auto Loans (2018)", and we can see in the author's research that he addresses how credit ratings develop and alter over the course of an economic cycle. He also researched about the internal or personal variables that credit scores are intended to account for, and he explores how external influences may also have an impact on them [4]. In Gharat Pranav Umesh's "A STUDY ON CREDIT RATING AGENCIES OF INDIA (2019–2020)", the author describes about the credit rating agencies in India. He also done a demonstrative difference between the agencies and how they work [5]. In Dr. E. Punarselvam, Mr. S. Gopi, and S. Kowsalya's "Free from CIBIL Score and Strong Your Financial (2020)", they gave an idea about how to maintain good civil score and get a good credit score. Mainly, they have made a web application for providing awareness over the CIBIL score and properly providing the various mechanisms to pay the due within the time [6]. To study the credit score prediction by the use of machine learning, we can find it in, Simon Williams's "Credit Card Score Prediction Using Machine Learning (2021)". The author used a bank loan database to check the connection of the borrower classification model and examined machine learning technology. This paper is each about how to predict credit score using machine learning [7]. From Sunitha. G and Dr. V. Venu Madhav's "An Empirical Study on Impact of Credit Rating on Credit Risk of Banks (2021)", we got an idea about credit rating and also got knowledge of what is credit scoring, impact on banks by credit scoring, and need of credit score. It helps to know about credit scoring agencies in India. It will also help to know about how credit ratings help in a country's development [8]. From Experian's "A REVIEW OF INDIA'S CREDIT ECOSYSTEM (2021)" report, we got a very clear overview of Indian credit ecosystem. Mainly there, they demonstrate about Indian economy system and Indian credit growth [9]. Besides that, ceicdata.com, bain.com, taxguru.in, pib.gov.in, un.org—these websites helped us to know about the current economical scenario of India. It helped us to be updated with India's GDP growth, number of new unicorns, population trajectory, unemployment rate, also about the loan percentage, etc. [10].

3 Result and Discussions

3.1 Economic Impact in India

Due to advancements in the Indian economy, it currently boasts one of the most rapidly expanding economies globally. The recent liberalization of foreign direct investment (FDI) has enabled India to leverage FDI across diverse sectors, generating employment opportunities and harnessing technology. The effective implementation of key government digitization initiatives has positioned the economy as a strong global competitor. This success not only contributes to the growth of existing sectors

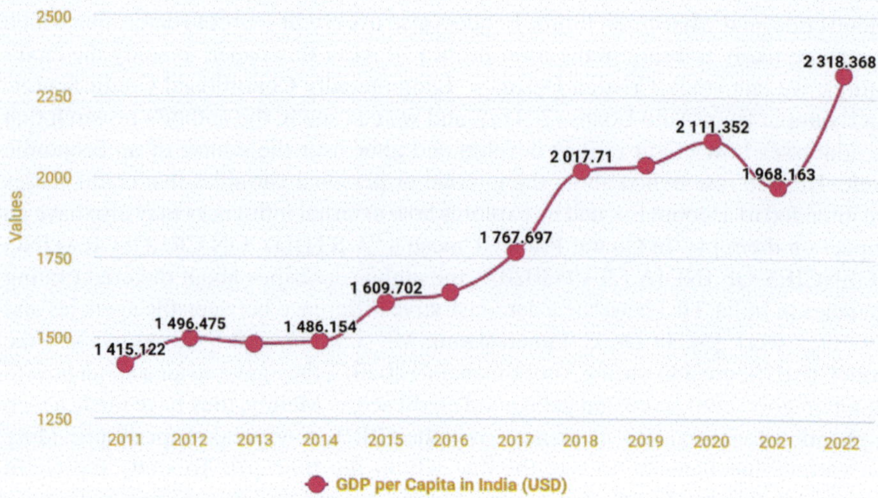

Fig. 1 Growth in GDP per capita in thousand [10]

like e-commerce and digital payments but also catalyzes the emergence of novel industries such as digital lending.

By cultivating and enhancing their abilities, the third-largest startup ecosystem in the world, India, has come a long way in terms of giving young entrepreneurs a level playing field. India raised 67 spots in the ease of doing business ranking over a four-year period (Fig. 1).

The proportion of people in India who are in working age has a significant negative impact on the economy which is dependent on population. Rising salaries and a low dependency ratio support the spending power of the working population. The country's consumption is growing as a result of the growing population (Fig. 2).

Government programs with a specific focus, such as Make in India, Pradhan Mantri Kaushal Vikas Yojna, and the Mahatma Gandhi National Rural Employment Guarantee Act (MNREGA), have assisted in addressing the grave employment gaps (Fig. 3).

Government initiatives like the Make in India initiative, the National Manufacturing Policy, the Credit Linked Capital Subsidy, and the Production Linked Incentives have benefited the manufacturing industry. The availability of more skilled workforce and a conducive investment climate are to blame for the persistent rise in profitability of manufacturing companies (Fig. 4).

From 2016 onwards, there has been a significant decrease in inflation, consistently staying within the targeted range. This suggests an enhancement in the overall quality of life and a stimulation of heightened economic productivity. However, the years affected by the pandemic, namely 2020 and 2021, deviate from this typical trend. Forecasts for the imminent years, 2022 and 2023, anticipate inflation returning to levels observed before the pandemic. The skillful management of inflation has led to a reduction in economic uncertainty, thereby fostering increased investment (Fig. 5).

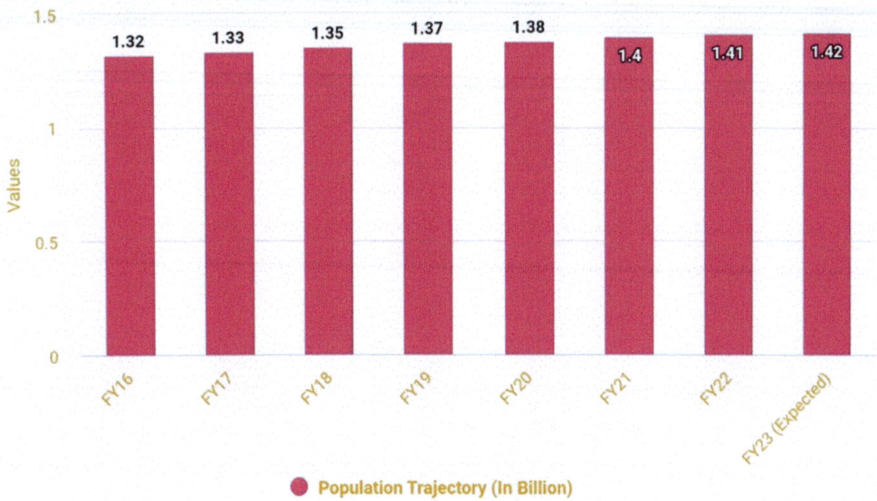

Fig. 2 Population trajectory [14]

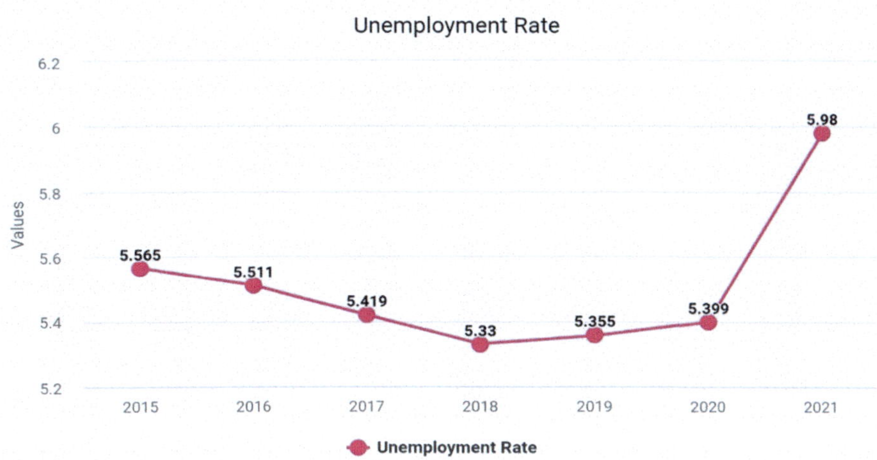

Fig. 3 Unemployment rate [13]

India has become a popular foreign direct investment (FDI) location worldwide. Since 2015, FDI equity inflows to India have increased, reflecting a rise in investor confidence that is a direct result of the country's strong reforms and investor-friendly policies.

From April to January 2021, a total of INR 5403.23 billion (USD 72.12 billion) in FDI was attracted to India. The statistics indicate that FDI equity inflow increased by 28% during the first ten months of FY 2020–21, reaching INR 4059.16 Bn (USD54.18 Bn), up from INR 3172.11 Bn (USD 42.34 Bn) during the same period the previous

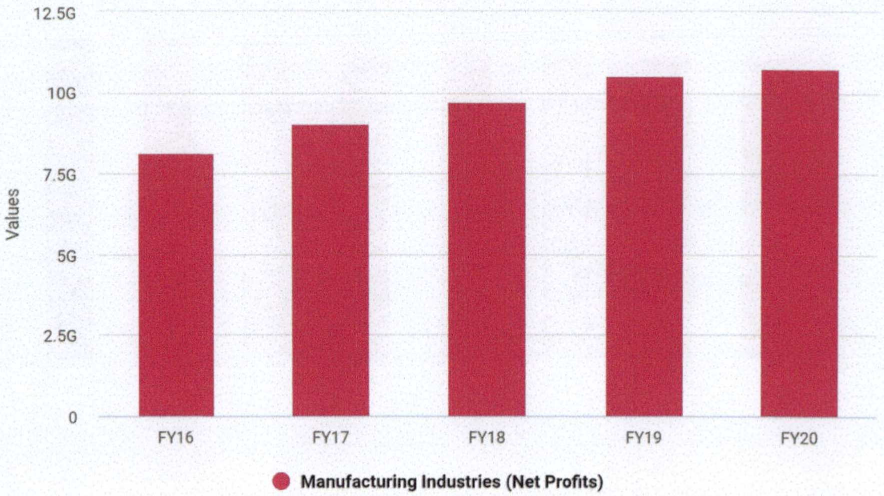

Fig. 4 Manufacturing industries (net profits) [15]

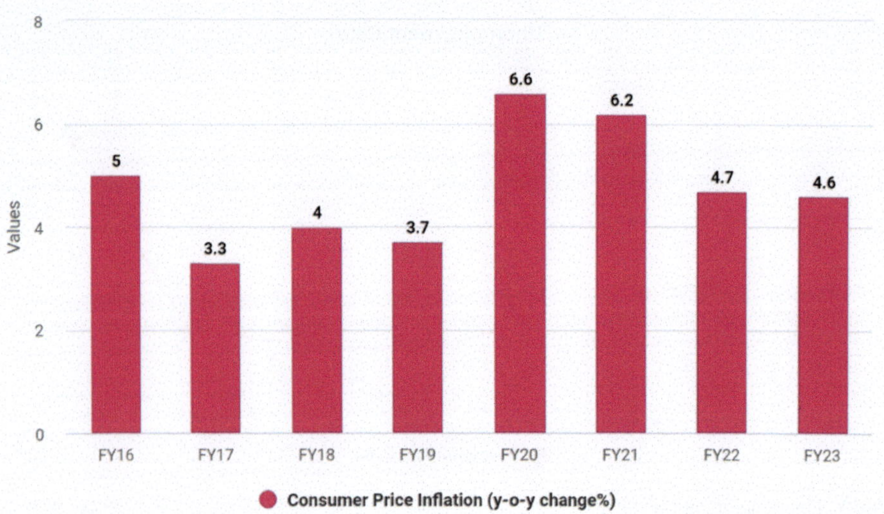

Fig. 5 Consumer price inflation [9]

year. While countries worldwide saw a significant reduction, FDI inflows to India persisted despite the pandemic (Fig. 6).

With INR 3,378.8 Bn (USD 45.1 Bn) in private equity, the largest in the previous ten years, India's investment momentum surged in 2019. BFSI, telecom, and real estate all notably contributed to the increase (Fig. 7).

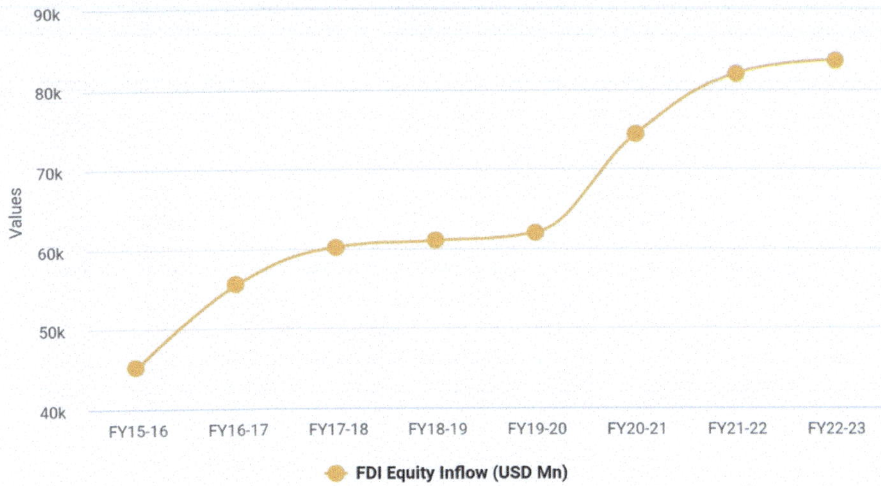

Fig. 6 FDI equity inflow [12]

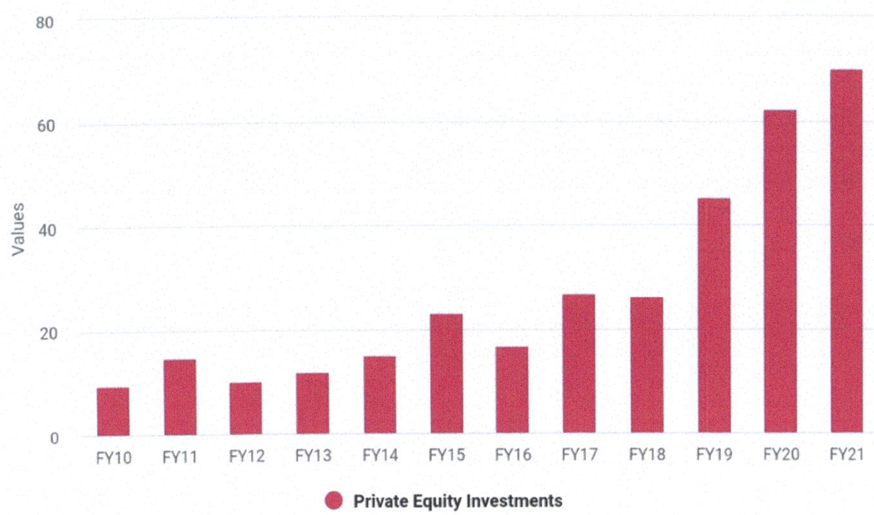

Fig. 7 Private equity investments [11]

With the support of Prime Minister Narendra Modi's flagship Startup India Initiative, India is on track to develop one of the largest and most active startup eco-systems in the world.

An unprecedented 10 new unicorns were created in the last year, signaling an increase in the number of Indian firms joining the exclusive club (Fig. 8).

The country's savings and lending system has become more formalized as a result of the large numbers of formerly unbanked people who have entered the banking

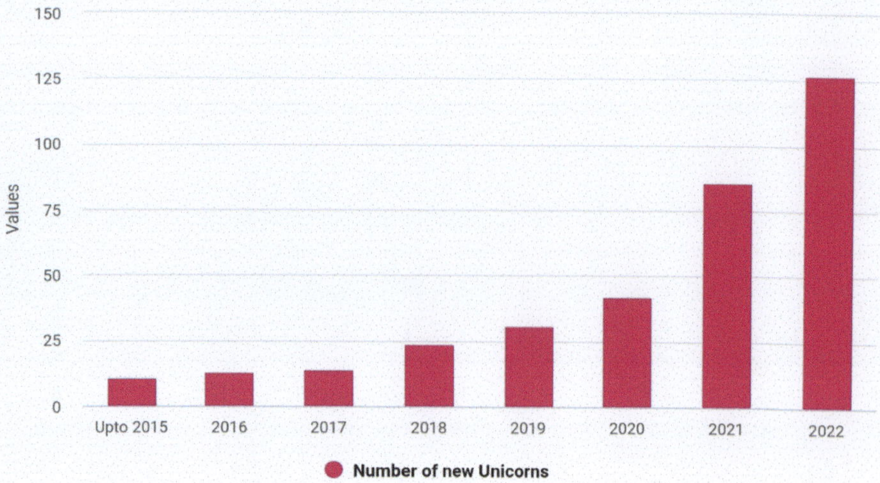

Fig. 8 Number of new unicorns [9]

sector. This generates new business for private lending institutions and presents a chance for financial technology companies to saturate the market of customers who have never used a bank.

The country's digital payment infrastructure has greatly improved over the past ten years as a result of decreased digital fraud risk and a rise in digital payment channels. With the ongoing expansion of the internet and rising confidence in electronic means of payment, digital lending services have been able to carve out a niche for themselves (Figs. 9 and 10).

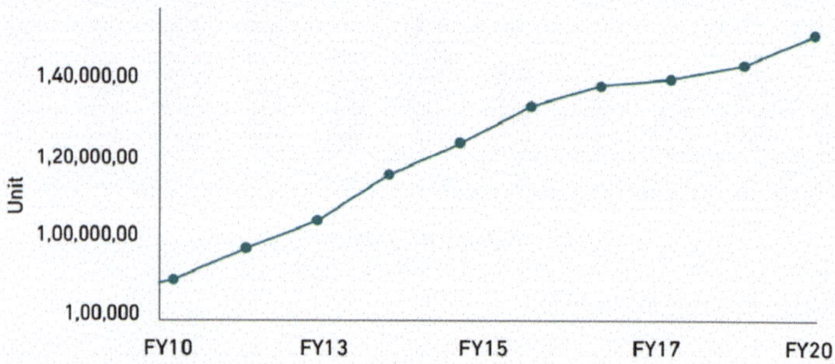

Fig. 9 Number of commercial banks [16]

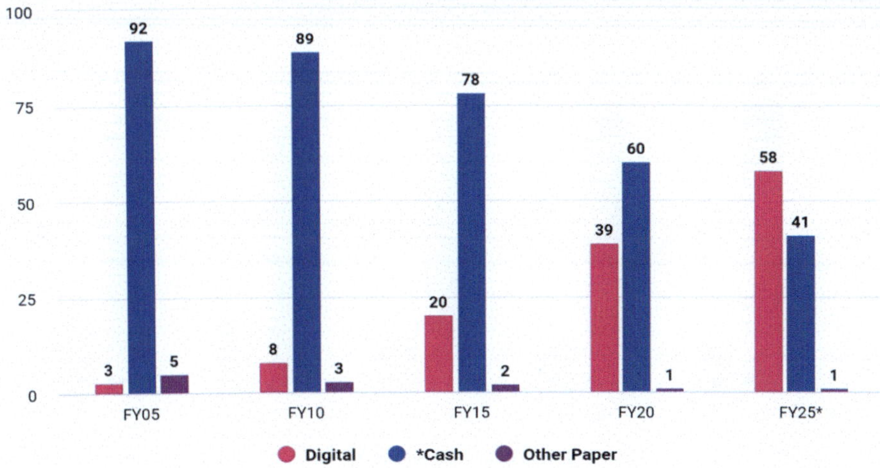

Fig. 10 Digital payments growth % [18]

3.2 *Credit Growth Trends*

Facilitating the transmission of monetary policy effects throughout the broader economy is crucial, particularly in financial systems like that of India, where banks play a significant role as financial intermediaries. Since 2019, the Reserve Bank of India (RBI) has pursued an accommodative monetary policy, with a specific emphasis on promoting credit growth, particularly among private lenders. In the consumer loan sector, significant product market sizes witnessed a notable 77% increase in 2020 compared to March 2017. Overall, both the total outstanding debt and the number of active loan accounts have consistently shown an upward trajectory at the market level. Between 2017 and 2020, overdue debt expanded at a Compound Annual Growth Rate (CAGR) of 21%, while the portfolio of loan accounts demonstrated an even higher CAGR of 26%.

3.2.1 Growth in Secured and Unsecured Products

Compared to secured products, which expanded at a CAGR of 17% from 2017 to 2020, the loan book for unsecured products increased at a rate of 38%. A loan that is unsecured is one that is simply supported by the borrower's creditworthiness and not by any kind of security like real estate or other assets.

It shows a consistent increase in secured loans over the years with total balance amount accounting to 76% of all credits in February 2021 (Fig. 11).

While there has been a consistent rise in secured lending, the proportion of loans in the unsecured category has surged to 71%, suggesting a growing trend among

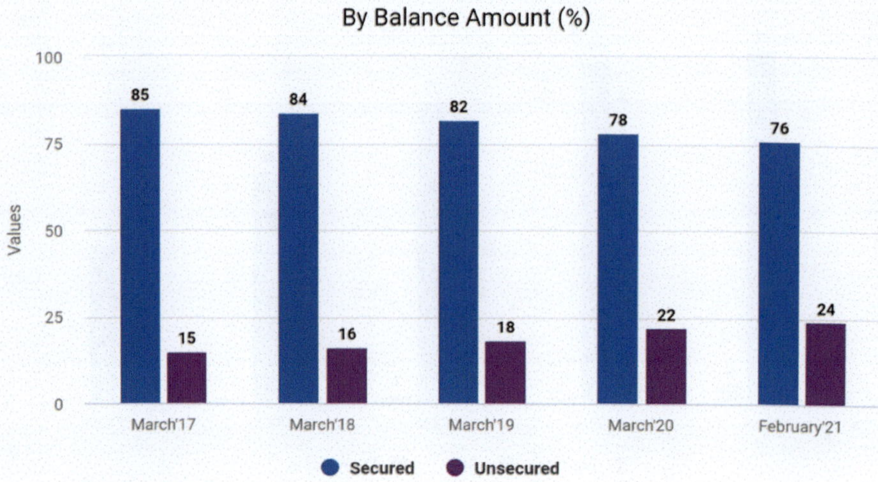

Fig. 11 Loan distribution (secured and unsecured) [9]

consumers to use credit for financing smaller expenses, such as consumer durables and vacations. This shift is noteworthy for lending institutions.

3.2.2 New Sourcing of Loans (Sanctions)

Unsecured loans were the main drivers of the jump in the number of new sanctioned loan accounts between FY18 and FY20, growing at a cumulative pace of 39%. The CAGR for these products was an astounding 49%. Two variables, firstly rising consumerism, are the main drivers of this loan rise in unsecured products. A larger portfolio of Non-banking Finance Companies (NBFCs) entering this market is the second factor on the supply side (Fig. 12).

There is a clear trend toward seeking smaller loan amounts, evident in the faster growth of loan accounts compared to balances. This shift is attributed to several factors, with key contributors being the rise in people's income profiles due to geographically dispersed economic growth and increased employment across various socioeconomic segments. Metropolitan areas have benefited from significant service migration to India, while government initiatives like MNREGA have notably boosted income levels in semi-urban and rural regions. This widespread economic activity has led to the emergence of new consumer segments with distinct credit needs. The demand for entry-level products and small-balance loans has risen, providing access to formal credit for a previously underserved group of consumers who either could not afford credit or were excluded due to insufficient income.

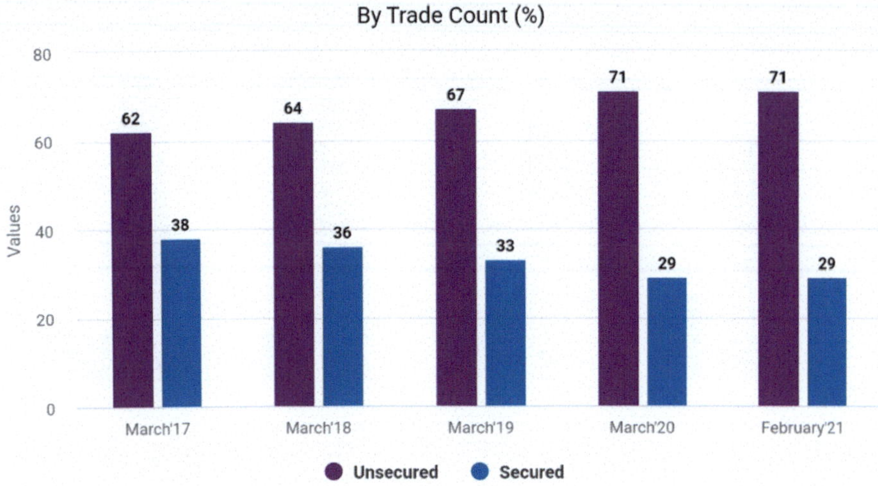

Fig. 12 Loan distribution (secured and unsecured) [9]

3.2.3 Product-Wise Distribution in New Sourcing (Sanctions) by Count

Consumer durables make up the largest portion of the sourcing in the graph above (Figs. 13 and 14). Consumer durables are a class of goods that do not need to be replaced regularly since they last a long time. They are also referred to as durables, durable goods, or consumer durables. When a person spends a significant amount of their income on durable, capital, or investment goods—items with longer economic useful lives—their wealth is conserved. Durables' purchases and new orders are closely monitored by economists, business owners, and investors as indicators of sustained economic growth (Figs. 15, 16, 17, 18, 19 and 20).

3.2.4 The Stratified Allocation of New Funding Sources in Semi-rural and Rural Areas is Poised to Drive the Subsequent Phase of Expansion

Another significant change has been the growing development of credit for lending to tier 3 and tier 4 markets. Given the concentration of the working population in metro areas, metros continue to be the largest lending markets; nevertheless, lending growth has been strongest in tier 3 and tier 4 markets.

* Share as of FY20 and CAGR from FY18 to FY20.

* Share and CAGR by sourcing amount.

Despite a change in customer borrowing preferences, Tier 1 and 2 remain the two largest credit markets. Low-ticket high-volume loan products increased significantly

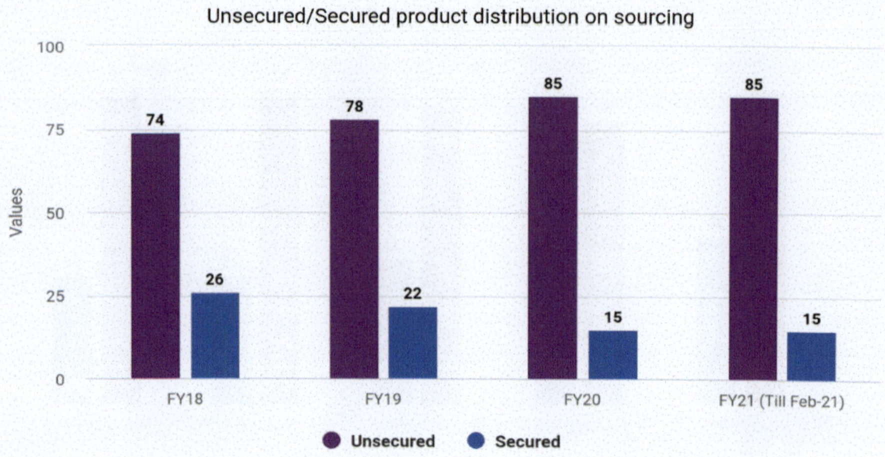

Fig. 13 Loan product distribution on sourcing (unsecured/secured; disbursements in that year; category) [9]

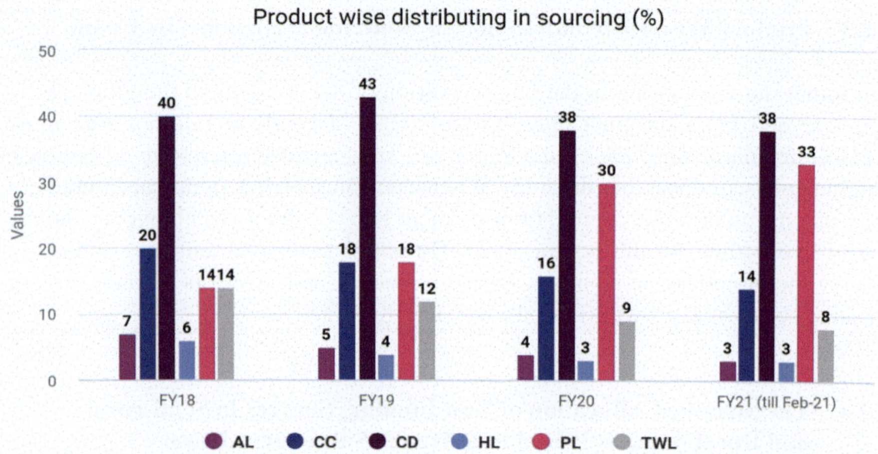

Fig. 14 Loan product distribution based on sourcing (category: AL, CD, PL, CC, HL, TWL; disbursements in that year) two-wheeler loan, consumer durables, credit cards, house loan, personal loan, and auto loans are all examples of loans [9]

in these markets. However, affordable goods like two-wheelers, entry-level cars, and reasonably priced homes helped the tier 3 and tier 4 markets flourish (Fig. 21).

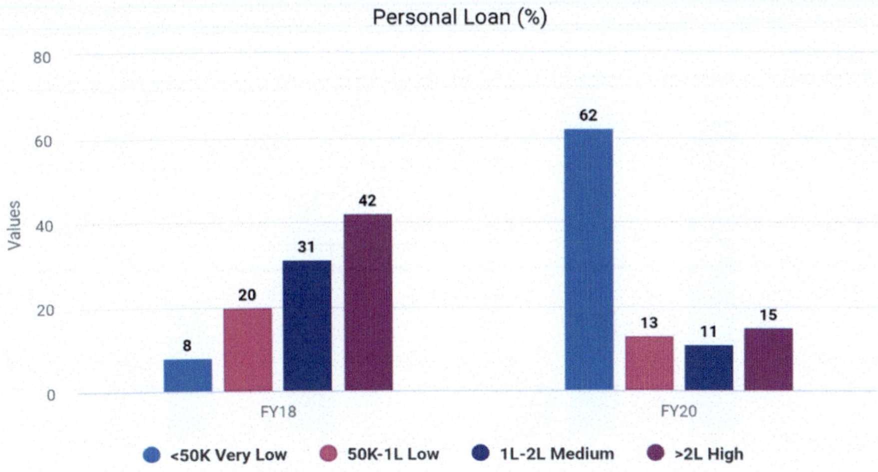

Fig. 15 Average loan disbursement ticket size for the years FY18 and FY20 (personal loan) [6]

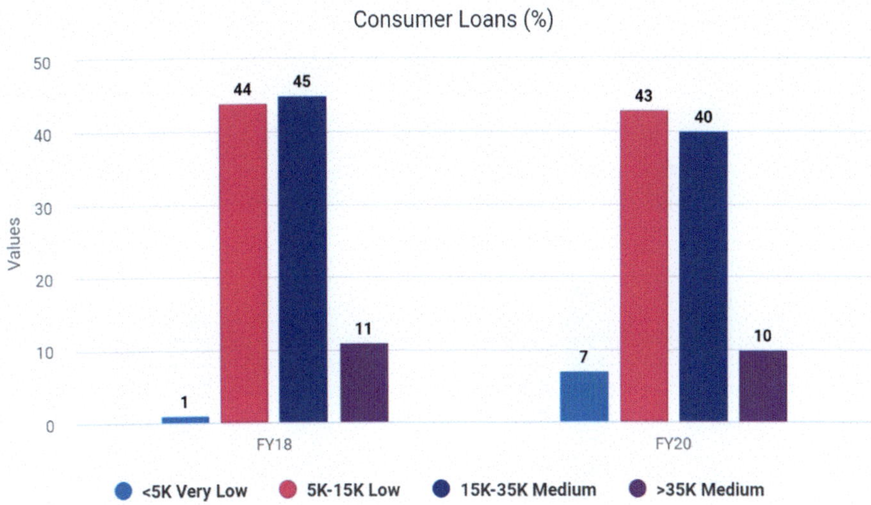

Fig. 16 Average loan disbursement ticket size for the years FY18 and FY20 (consumer loan) [6]

3.2.5 Source Distribution Through a Range of Lending Institutions

By increasing their lending activities, financial institutions have facilitated the extension of credit to diverse markets and product categories. Traditionally, the pursuit of credit expansion and financial inclusion goals in India has been led by a robust coalition of public and private banks, with the emergence of Non-banking Financial Companies (NBFCs) playing a pivotal role in propelling the growth of specific products, regions, and customer segments.

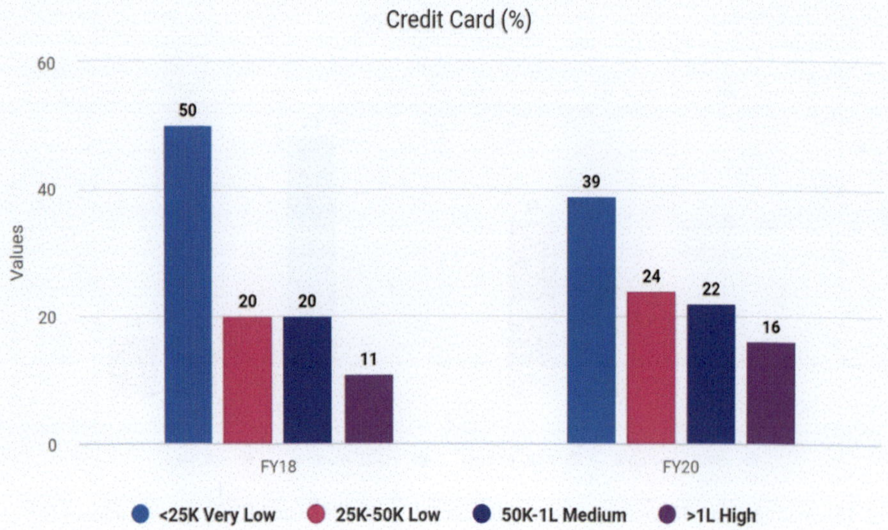

Fig. 17 Average loan disbursement ticket size for the years FY18 and FY20 (credit card) [6]

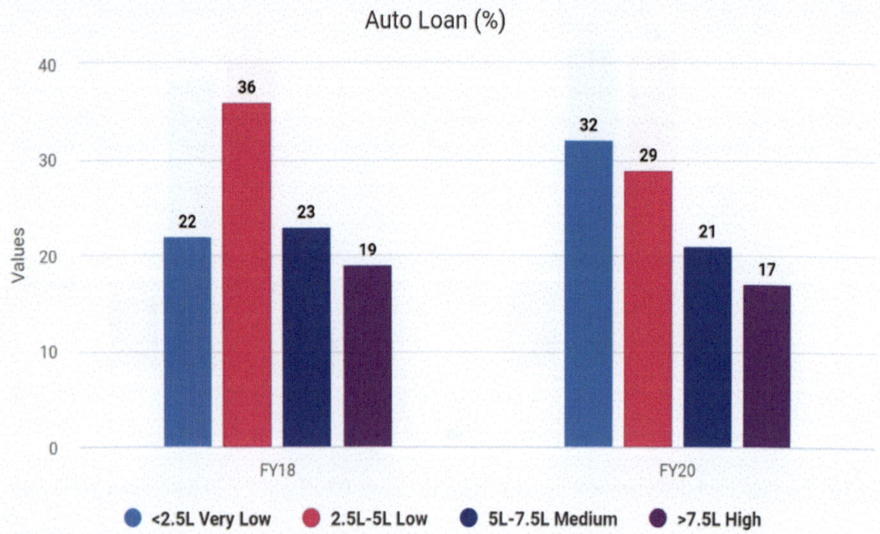

Fig. 18 Average loan disbursement ticket size for the years FY18 and FY20 (auto loan) [6]

In terms of the proportion of new loan acquisitions (by count), NBFCs dominate with a significant share of 68%, surpassing public banks (8%) and private banks (22%). In less structured markets, NBFCs have become a preferred choice due to distinctive business strategies that leverage a comprehensive understanding of client preferences.

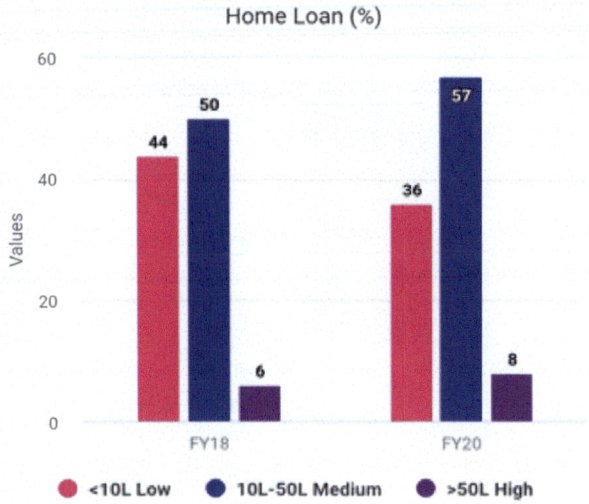

Fig. 19 Average loan disbursement ticket size for the years FY18 and FY20 (home loan) [6]

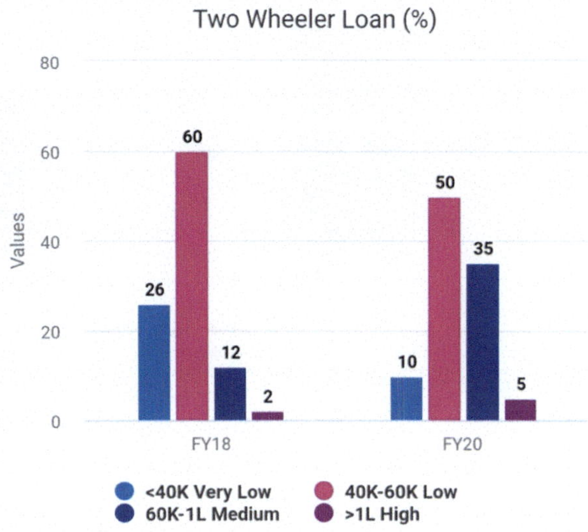

Fig. 20 Average loan disbursement ticket size for the years FY18 and FY20 (two-wheeler loan) [6]

Although NBFCs constitute a significant proportion of the overall number of sources, their sourcing remains limited due to the predominantly small size of the loans they acquire (Fig. 22). This limitation might also be attributed to the fact that NBFCs are still a relatively novel institution for the general public. Despite catering to a broader audience, the value of loans sourced by NBFCs is considerably smaller than that of public and private sector banks as of FY20. However, within a mere two-year span, there has been a five-percentage-point increase in their share, indicating an escalating acceptance of NBFCs (Fig. 23).

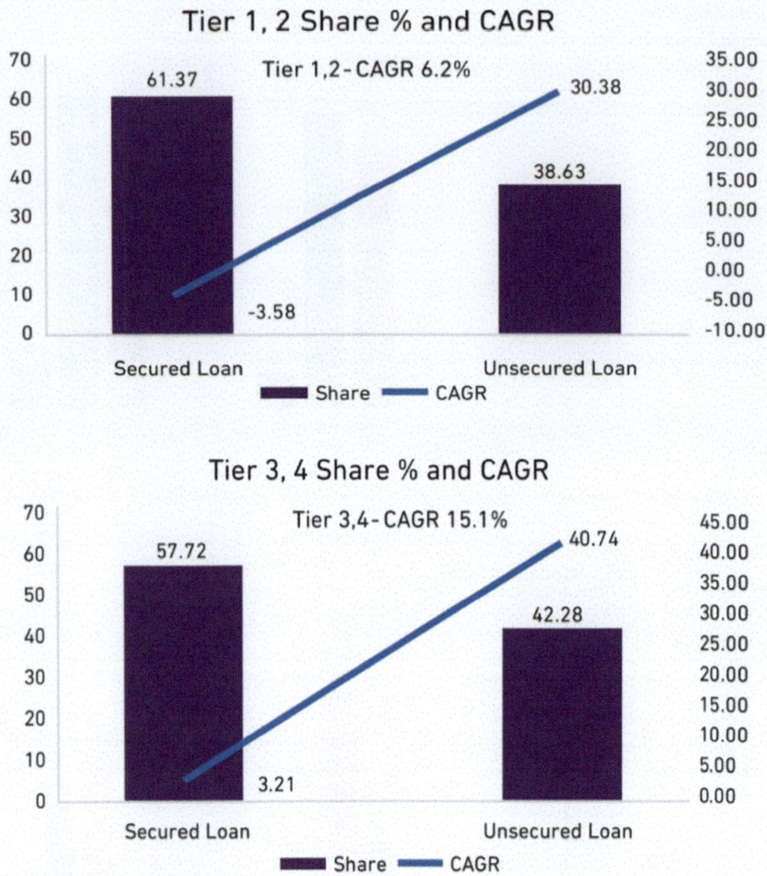

Fig. 21 Tier-wise growth [9]

As can be seen from the graph above (Fig. 24), the average ticket size for NBFCs has been the smallest of all years and has been steadily declining over time. NBFCs have found success in unorganized markets by focusing on customers with modest ticket requirements. The NBFCs' low ticket size business strategy and their sizable market share also demonstrate the sizeable market for low ticket lending in India.

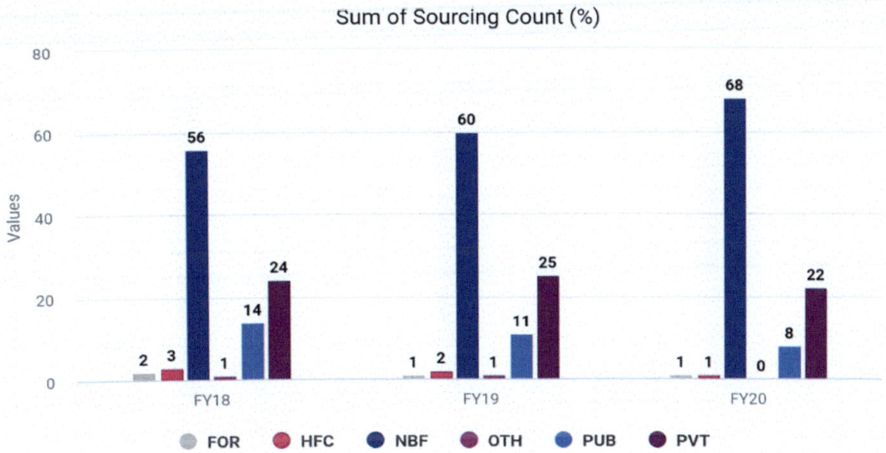

Fig. 22 Share of loans disbursed by type of lender (category: For, HFC, NBF, OTH, PUB, PVT, loans disbursed in that year) by percentage [9, 17]

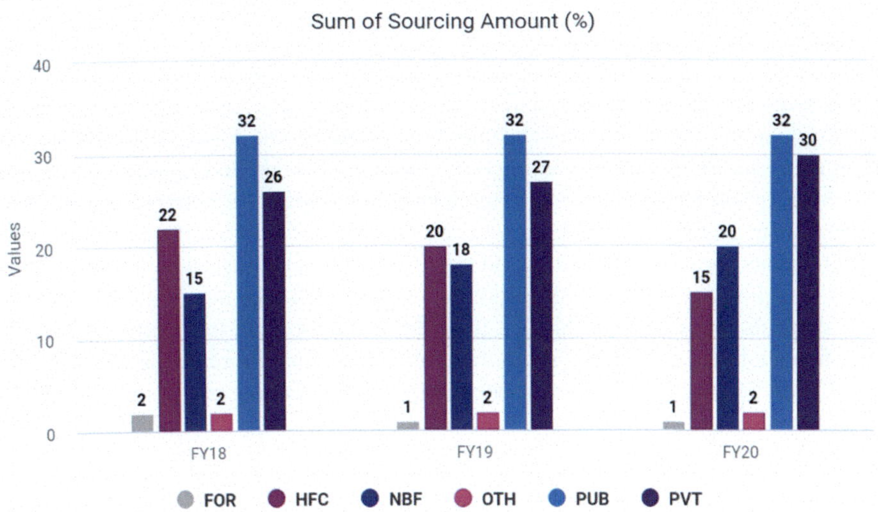

Fig. 23 Sum of sourcing amount [9, 17]

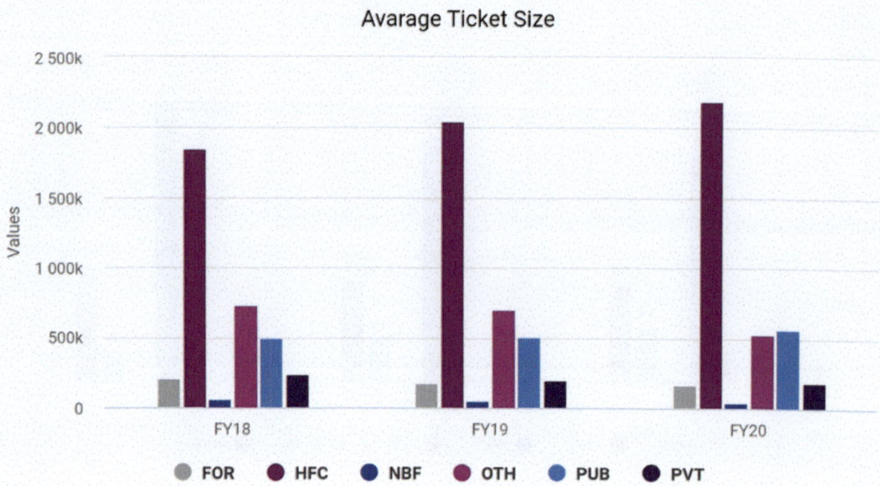

Fig. 24 Average ticket size loans from different lenders [9, 17]

4 Conclusion

In this world, everyone from small to large businessman, entrepreneurs, normal working man take loans for their business, car loans, home loans, and bike loans. So, in this case, credit score is an important indicator to take loans because lenders will check by your credit score of your capability to repay the loans. Loans will be credited from lenders by checking your credit score. If your credit score is good enough means above or similar to good range you will get loans when you need. But if you have low amount of credit score then you cannot take any loan. This credit score depends on your credit history, payment behavior, debt-to-income ratio, and other financial data to assign a score that represents how you repaying the debt on time. Credit scoring has become an integral part of the lending process as it enables lenders to make informed decisions based on data-driven analysis rather than relying solely on subjective opinions.

5 Future Scope

In this work, a detailed analytical survey report is done on credit scoring method. During literature review phase, it is found that machine learning algorithms help to predict credit score. In this work, it is shown how to calculate the credit score and an detailed analysis on if the credit score is low, how it can be increased. Future work is to create a model on how to increase the credit score which will help the consumers to take loans.

References

1. Singh C, Pemmaraju SB, Rohan D (2016) Economic growth and banking credit in India. IIMB WP N0 531
2. Kern AM (2017) Credit score analysis. Southern Illinois University Carbondale
3. Dr. Laxmanan S, Dr. Sankaramuthukumar S (2017) Awareness of credit score mechanism in India: a study with reference to credit information bureau India limited (CIBIL). ResearchGate Publication No 351979266
4. Cristian Deritis C (2018) Economically conditioned credit scores: an example from auto loans. Moody's Corporation
5. Umesh GP (2019–2020) A study on credit rating agencies of India. University of Mumbai
6. Dr. Punarselvam E, Mr. Gopi S, Kowsalya S (2020) Free from CIBIL score and strong your financial. Int J Eng Res Technol (IJERT) ISSN: 2278-0181
7. Williams S (2021) Credit card score prediction using machine learning. Int J Innov Sci Res Technol ISSN No:-2456-2165
8. Sunitha G, Dr. Madhav VV (2021) An empirical study on impact of credit rating on credit risk of banks. Indian J Financ Bank 7(1). ISSN 2574-6081 E-ISSN 2574-609X Published by CRIBFB, USA
9. Experian (2021) A review of India's credit ecosystem, experian.in
10. GDP capita, www.ceicdata.com/en/indicator/india/gdp-per-capita
11. Private equity investments, www.bain.com/insights/india-private-equity-report-2022/
12. FDI Equity, pib.gov.in/newsite/PrintRelease.aspx?relid=199821, pib.gov.in/PressReleasePage.aspx?PRID=1826946
13. Unemployment, www.un.org/development/desa/dpad/publication/un-desa-policy-brief-no-153-india-overtakes-china-as-the-worlds-most-populous-country/#:~:text=In%20April%202023%2C%20India's%20population,to%20grow%20for%20several%20decades, www.statista.com/statistics/271330/unemployment-rate-in-india/
14. Population Trajectory, www.un.org/development/desa/dpad/publication/un-desa-policy-brief-no-153-india-overtakes-china-as-the-worlds-most-populous-country/#:~:text=In%20April%202023%2C%20India's%20population,to%20grow%20for%20several%20decades
15. ManufacturingIndustries,www.ceicdata.com/en/india/manufacturing-industry-nic-2008-all-industries/manufacturing-industries-number-of-factories
16. Credit Growth in Commercial Bank, www.moneylife.in/article/
17. Avg ticket price, www.thehindubusinessline.com
18. Digital payments growth, community.nasscom.in/communities/digital-transformation/fintech/india-digital-payments-2020-launching-the-first-adoption-index-time-is-now.html

The Study of the Effectiveness and Sensitivity of Stochastic Divergence in Foreign Exchange Market

Mahamudul Hasan, Izazul Hoq Imran, Tropa Das, Nazib Al-Hoque, and Md. Al-Amin

Abstract In this research, we focus on stochastic divergence for the EUR/USD and AUD/USD currency pairs. It is a trading tool used in forex markets. Traders can use it to find trends and guess how prices might move over a range of time periods. One of the benefits of stochastic divergence is that it can show traders when the market might change direction or continue its current trend, which helps them make better decisions. But, all economic signs and stochastic divergence can sometimes give wrong indications about how the market is doing. This study looked at how well trading tactics based on stochastic divergence worked over three different time frames. One important thing that was looked at was the risk-reward ratio, which is an important factor in figuring out whether a trading plan will work or not. The study's results show both the pros and cons of using stochastic divergence in forex trading. Traders should be smart about how they use it and combine it with other technical and fundamental analyses.

Keywords Stochastic divergence · Technical analysis · Currency pairs · Trading strategies · Bullish divergence · Bearish divergence · Technical indicators · Break-even analysis

1 Introduction

This study examines the unusual thing of stochastic divergence within forex markets, with a specific idea on the currency pairings of EUR/USD and AUD/USD. Traders are able to enhance their ability to forecast future market patterns and anticipate divergences through the utilization of indicators, which include the comparison of price movements. The present study aims to evaluate the efficacy of stochastic divergence as a predictive instrument for the stock market, while also examining its responsiveness over varying time intervals [7]. This study analyzes the performance of a trading system on two currency pairs, EUR/USD and AUD/USD, and evaluates

M. Hasan (✉) · I. H. Imran · T. Das · N. Al-Hoque · Md. Al-Amin
CSE, East West University, Aftabnagar, Dhaka 1212, Bangladesh
e-mail: Munna09bd@gmail.com

S. Dutta et al. (eds.), *Cyber Intelligence and Information Retrieval*, Lecture Notes in Networks and Systems 1025, https://doi.org/10.1007/978-981-97-3594-5_22

its dependability across various trading situations [10]. The investigation primarily emphasizes the percentages of wins, losses, and break-even outcomes. This article aims to analyze the unusual thing of stochastic divergence in financial markets, with a specific focus on the processes of data collection, statistical methodologies, and the assessment of its effectiveness [12]. The findings of this study reveal notable patterns in the currency pairings of EUR/USD and AUD/USD, accompanied by an analysis of associated challenges, potential avenues for further investigation, and constraints encountered throughout the research process. The article underscores the significance of trading choices.

2 Literature Review

A well-known trading tool, stochastic divergence analyzes historical price data to forecast future price movements. This widely used instrument foresees market changes and trend reversals. In the stock market, it is a well-known instrument for making informed trading decisions. The investigations, hypotheses, and conclusions connected to stochastic divergence's use and efficiency in stock markets are summarized in this review [1]. The technique of stochastic divergence, which relies on the momentum oscillator, is employed to identify inconsistencies between price fluctuations and oscillator readings, therefore signifying shifts in market sentiment [2]. The indicator is employed in technical analysis to forecast bullish and bearish divergence between oscillator and price values, as well as to ascertain overbought and oversold levels [3]. The analysis of divergence patterns has significant importance in serving as signals for both entry and exit points in trading. Enhancing the reliability of these patterns may be achieved by cross-referencing them with various technical indicators [5]. In their study, authors conducted an analysis that yielded valuable findings pertaining to market reversals. Similarly, Brown and Jennings conducted a comprehensive investigation focusing on technical trading profitability, currency pairings, and moving averages [7]. A comprehensive grasp of these components is necessary for the effective implementation of technical trading strategies. Stochastic divergence, a prospective tool for forecasting, encounters computational obstacles and necessitates intricate algorithms to provide precise predictions [10]. The study conducted by [6] examines the efficacy and sensitivity of predicting price changes and trend reversals in stock markets [12]. The researchers take into account several elements such as time frames, currency pairings, and other information in their analysis.

3 Methodology

This study carefully collects and pre-processes historical pricing information for the currency pairs EUR/USD and AUD/USD over a range of time frames. A thorough analysis of bullish and bearish stochastic divergence is possible thanks to the data set's inclusion of transaction intervals, including 1-, 5-, 15-, 30-, and 60-min intervals [2]. Statistical investigations were conducted to evaluate the efficacy and responsiveness of stochastic divergence in stock markets. The study analyzed the percentage of wins, losses, and break-even trades for specific time periods and currency pairs. The average size of wins and losses was computed to assess risk-reward characteristics associated with stochastic divergence implementation [3]. A comparative study examined the statistical significance of stochastic divergence across currency pairs, using hypothesis testing to determine disparities in win, loss, and break-even percentages. A significant level of $\alpha = 0.05$ was employed for statistical analysis. This research on stochastic divergence focuses on historical data, ignoring real-time market circumstances and sentiment. It adheres to ethical guidelines by using publicly accessible historical pricing data and avoiding sensitive personal information. The study aims to evaluate the effectiveness of stochastic divergence as a trading strategy across different periods and currency pairings, using rigorous data collection, statistical analysis, comparative assessment, and hypothesis testing [9].

3.1 Technical Analysis and Indicators

Technical analysis looks at past data to guess how prices will change in the future [4]. The main focus is on finding patterns and trends in how market prices change. In this field, stochastic divergence is very important because it helps find future changes or reversals in trends by comparing price changes with analytical signs [7]. This method is very similar to the stochastic oscillator, which is a well-known technical sign in financial research. When the stochastic oscillator shows a trend that is different from the price movement of the underlying product, it means that the market mood has changed. When price movement and signal movement are not lining up, this method is very helpful for finding trade chances [8]. Stochastic divergence is a very important part of technical analysis because it can show when the market is about to turn around. A bullish or negative split could mean that present trends are weakening or that things are about to change quickly [14]. This study looks at how useful and sensitive stochastic divergence is in EUR/USD and AUD/USD pairs over a number of time periods.

3.2 Exponential Moving Average (EMA)

The stochastic divergence with the EMA technique includes studying divergences between the stochastic oscillator and price while utilizing an Exponential Moving Average (EMA) to validate the trend direction (Fig. 1). A buy signal is formed when the stochastic oscillator and the price diverge, and this is verified by the price being above the EMA, suggesting an upward trend. A sell signal is generated when there is a

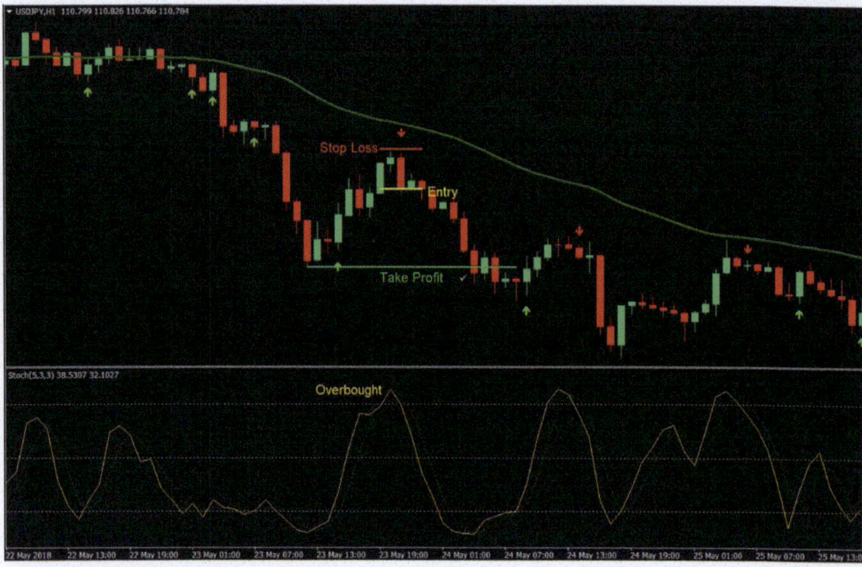

Fig. 1 EMA of stochastic

bearish divergence and the price is lower than the EMA, suggesting a downward trend. To maximize the chance of profitable trades, this technique combines momentum and trend-following features.

3.3 Calculation of Stochastic Oscillator

The stochastic oscillator measures momentum by comparing an asset's closing price to its price range over a given time frame, usually 14 days [3]. The following steps are involved in the calculation:

Step 1: Collect Historical Price Data: For each time period, compile the asset's peak, low, and closing prices. The standard notation for the time period is "n."

Step 2: Calculate the Lowest Low and Highest High: Using the historical price data, determine the lowest low and highest high values for the chosen time period "n." The lowest low and highest high each indicate the lowest and greatest prices that were experienced throughout the time [6].

Step 3: Calculate %K: The current closing price with respect to the price range throughout the chosen time period "n" is shown by the %K line. Following is the formula used to compute it:

$$\%K = \frac{\text{Closing Price} - \text{Lowest Low in } n \text{ periods}}{\text{Highest High in } n \text{ periods} - \text{Lowest Low in } n \text{ periods}} \times 100 \qquad (1)$$

Closing Price: The price at which the current time period ended. The lowest price seen during the last "n" times, or "low in n periods." Highest high in n periods: The highest high price recorded during the preceding "n" times.

Step 4: Plotting the Stochastic Oscillator: Plot the two lines on a chart after calculating %K and %D for each time period within the selected range. The %D line is frequently shown as a dotted line, whereas the %K line is generally shown as a solid line. From 0 to 100 are the possible values for the stochastic oscillator. A rating below 20 is often regarded as oversold, and one beyond 80 as overbought [14].

The stochastic oscillator helps traders spot bullish or bearish trends, find extreme readings, and use tools and studies to make smart trading decisions.

3.4 *Stochastic Divergence Indicator*

The stochastic divergence indicator predicts market trends by transforming erratic reactions between the price and stochastic curve. Traders typically use RSI and MACD for divergence trading, but the Stochastic indicator also provides helpful signals for swing trading opportunities on shorter time frames [5]. The stochastic divergence indicator detects trend reversals by generating trend lines on the indicator window and price chart. Graphical representations help traders identify price changes deviating from stochastic fluctuations. However, novice traders may struggle to identify atypical fluctuations (Fig. 2). Arrow indicators strengthen bullish/bearish divergences, enabling informed speculations on market reversals [11]. Divergence manifests when the price exhibits a contrary movement to the stochastic curve, such as a bullish divergence when the price establishes a lower bottom while the stochastic indicator indicates a higher low. To effectively discern trade signals and mitigate market fluctuations, it is crucial to incorporate stochastic value into the analysis when forecasting divergence trades. For instance, a stochastic value of 80 indicates oversold conditions, increasing the likelihood of short selling during oversold market situations, especially in the presence of a bearish divergence [12] (Fig. 2).

- A blue arrow heading up is seen in the indicator pane to indicate a positive divergence (Fig. 3).
- An oversold market state is indicated by the stochastic value staying close to 20.
- Exit long when a trendline or horizontal resistance limits the price [13].

- A red arrow on the indicator's plot indicates a bearish divergence.
- After the price fails to go below a trend line or horizontal support, exit short positions as the stochastic value hovers around 80, indicating an overbought market.

The presented algorithm is a stochastic divergence trading algorithm, which is used to determine trading signals based on the divergence between price action and the stochastic oscillator, which is a momentum indicator. To identify probable divergences, it examines the previous and current swing highs and lows in the price

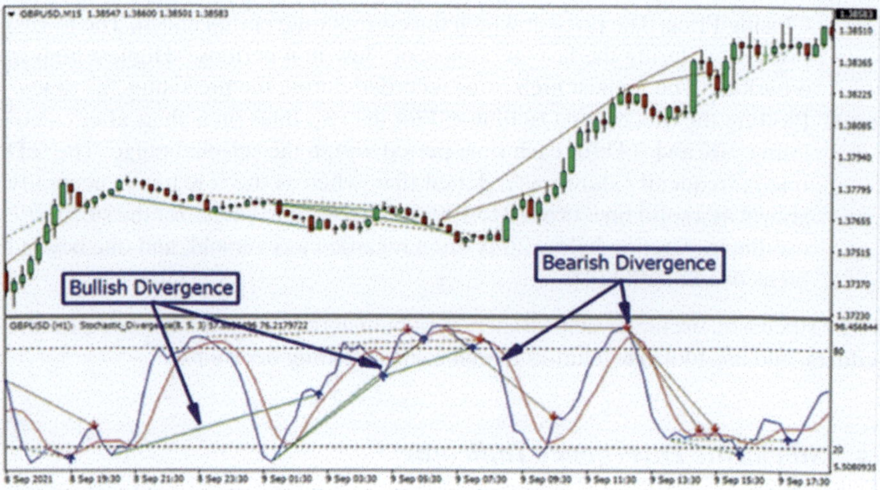

Fig. 2 Stochastic divergence indicators

Fig. 3 Stochastic divergence indicators: buy conditions

as well as the stochastic values. If a divergence is discovered and it corresponds to an Exponential Moving Average (EMA) crossover signal, the system provides a buy or sell signal. It is advised to wait for a clearer indication if the prerequisites for a purchase or sale are not satisfied. This algorithm appears to be a component of a trading system that employs both momentum and trend-following tactics. If both the EMA crossover and stochastic signals are bullish, a buy signal is created; if both are bearish, a sell signal is issued. If neither criterion is entirely satisfied, the algorithm recommends waiting for a better chance. At the conclusion, the final trade signal is returned.

Algorithm 1 Stochastic Divergence Trading Algorithm

```
Initialize variables:
prev_swing_high ← 0, prev_swing_low ← ∞, curr_swing_high ← 0, curr_swing_low ←
  ∞, ema_cross_over_signal ← 'none', Stochastic_signal ← 'none', trade ← 'wait', EMA [0]
  = Current EMA and EMA [1] = Previous EMA
for i = 0 to length(data) − 1 do
    if prev_high > curr_swing_high then
        prev_swing_high ← curr_swing_high
        curr_swing_high ← prev_high
    end if
    if prev_low < curr_swing_low then
        prev_swing_low ← curr_swing_low
        curr_swing_low ← prev_low
    end if
    stochastic_value ← calculate_stochastic(data[i])
    if (prev_stochastic_high > curr_stochastic_high and stochastic_value <
      curr_stochastic_high) or (prev_stochastic_low < curr_stochastic_low and stochastic_value >
      curr_stochastic_low) then
        stochastic_divergence_signal ← "divergence"
    end if
    if stochastic_divergence_signal is "divergence" then
        if prev_price_swing_high > curr_price_swing_high then
            stochastic_signal ← "buy"
        else if prev_price_swing_low < curr_price_swing_low then
            stochastic_signal ← "sell"
        end if
    end if
    if (stochastic_signal == "buy" and ema_cross_over_signal == "buy") then
        trade ← "buy"
    else if (stochastic_signal == "sell" and ema_cross_over_signal == "sell") then
        trade ← "sell"
    else
        trade ← "wait for better signal"
    end if
end for
```

4 Results and Findings

The analysis explores the effectiveness and sensitivity of stochastic divergence in EUR/USD and AUD/USD currency pairs, providing valuable insights into its performance as a technical analysis tool in real-world trading situations (Table 1).

The EUR/USD currency pair showed a win rate of 45% in the 1-min timeframe, with a 40% loss percentage and a break-even proportion of 15%. The 5-min data showed a 50%-win rate, 30% loss percentage, and 20% break-even proportion. According to the 15-min data, there were 45% wins, 30% losses, and 10% break-even games (Table 2).

The AUD/USD currency pair exhibits a win rate of 40%, a loss rate of 40%, and a break-even rate of 20%. These figures suggest that the pair has varying rates

Table 1 Trading performance by time frame for EUR/USD

Time frame (min)	Win (%)	Loss (%)	Break-even (%)
1	45	40	15
5	50	30	20
15	45	45	10
30	50	30	20
60	50	35	10
Total	48	36	16

Table 2 Trading performance by time frame for AUD/USD

Time frame (min)	Win (%)	Loss (%)	Break-even (%)
1	40	40	20
5	50	35	15
15	35	45	20
30	45	35	20
60	45	40	15
Total	43	40	17

of success and failure over different time periods. The aggregate success rate is at 45%, comprising a 40% failure rate and a 15% probability of achieving a break-even outcome.

4.1 Trading Balance Analysis

The investigation of trade balance resulting from stochastic divergence-based techniques in the EUR/USD and AUD/USD currency pairs is the focus of the study of the effectiveness and sensitivity of stochastic divergence in stock markets. To assess the success and risk involved with a particular investment or trading strategy, key metrics used in finance and investment analysis include the win-loss ratio and risk-reward profile. While the risk-reward profile offers insights into the relative profitability in relation to the accompanying risk, the win-loss ratio assesses the strategies' dependability in generating financial advantages. This thorough analysis offers insightful information on the use and sensitivity of stochastic divergence in stock markets.

The win-loss ratios and risk-reward profiles that are revealed by the EUR/USD exchange rate point to profitability in the US. The ratio of wins to losses in 60 s varies between 1.12 and 1.47, while the risk-reward profile in 1 min varies between 1.32 and 1.84. The 5-min win-loss ratio is in the range of 1.67–1.76, showing consistent profitable trades. The 15-, 30-, and 60-min durations show comparable trends in Fig. 4. The AUD/USD exchange rate reveals consistent trends in trade balances, with a win-

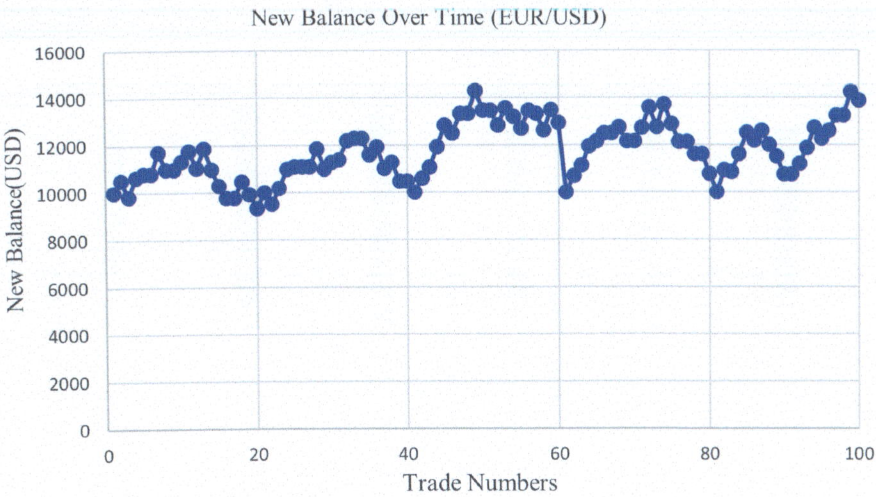

Fig. 4 Trading amount EUR/USD

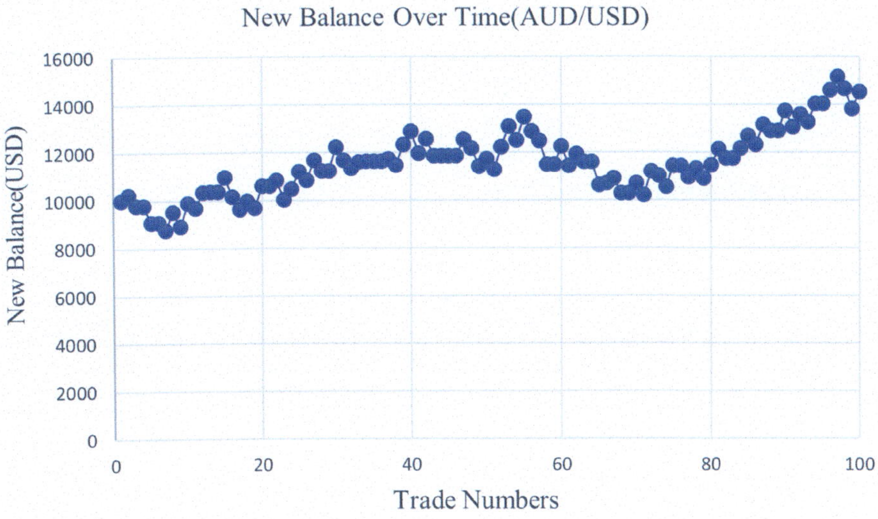

Fig. 5 Trading amount AUD/USD

loss ratio ranging from 1.00 to 1.03 within 60 s, suggesting equitable distribution of profits and losses. The risk-reward profile exhibited a range of 1.33–1.81, suggesting advantageous profitability based on risk. The win-loss ratio and risk-reward profiles showed stability in profit generation and diverse degrees of profitability in Fig. 5. The analysis of trading balance provides valuable insights into profitability and risk associated with stochastic divergence. Taking into account win-loss ratios and risk-reward profiles enables traders to develop successful strategies that guarantee ongoing prof-

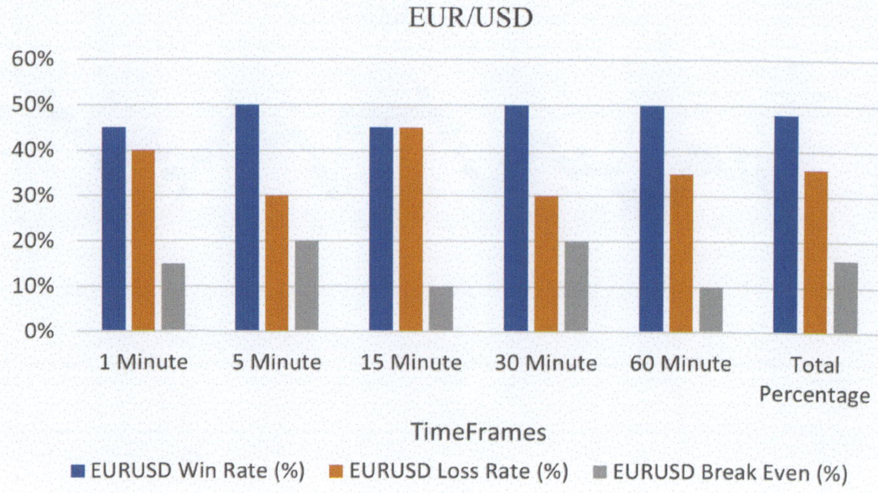

Fig. 6 Trading performance by time EUR/USD

itability and balance risk and reward factors. This aids traders in customizing tactics to suit their level of risk tolerance and desired level of profit.

4.2 *Experiment and Analysis*

The experimental part of this work used stochastic divergence-based trading strategies for the currency pairs EUR/USD and AUD/USD over various time periods. The researchers gathered historical price information and created trading plans that took bearish and bullish divergence indications into account. For every detected divergence signal, simulated transactions were run to determine the entry and exit positions based on the signal's direction. The usefulness of these strategies' response to various time periods and currency combinations is revealed by this study (Fig. 6).

The study shows how well and how easily stochastic divergence-based tactics can miss new market trends. The study found that 5-min intervals constantly had the best winning rates for both currency pairs. Shorter intervals, on the other hand, were better at spotting quick changes in price. Different time periods and currency pairs had different risk-reward profiles. The risk-reward profiles for shorter time periods were better. Short-term price changes and long-term trends can both be seen with stochastic divergence. Bullish divergence signals often mean that it might be a good time to buy, while bearish divergence signals mean that it might be a good time to sell (Fig. 7). Traders and buyers can use the study's findings in real life because the 5-min time frame works well for stochastic divergence-based methods, which lets them profit from short-term price changes.

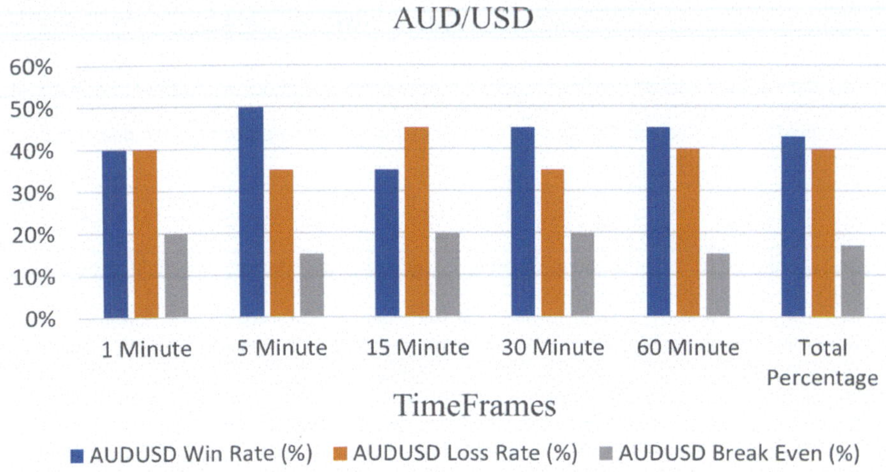

Fig. 7 Trading performance by time AUD/USD

5 Conclusion

The research looks at how well and how sensitively stochastic divergence works on the foreign exchange market, focusing on the EUR/USD and AUD/USD trades. Based on the stochastic oscillator, stochastic divergence is a trading method that helps traders guess when the market will change or go in a different way. Traders should use it along with other basic studies, though, because it doesn't always give correct signs. The study shows how important risk-reward and win-loss numbers are in stochastic divergence. In the future, we hope could improve the accuracy of predictions using stochastic divergence in forex trading by tweaking criteria and limits and adding more complex parameter changes and technical signs and we will also try to do research on this. Traders can make smart choices in the volatile foreign exchange market if they understand this approach and know how to control their risks.

References

1. Erdogan S, Okumus I (2020) Stochastic and club convergence of ecological footprint: an empirical analysis for different income groups of countries
2. Nascimento ADC, Cintra RJ, Frery AC (2012) Hypothesis testing in speckled data with stochastic distances. IEEE Trans Geosci Remote Sens
3. Barassi MR, Cole MA, Elliott RJ (2007) Stochastic divergence or convergence of per capita carbon dioxide emissions: re-examining the evidence
4. Fama EF, French KR (2012) The cross-section of expected stock returns. J Finance 47(2):427–465

5. Andersen TG, Bollerslev T, Diebold FX, Ebens H (2001) The distribution of realized stock return volatility
6. Lipton A, Sepp A (2008) Stochastic volatility models and Kelvin waves. J Phys A Math Theor 41
7. Murphy JJ (2011) Technical analysis of the financial markets: a comprehensive guide to trading methods and applications, 2nd ed
8. Stochastic Divergence Indicator for MT4. https://www.trendfollowingsystem.com/stochastic-divergence-indicator-for-mt4/?fbclid=IwAR1OsaWZaAAS_yLtiCkR3FVx3aqxrEsjSM_sjpCfdibcyYbWi8aTfUTYk3s
9. Jog V, Loh PL (2015) Information-theoretic bounds for exact recovery in weighted stochastic block models using the Renyi divergence
10. Brown C, Kim W (2018) An analysis of the stock market performance of different currency pairs. Int J Financ Stud 6(3):72
11. Neves W, Olivera C (2021) Stochastic transport equations with unbounded divergence
12. Stephens G, Dittman M (2017) The effects of time frames on technical trading profitability. J Bank Finance 85:21–41. https://doi.org/10.1016/j.jbankfin.2017.07.002
13. Lama A (2021) Stochastic divergence indicator for MT4. Trend following system. https://www.tradingview.com/script/nXRnJ0s3-Stochastic-RSI-Indicator/
14. Hayes A (2021) Stochastic oscillator: what it is, how it works, how to calculate. Investopedia. https://www.investopedia.com/terms/s/stochasticoscillator.asp

Data Science and Data Analytics

Conceptual Framework and Impact of Blended Education System on Employability Skills

Suplab Podder, Lindita Bicaj, and Debabrata Samanta

Abstract Employability skills are considered as the professional and organizational necessities in the modern challenging corporate world that demands skilled and efficient human resources to balance social ecosystem and corporate social responsibility. Blended Education System develops the social ecosystem that facilitates most effective education amenities for the diverse background of learners. The curiosity of the researchers and the professional obligation on social realization that leads to the development of conceptual framework and describe the employability skills. The present study is the empirical in nature and the research outcomes designate the responses, expert opinion, and systematic observation of stakeholders in the relevant domain. Principal outcomes of the research study reveal the social obligations of smart education system that develop the employability skills and independent business expansion skills. The present study is well connected to the futuristic research in the area of modern education, skills-based education, smart practices on work efficiency.

Keywords Blended Education System · Employability skills · Virtual simulation · Self-directed learning and institutional support system

1 Introduction

Modern organizations require the efficient and dynamic leaders to take responsibility, design the strategic plan and smart execution of strategies for innovative business practices. The dynamic leadership skills are developed through systematic

S. Podder
Postdoctoral Researcher, Rochester Institute of Technology, Europe, Kosovo

L. Bicaj · D. Samanta (✉)
Rochester Institute of Technology, Europe, Kosovo
e-mail: debabrata.samanta369@gmail.com

L. Bicaj
e-mail: lbicaj@auk.org

S. Dutta et al. (eds.), *Cyber Intelligence and Information Retrieval*, Lecture Notes in Networks and Systems 1025, https://doi.org/10.1007/978-981-97-3594-5_23

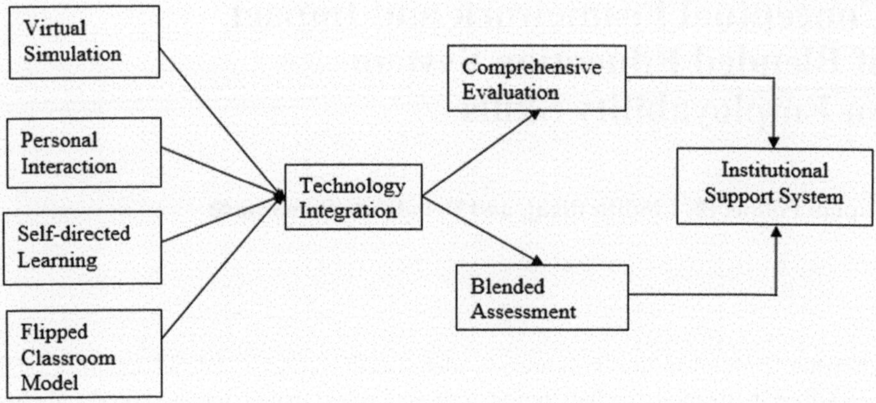

Fig. 1 Presents the conceptual framework of education system for developing skills and knowledge

learning and professional involvement in the better performance. Blended Education System is the proficient roadmap that gives the uniform opportunities for the learners from various geographical locations [1–3]. The system facilitates for the systematic learning environment that develops the employability skills and leadership skills in the later stage of professional development with the strong foundation for the practices of organization functions. Figure 1 presents the conceptual framework of education system for developing skills and knowledge.

2 Conceptual Framework

The conceptual framework is the technology-driven education structure that creates the favorable learning environment for all categories of learners in the society.

The conceptual framework of employability skills includes virtual simulation, personal interaction, self-directed learning, flipped classroom model, technology integration, blended assessment, comprehensive evaluation, and institutional support system. Blended educational laws for personalized learning experiences facilitate for education for everyone in remote locations [4–6]. Figure 2 presents the conceptual framework of employability skills for developing skills.

The conceptual framework for employability skills includes teamwork, leadership skills, time management, emotional intelligence, professionalism and work ethics, and digital literacy.

3 Objectives and Hypothesis Formulation

The objectives are given below:

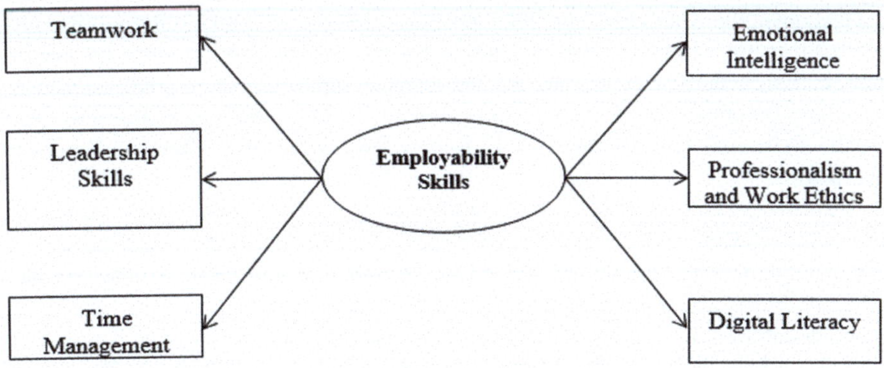

Fig. 2 Presents the conceptual framework of employability skills for developing skills

- To identify the element of conceptual framework of Blended Education System for developing employability skills.
- To identify the key elements of innovative organizational behavior [7, 8].
- To analyze the impact of impact of Blended Education System on employability skills and innovative organizational behavior.

Hypothesis Formulation

Null Hypothesis: H0 = There is no significant level of impact of Blended Education System on employability skills.

Alternative Hypothesis: H1 = There is a significant level of impact of Blended Education System on employability skills.

4 Research Methodology

The research study is the combination of descriptive and exploratory research design with the consideration of historical data and the data collected from the original sources [27, 28]. At first the questionnaire was designed with the help of Delphi technique; the pilot study was initiated to re-refine the questionnaire. The sample size is 400 from various educational institutions and government and private organizations [9–11]. The questionnaires were distributed and collect the responses from the respondents using snowball sampling method.

5 Results and Discussion

Analysis Related to Identify the Element of Conceptual Framework.

The conceptual framework includes virtual simulation, personal interaction, self-directed learning, flipped classroom model, technology integration, blended assessment, comprehensive evaluation, and institutional support system. Blended educational laws for personalized learning experiences [24–26]. **Analysis Related to Find out the Principal Components of Employability Skills for Effective Business Operations.**

The principal components of employability skills for effective business operations include the relevant issues of skills development and knowledge sharing in the area of time management and organizational skills, emotional intelligence, professionalism and work ethics, and digital literacy [12–14].

Understanding the Impact of BES on Employability Skills

Data analysis includes the Multiple Regression Analysis (MRA). The stakeholders from various educational institutions and government and private organizations responses regarding the real-time experience on skills development of young people.

Null Hypothesis: H0 = There is no significant level of impact of Blended Education System on Employability Skills and Innovative Organizational Behavior.

Alternative Hypothesis: H1 = There is a significant level of impact of Blended Education System on Employability Skills and Innovative Organizational Behavior.

The result of ANOVA (Regression Analysis) test indicates significant at 0.05 level (at 5% level of significance). The result from Table 1 specifies that all eight independent variables together have the positive impact on at least one dependent variable [15–17]. The t-test values with reference to virtual simulation, personal interaction, self-directed learning, flipped classroom model, technology integration, blended assessment, comprehensive evaluation, and institutional support system are significant at 0.05 level. The individual significant p values are less than 0.05 and 5% level of significance. Table 1 presents the summary of MRA.

6 Test of Hypothesis

The calculated F-value (Fstat) is 31.966, and the critical F-value (Fcritical) for this test can be obtained from the F-Table with numerator degree of freedom (8) and the denominator degree of freedom (391), Fcritical (8,391) = 1.938, since the (Fstat) is greater than (Fcritical). From the results of Multiple Regression Analysis, the p-value is 0.000 which is less than 0.05 at the 5% level of significance. Both the results (F-value and p-value) lead the same decision regarding the validity of regression model. So, the analysis can be concluded that the overall regression model is significant and valid. There is a significant relationship between the independent variables and at least one of the dependent variables. There is a significant level of impact of Blended Education System on employability skills [18–20].

Table 1 Presents the summary of MRA

Highlights of MRA						
Blended Education System	Coeff.	Std. Error	Std. Coeff.	Tolerance	t	p
Constant	4.302	0.609	0.000	-	7.064	0.00
Virtual simulation	0.054	0.029	0.116	0.256	2.829	0.00
Personal interaction	0.327	0.066	0.307	0.271	3.682	0.00
Self-directed learning	0.766	0.074	0.739	0.201	9.332	0.00
Flipped classroom model	0.314	0.060	0.304	0.304	10.225	0.00
Technology integration	0.342	0.054	0.386	0.275	7.314	0.00
Blended assessment	0.515	0.049	0.599	0.314	4.268	0.00
Comprehensive evaluation	0.175	0.044	0.248	0.263	8.968	0.00
Institutional support system	0.368	0.057	0.298	0.482	7.441	0.00
Source	Sum-of-Squares	df	Mean square	F-ratio	p	
Regression	430.136	8	53.767	31.966	0.00	
Residual	657.624	391	1.682			

7 Summary of Findings and Conclusions

The research study was conducted to understand the significant level of impact of Blended Education System on employability skills. The responses from various stakeholders like the employees from educational institutions and government and private organizations summarizes the positive impact of Blended Education System on employability skills [21–23]. All the stakeholders are witness of the systematic progress of social ecosystem and the practices of innovative organizational behavior incorporate world. When the blended learning system will be successfully implemented, we can experience the overall development of employability skills, favorable work culture that will lead to better corporate world for work and progress.

References

1. Askling B, Stensaker B (2012) Academic leadership: prescriptions, practices and paradoxes. Tert Educ Manag 8(2):113–125. https://doi.org/10.1023/A:1015612510179
2. Amba- Rao SC (2018) A Research on the effect of social media on tendency to entrepreneurship and business establishment (Case study: active iranian entrepreneurs in social media). J Bus Ethics 12(3):553–572. https://doi.org/10.28991/esj-2017-01115
3. Hedberg C, Malmborg F (2013) The global education and institution sustainability reporting in Swedish companies. Inst Soc Responsib Environ Manag 24(2):153–164

4. Carroll A (2015) Impact of institution social responsibility and higher education. J Bus Soc 38(3):268–295
5. Demeke M, Meschi M (2010) Technology adoption in the presence of constraints: the case of fertilizer demand in Ethiopia. Rev Dev Econ 7(2):58–70. https://doi.org/10.1111/1467-9361.00175
6. Daly H (1999) The allocation, distribution and scale towards an economics that is efficient, just and sustainable. J Ecol Econ 6(3):185–193
7. Elliot SR (2005) Sustainability: an economic perspective', resources conservations and recycling. J Econ 44(1):263–277. https://doi.org/10.5281/zenodo.1094141
8. Morton JF (2017) The impact of climate change on smallholder and subsistence agriculture. Proc Natl Acad Sci 14(2):680–685
9. Khan et al (2019) Embedded life cycle costing elements in green building rating tool. Civ Eng J 5(4):750–758. https://doi.org/10.28991/cej-2019-03091284
10. Abusalim N, Rayyan M, Jarrah M, Sharab M (2020) Institutional adoption of blended learning on a budget. I J Educ Manag 34(7):1203–1220
11. Adams D, Tan MHJ, Sumintono B (2021) Students' readiness for blended learning in a leading Malaysian private higher education institution. Interact Technol Smart Educ 18(4):515–534
12. Ayob HH, Daleure G, Solovieva N, Minhas W, White T (2021) The effectiveness of using blended learning teaching and learning strategy to develop students' performance at higher education. J Appl Res High Educ 12(2):345–362
13. Bedics BC, Hall MN (1987) Educating the generalist practitioner in the skills of public relations: a challenge for the eighties. J Soc Work Educ 23(1):58–63
14. Biswas M, Podder S, Shalini R, Samanta D (2019) Factors that influence sustainable education with respect to innovation and statistical science. Int J Recent Technol Eng 7(1):373–376
15. Chowdhury F (2020) Blended learning: how to flip the classroom at HEIs in Bangladesh? J Res Innov Teach Learn 13(2):228–242
16. Samanta D, Podder SK (2019) Level of green computing based management practices for digital revolution and New India. Int J Eng Adv Technol (IJEAT) 8(3):124–131
17. Fearon C, Starr S, McLaughlin H (2011) Value of blended learning in university and the workplace. Ind Commer Train 43(7):446–450
18. Torrisi-Steele G, Steve D (2013) The literature landscape of blended learning in higher education: the need for better understanding of academic blended practice. Int J Acad Dev 18(4):371–383
19. Gary L, Poole DL (1986) Educating leadership for effecting community change through voluntary associations. J Soc Work Educ 22(2):19–29
20. Ayala JS (2009) Blended learning as a new approach to social work education. J Soc Work Educ 45(2):277–288
21. Maile Cutri R (2011) Storied ways of approaching diversity: reconceptualizing a blended learning environment in a multicultural teacher education. Kitchen 13(2):195–216
22. Severson M (1998) Teaching mediation theory and skills in an inter disciplinary classroom. J Soc Work Educ 34(2):185–194
23. Podder SK, Samanta D, Gurunath R (2020) Impact of business analytics for smart education system and management functions. In: Sharma H, Govindan K, Poonia R, Kumar S, El-Medany W (eds) Advances in computing and intelligent systems. Algorithms for intelligent systems. Springer, Singapore. https://doi.org/10.1007/978-981-15-0222-4_45
24. Manral S, Samanta D, Podder SK (2020) Effective classroom activities on accounting using double entry system: the productive consequences. TEST Eng Manag 83(2):5448–5455
25. Podder SK, Samanta D (2022) Green computing practice in ICT-based methods: innovation in web-based learning and teaching technologies. Int J Web-Based Learn Teach Technol (IJWLTT) 17(4):1–18
26. Podder SK (2019) Information and communication technology in sustainable reporting towards paradigm shift in business. J Emerg Technol Innov Res 6(4):240–250

27. Podder SK, Karuppiah M, Thomas B, Samanta D (2022) Research initiative on sustainable education system: model of balancing green computing and ICT in quality education. Interdiscip Res Technol Manag (IRTM) 2022:1–5
28. Viebig C (2022) Blended learning in entrepreneurship education: a systematic literature review. Educ Train 64(4):533–558

Detection of DDoS Attacks in SDN Using Machine Learning Approaches: A Review

Saumitra Chattopadhyay, Ashok Kumar Sahoo, Sanjay Jasola, and Tanupriya Choudhury

Abstract Over the past several years, the Distributed Denial-of-Service attack has emerged as a big threat to Software-Defined Networks because of its frequent attack on SDNs. The DDoS attack, partially or fully, stops the genuine users of network services from accessing the network resources. A large amount of network resources is consumed to ensure the security of the SDN controllers. This factor leads to the poor performance of the SDN controllers in managing the entire network. The research and situation become tougher and stimulating as there are numerous categories of DDoS attacks against SDNs. In order to protect SDN networks from DDoS attack, one must understand the basic characteristics of SDN; network traffic behavior when DDoS attack occurs against SDN; and behaviors of DDoS attack. Different studies on DDoS attacks against SDN had shown many specific clear characteristics and features that can be used as indicators and in turn help us to identify DDoS attacks. With the help of machine learning techniques, one can detect DDoS attacks on SDNs. In this paper, various machine learning techniques which can be helpful in detection of DDoS attacks on SDNs are discussed. Also, a proposed solution is suggested at the end where the best machine learning (ML) with highest accuracy, precision, and recall will be used to create an inference engine for early detection of DDoS attack in SDN.

Keywords Software-Defined Network · DDoS · Machine learning · Deep learning · OpenFlow switch first section

S. Chattopadhyay (✉) · A. K. Sahoo · S. Jasola · T. Choudhury
Graphic Era Hill University, Dehradun, Uttarakhand 248002, India
e-mail: saumitroc@gmail.com

T. Choudhury
e-mail: tanupriya.choudhury@sitpune.edu.in

T. Choudhury
CSE Department, Symbiosis International (Deemed University), Symbiosis Institute of Technology, Pune, Maharashtra 412115, India

S. Dutta et al. (eds.), *Cyber Intelligence and Information Retrieval*, Lecture Notes in Networks and Systems 1025, https://doi.org/10.1007/978-981-97-3594-5_24

1 Introduction

The traditional network architecture is very complex and rigid in nature. It is a challenging job to control and manage it as the new applications and technologies are evolving day by day [1]. Information and communication technologies are changing at a rapid speed because of the heavy use of the Internet. Due to the integration of other related technologies like IoT, Wireless Sensor Network, mobile and cloud computing, etc., the user experiences are becoming better and better [2]. As stated above, due to the development of the Internet and integration of various technologies along with the core networking, the challenges have increased. It has become difficult to satisfy the demands of the modern-day networks like security, dependability, scalability, reliability, and other factors, and the SDN architecture [3, 4] is slowly becoming popular as the most desirable network architecture which can fulfill the modern applications. SDN is more responsive and flexible to manage and implement new applications in comparison with classical network design. The main objective of this paper is to explore and find out with extensive research and relevant datasets the best DDoS mitigation methods to identify the DDoS attacks at the earliest using machine learning algorithms.

2 Software-Defined Networks

SDN ideas are moving from the Datacenter's to the organizational networks and WANs and so involving several challenges. The concept of SDN is to simplify for the need of high capability, consistent and profitable networks that can handle cumulative traffic bulk, with constantly changing and separate applications for each organization.

A control plane is the part of the networking device that regulates how discrete devices inside a network cooperate with each other. The control plane protocols are similar to those used in routing scenario that is OSPF [5], BGP [6], and STP [7] and so these protocols regulate the ideal interfaces to forward packets to destination network. Although these routing, i.e control plane protocols, balance well enough and deliver an extreme of network flexibility, these create constraints. Routing protocols, either way, may only regulate the finest path over a network built on fixed metrics such as the bandwidth on particular interfaces or hop count. Similarly, the control plane cannot certainly have any discernibility into various applications that are running on the network and so accordingly these networks can be degrading application functioning. The SDN approach can have two different ways, control planes that are centralized or distributed control plane. Also, these two different architectural control planes can work together as a hybrid model, so there will be a common core SDN [8] controller for the same.

The data plane compatibility provides featured services such as NATing [9], ACLing [10], VPN [11], and Encryption [12]. These characteristics straightly influence how certain packets are being forwarded, also incorporating the packets that are being dropped. Still, a lot of these services are fixed and controlled by the static configuration of the routers or switches.

3 The SDN and Its Deployment Models

The SDN is described by the Open Network Foundation [13] as a breakthrough where the rigid bond between the forwarding and the control plane is broken and separated. Conventional switches or routers consist of only one plane where both the decision-making and forwarding tasks are performed. In a typical SDN model, the intelligence of a network is consolidated into a SDN controller, and also there exists a decoupling in between the control and the data plane. In the SDN model, a protocol known as OpenFlow [14] is present that stipulates connections running in the controller in between the control plane and the organization. The commitment of this model is dexterity and reduces the overall expenses of complete network.

To hold the potential for customer sections and its used cases, the classic SDN method may be useful, and it might not be best fit for some other customer sections and so is not deprived of its limits. An overall parting of whole forwarding plane from the control plane makes the network more efficient.

SDN architecture, removal from a unified controller, and so will be a fundamental removal from certain operations and designs from maximum networks today. Figure 1 shows the open network foundation model. The typical SDN method to network model presents probable confrontations in handiness, functioning, and security. Lastly, it should be extremely secure and also accepts certain applications which should be capable of adapting and modifying the primary network. The typical SDN model acquires benefits from the programmability of the networks. By using certain APIs [15], the network devices could be visible to the application layer [16]. Moreover, in a typical SDN model, one must decline the features of the native operation system, and so these abilities should be re-designed in the SDN controller.

For SDN placement and network programmability, the hybrid method [17] can acquire benefit of present characteristic groups inside the network operating system. It could be achieved without overall separation of the forwarding plane and the control plane.

This SDN model, which is hybrid in nature, permits various advantages of typical SDN model through a multi-domain controller, and still letting hosts to gain advantage from current capabilities of the network. Figure 2 shows hybrid SDN model. The network operation system can be more advanced than the overall advancement of the computer network.

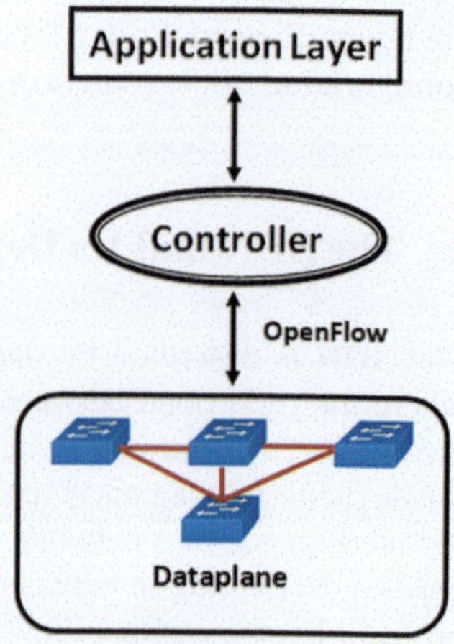

Fig. 1 Open network foundation model

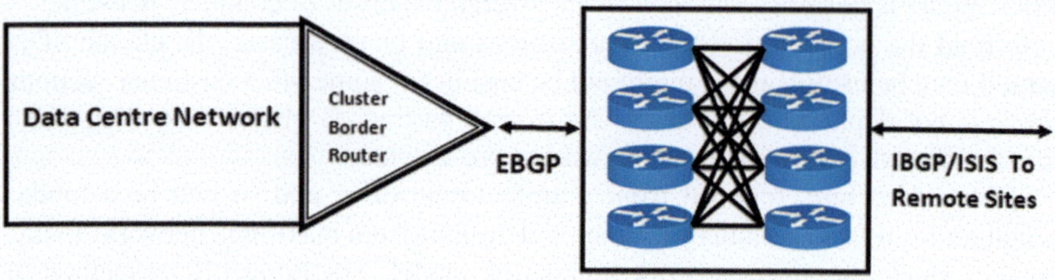

Fig. 2 Hybrid SDN model

4 Working of SDN

As a part of the advanced computer networks, the SDN is a novel concept. The important point of SDN comprises the concentration of network control in both domains, that is in single and multi-domain networks. The SDN controller controls and observes the performance of the networks. The decoupling of the control plane from the forwarding plane is an important goal of the SDN controller. Using an API, this decoupling is conceivable among the networking devices and the SDN controllers, for example, the OpenFlow Protocol. Further networks improvise into programmable networks, which allow the description of the performance of complete networks from the SDN controller and thus the likelihood to form higher network strategies that is security, routing, and load balancing.

Below, Fig. 3 represents three different network layers that the SDN comprises [13]. The initial layer is known as the application layer [16], which contains the applications or software which describes the performance of the domain network, commonly making use of the Representational State Transfer (RST) API. The RST API utilizes the HTTP or HTTPS [15] to permit distant applications to deliver commands to the controller and also recover data from the controller. At the second

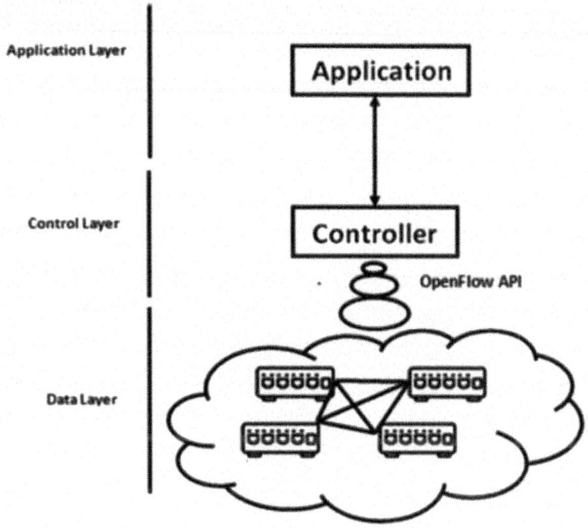

Fig. 3 SDN-configured network design

layer, that is the control layer, there exists an SDN controller that is integrated software based, liable for upholding the global view and managing the control plane of the system. The controller is also liable to give applications of an updated vision of certain network conditions. Finally, the data layer comprises switches, routers, or other devices to build a network liable for traffic forwarding. Figure 3 shows SDN-configured network design.

5 OpenFlow

To acquire communication between the switches, routers, or other networking devices and the SDN controller, an OpenFlow protocol is used. The OpenFlow protocol works as a method to genuinely calculate the cases and so to allow the decoupling of logical traffic flow from the actual traffic flow and so this allows the usage of a network router or switch for tests deprived of affecting with general traffic. This protocol permits changes in the certain switching tables of the switches, on some definite ports, by delivering instructions received by the SDN controllers. An OpenFlow switch can compare the packets besides an unlike header that permits more changing and adaptable packet flow commands than shared devices like switches or routers.

As soon as traffic reaches at an OpenFlow networking device, the header fields of the packets are obtained and paired besides the equivalent field's section entries of the switch table [13]. Suppose an entry to be matched is recognized, then suitable commands the switch applies along with a similar flow.

A distant SDN controller can operate a set of instructions by making use of the OpenFlow protocol, suppose—delete, update, or add entries from the switching tables. Pure and hybrid are the categorics that an OpenFlow network switch originates. Many commercial related switches existing nowadays are hybrid switches [3].

6 DDoS Attacks in SDNs

In the SDN architecture, the network traffic is managed and operated spontaneously as per the organization's dynamic needs. The DDoS [18] attack is one of the most popular and frequently occurring threats in SDN, especially if it is targeted at the controller of the SDN. The DDoS attack tries to stop authorized users from using network services or stop their access to network resources. The DDoS attacks are very difficult to detect with conventional security measures like Firewall, IDS, etc.; due to this, DDoS attack is one of the favorites for hackers for conducting attacks in SDN [19–21].

DoS attacks are usually segregated based on the protocol selected or in which plane (application or control) it is targeted [22]. Attacks targeted at the control layer frequently target UDP, ICMP, or TCP protocols so as to fully consume the target machine bandwidth. Attacks targeted at the application layer try to reject authorized clients of network services by fully engaging the main components of the device that provides the specific service to the client machines.

7 The Literature Review

A hybrid model has been designed by Ahuja et al. [23] using SVM classifier with random forest (SVM-RF) to classify the harmful traffic for detecting DDoS attacks. For an experiment on the SDN dataset, a novel feature extraction technique in the form of CSV file was used. The dataset has been trained using various machine learning algorithms. The experiment shows that using SVM-RF classifier, the SDN traffic detecting accuracy is 98.8% with a very marginal false alarming ratio.

In the paper, as described by Abdurrahman al [24], a new manner to improve the detection of DDoS attack on SDN-based networks was discussed. In their research, they focused on the entropy [25] value and the SDN controller for DDoS attack detection. In their new approach, they measured the randomness of the packets and classified them as normal traffic or attack type. Here they have used a two-step process, first measured the entropy value, and then predicted the attack. In their final result, they have proved that they were able to decrease the false positive rate in a scenario where there is a sudden flood of normal traffic.

A machine learning and deep learning model-based approach was suggested by Luong et al. [26] for creating a DDoS and defense system in SDN networks. Here deep neural network and machine learning models were used for investigation. As per the authors, the unified structure of SDN architecture encourages the problem of

DDoS attack, especially in SDN controller, and it can be minimized significantly by using the combination of machine learning and deep neural network models. Also, they have considered two different types of attack situations, one from outside the network and the other from the inside of the network.

Li et al. [27] proposed an innovative and effective method of programming the controller timing for identifying the DDoS attack in SDN. In normal conditions, the normalized time a switch has to wait, and the position of the attacked switch is used to schedule the controller. The experiments conducted by the researchers showed that the technique proposed decreases the delay and network establishment ratio drastically.

Wang et al. [28] emphasized for protecting the controller plane which is considered to be the most important component of SDN by presenting a concept of Safe Guard Scheme (SGS) in their work. In this, the main point stressed upon is by using the concept of multiple controllers placed in various places and grouped together as a cluster, which reduces the probability of DDoS attacks. Here, they have created two components in SGS, the first one focuses on anomaly traffic detection and the second one on controller dynamic defense. The anomaly traffic detection deals with the OpenFlow switches and segregates the fake traffic from genuine traffic by implementing a four-tuple feature vector. Here they have created a simulated environment and proved that the concept of SGS is very effective in detecting DDoS attack in SDN.

Tonkal et al. [29] used an algorithm using machine learning and Neighborhood Component Analysis for DDoS attack. The dataset used consists of 23 features which include important values like TCP, UDP, and ICMP data from genuine and fake traffic. The dataset had more than one lakh records and has features like packet per flow, byte count, duration in seconds, etc. The dataset is classified using kNN, Decision Tree, ANN, and SVM algorithms. It is concluded that decision tree has a better accuracy in detecting DDoS attack in SDN as compared with other machine learning algorithms.

From the above literature review, consideration of following problems associated with the existing DDoS attack detection is identified:

- Incapability to notice low-rate DDoS attack.
- Hefty load on the SDN controller to route, as devastating number of incoming packets within a very less time frame.
- Very high network resource usage.

On top of this, attack detection process is delayed due to illegal packets which put extra processing load on the SDN controller. Another problem area is specific methods which are used in case of reduced flow of traffic, which results in a high percentage of false positive rate. Also, till now, there is no proper approach to detect low-rate DDoS attacks with high accuracy and low false positive rate. As a result, an effective and high-performance technique of DDoS detection is required to evaluate, identify, and respond to DDoS attack incidents as early as possible. This will cause low or no negative effects in the SDN network functioning. The previous papers written till date are not able to give a convincing solution to questions like which dataset will be efficient and effective in finding the best solution for early and efficient DDoS attacks' detection in SDN at the same time consume less network resources. How are feature extraction methods to be integrated effectively in DDoS attack detection in SDN. Which methods and models (or framework) are best to identify the optimal features for DDOS attack detection in SDN. Which subset of features extracted from incoming packets in SDN can improve the classification performance of DDoS attack. How to implement the proposed framework in existing SDN to reduce the DDoS attacks.

8 A Proposed Solution

After reviewing all the existing papers above, a framework is proposed as depicted in Fig. 4 for early detection of DDoS attacks in SDN using ML algorithms. Firstly, the training data will be fed to the inference engine. After successful training of the proposed system, the testing data will be provided to the system. If the prediction is to accept, the incoming data packet will be only forwarded to the SDN controller/ OpenFlow switch for further processing in the SDN; otherwise, it will be dropped. Figure 4 shows the DDoS detection model in SDN.

The proposed model of the work is depicted clearly in Fig. 5. We will assess and create the dataset which will be relevant for SDN environment only. This dataset will be divided into two groups, training and testing data sets. The dataset will be fed to the proposed inference engine which will classify the packet entering as normal or attack type based on the various machine learning algorithms.

In the first phase of this, inference engine will do feature extraction to take only the most relevant features from the dataset. In the second phase, we will perform the feature selection to qualify the best parameters for classification of the incoming packets as normal or attack type.

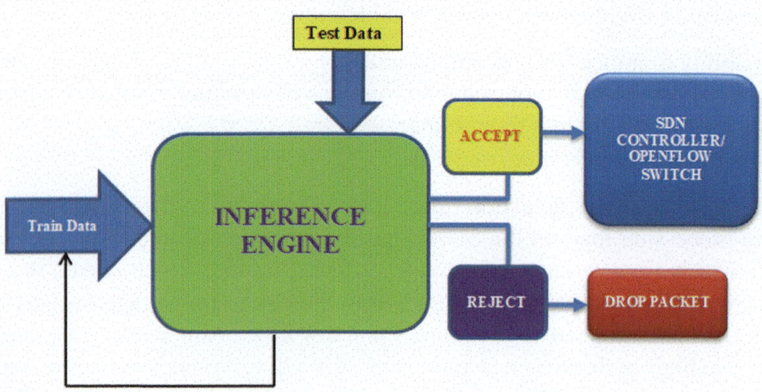

Fig. 4 DDoS detection model in SDN

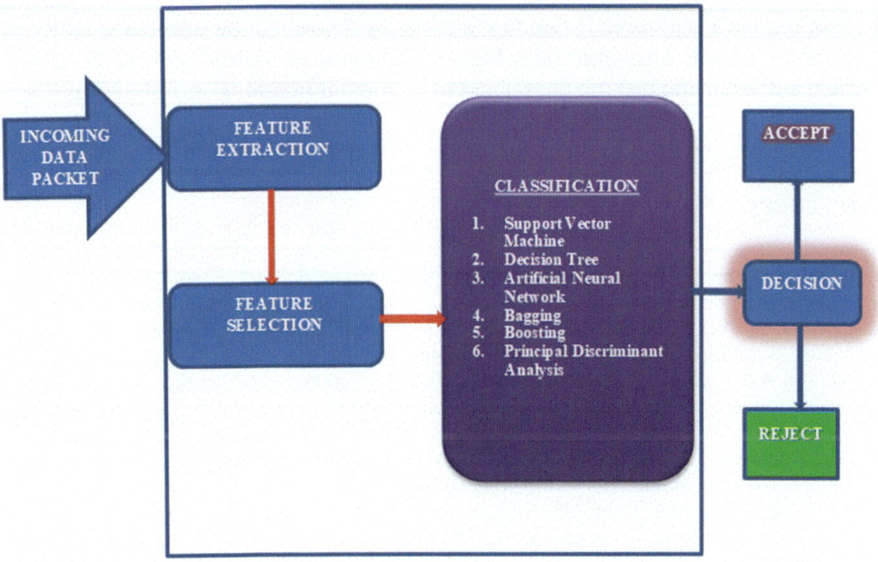

Fig. 5 Inference engine for DDoS attack in SDN

In the next phase, we will use the features selected using machine learning algorithms like SVM, decision tree, artificial neural network, bagging, boosting, and principal component analysis for training our model. Once trained, we will use the model for testing.

As stated above, the various machine learning algorithms will play a vital role in our proposed research work.

9 Conclusion and Future Work

Over the past several years, the DDoS attack has emerged as a big threat to SDN because of its frequent attack on the SDN networks. The DDoS attack, partially or fully, stops the genuine users of network services from accessing the network resources. A large amount of network resources is consumed to ensure the security of the SDN controllers. This factor leads to the poor performance of the SDN controllers in managing the entire network. The situation becomes tougher and stimulating as there are numerous categories of DDoS attacks against SDN. In order to protect SDN networks from DDoS attack one has to understand the basic characteristics of SDN; network traffic behavior when DDoS attack occurs against SDN; and behaviors of DDoS attack. Different studies on DDoS attacks against SDN had shown many specific clear characteristics and features that can be used as indicators and in turn help us to identify DDoS attacks.

Also, the proposed model stated in Section VI above will be implemented by us to find the best ML algorithm with highest accuracy, precision, and recall values. Also we will train and test the model and implement it in real-world SDN networks.

References

1. Chen J, Zheng X, Rong C (2015) Survey on software-defined networking. In: Second international conference on cloud computing and big data in Asia.Springer, Cham, pp 115–124
2. Leiner BM, Cerf VG, Clark DD, Kahn RE, Kleinrock L, Lynch DC, Wolff S (2009) A brief history of the Internet. ACM SIGCOMM Comput Commun Rev 39(5):22–31
3. Samson A, Gopalan NP (2016) Software defined networking: identification of pathways for security threats. In: Proceedings of the international conference on informatics and analytics, pp 1–6
4. Scott-Hayward S, Natarajan S, Sezer S (2015) A survey of security in software defined networks. IEEE Commun Surv Tutor 18(1):623–654
5. Kabir MH, Kabir MA, Islam MS, Mortuza MG, Mohiuddin M (2021) Performance analysis of mesh based enterprise network using RIP, EIGRP and OSPF routing protocols. Eng Proc 10(1):47
6. Dhir S, Agarwal A, Kapoor A (2022) Reduction of cryptography processing upward in BGP security protocols: an investigation. In: 2022 international mobile and embedded technology conference, pp 29–32
7. Willis P, Shenoy N (2019) A meshed tree protocol for loop avoidance in switched networks. In: 2019 international conference on computing, networking and communications pp 303–307
8. Rischke J, Salah H (2020) Software-defined networks. In: Computing in communication networks. Academic Press, pp 107–118
9. Fang S, Liu Q, Wu W (2021) HyperNAT: scaling up network address translation with smartNICs for clouds. In: 2021 IEEE global communications conference, pp 1–6
10. Chattopadhyay S, Sahoo AK (2021) A proposed approach to DDoS attacks detection on SDN using machine learning technique. In: Singh J, Kumar S, Choudhury U (eds) Innovations in cyber physical systems. Lecture notes in electrical engineering, vol 788. Springer, Singapore. https://doi.org/10.1007/978-981-16-4149-7_6
11. Izadi S, Ahmadi M, Nikbazm R (2022) Network traffic classification using convolutional neural network and ant-lion optimization. Comput Electr Eng 101:108024
12. Chuman T, Kiya H (2022) Security evaluation of block-based image encryption for vision transformer against jigsaw puzzle solver attack. In: 2022 IEEE 4th global conference on life sciences and technologies, pp 448–451
13. McKeown N, Anderson T, Balakrishnan H, Parulkar G, Peterson L, Rexford J, Turner J (2008) OpenFlow: enabling innovation in campus networks. ACM SIGCOMM Comput Commun Rev 38(2):69–74
14. Romanov O, Korniienko N, Burlaka H (2021) Construction of the SDN transport network model using the T-API interface. In: 2021 IEEE 4th international conference on advanced information and communication technologies, pp 220–224
15. Dhesikan S, Kathail P (2016) Broadcast and network integration using the network control (SDN) API. In: SMPTE 2016 annual technical conference and exhibition, pp 1–7
16. Abdulkarem HS, Dawod A (2020) DDoS attack detection and mitigation at SDN data plane layer. In: 2020 2nd global power, energy and communication conference, pp 322–326
17. Maity I, Misra S, Mandal C (2021) SCOPE: cost-efficient QoS-aware switch and controller placement in hybrid SDN. IEEE Syst J
18. Yungaicela-Naula NM, Vargas-Rosales C, Perez-Diaz JA (2021) SDN-based architecture for transport and application layer DDoS attack detection by using machine and deep learning. IEEE Access 9:108495–108512

19. Yan Q, Yu FR, Gong Q, Li J (2015) Software-defined networking (SDN) and distributed denial of service (DDoS) attacks in cloud computing environments: a survey, some research issues, and challenges. IEEE Commun Surv Tutor 18(1):602–622
20. Deore S, Patil A (2016) Survey denial of service classification and attack with protect mechanism for TCP SYN flooding attacks. Int J Eng Res Technol 3(5):1–4
21. Rajkumar MN (2013) A survey on latest DoS attacks: classification and defense mechanisms. Int J Innov Res Comput Commun Eng 1(8):1847–1860
22. Fenil E, Kumar PM (2022) Towards a secure software defined network with adaptive mitigation of DDoS attacks by machine learning approaches. In: 2022 international conference on advances in computing, communication and applied informatics, pp 1–13
23. Ahuja N, Singal G, Mukhopadhyay D, Kumar N (2021) Automated DDoS attack detection in software defined networking. J Netw Comput Appl 187:103108
24. Abdurohman M, Prasetiawan D, Yulianto FA (2017) Improving distributed denial of service (DDoS) detection using entropy method in software defined network (SDN). ComTech: Comput, Math Eng Appl 8(4):215–221
25. Liu Z, He Y, Wang W, Zhang B (2019) DDoS attack detection scheme based on entropy and PSO-BP neural network in SDN. China Commun 16(7):144–155
26. Luong TK, Tran TD, Le GT (2020) DDoS attack detection and defense in SDN based on machine learning. In: 2020 7th IEEE NAFOSTED conference on information and computer science, pp 31–35
27. Li S, Cui Y, Ni Y, Yan L (2019) An effective SDN controller scheduling method to defence DDoS attacks. Chin J Electron 28(2):404–407
28. Wang Y, Hu T, Tang G, Xie J, Lu J (2019) SGS: safe-guard scheme for protecting control plane against DDoS attacks in software-defined networking. IEEE Access 7:34699–34710
29. Tonkal Ö, Polat H, Başaran E, Cömert Z, Kocaoğlu R (2021) Machine learning approach equipped with neighbourhood component analysis for DDoS attack detection in software-defined networking. Electronics 10(11):1227

Development and Evaluation of an IoT-Based Gas Leakage Detection System Using Arduino Uno

Raj Kumar, Rubina Liyakat Khan, Ramveer Singh, Anupam Singh, Richa Vijay, and Danish Ather

Abstract The prime aims of the current study are to design, develop, and evaluate an Internet of Things (IoT)-based LPG leakage detection system utilizing the Arduino Uno microcontroller platform. The proposed system integrates various gas sensor detectors with Arduino Uno and establishes seamless communication with the IoT infrastructure to enable real-time monitoring and alerts. The study evaluates the system's performance in detecting LPG leakage events and assesses its reliability, sensitivity, and responsiveness. Further we explore the potential applications in domestic, industrial, and commercial environments of the built system. Additionally, the paper investigates the system's scalability, energy efficiency, and potential for integration into smart buildings and city infrastructures.

Keywords Internet of Things (IoT) · LPG gas leakage detection · Arduino Uno · Sensor integration · Smart building infrastructures

R. Kumar
Department of Computer Applications, Manav Rachna International Institute of Research and Studies (MRIIRS), Faridabad, India

R. L. Khan
The Applied College, Imam Abdulrahman Bin Faisal University, Dammam, Saudi Arabia

R. Singh
Department of CSE, Galgotias College of Engineering & Technology, Greater Noida, India

A. Singh
Department of Computer Science and Engineering, Graphic Era Hill University, Dehradun, India

R. Vijay
Amity University, Uttar Pradesh, Noida, India

D. Ather (✉)
Amity University in Tashkent, Tashkent, Uzbekistan
e-mail: danishather@gmail.com

S. Dutta et al. (eds.), *Cyber Intelligence and Information Retrieval*, Lecture Notes in Networks and Systems 1025, https://doi.org/10.1007/978-981-97-3594-5_25

1 Introduction

In recent years, the Internet of Things (IoT) has emerged as a ground-breaking paradigm with remarkable potential to revolutionize various sectors, including the safety and security industry. Among these sectors, the application of IoT in LPG leakage detection systems can significantly transform current approaches toward ensuring home safety, industrial safety, and environmental monitoring. Since the current solutions have been confined to conventional systems, which are fraught with multiple challenges like low efficiency, late detection, limited coverage, and lack of real-time alert mechanisms, this study provides a better alternative based on IoT using Arduino Uno.

The relentless advancements in technology have instigated a transformative shift in our daily life and business operations. This technology revolution has enabled interconnectivity among devices, empowering them to share and process information to accomplish complex tasks. One of the potential applications of IoT that could significantly enhance human safety pertains to LPG leakage detection systems. LPG leakage accidents are perilous incidents that have devastating consequences on both domestic and industrial fronts. These are primarily responsible for a significant number of accidents worldwide, leading to loss of life, property, and substantial economic damage. Traditional detection mechanisms for gas leaks have been quite successful in their application to some extent. Yet, the limitations of existing systems are becoming increasingly evident, particularly in the context of speed, area coverage, response time, and real-time communication as shown in Fig. 1.

Traditional LPG leakage detection systems rely on physical alarms or detectors, which often lead to delay in the detection process. They are designed to cover limited areas, failing to offer comprehensive detection solutions. Moreover, these systems

Fig. 1 LPG leakage accidents: example

lack real-time alert mechanisms and data analytics, which could offer significant advantages, such as rapid notification and preventative action. These gaps in the traditional system are where the IoT technology, coupled with the versatility of Arduino Uno, can truly shine.

IoT can provide a practical answer for these issues thanks to its real-time monitoring, data processing, remote access capabilities, and simple interaction with other technologies. By utilizing the power of IoT, we can increase detection speed, expand the coverage area, and enable immediate notifications and replies in the event of any leakage. IoT technology can be easily integrated and customized with Arduino Uno, a microcontroller board based on the ATmega328P, making it a suitable option for an improved LPG leakage detection system.

The current study was developed to address the drawbacks of current LPG leakage detection systems. It examines how the Internet of Things (IoT) and the Arduino Uno could transform present systems and provide a comprehensive and efficient detection approach. This study is strongly justified by the significance, urgency, and immense potential of the problem and the suggested solution. The primary objective of the project was to develop and evaluate an IoT-based LPG leak detection system using an Arduino Uno. It focuses on creating a system that can rapidly and accurately detect leaks while making advantage of IoT's data analytics and real-time alarming capabilities. The goal is to assess the system's performance in terms of speed, area coverage, and dependability in relation to that of more mature systems. The study also aims to assess the system's suitability for broader applications in various domestic and commercial settings.

Arduino Uno is anticipated to provide a number of benefits when used in the proposed system, including simplicity in programming, affordability, and adaptability in combining with other components and sensors. Because of its open-source design, it is a good fit for this study because it allows for ongoing improvements and adjustments. Additionally, the system's IoT connectivity guarantees that it can be quickly connected to the internet, enabling real-time data tracking, warnings, and remote access, considerably increasing the system's overall effectiveness.

2 Background and Motivation

Gas leaks are among the leading causes of both domestic and industrial accidents worldwide. These accidents not only lead to the loss of property but also endanger human lives. The severity of these accidents has necessitated the development of effective monitoring and early detection systems. Traditional detection systems, while having been moderately successful, exhibit limitations in terms of area coverage, detection speed, and response time. Additionally, they often fail to offer real-time data analysis and communication, which is critical for immediate action. These challenges have highlighted the need for more advanced, efficient, and comprehensive detection systems. The evolution of IoT technologies provides a promising platform to address these issues. IoT, with its real-time monitoring, data analytics, and remote access capabilities, coupled with the versatility of Arduino Uno, can significantly enhance the performance and effectiveness of LPG leakage detection systems.

The primary issue this research intends to address is the inadequacy of traditional LPG leakage detection systems. These systems, often marked by delayed detection, poor area coverage, and insufficient alert mechanisms, lead to significant response time, increasing the potential for catastrophic consequences. Furthermore, the lack of integration with modern communication and analytics technologies limits their overall effectiveness and the scope of preventive measures. In this context, the research will examine how an IoT-based system, using Arduino Uno, can overcome these limitations, providing an efficient and comprehensive solution for LPG leakage detection.

The study, in essence, strives to combine the potential of IoT and Arduino Uno in providing a comprehensive solution to a critical problem. By mitigating the limitations of the existing systems, the research aims to contribute substantially toward improving safety measures related to LPG leakages. Its potential impacts are far-reaching, including saving human lives, preventing property damage, and reducing economic losses caused by such accidents. The existing challenges posed by traditional LPG leakage detection systems call for innovative solutions. The application of IoT, combined with Arduino Uno, can pave the way for improved safety mechanisms. This research focuses on developing and evaluating such a system, contributing significantly to the domain of safety and security, and potentially saving numerous lives and properties from devastating accidents.

The motivation for the development and evaluation of an IoT-based LPG leakage detection system using Arduino Uno stems from a complex convergence of technological innovation, safety requirements, economic considerations, and societal demands. In the context of a rapidly growing technological landscape and an ever-increasing need for safety, the push for smarter solutions to old problems becomes paramount. This convergence creates a fertile ground for the innovation of a system designed to detect LPG leakage efficiently and effectively. This motivation is deeply rooted in several key factors as shown in Fig. 2.

2.1 Increasing Safety Concerns

LPG leakage, especially in household and industrial environments, is a common and often overlooked hazard. Leaks of combustible gases such as LPG, methane, and natural gas can lead to explosions, fires, and even loss of life. The current methods of gas detection are often slow, manual, and lack the real-time responsiveness needed to prevent accidents. An automated system that can provide continuous monitoring and instant alerts would significantly enhance safety measures in various settings.

Fig. 2 Motivation of the study

2.2 *Economic Considerations*

Traditional LPG leakage detection methods may involve costly equipment and manual inspection. By employing Arduino Uno and IoT, the proposed system offers a cost-effective solution. Arduino is not only affordable but also widely accessible, making it an attractive option for mass deployment. Additionally, real-time monitoring could prevent accidents, thus avoiding costly damages and litigation, further underlining the economic motivation.

2.3 *Technological Advancements*

With the advent of the Internet of Things (IoT), devices are now able to communicate and interact over the internet. This connectivity enables the seamless integration of sensors and devices to monitor and control various aspects of our environment. The Arduino Uno, an open-source microcontroller platform, offers a flexible and affordable means to build IoT devices. Leveraging these technological advancements for LPG leakage detection aligns with the trend toward smarter, interconnected systems that can enhance our quality of life.

2.4 Scalability and Integration

The flexibility of Arduino and IoT allows for scalability and integration into various applications, from domestic kitchens to large industrial plants. This adaptability encourages development, as the technology can be applied in numerous contexts, broadening its impact and utility.

2.5 Societal Expectations

As our society becomes more technology-driven, there is an expectation that solutions to common problems will become smarter and more efficient. This societal pressure adds to the motivation to develop solutions like the IoT-based LPG leakage detection system. Undetected LPG leakages can also have serious environmental consequences. Leaking gases can contribute to air pollution and greenhouse gas emissions. An effective detection system helps in minimizing these environmental impacts by providing early detection and facilitating immediate corrective action.

2.6 Global Reach and Social Impact

The universal applicability of such a system means that it could have a global reach. In developing regions, where safety systems might be lacking, affordable and efficient solutions like this could make a significant social impact.

2.7 Energy Efficiency

In addition to detecting leakage, intelligent monitoring can contribute to energy efficiency by ensuring that gases are utilized optimally, without waste. This aligns with broader global goals of energy conservation and sustainability.

2.8 Research and Development Opportunities

The field of IoT and sensor integration is still ripe for exploration. The development and evaluation of this system contribute to the broader body of research and provide opportunities for further innovation and refinement.

Based on above, the motivation to develop and evaluate an IoT-based LPG leakage detection system using Arduino Uno is multifaceted, driven by technological innovation, safety imperatives, economic considerations, environmental stewardship, regulatory compliance, societal expectations, and the potential for global impact. This convergence of factors makes the pursuit of such a system not only technologically exciting but also socially relevant and economically viable. It represents a step toward a future where our living and working environments are safer, smarter, and more responsive to our needs and the needs of our planet.

3 Literature Review

The application of the Internet of Things (IoT) in the safety and security industry is a rapidly developing field of interest for researchers worldwide. Coming section presents an overview of relevant studies and concepts pertinent to the topic.

In an early attempt, Al-Fuqaha provides an excellent starting point, laying down the foundational knowledge of IoT. The authors delve into the core components of IoT, emphasizing its capabilities in data collection, processing, and transmission. They also shed light on its potential in safety and security applications, which aligning well with the focus of current study [1].

A study conducted by Liu demonstrates a practical application of IoT in safety monitoring [2]. The study focuses on developing an IoT-based real-time safety monitoring system, demonstrating IoT's potential in safety-related applications. In conclusion the study underlines the advantages of IoT-based systems, such as real-time monitoring and swift response, which are integral to this study as well.

Arduino's utility as a platform for implementing IoT-based systems has been explored in depth by many. For instance, Shrouf gives a clear account of Arduino's usage in developing smart factories using IoT [3]. They highlighted the platform's flexibility and its ability to facilitate a seamless interconnection of physical and digital systems. The insights from their research can be invaluable in understanding the potential of Arduino Uno in our proposed system.

Literature specifically addressing the integration of IoT and Arduino for safety applications is relatively scant. However, some studies provide a useful precedent for our work. A case in point is the work by Shaikh [4], wherein the authors developed an IoT-based fire detection system using Arduino. Their results demonstrate the system's effectiveness, providing a basis for our research's feasibility. In the same line, Mokhtar developed an IoT-based fire alarm system using Arduino, showing a real-world application that mirrors our research context [5]. Their successful implementation and the demonstrated effectiveness of the system underscore the feasibility of our proposed study.

Evaluation methodologies for IoT-based systems have also been studied. Botta discuss various approaches for evaluating the performance of IoT-based systems, with an emphasis on performance indicators like efficiency and reliability [6]. These evaluation techniques can be instrumental in assessing IoT technology that forms the backbone of our proposed research model. A comprehensive understanding of its architecture, capabilities, and potential applications is therefore crucial. Gubbi presents a detailed exploration of the IoT paradigm, which includes its key building blocks, potential applications, and associated challenges [7]. Their discussion on IoT's role in developing smart environments provides a valuable basis for our proposed system.

Applying IoT to safety and security, especially in LPG leakage detection, has been an area of interest in recent studies. For instance, Liu in [2] built an IoT-based safety monitoring system, demonstrating the practical utility of IoT in such applications. Furthermore, research by Vishwakarma and Agrawal [8] specifically focuses on the development of an IoT-based LPG leakage detection and alerting system. Their work affirms the feasibility of IoT in this domain and inspires our research methodology.

The Arduino platform, especially the Arduino Uno, has been widely recognized for its adaptability and ease of use in various IoT applications. Banzi and Shiloh provide a comprehensive guide on Arduino, including its varied uses in creating interactive projects [9]. Their work offers practical insights into using Arduino Uno for our proposed system.

Performance evaluation of IoT-based systems has been a central theme in recent research works. For example, Ray discusses various performance indicators and evaluation methodologies for IoT systems. The discussed techniques provide a framework for assessing the performance of our proposed LPG leakage detection system [10].

4 Proposed Methodology

4.1 System Architecture and Components

The proposed IoT-based LPG leakage detection system as in Fig. 3 using Arduino Uno would consist of several integral components, each serving a unique purpose in the system.

Sensors. The system would primarily require gas sensors detectors. Suitable sensors might include MQ-series gas sensors (like MQ2, MQ5, and MQ9 for detecting different gases such as methane, LPG, and carbon monoxide). For the current study we propose to use MQ2 sensor to deduct the LPG leakage.

Arduino Uno. This microcontroller board serves as the central processing unit. It collects data from sensors and communicates with the IoT platform.

ESP8266 Wi-Fi Module. This Wi-Fi module enables the Arduino Uno to connect with the internet and send data to the IoT platform.

IoT Platform. An IoT platform like ThingSpeak, Blynk, or AWS IoT is used for data collection, processing, visualization, and triggering alerts in real-time. We rely on ThingSpeak for the current work for its own benefits.

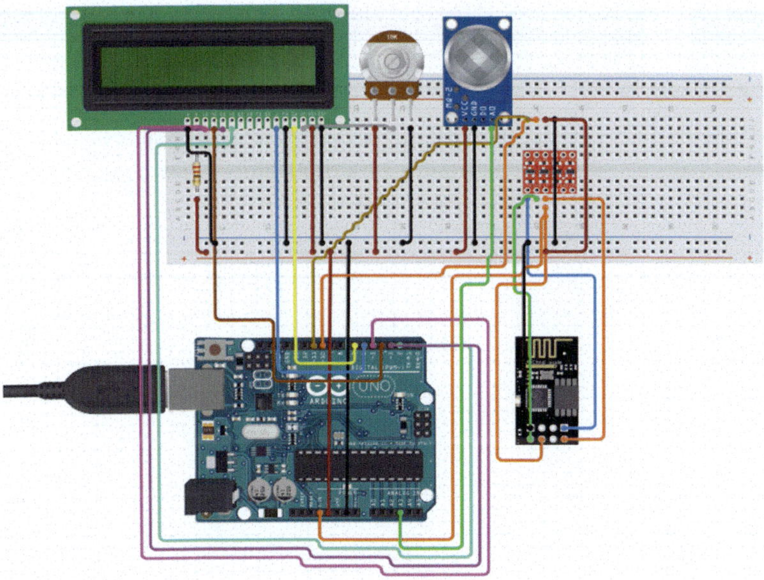

Fig. 3 Implementation circuit of LPG leakage system

Alarm System. The system includes a buzzer or similar alarm system to alert the local premises in case of a detected leak or fire. An appropriate alarm is being adapted.

Integration of these sensors with Arduino Uno involves connecting the sensor output to the analog input pins of the Arduino Uno. The Arduino board processes the sensor data and communicates it to the IoT platform via the ESP8266 Wi-Fi module as shown in Fig. 4.

4.2 Communication Protocols and IoT Connectivity

The communication between the Arduino Uno and the IoT platform occurs over Wi-Fi, using the ESP8266 Wi-Fi module. This communication employs standard protocols like HTTP or MQTT.

Hypertext Transfer Protocol (HTTP) is widely used due to its simplicity. However, it can be data-intensive and may not be suitable for continuous real-time monitoring.

Message Queuing Telemetry Transport (MQTT) is a lightweight publish-subscribe-based messaging protocol designed for low-bandwidth, high-latency networks, making it ideal for IoT applications. MQTT is a better choice for our system due to its low power consumption and real-time capabilities.

The Arduino Uno, with its Wi-Fi module, connects to the internet and sends data to the IoT platform, which can then be accessed from any internet-connected

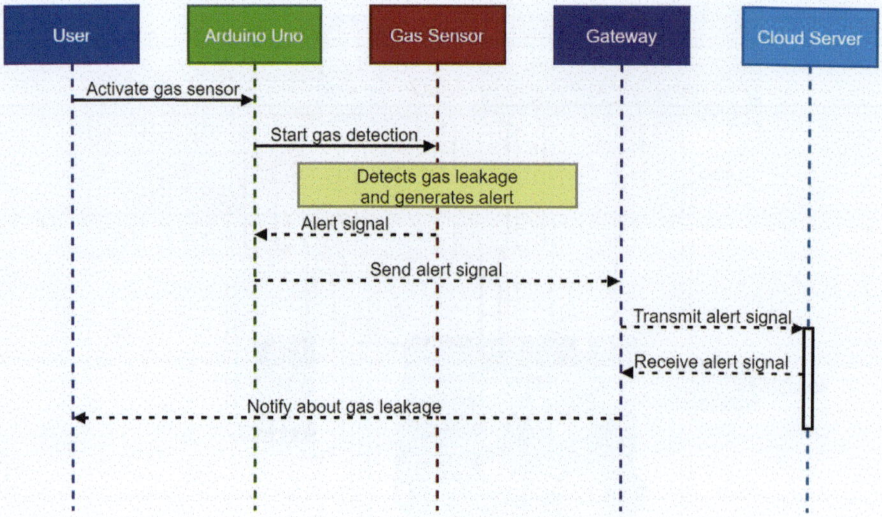

Fig. 4 Process flow

device. The IoT platform also allows for real-time data visualization and can be configured to send alerts or notifications based on specific conditions, such as when gas concentration exceeds a certain level, or a fire is detected.

5 Evaluation and Results

In the table above, "ppm" stands for parts per million, a common measure of gas concentration. The "Status" column indicates the system's response based on the detected concentration: "Normal" for safe levels, "Warning" for elevated but not immediately hazardous levels, and "Alarm" for potentially dangerous levels. The specific thresholds for each status may vary and should be set based on relevant safety guidelines.

An analysis of the data collected over a 30-min interval shows a positive correlation between the expected LPG concentration (ppm) and the actual reading obtained from the MQ2 sensor as shown in Tables 1 and 2.

At the 0-min mark, both the expected and actual sensor readings indicated a normal status with no presence of LPG. As time progressed and the LPG concentration increased, the MQ2 sensor readings similarly escalated, showing the sensor's responsiveness to changes in LPG concentration levels.

At the 5-min interval, the expected concentration was 300 ppm, and the sensor reading was 290 ppm. This shows a minimal discrepancy of 10 ppm between the expected and recorded values. Similarly, at the 10-min mark, the sensor reading was 510 ppm, which is quite close to the expected concentration of 500 ppm.

Table 1 LPG cooking detection at different time intervals: expected

Time interval	Reading using MQ2 sensor	Status
0	0	Normal
5	200	Normal
10	500	Normal
15	800	Normal
20	1200	Warning
25	1500	Warning
30	2000	Alarm

Table 2 LPG cooking detection at different time intervals: actual

Time interval	Reading using MQ2 sensor	Status
0	0	Normal
5	290	Normal
10	510	Normal
15	820	Normal
20	1230	Warning
25	1450	Warning
30	1900	Alarm

At the 15-min mark, the LPG concentration was expected to be 800 ppm, and the recorded sensor reading was 820 ppm. The slightly higher sensor reading could be attributed to sensor sensitivity or external environmental factors.

During the 20-min and 25-min intervals, the status shifted from "Normal" to "Warning", indicative of elevated LPG levels. The MQ2 sensor readings for these time intervals were 1230 ppm and 1450 ppm, respectively, demonstrating a minor difference from the expected values.

At the 30-min interval, when the status reached "Alarm" indicating hazardous LPG levels, the expected concentration was 2000 ppm, and the MQ2 sensor recorded a value of 1900 ppm. This shows that the sensor responded correctly to the critical increase in LPG concentration, even though the reading was slightly less than expected.

Overall, the MQ2 sensor showed a consistent response to changes in LPG concentration levels as represented in Fig. 5. Even though there were minor discrepancies between the expected and actual sensor readings, these variations were within an acceptable range. Hence, the MQ2 sensor appears to be reliable and effective in detecting and tracking changes in LPG concentrations in this hypothetical scenario.

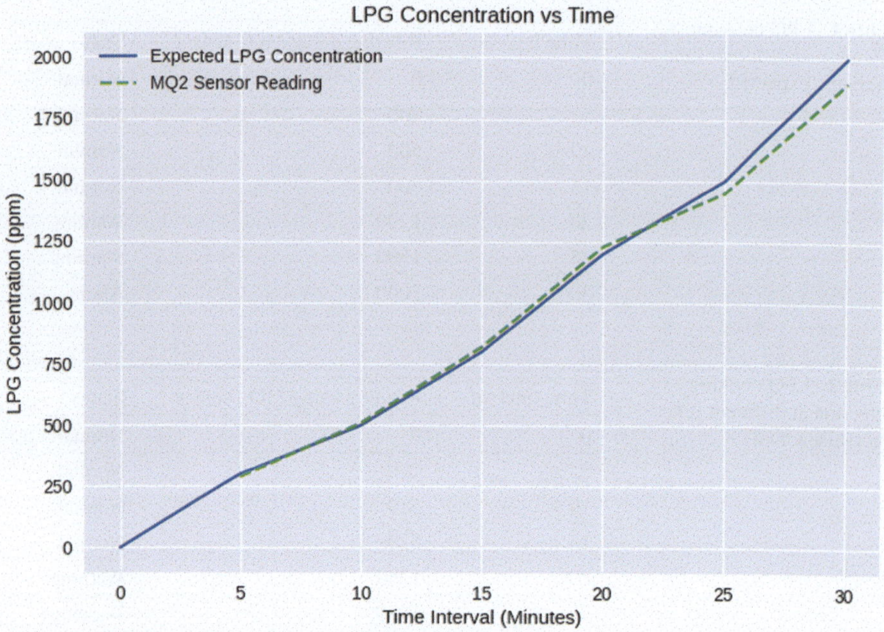

Fig. 5 Comparison of expected versus actual readings

6 Conclusion

In conclusion, our analysis demonstrates the effective performance of the MQ2 sensor in tracking changes in LPG concentrations. The sensor readings closely mirrored the expected concentration levels, with only minor discrepancies. These findings suggest that the MQ2 sensor, coupled with an IoT framework and Arduino Uno, could provide a robust solution for LPG leakage detection. This research paves the way for developing advanced, real-time monitoring systems to enhance safety in environments susceptible to gas incidents, contributing to a safer and more secure society.

7 Future Work

In the future we plan to expand on this research by integrating more sensor types into the system to detect a wider range of gases. It would be beneficial to explore the system's performance under varying environmental conditions, such as different temperatures and humidity levels. This could lead to improvements in the sensor's calibration process. Additionally, exploring other IoT platforms and communication protocols could lead to optimized performance and increased system robustness.

Further, we would like to test our system in real-world settings, across residential and industrial environments, to validate its efficacy and make necessary modifications to ensure its practical applicability.

References

1. Al-Fuqaha A, Guizani M, Mohammadi M, Aledhari M, Ayyash M (2015) Internet of things: a survey on enabling technologies, protocols, and applications. IEEE Commun Surv Tutor 17(4):2347–2376
2. Liu H, Yang B, Liu H, Liu X (2018) Research on safety monitoring system of coal mine based on the internet of things. IOP Conf Ser Earth Environ Sci 170(4):042074
3. Jain A, Sarkar A, Ather D, Raj D (2022) Temperature based automatic fan speed control system using Arduino. SSRN Electron J. https://doi.org/10.2139/ssrn.4159188
4. Khan RL, Priyanshu D, Ather D, Allataifeh H (2022) An implementation of internet of things-based live temperature and humidity monitoring system. In: Proceedings of the 2022 11th international conference on system modeling and advancement in research trends, SMART 2022. https://doi.org/10.1109/SMART55829.2022.10046723
5. Mokhtar M, Eldin AS, Aborokbah A, Alotaibi A (2018) Fire alarm system based on Arduino UNO and IoT. In: 2018 IEEE 7th international conference on computer, control and communication (IEEE-IC4), Karachi, 2018, pp 1–5
6. Botta A, de Donato W, Persico V, Pescapé A (2016) Integration of cloud computing and internet of things: a survey. Futur Gener Comput Syst 56:684–700
7. Agarwal AK, Ather D, Astya R, Parygin D, Garg A, Raj D (2021) Analysis of environmental factors for smart farming: an internet of things based approach. In: 2021 10th international conference on system modeling and advancement in research trends (SMART), 2021, pp 210–214
8. Vishwakarma S, Agrawal S (2019) IoT based gas leakage detection system. In: 2019 Amity international conference on artificial intelligence (AICAI), Dubai, United Arab Emirates, 2019, pp 909–914. 2(5):99–110
9. Banzi M, Shiloh M (2014) Getting started with Arduino: the open source electronics prototyping platform. Maker Media, Inc.
10. Ray PP (2016) A survey on internet of things architectures. J King Saud Univ Comput Inf Sci 30(3):291–319

Optimized Depth-Wise Separable Convolution Neural Network for Lung and Colon Cancer Detection

Amit Seth and Vandana Dixit Kaushik

Abstract In this study, we develop lung disease data classification using deep learning technique. There are three stages in the proposed method namely, preprocessing, segmentation, and classification. Initially, CT images are collected and preprocessed. A modified median filter is used for preprocessing. Then, the preprocessed images are fed to the segmentation process. In order to segment the image, K-means clustering is used. Then, this segmented region is fed to the input of an optimal depth-wise segmented CNN classifier to classify the image as a normal or cancerous image. To optimize the hyperparameter of the classifier Coati Optimization Algorithm (COA) is used. The suggested approach performs better, with a 95.07% accuracy rate when evaluated using several measures.

Keywords Lung cancer · Colon cancer · Modified median filter · K-means · DSCNN · Coati optimization algorithm

1 Introduction

A cancerous disease is one that damages the organs in the body. It covers a wide range of diseases that cause damage to the organs in the body [1]. Infection and metastasis are caused by undifferentiated cells that grow too fast and spread to neighbouring tissues. The most common cause of cancer-related death is metastasis. It has been observed that the prevalence of cancer has increased significantly over time as a result of increasing susceptibility to specific cancers with age [2]. Colorectal cancer affects both the colon and the rectum and is one of the most common causes of death. As a leading cause of death, lung cancer (12.2%) is the most commonly recognized

A. Seth (✉) · V. D. Kaushik
Department of Computer Science and Engineering, Harcourt Butler Technical University, Kanpur, Uttar Pradesh, India
e-mail: 200304004@hbtu.ac.in

V. D. Kaushik
e-mail: vdkaushik@hbtu.ac.in

S. Dutta et al. (eds.), *Cyber Intelligence and Information Retrieval*, Lecture Notes in Networks and Systems 1025, https://doi.org/10.1007/978-981-97-3594-5_26

malignancy, however colon cancer (10.7%), which affects the colon and rectum, is also a major cause of death [3].

Globally, it is among the main causes of death, even though its impact is unevenly distributed. There are several approaches to identifying cancer symptoms [4]. Data is being stored in archives at an ever-increasing rate due to technological advancements. Using traditional methods, it is difficult to analyze a huge quantity of information and extrapolate the results. With the increasing availability of health data, scientists have an opportunity to enhance existing approaches in order to conduct a more comprehensive analysis of clinical cases [5]. Biomedical data is commonly analyzed using machine learning and deep learning technologies. This study proposed a lung disease data classification using a deep learning technique. This article is organized as follows: Sect. 2 reviews some Lung and colon cancer detection models. K-means–DSCNN is used to automatically segment and classify images for multiple diseases as outlined in Sect. 3. Section 4 discusses the effectiveness of K-means–DSCNN. The results of the experiment are presented in Sect. 5 of the paper.

2 Literature Survey

As an alternative to the techniques used now for cancer detection, Mehmood et al. [6] proposed an accurate and computationally efficient model for lung and colon cancers. The training and validation process of this study was conducted using a large dataset of lung and colon histopathology images. The dataset consisted of 25,000 images of histopathology that were divided into five classes according to their histology. The four layers of an already-trained neural network (alexnet) were altered prior to training on the dataset. With an overall accuracy of 89%, the initial classification results were promising for all but one class of images. A contrast enhancement technique was used to improve the image quality of the underperforming class, rather than performing image enhancement techniques on the entire dataset, in order to improve the overall accuracy and computational efficiency of the model. Previous forecast rates did not improve. Srinivasan and Asunda [7] used deep learning to locate cancerous lung nodules. An advanced feature extraction technique called Histogram of Oriented Gradients (HoG) was used here, along with wavelet-based features, system-invariant feature transformation (SIFT), a local binary pattern (LBP), and Zernike Moments. To select the best feature, Fuzzy Particle Swarm Optimization (FPSO) algorithm was used after extracting texture, geometric, volumetric, and density features.Following this, deep learning was used to classify the features. The computational complexity of CNN was reduced by a new FPSOCNN algorithm. Compared to other techniques, FPSOCNN performed better in the experiments. However, it needs further improvement in terms of accuracy in classifying pulmonary nodules.

Based on histopathological images, Chehade et al. [8] proposed a method to establish a computer-aided diagnostic system that is capable of accurately classifying five types of colon and lung cancer tissuesIn order to classify histopathological images of lung and colon cancers obtained from the LC25000 dataset, five models were used: XGBoost, SVM, RF, LDA and MLP. Since machine learning models are based on feature engineering, they are able to provide a more accurate description of the classification model. Based on the results obtained from the experiments, machine learning models were found to produce satisfactory results and to be highly accurate in the detection of lung and colon cancer subtypes. The selection of features in this method needs to be improved. Bhatia, Sinha and Goel [9] considered preprocessing techniques to highlight cancer-prone lung areas and extract features. For several classifiers, features were provided. A CT scan can be predicted as cancerous based on XGBoost and random forest, as well as the combined use of individual predictors. Compared to previous attempts, the Lung Image Database Consortium image collection (LIDC-IRDI) achieved an accuracy of 84%. The accuracy of this work needs to be improved.

Shafi et al. [10] used data from colon cancer patients to analyze and predict using a machine learning approach based on random forest classification. In a random forest classifier, this method combined "mean reduction accuracy" and "mean reduction Gini" to make fast computations. Experimental results demonstrated that this model was effective and favorable. It wasn't advanced enough to use sophisticated feature selection methods. Singh and Gupta [11] proposed an efficient method for detecting and classifying lung cancer-related CT scan images as benign or malignant. Using supervised learning algorithms, these images were first processed using image processing techniques. A combination of texture features and statistical features was extracted, and various extracted features were then fed into a classifier. There were seven different classifiers used, including k-nearest neighbors, support vector machines, and decision trees. To train and test these classifiers, 15750 clinical images containing 6910 benign and 8840 malignant lung cancer images were used. Compared to other classifiers, the MLP classifier achieved an accuracy of 88.55%. The accuracy of this work needs to be improved.

3 Proposed Lung and Colon Cancer Detection Model

The purpose of this work is to propose a deep learning-based multi-level lung and colon cancer detection technique that combines DSCNN with K-means clustering. Figure 1 shows the overall structure of the proposed work. In the initial stage, CT images are obtained from the dataset and preprocessed using a median filter. Then, the preprocessed image segmented using the K-means clustering algorithm. After that, the segmented image given as input to the DSCNN classifier, which classifies the segmented image as normal and diseased image. The hyper parameter of the DSCNN is optimized using COA optimization algorithm.

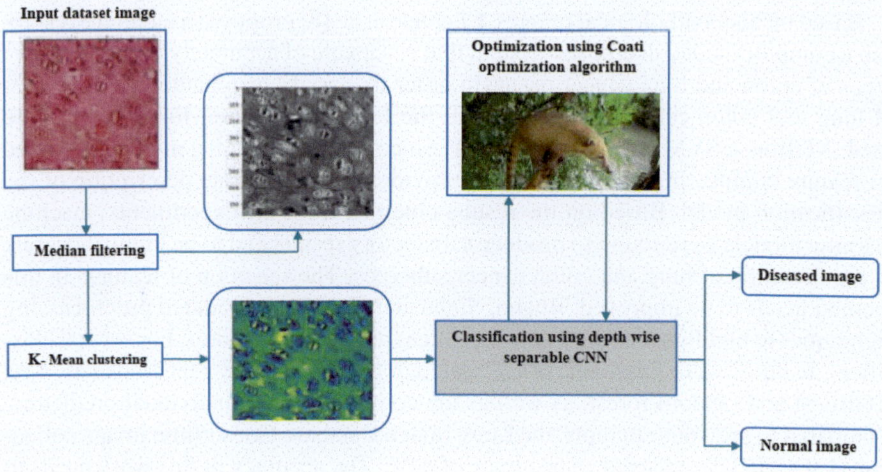

Fig. 1 Architectural diagram for proposed lung and colon cancer detection

3.1 Modified Median Filter-Based Preprocessing

Median filters are most effective when noise levels are below 0.2. When the noise is suppressed, they ensure that details on the screen will not disappear. The adaptive median filter is designed with a W_{ld} window, as with any other filter. In the adaptive median filter, the window size can be modified and, regardless of the modification, the result of the conducted operation will remain one and single. The center pixel value of the window can be used to adjust the result. Then, the preprocessed image is given as input to a K-means clustering technique for segmentation.

3.2 Segmentation Using K-Means Clustering

A clustering algorithm is a technique commonly used for segmenting data into groups defined by the user. K-means clustering is one of the most widely used and popular clustering algorithms because of its simplicity and low computational requirements. Clustering with k-means employs an iterative procedure in which n observations are iteratively assigned to one of k clusters defined by centroids. The number k is selected before the algorithm is run.

It may be assumed that $Y = \{y_1, y_2, y_3, \ldots, y_j\}$ is a set of j data points and $C = \{c_1, c_2, c_3, \ldots, c_i\}$ is a set of cluster centers. The n data points should be clustered into m cluster centers. According to Eq. (1), K-means clustering has an objective function.

$$N = \sum_{m=1}^{i} \sum_{n=1}^{j} \|y_m - c_n\|^2 \tag{1}$$

where N is the objective function, j represents the number of data, c_n represents the nth cluster center, and y_m represents the data point. As a result of minimizing this objective function, optimal centers are obtained. It is important to note that the results of K-means clustering are highly dependent upon the initial cluster center. As a result, it is important to select cluster centers carefully.

Since k-means operates on 1D data, the 3D lung segmentation data must first be transformed into a vector. Using cluster number 3, the vector is clustered using the k-means algorithm. Cluster 3 is selected because the dissected lung contains soft tissue, background, and tumor bronchi. A tumor feature is selected from the clustered vector that has the lowest probability, which includes the airways as well.

3.3 Classification Using Depth-Wise Separable Convolution Neural Network

After segmentation, the DSCNN classifier is used to classify the clustered image as normal and abnormal. Laurent Sifre [14] developed depth-wise separable curves for the purpose of image classification. There are two types of factorized curves: DS separable curves and DS singular curves. Figure 2 illustrates structure of DSCNN. The former divides a fixed cycle into depths and points. A single filter is used for each input channel in Deep Curve to perform lightweight filtering. In order to linearly combine the input channels, point-wise convolution is followed by 1×1 convolutions. The DS algorithm provides a convolution with a factorized convolutional function of 3 layers, a spatial filter of 1 layer and a phase filter of 1 layer. Using DS convolution instead of standard convolution, one layer can be used for spatial filtering and another layer for concatenation. Redundant computation and mode size can be greatly reduced by DS convolution.

An in $H \times W \times I_{in}$ input feature map M is received by the standard convolution layer, which then executes a in out $K \times K \times I_{in} \times I_{out}$ convolution kernel k to produce an out $H \times W \times I_{out}$ output feature map o, where H and W represent the height and width of input feature maps, K represents the square kernel dimension, in I_{in} and out I_{out} are represent the number of input channels and the number of outcome channels, respectively. As a result of zero padding, the resulting feature map has the same spatial extent as the input feature map. The calculation difficulty of a normal convolutional algorithm is $K^2 \cdot I_{in} \cdot I_{out} \cdot H \cdot W$ (the multiplication and accumulation function counts).

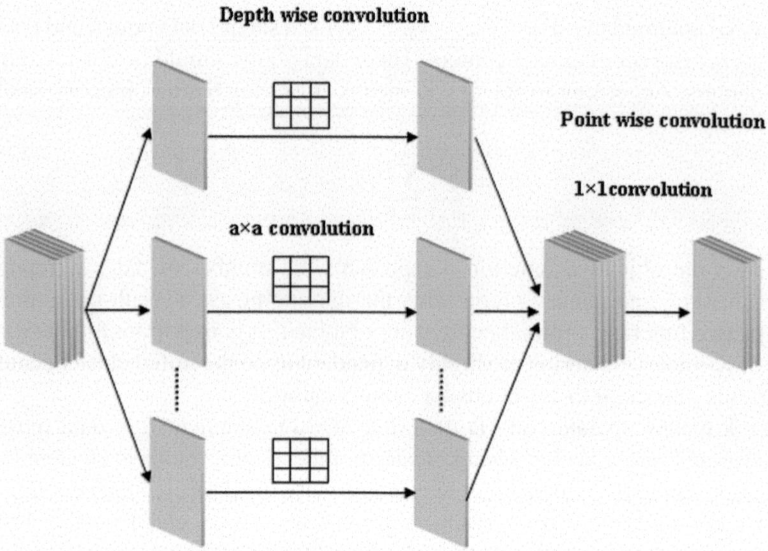

Fig. 2 Structure of DSCNN

In this method, depth-wise convolutional algorithms are used to filter input feature maps across all channels. A depth-wise convolutional function can be described by Eq. (2).

$$D(z, y, n) = \sum_{i=1}^{K}\sum_{i=1}^{K} k(u, u, n) \times M(z + u - 1, y + u - 1, n) \tag{2}$$

k refers to the length of the convolutional kernels in depth of size $k \times k \times I_{in}$. The rth filter in k is applied to the rth channel in M in order to generate the rth channel in D. A point-wise convolution is constructed by using depth-wise convolutions using 1×1 convolutions to construct novel features. A point-wise convolution can be expressed by the Eq. (3).

$$o(z, y, l) = \sum_{n=1}^{I_{in}} D(z, y, n) \times p(n, l) \tag{3}$$

This convolutional kernel has a size of $1 \times 1 \times I\text{in} \times I\text{out}$. By modifying m, the outcome feature map is modified in terms of the number of channels. It is similar to the dense 1×1 convolutional function which does not impose any restriction on locality, so the parameter does not need to be rearranged in memory. After this, a highly optimized matrix multiplication technique is applied directly. It is difficult to express a DWS convolutional calculation as Eq. (4):

$$I_d = K^2 \cdot I_{in} \cdot H \cdot W + I_{in} \cdot I_{out} \cdot H \cdot W \quad (4)$$

Fundamentally, it is an estimate of depth and point that convolutional takes into account. The calculation is simplified by a DS convolution factor, which can be written as (5).

$$\eta = \frac{I_d}{I_{std}} = \frac{K^2 I_{in} HW + I_{in} I_{out} HW}{K^2 I_{in} I_{out} HW} = \frac{1}{I_{in}} + \frac{1}{K^2} \quad (5)$$

There are many cases in which m is extremely significant; therefore the factor η is approximately similar to $1/k^2$. In this paper, 3×3 DS convolutional is used, which reduces computation complexity and parameter count by 7–8 times. The hyper-parameters of this DSCNN are optimized using the following Coati Optimization Algorithm.

3.3.1 Hyper Parameter Optimization Using the COA Algorithm

In this section, we present a mathematical description of the steps of the Coati Optimization Algorithm (COA). This species is a member of the Procyonidae family and is found in the genera Nasua and Nasuella, and is also known as the Cotymundis. There is a diurnal species of this species of mammals found in the Mexico, southwest United States, South America, and Central America. As a signaling and balancing device, they have a long, non-prehensile tail that acts as a signaling and balancing device with an elongated, flexible nose, small ears, black paws, and an elongated, flexible nose. From head to tip of tail, the adult coati measures 33–69 cm, which is as long as their body. An average coati weighs between two and eight kilograms and stands 30 cm tall at the shoulder. There is no limit to the size of males, and their fangs are sharp and large. Males can reach twice the size of females. South American coatis and white noses are measured in these measurements. Among the two coati species, the mountain coati is the smallest. As a result of their intelligent behavior, iguanas apply strategies to attack other iguanas and to escape from predators. A population-based metaheuristic is applied to the COA approach, which considers the coatis as members of a population. Developing the proposed COA method is based on the simulation of natural coati behaviors.

Step 1: Initialization

A population-based metaheuristic is central to the COA approach, which considers coatis as members of a population within the system. The initialization process is an essential component of every optimization problem. In this case, solutions are defined in terms of height (h), weight (w), spatial dimensions (d), and count of input (i) and output (o) channels. The solution refers to the crocodile. ACO initializes its solutions randomly at the beginning. In the beginning, crocodiles are formed from these random solutions. Equation (6) gives the initial form of the solution.

$$\{Y_i\} : Y_{i,j} = LB_j + R \cdot (UB_j - LB_j), \quad i = 1, 2, 3, \ldots, m, \quad j = 1, 2, 3, \ldots n \tag{6}$$

$$y_{i,j} = [h_j, w_j, d_j, i_j, o_j] \tag{7}$$

where represents the *i*th coati's position, *m* represents number of coatis, *R* is the random real number with the interval of [0, 1] and upper bound and lower bound of the *j*th decision variable, respectively.

Here, Y_i is the position of the *i*th coati, *m* is the number of coatis, *R* is the random real number with the interval [0, 1], and respectively UB_j and LB_j are the upper and lower bounds of the *j*th decision variable.

Step 2: Objective Function

After initialization, there is a need to determine the objective function of the solution. The fitness of each solution is determined based on accuracy. It can be defined using the Eq. (8),

$$Fitness = \max(accuracy) \tag{8}$$

Step 3: Update the Solution

As part of the optimization process in COA, two natural behaviors of coatis are modeled, including their strategy when attacking iguanas (exploration) and their defense strategy when escaping from predators (exploitation). Based on these two different phases COA population is updated.

Phase1: Exploration

Simulated locust populations in the search space are updated according to their attack strategy against iguanas. Several coatis climb a tree and approach an iguana in an attempt to scare it away, while others wait under the tree until the iguana falls to the ground and attack the animal. In global search for problem-solving solutions, COA shows the capability of coatis to change positions in the search space in order to explore new information.

COA assumes that the iguana is the most successful member of the population based on its COA design. It is believed that the remaining half of the coatis will wait until the iguana falls to the ground before climbing up the tree. Thus, Eq. (9), a mathematical model, simulates the position of the coatis as they rise from the tree

$$Y_i^{P1} : Y_{i,j}^{P1} = y_{ij} + R \cdot (IG_j - I \cdot y_{ij}), \text{ for } i = 1, 2, 3, \ldots, \left\lfloor \frac{M}{2} \right\rfloor \text{ and } j = 1, 2, 3, \ldots, n. \tag{9}$$

A random position is assigned to the iguana (IG) in the search space. Equations (10) and (11) are used to simulate coatis moving in the search space given a random position on the ground.

$$IG^g : IG_j^g = LB_j + R \cdot (UB_j - LB_j),\ j = 1, 2, 3, \ldots, n, \tag{10}$$

$$Y_i^{P1} : y_{i,j}^{P1} = \begin{cases} y_{i,j} + R \cdot (IG_j^g - I \cdot y_{i,j}), & Fitness_{IG^g} < Fitness_i, \\ y_{i,j} + R \cdot (y_{i,j} - IG_j^g), & \text{else, for } i = \lfloor \frac{M}{2} \rfloor + 1,\ \lfloor \frac{M}{2} \rfloor + 2, \ldots, m \\ & \text{and } j = 1, 2, 3, \ldots, n \end{cases} \tag{11}$$

Providing that the new positions for each coati improve the value of the objective function, the new positions for each coati are acceptable for the update process. Otherwise, the coati will remain in its previous position. In the case of $i = 1, 2, ..., M$ in the simulation, the following update condition applies (Eq. (12)).

$$Y_i = \begin{cases} Y_i^{P1}, & Fitness_i^{P1} < Fitness_i, \\ Y_i, & else. \end{cases} \tag{12}$$

where Y_i^{P1} represents the recent position calculated for the ith coati, $y_{i,j}^{P1}$ represents the new position's jth dimension, $Fitness_i^{P1}$ is the objective function value, IG represents the position of an iguana in a search space, $Fitness_{IG^g}$ represents the objective function, IG_j represented its jth dimension, I represents the randomly selected integer from $\{1, 2\}$, IG^g represents randomly selected position of the iguana on ground, IG^g represents its dimension, and $\lfloor \cdot \rfloor$ represents the supreme integer function.

Phase 2: Exploitation

It is based on the natural behavior of coatis when interacting with predators and escaping from them that the second phase is based on after determining the location of coatis in the search space. When a predator attacks the coati, it escapes from its position. The activities of Coati in this strategy have placed it in a safe position close to its present location, providing evidence of its ability to exploit local resources.

For the purpose of simulating this behavior, Eqs. (13) and (14) generate a random position near the position of each coati.

$$LB_j^{local} = \frac{LB_j}{t},\ UB_j^{local} = \frac{UB_j}{t},\ \text{where } t = 1, 2, 3, \ldots, T. \tag{13}$$

$$Y_i^{P2} : y_{i,j}^{P2} = y_{i,j} + (1 - 2R) \cdot (LB_j^{local} + R \cdot (UB_j^{local} - LB_j^{local})),\ i = 1, 2, 3, \ldots, M,\ j = 1, 2, 3, \ldots, n, \tag{14}$$

When the new position improves the objective function that Eq. (15), it is acceptable to calculate the new position.

$$Y_i = \begin{cases} Y_i^{P2}, \ Fitness_i^{P2} < Fitness_i, \\ Y_i, \quad else, \end{cases} \tag{15}$$

Based on the second phase, the new position calculated for the ith coati is Y_i^{P2}, $y_{i,j}^{P2}$ represents its jth dimension, $Fitness_i^{P2}$ represents the objective function, and t represents the iteration counter.

Step 4: Termination

COA iteration is conducted after completion of the first and second phases in order to update all nodes in the search space. Based on Eqs. (9) to (15), the population is updated through a number of iterations of the algorithm. Once the run is complete, the COA algorithm returns the best solution obtained during all iterations.

4 Result and Discussion

This study proposes a lung and colon cancer prediction approach using a combination of K-means and DSCNN algorithms. An Intel Core i7 processor with 8 GB of memory running Windows 10 was used to simulate the proposed method. The experimental used images are collected from https://www.kaggle.com/datasets/andrewmvd/lung-and-colon-cancer-histopathological-images. The dataset consist of 25,000 images and five different classes. The performance of proposed approach is analyzed based on the different metrics namely, accuracy, sensitivity, specificity, precision, F-score, dice sore, (DS), and Jaccard index (JI).

4.1 Experimental Results:

As shown in Table 1, the lung squamous cell carcinoma disease was classified according to a comparative classification system. Here, the DS and the JI of our proposed method is 0.915 and 0.862. As we see in Table 1, it is higher than K-mean & ANN and SVM & CNN. A comparison was made between the proposed method and two existing methods: K-means and artificial neural networks (K-means & ANNs), convolution neural networks and support vector machines (CNNs & SVMs).

On the basis of the DS and the JI, the segment result is compared. This results in better results than the existing techniques as a result of our proposed technique.

Table 1 Comparative result evaluation for the proposed and existing method

Techniques	Input image	Segmented image	DS	JI
Proposed K-means & DSCNN	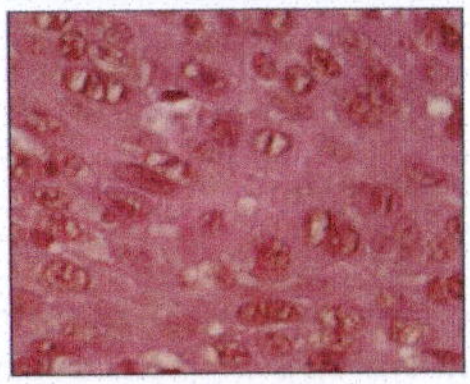	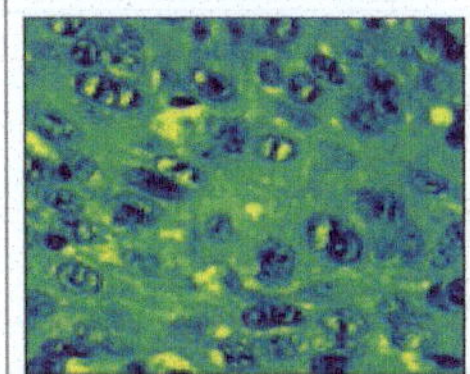	0.915	0.862
K-means & ANN	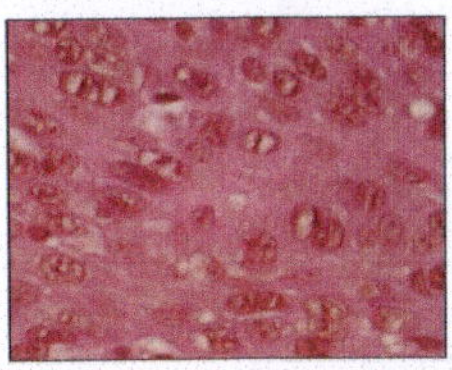	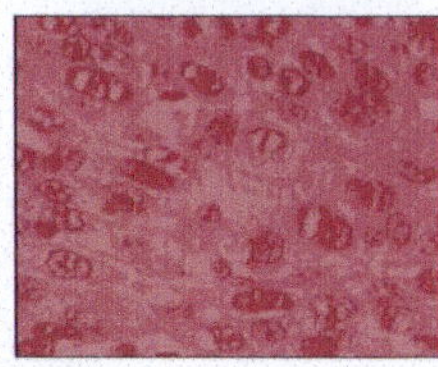	0.632	0.461
CNN & SVM	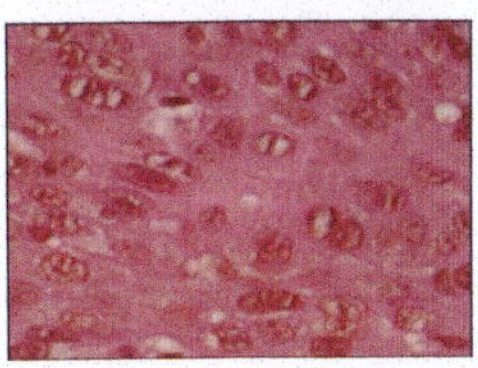	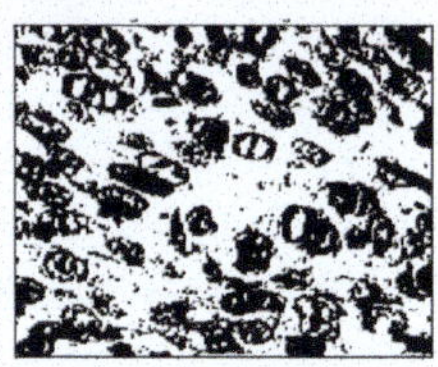	0.738	0.421

Figure 3 shows the training and validation accuracy and loss results, ROC curve and confusion matrix for the proposed work. Figure 4 represents the quantitative evaluation results for the proposed work. Our proposed work achieves 95.07% accuracy, 96.28% sensitivity, 93.90% specificity, 93.79% precision, and 95.02% F-scores of measurements medians. The proposed K-means & DSCNN technique improved the accuracy by 7.63% and 4.27% over the existing K-means & ANN and CNN & SVM techniques. Similarly, the sensitivity of our proposed work is improved by 6.68% from K-means & ANN and 4.45% from CNN & SVM. The specificity of the proposed work is improved by 8.49% and 4.87% over the existing K-methods & ANN and CNN & SVM techniques. The precision of our proposed work is improved by 8.73% from K-means & ANN and 4.33% from CNN & SVM. Also, the F1-score of the proposed work is improved by 7.65% from K-means & ANN and 4.39% from CNN & SVM. The comparative metrics evaluation indicates that the results obtained by our proposed method are superior to the results obtained by existing techniques.

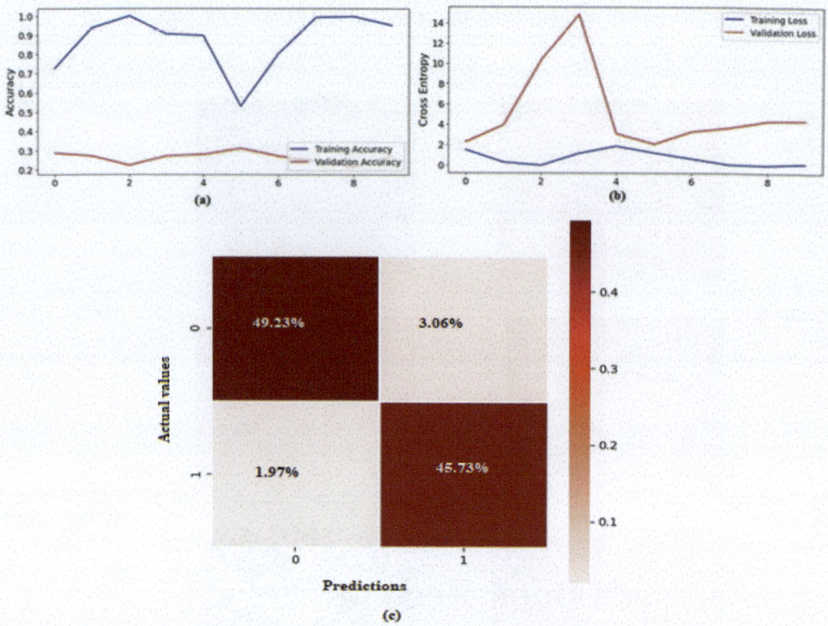

Fig. 3 **a** Training and validation accuracy, **b** training and validation loss results, and **c** confusion matrix for the proposed work

Fig. 4 Metrics evaluation for the proposed and existing techniques

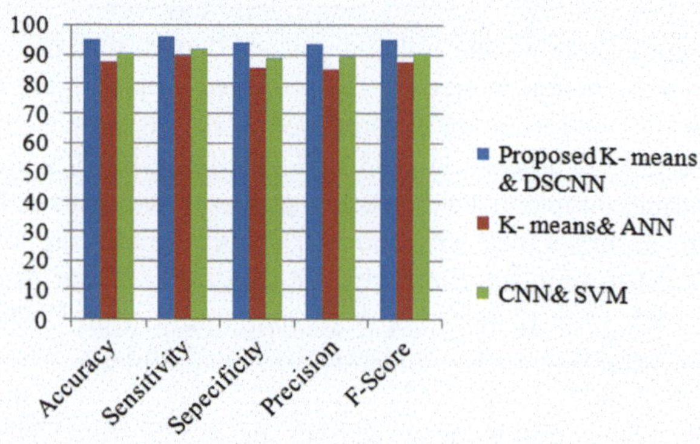

5 Conclusion

The aim of this study is to propose a lung and colon cancer detection method using K algorithms and DSCNN. Initially, CT images are collected and preprocessed using a modified median filter. After that, the preprocessed image is segmented using K-means technique. From this technique, we obtained dissected lung nodules. Then, these segmented nodes are classified using DSCNN technique. Hyper parameters of DSCNN are optimized using COA algorithm. The proposed work is implemented in Python 3 framework and achieves 95.07% accuracy, 96.28% sensitivity, 93.90% specificity, 93.79% precision, and 95.02% F-score. We have found that the DSCNN cannot learn properly in training if the input image is already small. In order to improve the detection of lung and colon cancer in the future, we intend to use a different deep learning method.

References

1. Giri B, Dey S, Das T, Sarkar M, Banerjee J, Dash SK (2018) Chronic hyperglycemia mediated physiological alteration and metabolic distortion leads to organ dysfunction, infection, cancer progression and other pathophysiological consequences: an update on glucose toxicity. Biomed Pharmacother 107:306–328
2. Singal AG, Lampertico P, Nahon P (2020) Epidemiology and surveillance for hepatocellular carcinoma: new trends. J Hepatol 72(2):250–261
3. Talukder MA, Islam MM, Uddin MA, Akhter A, Hasan KF, Moni MA (2022) Machine learning-based lung and colon cancer detection using deep feature extraction and ensemble learning. Expert Syst Appl 205:117695
4. Singhi AD, Koay EJ, Chari ST, Maitra A (2019) Early detection of pancreatic cancer: opportunities and challenges. Gastroenterology 156(7):2024–2040
5. Syed L, Jabeen S, Manimala S, Elsayed HA (2019) Data science algorithms and techniques for smart healthcare using IoT and big data analytics. In: Smart techniques for a smarter planet: towards smarter algorithms, pp 211–241
6. Mehmood S, Ghazal TM, Khan MA, Zubair M, Naseem MT, Faiz T, Ahmad M (2022) Malignancy detection in lung and colon histopathology images using transfer learning with class selective image processing. IEEE Access 10:25657–25668
7. Asuntha A, Srinivasan A (2020) Deep learning for lung cancer detection and classification. Multimed Tools Appl 79:7731–7762
8. Hage Chehade A, Abdallah N, Marion JM, Oueidat M, Chauvet P (2022) Lung and colon cancer classification using medical imaging: a feature engineering approach. Phys Eng Sci Med 45(3):729–746
9. Bhatia S, Sinha Y, Goel L (2019) Lung cancer detection: a deep learning approach. In: Soft computing for problem solving: SocProS 2017, vol 2. Springer, Singapore, pp 699–705
10. Shafi ASM, Molla MI, Jui JJ, Rahman MM (2020) Detection of colon cancer based on microarray dataset using machine learning as a feature selection and classification techniques. SN Appl Sci 2:1–8
11. Singh GAP, Gupta PK (2019) Performance analysis of various machine learning-based approaches for detection and classification of lung cancer in humans. Neural Comput Appl 31:6863–6877
12. Sirinukunwattana K, Raza SEA, Tsang YW, Snead DR, Cree IA, Rajpoot NM (2016) Locality sensitive deep learning for detection and classification of nuclei in routine colon cancer histology images. IEEE Trans Med Imaging 35(5):1196–1206
13. Fahami MA, Roshanzamir M, Izadi NH, Keyvani V, Alizadehsani R (2021) Detection of effective genes in colon cancer: a machine learning approach. Inform Med Unlocked 24:100605
14. Sifre L, Mallat S (2014) Rigid-motion scattering for texture classification. arXiv:1403.1687

Often Overindulge-Based Personality Classification Used in Neural Network Compared with Random Forest to Attain Accuracy

G. Dharma Teja, A. Selvakumar, and S. Ramesh

Abstract Aim: Personality classification based on overindulgence behavior has gained significant attention in recent research due to its potential implications for understanding human behavior and improving mental health interventions. In this study, we propose a methodology for personality classification using Neural Network compared with Random Forest to get better accuracy. Materials and Methods: Two techniques of Neural Networks and Random Forests (RF) have been applied. Total 40 samples are considered for the both algorithms. The purpose of calculating the accuracy %, the simulation results were repeatedly inspected and assessed. Results: Neural Network (NN) is more accurate than Random Forest (RF) (83.17%), with a score of 77.18%. Based on a sample T-test with a significance value of $p = 0.001$ ($p < 0.05$), the research study has no significance between the two groups. Based on the results, Neural Network accuracy was much superior to Random Forest accuracy based on the findings.

Keywords Visual activity · Novel neural network · Dynamic picture · Extract · Health · Personality

G. Dharma Teja · A. Selvakumar
Department of Computer Science and Engineering, Saveetha School of Engineering, Saveetha Institute of Medical and Technical Sciences, Saveetha University, Chennai, India
e-mail: gaduputidharmateja19@saveetha.com

A. Selvakumar
e-mail: selvakumara.sse@saveetha.com

S. Ramesh (✉)
Department of Electronics and Communication Engineering, Saveetha School of Engineering, Saveetha Institute of Medical and Technical Sciences, Saveetha University, Chennai, India
e-mail: rameshshunmugam.sse@saveetha.com

S. Dutta et al. (eds.), *Cyber Intelligence and Information Retrieval*, Lecture Notes in Networks and Systems 1025, https://doi.org/10.1007/978-981-97-3594-5_27

1 Introduction

A person's personality determines how they think, behave, and how they interpret their environment. The classification of personalities is crucial because it can provide a foundation for job assignments, especially in high-risk professions like the police force. Numerous characteristics of individual qualities exist, but not all of them can be utilized to describe a person's personality Radzi et al. [1] [2]. Modeling dominance effects, role identification, emerging leader recognition, leadership style prediction, and the classification of personality traits are a few applications for automatic nonverbal behavior analysis. The significance of nonverbal characteristics (NFs) that are based on VA and typically derived from audio or video or coupled with them. There are many methods to assess someone's novel VA. For instance, weighted motion energy pictures were then used to extract motion from both the head and the body when optical flow were employed for identifying head activity and image differencing to detect health body activity. A different suggestion was made: a 2-dimensional (2D) body position conveying Visual activity Yousuf [3–6], (Department of Computer Science and Engineering Jamia Hamdard New Delhi, and India 2018). Their approach dramatically enhanced activity identification and used recently discovered dynamic picture representation to condense the spatiotemporal combining the video's information with one picture. With the purpose of extract short-term visual activity, which has never been used in novel nonverbal behavior analysis, to alter that method in this study Baumert et al. [7]. Their approach dramatically enhanced activity health identification and used recently discovered dynamic picture representation to condense the spatiotemporal the video's content into a single image. For nonverbal behavior analysis, short-term VA extract, which has never been used, to alter that method in this study. Moreover, Visual activity (VA)-based nonverbal behavior analysis might be based on their development. Recently, suggested deep VA-based NFs generated by a hybrid CNN model were used to categorize low and high extraversion in small group meetings, predict leadership style, and identify rising leaders. These deep VA-based NFs outperformed many multi modal NF combinations for leadership applications, but they did not perform well enough to be categorized as low or high extroverts Luo et al. [8]. This demonstrates the continued demand for a more advanced technique to extract deep Visual activity (VA)-based NFs. Psychological assessments, Human resources, Marketing and advertising, Personalized recommendations Polzehl [9].

2 Literature Survey

The concept of personality classification based on overindulgence has been a topic of interest in the literature. Several studies have explored the use of different methods, which includes Neural Networks and Random Forests, to improve the accuracy of personality classification in this context. To the finest knowledge, this is the first attempt to find novel key-dynamic images in the social and emotional health computing fields using a deep learning model. Using key-frames and spatiotemporal saliency. This is also the first research to mix CNN + LSTM architecture with dynamic pictures. Because dynamic pictures are composites of numerous RGB images, video input is created with dynamic images is far smaller than the original movie, allowing training to be finished more rapidly. To improve classification accuracy, a modification is also made to the novel covariance-based feature encoding method suggested Allport et al. [10] Two datasets one made up of vlogs and the other of small group meetings are used to assess the technique for automatically classifying personality traits. As it may include several individuals communicating and personality are concurrently expressed through a variety of nonverbal channels including health, voice, body and face, etc., as evidenced by the nonverbal communication literature, computational analysis in personality traits is a difficult research issue. Yet in this study, Visual activity VA is the only area of attention.

The research gap in "Often Overindulge-Based Personality Classification" could be identifying the factors that contribute to overindulgence in personality classification and exploring alternative methods to reduce such tendencies. It could be beneficial to avoid utilizing NFs collected from all frames when doing a successful nonverbal behavior analysis. (in this paper, this is referred to as the "holistic method"), but rather discover a means to identify the frames that include more novel discriminative information, or "key-frames," and extract NFs solely from them Matthews et al. [11], a deep learning model Ramesh et al. [12] is employed in this study to search for key-frames in order to verify this supposition. Just the frames necessary for the extract of Visual activity (VA)-based NFs with noticeable motion were used, while other frames were left out of the analysis in order to identify emergent leaders. This approach is reasonable if there are still portions of the movie that produce frames that are remarkably similar to one another and nearly identical NFs, possibly leading to redundancy and complicating the classification process Mahrenholtz [13].

3 Materials and Methods

The proposed work is conducted in the Saveetha School of Engineering's Artificial Intelligence Laboratory. There were two groups employed in the investigation. A decision tree algorithms are used in first group, and a Random Forest techniques are used in second group. Total of 20 samples in each group. The computation is performed using G-power 0.8 with alpha and beta values 0.05, 0.2 with a confidence interval at 95% Kane et al. [14]. Two algorithms accuracy are compared in this research work.

The dataset was gathered from the Kaggle website (Kaggle: Your Machine Learning and Dat…). This dataset contains names, rank, platform. There are 40 samples collected for each of the groups. Testing data and training data are separated from the dataset. Twenty samples are taken for training purposes and twenty for testing purposes. To predict the accuracy values, the datasets are separated, and the technique is then fitted to the train and test sets. Due to the wide range of image-making tools that MATLAB (2013) and Windows 10 - i4 core have developed, MATLAB is the primary tool used in this work.

3.1 Neural Network Algorithms

Computers are taught to process data in a manner fashioned after the human brain using a technology known as a novel neural network. A type of machine learning is called machine learning. Uses tiered topology and linked nodes or neurons to replicate the human brain. Computers may use this to build an adaptive system that allows them to continuously learn from mistakes and get better. In an effort to more precisely handle complicated issues, such as synthesizing documents or recognizing faces, artificial novel neural networks.

3.2 Neural Network Algorithms

Step 1: Import the necessary packages, including pandas, numpy.
Step 2: Use csv to read the necessary data.
read () function in Pandas.
Step 3: Data preparation.
Step 4: Setting index as data values.
Step 5: Import Library Classification and MLP models.
Step 6: List the sample size.
Step 7: Producing the distinct data.
Step 8: Now split data as train and validation sets to assess the model's efficacy.
Step 9: Now, plot the outcomes from the trained data and the validation set.

3.3 Random Forest Algorithm

For the purpose of solving classification and regression problems, the Random Forest algorithm, a well-known supervised machine learning technique, is utilized. A forest is made up of many distinct kinds of trees, and the forest is more vigorous the more types of trees there are. Similar to this, the number of trees in a Random Forest algorithm enhances the algorithm's accuracy and ability to solve problems. The classification technique Random Forest uses a number of decision trees on different subsets of the input data to raise the expected accuracy of the dataset. It is based on the concept of ensemble approaches, which include combining a number of classifiers to tackle a complex issue and improve the performance of the model.

3.4 Random Forest Algorithm

Step 1: Import the algorithm's necessary libraries, such as csv, pydh.
Step 2: Utilize a csv file to read the necessary data.
read() function in Pandas.
Step 3: Data preparation.
Step 4: Identify the explanatory variables.
Step 5: Describe the target variable.
Step 6: Data separates as two categories train and test.
Step 7: Use RF for the sequential and imported models.
Step 8: Reliability of classifiers.
Step 9: Execution of the strategy.
Step 10: Plot strategy returns versus actual outcomes.

3.5 Statistical Analysis

SPSS software is used to statistically analyze novel Neural Network and Random Forest. Independent variables include the image, objects, distance, frequency, modulation, amplitude, volume, and decibels [13]. The independent variables are images and things. An independent T-test analysis is conducted for both procedures to ascertain correctness.

4 Discussion

Neural Network and Random Forest both were implemented to predict the prevent from personality questionnaire classification and to improve the accuracy of the existing model and health from the obtained results in the article it is concluded that the novel Neural Network ($p = 0.001$) is more efficient and accurate in prediction compared with Random Forest for the larger datasets ("Pneumonia Classification of Thorax Image Using Convolutional novel Neural Network"). The classification of personality traits can also be viewed. The objective of this study is classification rather than regression, hence the characteristics must be binarized into low and high levels Tyrer [15]. Baselines for each participant were used to guide this.

In order to perform regressions on subsequent projects, the given approach can however be adjusted so as to take account of continuous Value Added Notes. The downside of the suggested strategy is that a lot of dynamic picture are needed as input [2]. Nevertheless, in later significance testing, the quantity of novel key-dynamic pictures Visual activity would be trained as a parameter. The data used in this study was adequate in statistical analysis to show that the suggested technique works effectively Millon et al. [16]. Human health prediction and analysis research is in high demand right now. This system's capacity to forecast a candidate's personality has facilitated numerous scenarios.

The domains include things like hiring procedures and medical advice. When questionnaires are used to evaluate personality traits, behavioral characteristics of respondents may be revealed Livesley et al. [17]. The limitations and consider a comprehensive and holistic approach while using personality classification based on overindulgence or any other single factor. It's important to use such classifications with caution and in conjunction with other validated personality assessment tools and approaches to gain a more accurate understanding of an individual's personality. Identifying a correct personality recognition method based on surveys is the primary objective of this study. Additional analyses have been carried out during the meetings in order to classify low and high extraversion health. In particular, the essential selection element of the recommended approach proved to be more effective in selecting key-frames for most popular social and emotional computing techniques which used all dynamic pictures as well as selected random and uniform static images.

5 Results

Random Forest method mean accuracy was 77.18%, and the Neural Network technique mean accuracy was 83.17%. The various parameters for both groups are shown in Table 1. For RF and NN, the accuracy loss has been computed. NN exceeds the Random Forest in terms of accuracy (83.17%) and loss (16.83%), according to a comparison of the two groups. The statistical analysis of Random Forest and novel Neural Network using various test datasets is shown in Table 2. The table shows that when compared to RF algorithm, NN algorithm produces greater accuracy. Additionally, RF has a much higher loss than NN. The NN method outperforms the Random Forest approach in terms of performance.

$$\text{Accuracy} = \text{TP} + \text{TN}/\text{TP} + \text{TN} + \text{FP} + \text{FN} \tag{1}$$

Table 1 Group Statistical Analysis of NN and RF. Mean, standard deviation, and standard error mean are obtained for 20 samples. NN has higher mean accuracy and lower mean loss when compared to RF algorithms

	Group	N	Mean	Std. deviation	Std. error mean
Accuracy	Neural network algorithm	20	83.1715	5.81366	1.29997
	Random forest	20	77.1860	5.63932	1.26099

Table 2 Accuracy and loss analysis of neural network (NN) and random forest (RF) algorithm obtained for each iteration while evaluating the dataset for various test sizes

S. no	Neural network	Random forest
1	74.77	69.21
2	75.33	69.65
3	76.74	69.84
4	77.21	70.64
5	77.84	71.46
6	78.53	72.85
7	79.66	73.22
8	80.39	74.65
9	81.87	75.76
10	82.44	76.34
11	83.62	77.65
12	84.32	78.25
13	85.69	79.62
14	86.79	80.89
15	87.42	81.64
16	88.64	82.37
17	89.23	83.55
18	90.54	84.67
19	91.76	85.34
20	92.64	86.12

where TP = True Positive TN = True Negative FP = False Negative Loss: A scalar worth that endeavor to limit during the preparation of the model. The statistical analysis of significant values for both groups is shown in Table 3. The difference between the two groups is minuscule, being 0.004 for loss and 0.001 for accuracy. Therefore, NN is superior to RF.

Figure 1 shows the mean accuracy and loss of the Neural Network and Random Forest (RF) models (NN). The findings demonstrated that the NN has superior accuracy (83.17%) and loss (16.83%) in comparison to the Random Forest's accuracy and loss. However, compared to the Logistic Neural Network, Random Forest has a lower mean error.

Table 3 Independent sample T-test: neural network is insignificantly better than random forest with p value 0.001 ($p < 0.05$)

		Levene's test for equality of variances		T-test for equality means with 95% confidence interval						
		f	Sig.	t	df	Sig. (2-tailed)	Mean difference	Std. error difference	Lower	Upper
Accuracy	Equal variances assumed	0.006	0.938	3.305	38	0.001	5.98550	1.81109	2.31915	9.65196
	Equal variances not assumed			3.305	37.965	0.001	5.98550	1.81109	2.31904	9.65196

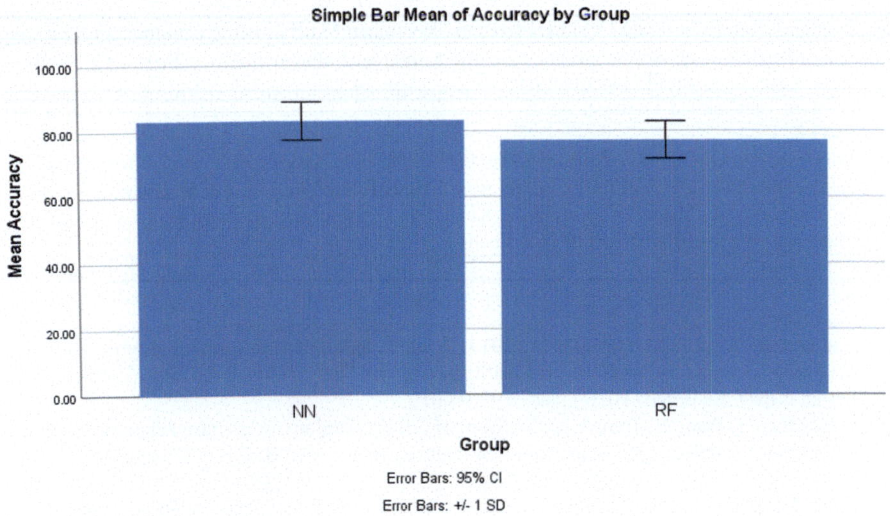

Fig. 1 Comparison of neural network and random forest. Classifier in terms of mean accuracy and loss. The mean accuracy of the neural network is better than random forest. Classifier; standard deviation of neural network is slightly better than random forest. X axis: neural network versus random forest classifier and Y axis: mean accuracy of detection with ±SD

6 Conclusion

In this study, data was analyzed using both Neural Network and Random Forest means clustering methods. The results showed that the accuracy of the Neural Network was 83.17%, while the accuracy of Random Forest was 77.18%. Based on the analysis, it was found that the Neural Network outperformed the Random Forest in terms of accuracy.

References

1. Radzi NHM, Mazlan MS, Mustaffa NH, Sallehuddin R (2017) Police personality classification using principle component analysis-artificial neural network. In: 2017 6th ICT international student project conference (ICT-ISPC). https://doi.org/10.1109/ict-ispc.2017.8075348
2. Ro E, Stringer D, Clark LA (2012) The schedule for nonadaptive and adaptive personality: a useful tool for diagnosis and classification of personality disorder. The oxford handbook of personality disorders. https://doi.org/10.1093/oxfordhb/9780199735013.013.0004
3. Yousuf S, Department of Computer Science and Engineering Jamia Hamdard New Delhi and India (2018) Analytical study of image classification using deep learning. Int J Adv Res Comput Sci. https://doi.org/10.26483/ijarcs.v9i2.5802

4. Subbulakshmi B, Nirmala Devi M, Sriram S, Arvindhan M (2023) A hybrid machine learning model for house price prediction. In: Deepak BBVL, Bahubalendruni MVAR, Parhi DRK, Biswal BB (eds) Intelligent manufacturing systems in industry 4.0. IPDIMS 2022. Lecture notes in mechanical engineering. Springer, Singapore. https://doi.org/10.1007/978-981-99-1665-8_35
5. Munshi A, Arvindhan M, Thirunavukkarasu K (2021) Random forest application of Twitter data sentiment analysis in online social network prediction. In: Mangla M, Sharma N, Mittal P, Wadhwa VM, Thirunavukkarasu K, Khan S (eds) Emerging technologies for healthcare. https://doi.org/10.1002/9781119792345.ch12
6. Arvindhan M, Anand A, Wassey MA (2022) An optimal system on data challenge with distributed data management on cloud, fog and edge computing. Multimed Comput Syst Virtual Real 129
7. Baumert A, Schmitt M, Perugini M (2019) Towards an explanatory personality psychology: integrating personality structure, personality process, and personality development. Pers Individ Differ. https://doi.org/10.1016/j.paid.2019.04.016
8. Luo G, Zhu Y, Wang R, Tong Y, Lu W, Wang H (2020) Random forest-based classification and analysis of hemiplegia gait using low-cost depth cameras. Med Biol Eng Comput 58(2):373–382
9. Polzehl T (2014) Personality in speech: assessment and automatic classification. Springer
10. Allport FH (1921) Personality traits: their classification and measurement
11. Matthews G, Deary IJ, Whiteman MC (2009) Personality traits. Cambridge University Press
12. Ramesh S, Sasikala S, Gomathi S et al (2022) Segmentation and classification of breast cancer using novel deep learning architecture. Neural Comput Appl 34:16533–16545. https://doi.org/10.1007/s00521-022-07230-4
13. Mahrenholtz O (1970) New approaches to personality classification. Columbia University Press, New York
14. Kane SP, Phar, BCPS (2023) Sample size calculator. Accessed 18 April 2023. https://clincalc.com/stats/samplesize.aspx
15. Tyrer P (2020) Classification of complex disorders is a challenge solved by simplicity: commentary on controversies in the classification and diagnosis of personality disorders. The Cambridge handbook of personality disorders. https://doi.org/10.1017/9781108333931.022
16. Millon T (1994) Personality disorders: conceptual distinctions and classification issues. In: Personality disorders and the five-factor model of personality. https://doi.org/10.1037/10140-032
17. Livesley WJ, John Livesley W (2011) The current state of personality disorder classification: introduction to the special feature on the classification. J Pers Disord. https://doi.org/10.1521/pedi.2011.25.3.269
18. Pneumonia classification of thorax image using convolutional neural network. https://doi.org/10.25139/inform.v5i2.2707.g1813

Accurate Analysis of Blast Disease Prediction in Rice Crop Using Artificial Neural Network Algorithm Compared with Random Forest Algorithm

M. Tharun Kumar Reddy, S. Ramesh, and A. Selvakumar

Abstract The proposed work is to find the blast disease in rice crops using artificial neural network compared with random forest algorithm. To predict the rice blast illness, an artificial neural network and random forest are used with different training and testing splits. In this instance, each of the two groups' 169 samples was used in an analysis that required a total of 10 iterations. The ClinCalc program is used as a tool to calculate the setup's correctness for supervised learning. With Gpower setting parameters of 0.05 and 0.80, the test's average Gpower is roughly 80%. Identifying blast disease detection in rice crops can be used to classify the presence or absence of the disease based on input features such as crop characteristics, environmental factors, and symptoms. With an independent sample T-test value of $p = 0.001$, artificial neural network (ANN) (91.8100%) outperforms random forest (RF) (78.1000%) in terms of both accuracy and loss, which is considered to be statistically significant. In comparison with the random forest (78.10%), artificial neural networks (91.81%) are more accurate for the selected dataset.

Keywords Novel artificial neural network · Random forest · OpenCV · Significance value · Diseases · Plants · Crops

1 Introduction

In order to properly identify objects, this research suggested an appropriate framework that included enhancement, filter, colour segmentation, and colour feature for classification phases [1]. This study offers a workable technique that colours both the input and the output images. This study considered a number of image processing

M. Tharun Kumar Reddy · S. Ramesh (✉) · A. Selvakumar
Department of Electronics and Communication, Saveetha School of Engineering, Saveetha Institute of Medical and Technical Sciences, Saveetha University, Chennai, India
e-mail: rameshshunmugam.sse@saveetha.com

M. Tharun Kumar Reddy
e-mail: tharunkumarm19@saveetha.com

S. Dutta et al. (eds.), *Cyber Intelligence and Information Retrieval*, Lecture Notes in Networks and Systems 1025, https://doi.org/10.1007/978-981-97-3594-5_28

phases, including enhancement, noise reduction, colour picture segmentation, and colour crop characteristics for identification [2]. This study offers an innovative method for classifying diseases via image processing that is based on the fraction of the affected area's RGB value. The main benefit of identifying rice plant illnesses early is that farmers may take preventative action, which leads to increased yields and better-quality rice. This is especially true of diseases that harm the leaves of rice plants. An average of 143 research articles has been published in IEEE Xplore, and 47 papers have been published in science direct as a result of the research on identifying blast disease in rice. This study proposed an autonomous deep learning-based system for localising rice blast diseases [3] and for classifying images of blast diseases as either healthy, moderate, or susceptible [4]. In this work, the damaged area of the rice plant is segmented from the grayscale photographs using the k-means method, and the illness is then classified using the support vector machine (SVM) [5]. Software system is created to detect the disease of brown spot and leaf blast. The system was created using the YOLO algorithm. The goal is to create an expert system to help farmers in the state of Meghalaya deal with rice illnesses. After consulting with subject matter experts and engaging with local farmers via a plRandom Forested interview schedule, the most common rice illnesses in this region were identified and confirmed. With the present system, the research gap was identified with subpar accuracy.

2 Literature Survey

The aim of this work is based on option posters by players and upvoting-based prediction and suggestions to enhance performance of the server using neural network algorithms compared making use of random forest techniques to improve accuracy using machine learning in which the neural network algorithms are more accurate than the algorithm of random forests; the staff has a wealth of knowledge and research expertise, which has resulted in publications of the highest calibre; its proposed solution is to get more accuracy with less information in order to achieve these results that give more accuracy by the chosen algorithm, as the existing literature survey shows that predicting discord server gaming performance through the submission of game players' suggestions gets accurate values of accuracy. This method's assessment and outcome are mostly based on gaming server data, which necessitates longer training time economic development. The purpose of this project is to use novel neural network algorithms to better correctly predict.

3 Materials and Methods

This research experimental setup has taken place in the Machine Learning Laboratory at Saveetha School of Engineering, Saveetha Institute of Medical and Technical Sciences (SIMATS), Chennai. The Gpower software has calculated the sample size by comparing the artificial neural networks and random forest. Two groups are picked in order to compare their methods and outcomes. For this investigation, a total 10 sets of samples from each group were selected. Two algorithms, novel artificial neural network and random forest, are used in technical analysis software [6]. The sample size was determined to be 10 for each group using the Gpower 3.1 program (Gpower setting parameters: = 0.05 and power = 0.80).

The proposed task is planned and executed using Python OpenCV software. Deep learning was tested on the Windows 10 operating system. The hardware configuration included an Intel Core i7 processor and 4 GB of RAM, and a 64-bit operating system was used. While the code is running, the dataset is being worked on in the background to finish an output process and ensure accuracy.

Artificial Neural Network

The ANN algorithm belongs to the supervised classification subfield of machine learning [7]. As implied by the name, this algorithm produces a forest made up of numerous individual trees. The forest looks stronger when there are more trees there. Like previous classifiers, the novel artificial neural network classifier gets more accurate as the size of the forest's tree population increases. The same apache of decision construction based on information gain or the Gini index will not be used to model a huge number of decision trees to produce a forest.

Random Forest

Computer systems that are modelled on the biological neural networks that make up human brains are known as artificial neural networks, sometimes known as neural networks or neural nets. Artificial neurons are a collection of interconnected nodes, or units, that make up a random forest. These synthetic neurons are somewhat similar to the neurons in the human brain. Each link has the ability to communicate with other neurons, just like synapses do in a real brain. A synthetic neuron receives impulses, interprets them, and then transmits messages to other neurons that are interconnected with it. The pieces that connect are called edges. The weights of the majority of neurons and edges evolve as learning progresses. Depending on the weight, a connection's signal strength increases or decreases. A signal is only sent by a neuron when the overall signal exceeds a certain level, which is known as a "Threshold" in neurons.

Statistical Analysis

Using the data from the Excel sheet, an independent sample T-test was performed to compare the novel artificial neural network and random forest algorithm. Independent variables include the image, objects, distance, frequency, modulation, amplitude, volume, and decibels. The dependent variables are images and things. An independent T-test analysis is conducted for both processes to ascertain accuracy.

4 Results

The recommended artificial neural network and random forest were performed in Anaconda Navigator with a sample size of 10 at various points. Table 1 displays the novel artificial neural network's projected accuracy and loss. The estimated accuracy and loss of random forest are shown in Table 2. These ten data samples are used for each strategy in order to get statistically comparable loss levels and associated statistical statistics. According to Table 3's findings, the novel artificial neural network's mean accuracy was 91.8100% while random forest's was 78.100%. Given standard deviations of 2.19564 for the artificial neural network and 1.69050 for the random forest, the artificial neural network has a mean value that is higher than the random forest, an inconsequential level of 0.242. Table 4 represents independent sample T-test value of $p = 0.242$, all findings for the random forest and novel artificial neural network. Figure 1 contrasts random forest with artificial neural network in terms of mean accuracy and loss.

For the novel artificial neural network, the corresponding values for mean, standard deviation, and standard error mean are 91.8100, 2.19564, and 0.69432. Additionally, random forest's mean, standard deviation, and standard error mean values are 78.1000, 1.69050, and 0.53458, respectively. In contrast, the mean, standard deviation, and standard error mean loss values for the artificial neural network are,

Table 1 Predicting accuracy of rice leaf disease detection using artificial neural networks (mean accuracy = 91.81%)

Iterations	Accuracy (%)	Loss (%)
1	89.80	10.20
2	93.92	6.08
3	91.95	8.05
4	88.27	11.73
5	93.79	6.21
6	92.22	7.78
7	91.03	8.97
8	95.47	4.53
9	89.75	10.25
10	91.90	8.10

Table 2 Predicting accuracy of rice leaf disease detection using random forest (mean accuracy = 78%)

Iterations	Accuracy (%)	Loss (%)
1	78.2	44.41
2	78.3	47.21
3	73.5	48.32
4	78.4	45.63
5	79.3	44.36
6	78.6	49.89
7	79.3	51.01
8	78.3	50.31
9	77.9	49.36
10	79.2	50.48

Table 3 Group statistical analysis of artificial neural network and random forest. Mean, standard deviation, and standard error mean are obtained for ten samples. Artificial neural network has higher mean accuracy and lower mean loss when compared to random forest

	Group	N	Mean	Std. deviation	Std. error mean
Accuracy	Artificial neural network	10	91.8100	2.19564	0.69432
	Random forest	10	78.1000	1.69050	0.53458
Loss	Artificial neural network	10	8.1900	2.19564	0.69432
	Random forest	10	48.0980	2.54631	0.80521

respectively, 8.1900, 2.19564, and 0.69432, while those for the random forest are, respectively, 48.0980, 2.54631, and 8.0521. The group statistics value, mean, standard deviation, and standard error mean are given for each of the two algorithms. The measures of loss for the two algorithms artificial neural network and random forest are represented graphically and categorised. This shows that the ANN shows a classification accuracy of 91.8100%, more than random forest's accuracy of 78.100%. Table 5 shows comparison of the artificial neural network and random forest with their accuracy.

5 Discussion

In this study, the insignificance value obtained with an independent sample T-test value of $p = 0.001$ implies that ANN appears as best than random forest. The artificial neural network's accuracy is analysed as 91.8100%, whereas the accuracy of the natural neural network is 78.100%. Blast, bacterial leaf blight, and tungro are three rice diseases studied in this research. Three classes of 240 photographs were taken. Our experimental data analysis of DenseNet201 has 96.09% accuracy compared to existing models. Same dataset is used for simple computational neural networks,

Table 4 Independent sample T-test: artificial neural network is significantly better than random forest with p value 0.001 (two-tailed, $p < 0.05$)

		Levene's test for equality of variances		T-test for equality means with 95% confidence interval						
		f	Sig	t	df	Sig. (two-tailed)	Mean difference	Std. error difference	Lower	Upper
Accuracy	Equal variances assumed	1.462	0.242	15.646	18	0.001	13.71000	0.87628	11.86901	15.55099
	Equal variances not assumed			15.646	16.896	0.001	13.71000	0.87628	11.86035	15.55965
Loss	Equal variances assumed	1.462	0.242	–15.646	18	0.001	–13.71000	1.06323	–15.55099	–11.86901
	Equal variances not assumed			–15.646	16.896	0.001	–13.71000	1.06323	–15.55965	–11.86035

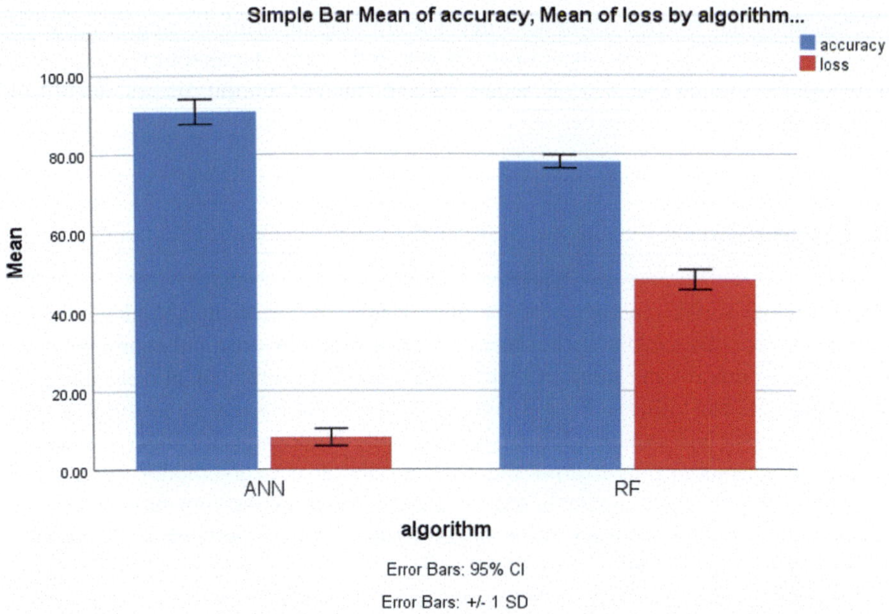

Fig. 1 Comparison of artificial neural network and random forest. Classifier in terms of mean accuracy and loss. The mean accuracy of artificial neural network is better than random forest classifier; standard deviation of artificial neural network is slightly better than random forest. X axis: artificial neural network versus random forest classifier and Y axis: mean accuracy of detection with ±1 SD

Table 5 Comparison of the artificial neural network and random forest with their accuracy

Classifier	Accuracy (%)
Artificial neural network	91.8100
Random forest	78.1000

which has 62.20% accuracy [8]. Various sensors, including a camera, a temperature sensor, and a moisture sensor, were used to create the proposed model on a Raspberry Pi 3b+. To train the suggested model, a CNN classifier was used. With a 99.7% total accuracy rate, this model was capable of identifying the condition (Tawde et al. 2021). This study used digital image processing to remove subjectivity in rice plant disease inspection. About 70% of the network's 134 illness images were utilised for training, 15% for validation, and 15% for testing. The program will present disease-specific strategy alternatives after processing. Program accuracy was 100%. This technique can recognise the five most prevalent diseases that affect rice leaves: bacterial leaf blight, sheath blight, bacterial leaf blast, brown spot, and tungro. Extraction and illness classification require textural and statistical features. This approach has 92.77% accuracy and efficiently identifies rice leaf diseases.

This study has several drawbacks because training an artificial neural network, especially with big datasets, takes a very long period. The system should be expanded to incorporate more objects while requiring less time to train the dataset, according to the study's future objectives.

6 Conclusion

While random forest's accuracy score is 78.100%, the artificial neural network's accuracy score is 91.8100%. According to the investigation, artificial neural network (91.8100%) outperforms random forest (78.1000%) in terms of performance. In conclusion, the blast illness in rice crops may be detected using both artificial neural network (ANN) and random forest (RF) models. Nonetheless, it seems that the random forest model outperforms the artificial neural network model in terms of accuracy and predictive power based on a comparison of the two models for the selected dataset. The accuracy value of the artificial neural network is 91.8100%, whereas the accuracy value of random forest is 78.1000%. Based on the analysis, artificial neural network (91.8100%) performs better than random forest (78.1000%).

References

1. Shahriar Sazzad TM, Anwar A, Islam S, Mila SA, Jannat Oishee S, Anjum A (2020) A computer based image processing approach to identify rice blast. In: 2020 2nd international conference on sustainable technologies for industry 4.0 (STI), pp 1–5. https://doi.org/10.1109/STI50764.2020.9350507.2
2. Shahriar Sazzad TM, Anwar A, Hasan M, Ismail Hossain Md (2020) An image processing framework to identify rice blast. In: 2020 international congress on human-computer interaction, optimization and robotic applications (HORA), pp 1–5
3. Ramesh S, Tamilselvi M, Ramkumar G, Anitha G, Nirmala P (2022) Comparison and analysis of rice blast disease identification in greenhouse controlled environment and field environment using ML algorithms. In: 2022 international conference on advances in computing, communication and applied informatics (ACCAI), Chennai, India, pp 1–5. https://doi.org/10.1109/ACCAI53970.2022.9752538
4. Tosawad T, Kasetkasem T, Laungnarutai W, Phatrapornnant T, Kumazawa I (2022) On the use of class activation map on rice blast disease identification and localization. In: 2022 19th international conference on electrical engineering/electronics, computer, telecommunications and information technology (ECTI-CON), pp 1–4
5. Siddarajamma B, Praveen Kumar MS, Sivaprasad N, Boranna R (2022) Machine learning-based approach to detect and classify rice blast disease. In: 2022 IEEE 2nd Mysore sub section international conference (Mysore Com), pp 1–6
6. Mandal N, Adak S, Das DK, Sahoo RN, Mukherjee J, Kumar A, Chinnusamy V et al (2023) Spectral characterization and severity assessment of rice blast disease using univariate and multivariate models. Front Plant Sci 14(February):1067189
7. Ramesh S, Vydeki D (2020) Rice disease detection and classification using deep neural network algorithm. In: Sharma DK, Balas VE, Son LH, Sharma R, Cengiz K (eds) Micro-electronics

and telecommunication engineering. Lecture notes in networks and systems, vol 106. Springer, Singapore. https://doi.org/10.1007/978-981-15-2329-8_56

8. Rukhsar, Upadhyay SK (2022) Rice leaves disease detection and classification using transfer learning technique. In: 2022 2nd international conference on advance computing and innovative technologies in engineering (ICACITE), pp 2151–2156

Design of Deep Learning Methodology for AES Algorithm Based on Cross Subkey Side Channel Attacks

Ammar Abdulhassan Muhammed, Hassan Jameel Mutasharand, and Amjed A. Ahmed

Abstract Recent Deep-Learning Side-Channel Attacks (DLSCAs) utilize networks of neural that have been skilled over part of tracings that merely includes actions relevant to subkeys that were targeted. These attacks are known as "side-channel attacks." However, due to the restricted number of training traces that are available for deep learning models, such as in the case of the ASCAD database, there is a risk of overfitting occurring. A technique known as "data augmentation" is an example of a data-level method. This technique makes use of additional traces that have been synthetically altered to act as a regularizer and provide deep learning models with a greater capacity for generalization. Within the scope of this investigation, we provide a cross-subkey training approach that acts as an addition to traces. In order to train deep learning models, we use segments for the other 15 subkeys of AES-128 in addition to a trace segment that includes the SBox operation of the target subkey. We show that developing a network model with a mixture of numerous subkeys is superior to the more conventional method of training a network model with a single subkey by using two well-known datasets.

A. A. Muhammed (✉) · H. J. Mutasharand · A. A. Ahmed (✉)
Department of Computer Techniques Engineering, Imam Al-Kadhum College (IKC), Baghdad, Iraq
e-mail: amar.abdalhasan@alkadhum-col.edu.iq

A. A. Ahmed
e-mail: amjedabbas@alkadhum-col.edu.iq

H. J. Mutasharand
e-mail: hasan.jamel@alkadhum-col.edu.iq

A. A. Ahmed
Center for Cyber Security, Faculty of Information Science and Technology, Universiti Kebangsaan Malaysia (UKM), 43600 Bangi, Malaysia

S. Dutta et al. (eds.), *Cyber Intelligence and Information Retrieval*, Lecture Notes in Networks and Systems 1025, https://doi.org/10.1007/978-981-97-3594-5_29

1 Introduction

Deep neural network (DNN) [1] also known to be as deep learning is a technique to machine learning that were used in the development of an image-recognition system that can be utilized by security cameras or autonomous vehicles. In addition to this, it has been implemented in applications that are concerned with security, like sensing potentially harmful electronic-mails.

Traditional side-channel attacks [2] (SCAs), like template attack (TA), differential power analysis (DPA) and correlation power analysis (CPA), needs analysis of signals having with statistical computing concerning waveform of waveform based on implementation knowledge of the target cryptographic algorithm. These attacks are known together as power analysis techniques. Studying cryptographic circuits with SCA countermeasures like clock jitter or/and arbitrary number masking also requires a high degree of skill. This is because both of these countermeasures are difficult to analyze.

It has been recommended to use deep learning SCAs, also known as DL-SCAs, which combine SCAs with deep learning [3]. It allows attacks on cryptographic circuits that are protected by typical SCA countermeasures despite the fact that the attacker is unaware of the countermeasures.

Both profiled and non-profiled categories are available for SCAs to choose from. Side-channel data obtained from a freely controlled device (the profiling device) that is analogous to the device being attacked (the target device) is used in profiled attacks in order to construct a leaky model. After then, attack were executed through utilizing both developed leakage model along with side-channel data received through gadgets that is being targeted. The term "premium" refers to a kind of attack that is used to describe a product. Attacks that do not use profiles depend only on the data that can be acquired from the side channels of the target device. Examples of attacks that are not profiled include CPAs. DL-SCAs are normally targeted by profile assault; however, DL-SCAs non-profiled were available for first time in 2018 [4].

2 Literature Review

In 2016, Maghrebi employed layers of convolution primarily in DL-SCAs contrary to an AES implementation of software that was insulated by masking countermeasures. This demonstrated the efficacy of layers of convolution comparatively to SCAs [5]. The DL-SCAs were carried out by a group of researchers from Californian University California, Los Angeles, they were able to show the usefulness of the procedure by using a range of different methodologies. After doing some permuting in ascertaining mask and offset settings, a CPA versus DPA Contest V4, which is an information set of AES software deployment was carried out by Hou, that was secured by the Rotating Sboxes Masking (RSM) countermeasure, and was successful in attacking it [6]. In addition, Gilmore made use of a multilayer perceptron (MLP) so as to successfully attack DPA Contest V4 [7].

Prouff carried out profiled DL-SCAs in the year 2018 and distributed ANSSI SCA Database (ASCAD), a dataset of public domain belonging to AES software deployment secured through re-processing of masking countermeasures [8]. This was done in order to determine whose parameter of DNN, like amount of filters and convolutional layers, are appropriate concerning the DL-SCA.

To counter non-profiled attacks (non-profiled DL-SCAs), around 2018, Timon created an attack that employs deep learning; moreover he was successful in demonstrating attack's efficiency over an AES software deployment that was secured by the ASCAD's masking countermeasure and random jitter. Timon was able to break the AES software implementation. Following an investigation of the mask's leakage point and the masked intermediate values, it was discovered that a single partial key was sufficient to mount a successful attack on the masking countermeasure as well as the restricted points of interest (PoI) [9]. Additional relative research over non-profiled DL-SCAs includes strategy that leverages side-channel data out of 1D tracings that has been converted into format of 2D images and attack pertaining to concealed countermeasure that depends on correlated noise production. Both of these techniques were featured in other publications.

3 Methodology

3.1 *Deep-Learning-Based Side-Channel Attack*

A part of machine learning is Deep learning [10] exploiting network of deep neural in accomplishing goals of learning through experiences, comprehending information at input in the context of a hierarchical methods. SCAs that are based on deep learning are much more effective than traditional cryptanalysis because deep learning methods are particularly adept in eliciting characteristics from unprocessed information [11]. In most cases, a deep learning side-channel attack will consist of following two phases.

In phase of profiling, attacks [12] wishes in making use of deep learning model in creating leaking profile through exploiting huge collection of power tracings (T) that were collected through gadgets of profiling, in which amount of training tracings is "m". This will allow attacker to determine how much power is being leaked. Every tracing Ti are distinguished through information computed at point of attack $lT_i \in L$, where $L = 0, 1, 255$, and from which the subkey can be obtained using some input that is already known (for instance, ciphertext, plaintext).

Method of constructing a neural network could be observed like mapping: $N : R^m \rightarrow I^L$ where outcome remains a score vector $S = N(T) \in I^L$. Element s_j containing measure j belonging to S shows possibility where $l(T) = j$.

During attack phase, attacks will use skilled deep learning scheme in classifying recorded tracing coming out of attacked gadgets in order to get vector of score. Attack has capability of discovering ith subkey $k_i = j$, containing greatest probability over S. For the authentic subkey, the sign $k*$ is what we use. After entering $k_i = k_j*$ the subkey can be successfully obtained. When calculating the neural network's classification error, we make use of cross-entropy [13] considered as loss function along with RMSprop (Root Mean Square prop) as the optimizer.

$$k_i = \underset{0 \leq j \leq 255}{\arg\max} \tilde{S_j} . \tag{1}$$

3.2 Power Traces Composition

The power-dependent side-channel attacks take use of reality where it is possible for target device's power consumption to fluctuate throughout the execution of an encryption process, depending on the data and processes that are being input. Consequently, parts of a power consumption trace that depend on data and processes, in that order, are two parts that are most interesting to look at. In addition, since electronic noise is a component of noise, using same apparatus to carry out an execution having an identical input information would need a various quantities of energy each time. The constant component, also known as "const," is amount of power consumption that is created by transistor switching processes, that are not based on input information. It is abbreviated as "const." As a direct consequence of this, becomes feasible in characterizing every junction on power tracing as combination of such parts [14].

$$\rho_{\text{total}} = \rho_{\text{data}} + \rho_{\text{op}} + \rho_{\text{noice}} + \rho_{\text{const}} \tag{2}$$

4 Cross-Subkey Attacks

Figure 1 provides a high-level summary of the training procedure for a cross-subkey model. During this procedure, a number of subkeys collaborate with one another to enhance the capacity of the target subkey to generalize.

Figure 2a, b shows second and first S-Box execution in primary turns as power traces which were taken out of 8-bit microcontroller deployment of AES. If same data is handled, traces seem quite similar. Processing can be done on either (a) same data or (b) different data. Figure 3a, b shows entire POI of subkeys in most recent AES cycle with AES_GPU.

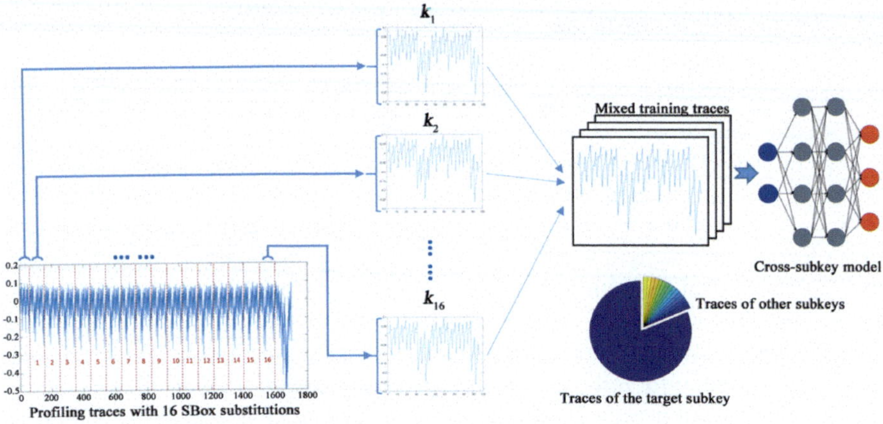

Fig. 1 Training process for the cross-subkey model that makes use of the mixed profile set

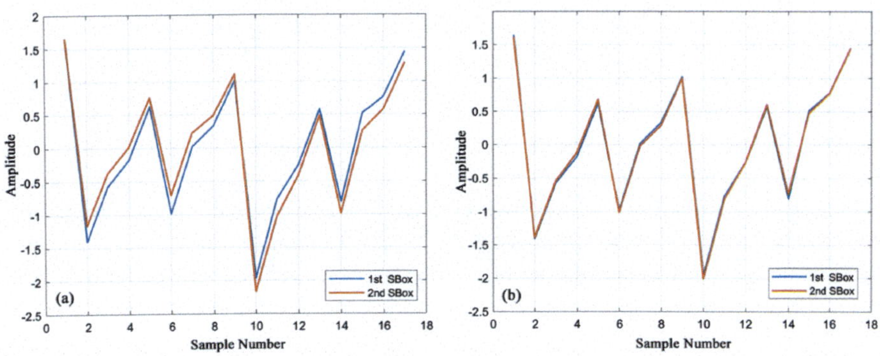

Fig. 2 Second and first S-Box execution in primary turns as power traces which were taken out of 8-bit microcontroller deployment of AES. If same data is handled, traces seem quite similar. Processing can be done on either **a** same data or **b** different data

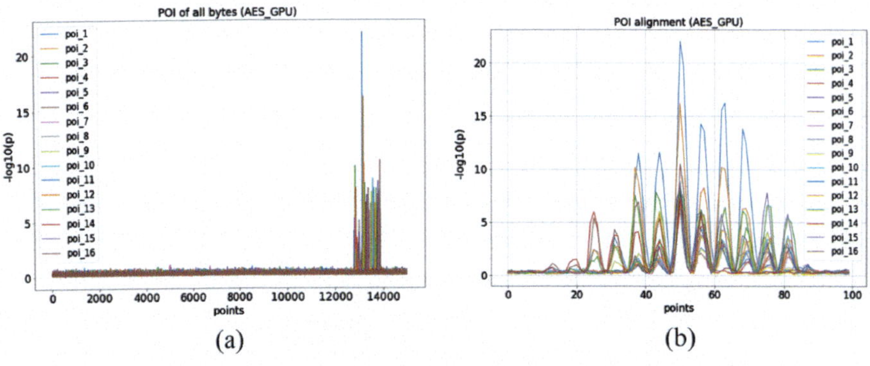

Fig. 3 **a** POI alignment; **b** entire POI of subkeys in most recent AES cycle with AES_GPU

4.1 Trace Augmentation

Deep learning strategies have shown to be very effective in a wide range of side-channel attack scenarios. However, as a result of insufficient training data, deep learning models often resort to overfitting. Overfitting is the phenomenon that occurs if network skills itself about a function that has extremely higher variation for accurately representing training information [15]. Regrettably, a large number of attackers cannot have access to significant volumes of profiling information. As an example, they can't have total regulation over devices of profiling and can be restricted traces amount that they are able to capture. Single information-level approach regarding issue of limited training information is known as data augmentation. This technique works by attempting for utilizing extra synthetically altered tracings in behaving as frequenter and assist in preventing over-fitting if skilling models over the context of side-channel attacks. As a result of the fact that instructions in software implementations of AES are carried out one at a time [16], leakage is time-dependent. The creation of a leaky profile between traces and the target subkey has become a common technique for exploiting software that uses AES as a result of this. The encryption procedure for 8-bit microcontrollers and microprocessors often involves doing each byte individually. Because data-based constituent and execution based parts as provided through Eqs. 2, 3 remains identical, power tracings of these two operations can be similar when identical information is computed through two different S-Box replacements. The power tracings that were captured during second and first S-Box execution in primary cycle of an 8-bit AES implementation on a microcontroller are shown in Fig. 2. When the identical data is processed by two different SBox replacements, it's possible to see that the power traces look remarkably similarly to one another. Due to this, pertaining to set of training, that only consist of traces concerning to target subkey, we can use select number of traces that are concerned to un-target subkeys behaves as regularizer. This is augmentation of data pertaining to certain subkey that is being done so that the model can be improved so that it has a greater ability to generalize.

4.2 Training of Cross-Subkey Model

Through Fig. 1, traces that consist of 16 S-Box processing from primary round is first distributed in 16 subtracings. Both label l_i, which refers to outcome of ith S-Box operation, and the label p_i, which refers to ith byte of plaintext, are used to identify the ith subtrace.

$$l_i = SBox(p_i \oplus k_i) \tag{3}$$

As part of profile process, tracings were split in 16 subtraces through examining Point of Interest (POI), every subtrace is assigned a label based on the output from the relevant SBox. The majority of the time, the best way to learn about a new technology is to see how it works. By include a tiny number of subtraces that reflects other 15 S-Box execution in training sets, we take cross-subkey training one step further. The percentage of subtraces belonging to targeted subkey in relation to whole train set is defined through x [0, 1]. As a result, there are 1× additional subkeys over training set. The training set for average has an equal distribution of the remaining 15 subkeys. Following that, as shown in Fig. 1, on this mixed trained set deep learning model is trained. The key is final k_i of last turns of AddRoundKey, point of attack is input of terminating round of S-Box. Few data in labeled l_i would get modified too, changing plaintext p_i, to ciphertext c_i.

$$l_i = SBox^{-1}\left(c_i \oplus k_i^{\text{last}}\right) \tag{4}$$

During attack phase, trained cross-subkey model is used. This allows for identification of traces and retrieval of target subkey.

5 Experimental Setup

The two best called dataset concerning side-channel attacks are initially described in this section's opening paragraph. Then, we demonstrate the way we skills deep learning model, assessing efficiency of attack.

Two databases that we utilized in our tests are summarized in Table 1. In primary repository, AES_GPU [17], electromagnetic tracings through NVIDIA GeForce GT620 GPU deployment of AES are recorded. A non-profiling attack to retrieve the key in [18] can need up to 11,000 traces. The attack efficiency is then further increased by Guang et al. [19] who suggest an autoencoder convolutional denoiser and recover a subkey using only 50 traces. In conclusion, the 39,511 total traces in the AES_GPU database, each containing 15,001 samples, are shown in the table below. 3.5 K randomly selected traces are used for validation along with 35 K training traces in our trials. In addition, testing involves 4511 traces.

A masked 8-bit AVR version of AES-128 is implemented in the second database, known as ASCAD, where traces are recorded using an electromagnetic probe. 100 K samples of each trace are included in the dataset, which consists of 60 K traces. According to article [20] for more information. The testing set consists of 10 K fixed-key traces, whereas sets of training has 5 K traces randomly selected for validation 50 K variable-key traces in our trials.

5.1 Model Structure

Table 2 provides a structural representation of the two deep learning models—one for the AES_GPU database and the other for the ASCAD database—that we used in the course of our research. Existing studies make use of a variety of different kinds of deep neural networks in order to counteract various forms of attacks. Multiple layer perceptrons (MLPs) looks adjustable design in situations in which tracing were least distorted with good synchronization [16]. As a consequence of this, MLPs for the AES_GPU repository were used by us, which consists of four dense layers that are interconnected in every way as shown in Table 1. In either case, the optimizer is configured to RMSprop, and the Rectified Linear Unit (ReLU) is used in its role as the activation function.

Table 1 Our experimental two databases deployed: a summary

Databases	AES_GPU dataset	ASCAD dataset
#Training trace set	32,000	45,500
#Validation trace set	4100	6000
#Testing trace set	4411	9000

Table 2 CNN and MLP network framework in our research

<table>
<tr><th>Target</th><th>I/P layers</th><th>Convolutional layers</th><th>Dense layers</th><th>O/P layers</th></tr>
<tr><td rowspan="2">AES_GPU dataset</td><td rowspan="2">100</td><td>-</td><td rowspan="2">64|128|128|256</td><td rowspan="2">256</td></tr>
<tr><td>Size × filter = 11 × 64</td></tr>
<tr><td rowspan="2">ASCAD dataset</td><td rowspan="2">300</td><td>Size × filter = 11 × 128</td><td rowspan="2">512</td><td rowspan="2">256</td></tr>
<tr><td>Size × filter = 11 × 256</td></tr>
</table>

DLSCAs have made use of Convolutional Neural Networks (CNNs), which are particularly effective when it comes to the extraction of features. For example, CNNs have been successfully used in context of side-channel attacks for bypassing jitter-based defenses and circumvent the problem of trace misalignment [17]. CNNs were also successful in breaking Protected AES [18]. In light of the fact that the implementation is a cloaked AES, we make use of CNNs in order to classify the traces included in the ASCAD database. We carry out an experiment with the goal of identifying the most effective CNN model for the four convolutional layers and one dense layer that are included inside the ASCAD database.

We use the identical power model in our studies, which makes the assumption that power usage is proportionate to the amount of information computed on point of attack. Neural network's output size is determined by its power model. Size of outcome pertaining to model is configured at 256 when information computed at attack point remains a single byte.

5.2 *Setup of Training*

It was known by us that augmentation of data rises training data number through making minute adjustments to trained tracings that already exist. But, if there is a significant amount of change made to the training data, the neural network can get confused. As a consequence of this, it is currently a challenging task to ascertain the optimum amount of augmenting traces that should be included in the training set. In order to train sixteen different deep learning models, we construct sixteen distinct training sets for each database. Each of these training sets has a unique number of augmenting traces. The procedure for constructing training sets is shown in Fig. 1, which can be seen below. The term "set" refers to each of these individual workout repetitions, of which there can be anywhere from one to sixteen. Assume that there are x training traces in the database, and that each training trace is divided into 16 segments, as shown in Fig. 1. Each of these segments is associated to a separate subkey. As a consequence of this, there need to be a total of sixteen trace segments. The training set that was used to train the model for the 1st-subkey consists of X segments for the first-subkey, and Y segments for the other subkeys.

5.3 Estimation Metric

Entropy of Partial Guessing, also known as PGE, Success Rate also known as SR, are two common metrics that are used in SCAs to assess the performance of models. The candidate for the a subkey is denoted by the letter k', and the subplaintext that is linked with the a subkey is denoted by the letter p. Subkey's rank are denoted by amount of subkey measures that have larger possibility compared to value of k.

$$\mathrm{Rank}(k) = |k' \in \kappa : \Pr\left[k|T, p\right] < \Pr\left[k'\left[T, p\right]\right] \tag{5}$$

Calculating an expected rank among all possible subkeys allows the PGE to make an estimate of the number of subkey candidate options that need to be investigated before an attack can be considered successful.

$$\mathrm{PGE}(k) = \mathop{\mathbb{E}}_{k \in \kappa} (\mathrm{RANK}(k)) \tag{6}$$

The access that the attacker has to the device that the victim is using may, however, in certain cases be fairly limited. For the whole of the attack stage, for instance, the attacker can be able to capture just a single trace. With such particular condition, rate of success is additionally appropriate threshold for evaluating. It calculates the usual chance that PGE $(k) = 0$ of categorizing a single trace by applying the deep learning model, which is equivalent to the predicted likelihood about success of attacking on a single round [19, 21, 22].

6 Result of Experimental

With section at start, experimental outcome that were obtained from our cross subkey prototype testing on AES_GPU repository was provided by us. After that, information is entered into ASCAD repository. Leakage sensing method was used by us [20] in order to locate point of interest (POI) for every subkeys.

Due to the fact that traces from AES_GPU were synchronized pertaining to most recent turns of AES, AES_GPU's attack point has been configured to S-Box inputting of that turns. Figure 3a demonstrates outcomes of leakage sensing process that made use of the attack point that was discussed earlier; more specifically, we identify segments for each subkey. In the course of our research, one hundred samples have been collected from each trace section. [13081:13181] is the precise trace interval that should be used for the first SBox operation. The synchronization process that we use for the different subkeys is shown in Fig. 3b. We establish 16 training sets by using the training technique described in experiment 2, and we refer to these sets as 1, 2, 16. Each training set contains a total of 31.5 thousand traces, of which only 3.5 thousand of the primary-subkey tracings are chosen for authentication. In test set, there are a total of 4511 traces related to the first subkey. After that, the models 1, 2, 16, MM, and so on are each trained separately on their own training sets. The maximum number of epochs that can occur is set at 500, and the batch size for training is 256. The accuracy of these 16 models in performing a single-trace attack is outlined in Table 3. It has come to our attention that the model M15 is capable of achieving the greatest possible single-trace attack accuracy in order to get the initial subkey.

Figure 4 displays, using model M15, the PGE findings for recovering the first subkey of the AES_GPU. These results were obtained using the PGE. As can be seen below, the model requires a total of just 17 traces in order to recover the subkey. As evident from Table 4, application of cross-subkey model yields 66% rise in accuracy in comparison to earlier research that was made public. In addition, we investigate whether or not adding disturbed signals to train tracings would contain identical effects like synthetically changing them. What we end up doing is coming up with a training set that is composed of "*x*" traces that were connected to primary subkey.

Table 3 16 models' accuracy results for the test set (AES_GPU)

Model number	Accuracy (%)	Model number	Accuracy
M_1	1.0	M_9	2.0
M_2	0.8	M_{10}	2.0
M_3	1.2	M_{11}	2.0
M_4	1.7	M_{12}	2.0
M_5	1.7	M_{13}	2.0
M_6	1.8	M_{14}	2.0
M_7	1.9	M_{15}	2.3
M_8	2.0	M_{16}	2.0

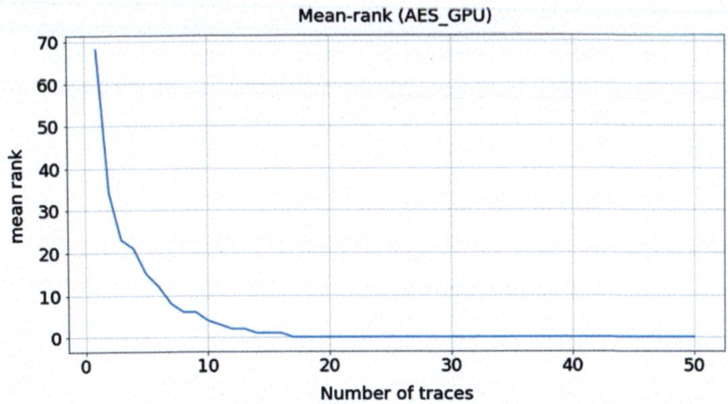

Fig. 4 PGE outcome of deploying M15 model in recovering primary subkey (AES_GPU)

After that, we added "y" traces of white Gaussian noise consisting random labels that run from 0 to 255 and have a mean and standard deviation of 0 and 0.01 correspondingly for each variable. The fact that we are able to generate a novel data-set known as $x/(x + y) = 15|16$ by making use of noisy traces rather than any other byte traces is evidence that the cross-subkey model is effective. This is shown by the fact that we did so. The trained model only has an accuracy of 0.6% when applied to the test set.

Table 4 Summary of existing works (AES_GPU)

Previous works	Number of traces for a successful attack
[19]	650
[20]	50
Our model	17

7 Conclusion

With concerned research, suggestion of cross-subkey deep learning side-channel attack was done by us that takes use of the additional intentionally altered power tracings like augmentation of data to produce prototypes having greater capacity of generalization. This attack uses power traces to construct framework having higher generalization capability. Network model precision was demonstrated by us in which accuracy can be improved by appending additional data subkeys to targeted subkey train set. Consequently, it will allow network model to better predict the data on the test set. This study demonstrates that cross-subkey training is successful by employing two datasets that are well known; yet, the analysis of the links that exist between the different subkeys still has a great deal of exciting questions that remain unresolved. As was covered in the sections that came before this one, the connection that exists between the various subkeys can be investigated in a variety of different ways. The coherence of the field as a whole, as well as the degree of confidence in the results, will ultimately improve as a result of these orientations.

References

1. Hu F, Wang J, Wang W, Ni F (2022) Open Access Libr J 9(1):1–15
2. Hu F, Ni F (2022) Presented at the 2022 international conference on cyber warfare and security (ICCWS)
3. Hu F, Wang H, Wang J (2021) IACR Cryptol ePrint Arch 2021:1328
4. Wang C, Dani J, et al (2023) Presented at the 2023 IEEE international symposium on hardware oriented security and trust (HOST)
5. Hu F, Wang H, Wang J (2022) Cryptol ePrint Arch
6. Hu F, Wang H, Wang J (2022) Sci Rep 12(1):6254
7. Ghandali S, Ghandali S, Tehranipoor S (2021) IEEE Access 9:136448–136458
8. Perin G, Wu L, Picek S (2022) IACR Trans Cryptogr Hardw Embed Syst 2022(4):828–861
9. Ou Y, Li L (2022) Front Comp Sci 16:1–11
10. Masure L, Strullu R (2023) J Cryptogr Eng 13(2):129–147
11. Fernández Ortiz C (2022) Bachelor thesis, Universitat Politècnica de Catalunya
12. Do NT, Hoang VP, Doan VS, Pham CK (2023) IET Inf Secur 17(3):377–393
13. Picek S, Perin G, Mariot L, Wu L, Batina L (2023) ACM Comput Surv 55(11):1–35
14. Wang C, Ninan M et al (2023) Proceedings of the 16th ACM conference on security and privacy in wireless and mobile networks. Association for Computing Machinery, Guildford, United Kingdom, pp 227–238
15. Chang L, Wei Y, He S, Pan X (2022) Appl Sci 12(16):8246
16. Ito A, Saito K et al (2021) IEEE Trans Inf Forens Secur 16:3790–3802
17. Kwon D, Hong S, Kim H (2022) IEEE Access 10:5957–5967
18. Hettwer B et al (2020) J Cryptogr Eng 10:135–162
19. Arpaia P, Caputo F, Cioffi A, Esposito A, Isgrò F (2023) IEEE Trans Instrum Meas 72:1–8
20. Cao P, Zhang H, Gu D, Lu Y, Yuan Y (2022) Proceedings of the 59th ACM/IEEE design automation conference. Association for Computing Machinery, San Francisco, California, pp 691–696
21. Sasank TS et al (2021) J Phys Conf Ser 1879 032124
22. Mitra A, Gochhait S, Obaid AJ, Alkhafaji MA (2023) A strategic data protection plan for the healthcare industry-a review. In: 2023 8th international conference on communication and electronics systems (ICCES), Coimbatore, India, pp 1784–1788. https://doi.org/10.1109/ICCES57224.2023.10192830

Content-Based Book Recommender System Using Supervised Learning

Mayur Rahul, Akhilesh Singh, Shekhar Verma, Himanshu Shukla, and Vikash Yadav

Abstract Online shopping in India has already got more attention in the last few decades. The Indians are more interested in online shopping which has changed all the concepts of the market scenario. There are so many online shopping websites available nowadays like Flipkart, Amazon, Junglee, Snapdeal, etc. With the increase in the number of traders and buyers, efficient and effective techniques are used to manage the vast amount of data formed day by day. Recommender systems give a platform to filter out the given data and able to provide relevant information to the end users. The methods used in the recommender systems are content-based, collaborative-based, and demographic-based but there are some issues that fail in producing efficient information for the end users. Thus, it is very important to find the most important features that are capable of optimizing recommender systems. This paper proposed a content-based book recommender system using PCA and SVM with the help of two publicly available datasets like Amazon Book Reviews (AB) and BookCrossing (BX). The results of the proposed recommender system prove the superiority of the system.

Keywords Recommender systems · Demographic · Collaborative · Content · PCA · SVM

M. Rahul (✉) · A. Singh · S. Verma · H. Shukla
Department of Computer Application, School of Engineering and Technology, CSJM University, Kanpur, Uttar Pradesh, India
e-mail: mayurrahul209@gmail.com

S. Verma
e-mail: shekharverma@csjmu.ac.in

H. Shukla
e-mail: himanshushukla@csjmu.ac.in

V. Yadav (✉)
Department of Technical Education, Government Polytechnic Bighapur Unnao, Manpur, Uttar Pradesh, India
e-mail: vikas.yadav.cs@gmail.com

S. Dutta et al. (eds.), *Cyber Intelligence and Information Retrieval*, Lecture Notes in Networks and Systems 1025, https://doi.org/10.1007/978-981-97-3594-5_30

1 Introduction

In present time of data and information profusion, internet customers are unable to choose from available services and products. There is a demand of efficient recommendation system that formulates personalized recommendations. The basic objective of recommender system is to provide framework for asking familiarity suggestions when anyone chooses a book, movies, restaurant, and so on. A recommender system estimates how likely user is interested in a given item that may be not familiar to him. Concerning to create suggestions, a recommender system normally requires user feedback, items, and data on given item. After recommendations, user response on given item is found either implicitly or explicitly. The system accumulates the user response in database and uses it for further suggestions [1].

The applications of recommender systems in e-commerce have many advantages to both consumers and sellers. The major objective is to give their items obtainable to the customer and to get user loyalty and satisfaction. Their goal could be obtained when user get accurate suggestion of the items as per need. Sellers can also be able to make gain with the help of RSs without any problem. The users would be able to receive lists of items that can be very useful. They were also able to save money, efforts, and time and also able to find the items which they appreciate.

The main advantages of recommender system go beyond the limit of e-commerce framework. In the given paper, we introduced book recommendation system. Such types of suggestion could be useful in schools, e-learning, and libraries portals. The expansion of e-book produces reader's access to expensive books with very small effort. We will presume that the book reading is spreading more frequently regions around the world. The number shows the converse meaning because nowadays students' interests in reading the books is falling day by day.

Furthermore, the vast quantity of accessible books creates the benefits of RSs a compulsory thing. Google found 129,864,880 in the year 2010, the count increased above to about one billion new books, and changed books have been issued in USA, China, and UK alone. This is very important goal to suggest the perfect fit regarding books for given user in this large publication.

Many researchers have showed the advantages of reading books. On contrast between non-readers and readers of books, latter have been found with better mental health and also happier in life [2, 3]. These data must assist the declaration given by National Institute of Child Health and Human Development "Reading is primary thing compulsory for a successful, fruitful and happy life".

Fiction is found to simulate great social interaction [4]. The fiction maps with a huge ability for social support and empathy. It is the biological effect in connectivity of brain. Therefore, it is so valuable attempt to apply the concepts of artificial intelligence to create people attentiveness by suggesting the right choice of the books. People rely on recommender systems that show user will be interested in books entitled "people bought this also bought this" than book entitle suggested by the book shop staff.

We found that from the survey of so many papers that we need a paper which is capable of suggesting some books to the user. Many papers specify the specific problems of recommender systems so far. They shall be able to summarize different trends, technique, and possible research challenge in this field. This paper proposed a content-based book recommender system using PCA and SVM with the help of two publicly available datasets like Amazon Book Reviews (AB) and BookCrossing (BX). The main objectives of this research are:

(1) We have introduced novel content-based book recommendation systems using new method.
(2) We have applied PCA and SVM techniques to enhance the outcomes and therefore minimize error rate.
(3) We are able to enhance the productivity of SVM by adding PCA.
(4) Our introduced method is assessed using free and standard Amazon Book Reviews (AB) and BookCrossing (BX) datasets.
(5) We used computational time, average similarity, Dunn index, MAE, T-value, RMSE, and SD to inspect the achievement of framework.
(6) Our introduced technique has better MAE of 0.74 and better simulation time of 123.65 with BX, which indicates the better performance when compared with AB datasets.

This paper is divided into 5 sections, in which Sect. 2 talks about some related work, Sect. 3 elucidates proposed methodology and evaluation measures, Sect. 4 details experiments and results, and finally in Sect. 5, conclusions and future work.

2 Related Works

This section explains the various algorithms applied to suggest item to the users: content-based, collaborative-based, demographic-based, association rule-based, context-aware, and social recommendation systems.

Collaborative filtering depends on available ratings of user to estimate other user's preferences. It can be categorized into two categories. Item-based calculates the similarity factor between rated items while user-based computes the similarity factor between active users [5]. To create suggestions, it only requires a user-item rating matrix, so it is very easy to create. Due to sparse issue in matrix especially in the case of new users and items, it forms cold start problem. This sparsity problem might produce bad recommendations. It also has two disadvantages. The shilling attack issue happens when item contains false ratings in the form of furtherance. The gray sheep issue comes when system suggests to a user who has liked completely opposite than other users [6]. The commonly used techniques in CF are slope algorithm and KNN. The slope simply uses the technique of regression to predict the user ratings, while KNN searches for user/item with dimensions similar to the preferred target user/item [7]. It is widely used in book recommender systems.

In content-based filtering, the content of given items can be represented by bag of words, ontology, and space model in the form of author and genre of book [8]. It is a classification technique which learns the similarities and patterns in buying history of user to suggest his future likes. The contents may be referred by its summary, title, and outline; metadata includes publisher, author, publication year, page number, genre, etc. The previous literature shows the list of various machine learning techniques that are used for suggestion of items. These techniques analyze the textual data of user suggested item and are able to find identical features among these items. Content-based avoids many issues of collaborative filtering including new item issue. It does not depend on the community ratings. It also gives efficient suggestions that depend on the item's attributes. Still, content-based technique faces many problems due to inadequate information of the items [9]. The content also required transformation of unstructured text to meaningful data or attributes using natural language processing. The over specialization is another issue with CB that shows the diverseness of the suggestions.

Demographic-based RS connected with items with user's classes having gender, country, age, etc. These recommender systems overcome the issue of new users but might disturb the privacy. The demographic-based recommender system is widely used in the book recommender system [10]. On the other hand, social RS makes use of the relationship, tag, and other features found on social media to make recommendations [11]. The social RS makes use of other recommender systems to suggest an item. Some used trust-based technique, and some used the concept of tag to suggest the book.

The context-aware RS was applied in book RS. This method is based on the current situation of the books for example location and time, to make suggestions [12]. The other author proposed the book RS based on association rules [13]. Hahsler et al. proposed an approach based on the patterns found in the huge database of library records. If two given books are interrelated, then appearance of one or the other in any records gives appearance of other book. To assess the dependability of technique, one calculates support and confidence. The regulation $(X, Y) \rightarrow Z$ represents the quantity of records which contains X and Y containing Z as confidence while support is the no. of transactions that contains X and Y [14].

The research domain of the recommender system is very big and active. Most of the systems are generic in nature and not limited to the particular type of item. The current RS can be applied to book RS. For example Amazon used the same technique for all items available in its database [15].

The other systems also depend on context, metadata, and demographic data [16]. An item-based recommender system computes the Euclidean and cosine length in users x authors and users x books using matrices. The book RS and author RS have been calculated by LitRec dataset. The suggestion completion was optimal when 90% of book RS and 10% of author RS were combined. Vaz et al. introduced the concept that was based on computation of relevance of ratings in item CF [17]. All the experiments have been conducted on LitRec dataset and found the large prediction error using only latest ratings and neglecting previous ratings. It was constructed by the author for better quality suggestions that required each ratings of items and not the latest ratings.

Kapusuzoglu et al. proposed a book RS with the help of book information using enlarged ontology [18]. If the ontologies have attribute author, then extended ontology will contain all the publications and awards of the author. The system represents the book as relational database. The extended form of Euclidean and cosine similarity metrics is introduced to manage the formed ontologies. All the simulations were running 2791 server logs of online book portal. The author was also able to prove that extended ontologies enhance the accuracies of suggestions in contrast with existing ontologies, mainly with the introduced Euclidean length.

Rajpurkar et al. introduced the combination content-based and collaborative filter to enhance the quality of recommendation [13]. The author has introduced the concept of inserting content-based RS into collaborative filter RS. The author has developed various association rules. The system retrieves book that found in transaction using user-preferred items. The author is able to give the recommendation by using intersection between recommendation results and association rules.

Pathak et al. introduced the hybrid system that combines the three RS, i.e., context-aware, CF, and CB RS [12]. In starting of system, CB recommends book on the basis of predefined books, from which maximum ranked book is suggested. The performances of their hybrid systems are better than all systems individually. Mikawa et al. introduced a SVM-based RS that is used to identify the age and gender of the people present in the library. The SVM is applied to various images that are taken by camera placed in the entrance. The book suggestion was given on the basis gender and age.

3 Proposed Methodology

In this section, we are introducing recommender system using PCA, sampling, and SVM methods. The SVM is chart-topping method under supervised method. The PCA is the main technique in our paper because it is more efficient and gives promising results when compared to recent methods.

In Fig. 1, the block diagram of our introduced book recommendation system is depicted. The population from AB and BX datasets is used. The sampling is a method which is used to extricate some important information from given datasets. Sampling is applied because conducting experiment on entire dataset is too costly. Sampling method applied in the given methodology is random sampling with replacement. The objective using random sampling is that each item has similar chances and with replacement means it is deleted from given population.

In recommendation system, datasets are having characteristic with large parameter space which contains large information. The characteristic of each object has minimum importance in big characteristics space. The length of linking every feature points and densities is less effective in classification of large information space. In matrix of billions of columns and rows, uttermost of data are found to be zero. To

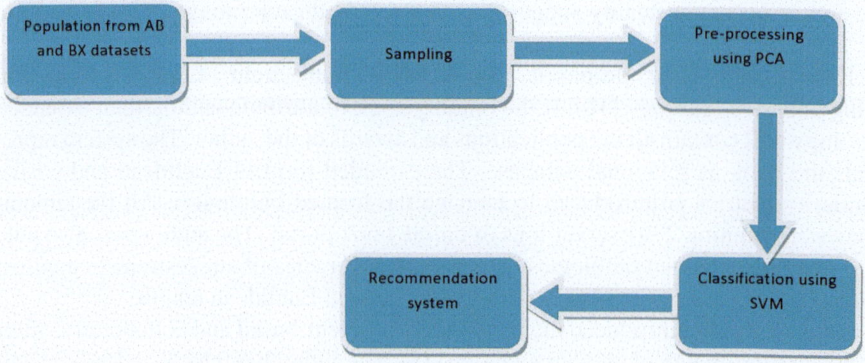

Fig. 1 Proposed methodology

minimize this ineffective data, we have applied PCA for minimization of data space. PCA is the technique to compute patterns in large database. PCA investigation is done in 2D magnitude of vector space with Gaussian technique. The two main parts are found after compacting of information. The first part consists 90% of data, so deleting second part produces only 10% of whole information reduced from given dataset. SVM is a supervised learning technique. The principal objective of method is to label an item to given group so that items in group seem to be more similar than that of rest of the groups. This technique is able to produce more significant groups in database.

4 Experiments and Results

In the given section, we demonstrate rigorous investigation of our simulation outcomes received from our introduced book recommender systems using AB and BX. We conduct all simulations in computer that have structure of I9 processor, 8 GB RAM, and Python 3.10.4 environment.

We have considered Amazon Book Reviews (AB) and BookCrossing (BX) datasets available freely online for investigation of outcomes of our introduced book recommender system. The limitation on every user is to rate at least 20 books. We have used different assessment measures in beneficial to assess the achievement of our book recommendation system. We are able to compute Dunn index, MAE, SD, MAE, RMSE, T-value, processing time, and average similarity to investigate the achievement of our introduced recommendation system.

We have collated our new introduced work outcomes with AB and BX datasets as depicted in Table 1. The MAE of our introduced work with BookCrossing (BX) is 0.74, which is far better as compared to Amazon Book Reviews. The outcomes are comparatively not good with AB in contrast with BX datasets.

Table 1 Comparison of different metrics-I

Dataset	RMSE	MAE	SD	T-value
Amazon Book Reviews (AB)	3.35128	1.85	0.34336	7.82321
BookCrossing (BX)	1.31763	0.74	0.12454	2.91232

Table 2 Comparison of different metrics-II

Dataset	Dunn index	Average similarity
Amazon Book Reviews (AB)	0.31123	0.89
BookCrossing (BX)	0.11643	0.86

Table 3 Comparison of processing time

Method	Processing speed in sec
Amazon Book Reviews (AB)	135.23
BookCrossing (BX)	123.65

The datasets explained in preceding table are again contrasting regarding similarity index and Dunn index as depicted in Table 2. The average similarity and Dunn index are 0.86 and 0.11643 for BX, which is significantly good than AB dataset. The outcomes in table specify better achievements of our introduced work with the BX dataset. In Table 3, matching in terms of processing time is shown. The outcome with BX is significantly good than AB dataset.

5 Conclusion and Future Works

Recommender system is chart-topping in a corporation to suggest items to user group and individual user from millions of item in area of online trading and additional type of corporations. They are capable of suggesting the etiquette of present user to produce latest items and to investigate running items. We are capable of introducing effective and reliable content-based book recommender system. Our outcomes also proved that achievement of our introduced recommendation system works with BX is better than AB dataset. Our introduced research regarding processing speed with AB is significantly better than BX dataset. In the future prospects, we will add some new feature extrication, machine and deep learning algorithms to enhance the performance of recommendation systems. We will further apply introduced work to large data and try to maximize its efficiency and privacy.

References

1. Ricci F, Rokach L, Shapira B (2011) Introduction to recommender systems handbook. In: Ricci F, Rokach L, Shapira B, Kantor BP (eds) Recommender systems handbook. Springer, US, Boston, pp 1–35
2. Mikawa M, Izumi S, Tanaka K (2011) Book recommendation signage system using silhouette-based gait classification. In: 2011 10th international conference on machine learning and applications and workshops (ICMLA), vol 1, pp 416–419
3. Hill K (2013) The arts and individual well-being in Canada. [Online; posted 13 February 2013]
4. Mar RA, Oatley K (2008) The function of fiction is the abstraction and simulation of social experience. Perspect Psychol Sci 3(3):173–192
5. Sarwar B, Karypis G, Konstan J, Riedl J (2001) Item-based collaborative filtering recommendation algorithms. In: Proceedings of the 10th international conference on World Wide Web, WWW'01. ACM, New York, pp 285–295
6. Su X, Khoshgoftaar TM (2009) A survey of collaborative filtering techniques. Adv Artif Intell 4:2–4:2
7. Lee J, Sun M, Lebanon G (2012) A comparative study of collaborative filtering algorithms. CoRR. arXiv:1205.3193
8. Lops P, Gemmis M, Semeraro G (2011) Content-based recommender systems: state of the art and trends. In: Ricci F, Rokach L, Shapira B, Kantor BP (eds) Recommender systems handbook. Springer, US, Boston, pp 73–105
9. Pazzani MJ, Billsus D (2007) Content-based recommendation systems. In: Brusilovsky P, Kobsa A, Nejdl W (eds) The adaptive web: methods and strategies of web personalization. Springer, Berlin Heidelberg, Berlin, pp 325–341
10. Sase A, Varun K, Rathod S, Patil D (2015) A proposed book recommender system. Int J Adv Res Comput Commun Eng 4(2):481–483
11. Tang J, Hu X, Liu H (2013) Social recommendation: a review. Soc Netw Anal Min 3(4):1113–1133
12. Pathak D, Matharia S, Murthy CNS (2013) NOVA: hybrid book recommendation engine. In: 2013 IEEE 3rd international advance computing conference (IACC), pp 977–982
13. Rajpurkar S, Bhatt D, Malhotra P (2015) Book recommendation system. IJIRST–Int J Innov Res Sci Technol 1(11):314–316
14. Hahsler M, Grün B, Hornik K (2005) arules – A computational environment for mining association rules and frequent item sets. JSS J Stat Softw 14(15):1–25
15. Linden G, Smith B, York J (2003) Amazon.com recommendations: item-to-item collaborative filtering. IEEE Internet Comput 7(1):76–80
16. Vaz PC, Martins de Matos D, Martins B (2012) Improving a hybrid literary book recommendation system through author ranking. In: Proceedings of the 12th ACM/IEEE-CS joint conference on digital libraries, JCDL'12. ACM, New York, pp 387–388
17. Vaz PC, Ribeiro R, Martins de Matos D (2013) Understanding temporal dynamics of ratings in the book recommendation scenario. In: Proceedings of the 2013 international conference on information systems and design of communication, ISDOC'13. ACM, New York, pp 11–15
18. Kapusuzoglu H, Gunduz Öguducu S (2011) A relational recommender system based on domain ontology. In: 2011 international conference on emerging intelligent data and web technologies (EIDWT), pp 36–41

Texture-Based Garments Defect Detection Method Using Machine Learning

Mayur Rahul, Shekhar Verma, Akhilesh Singh, Sarvachan Verma, and Vikash Yadav

Abstract Fabric defects identification is significantly important in textile industry. Various activities are involved in detecting fabric quality. The total profits of the industries depend on the defects on the quality of fabrics that produces loss. Existing defect identification methods are conducting in different organizations with the help of human defect inspectors who draw manually defect pattern. However, these detection techniques have few drawbacks like complication, inaccuracy, negligence, tediousness, and exhaustion in addition to time-consuming which causes decrease performance in failure finding. For the sake of overcoming these problems, a framework that depends on processing of image has been introduced to efficiently and effectively identify the defects of fabrics. The given introduced work consists of image segmentation, morphological methods, feature extraction, and classification. We have used edge detection for image segmentation, ICA for feature extraction, and SVM for image classification. The results achieved from the given framework are superior than existing framework. The experiments have been conducted in Python environment using TILDA datasets.

Keywords Texture-based · Machine learning · SVM · ICA · TILDA

M. Rahul (✉) · S. Verma · A. Singh
Department of Computer Application, School of Engineering & Technology, CSJM University, Kanpur, Uttar Pradesh, India
e-mail: mayurrahul209@gmail.com

S. Verma
e-mail: shekharverma@csjmu.ac.in

A. Singh
e-mail: akhileshsingh@csjmu.ac.in

S. Verma
Ajay Kumar Garg Engineering College, Ghaziabad, Uttar Pradesh, India
e-mail: sarvachan.verma@gmail.com

V. Yadav (✉)
Department of Technical Education, Government Polytechnic Bighapur Unnao, Manpur, Uttar Pradesh, India
e-mail: vikas.yadav.cs@gmail.com

S. Dutta et al. (eds.), *Cyber Intelligence and Information Retrieval*, Lecture Notes in Networks and Systems 1025, https://doi.org/10.1007/978-981-97-3594-5_31

1 Introduction

Image classification and segmentation models are applicable in different domains such as organization-based issues [1–4]. Garments are contemplated as one of the important needs of human beings, and garments history is very old as human civilization. The garments are contemplated as main components for human clothes which is also very useful in various industries' products [5–8]. The natural products like nylon, polyester, cotton, and wool could be used to manufacture textiles garments. Various types of tools are used in textile organization to create garments, and issues have been detected using inspection method. Conventionally, inspection method is executed by applying manual people attempt to determine the standard of garments [6, 9–11]. The cost of the garment is evaluated in market based on quantity of defects in the garments, and it is directly proportional to the quantity of defects [11, 12].

The production is stopped, if an issue is detected in garments, and the detail about occurred issue is recorded with its corresponding location by operator of machine. The primary drawback found during manual investigation is: (1) the training of particular garments inspector is required; (2) ignoring small defects while considering only major defects due to carelessness of human; (3) various human attempts have been required to identify defects in garments; and (4) it is very complicated for garment inspector to target on production that leads to the less production of the organization. Only 60–70% accuracy is found in identifying defects in garments that could lead to produce high cost of garments in the markets [3].

The above issue is very critical and complex for the garment factory to overcome. Due to these issues, we have to select an efficient defect detection framework that can reduce the cost of manufacturing garments. The simulation-based framework is used for this issue, and garments products are processed using the given process and could produce higher rate of accuracy. The concept of image processing and computer vision plays an important role in garment industry for the detection of garments defects [9]. The checking of garments is done during the production process and said to be real-time application (see Fig. 1). The real-time processing of defect detection is done using digital image processing and computer vision. According to different papers published so far, more than 70 defects have been found in industries. The different issues in garments are categorized as critical, major, and minor issues based on the condition of issues [13].

Some of regularly occurring primary defects are smash, cracks, snarls, double pick, bad selvedge, oil marks and stains, loose threads, knots, stitches, holes, float, end outs, and broken pick. Hole is identified as major issue in garments caused by broken needle or defects in machine. Oil marks are obtained due to uncontrolled oil in machine. Collective netting comes under the minor effect due to the combination of more than one broken threads. The appearance of broken thread is due to the breaking of warp yarn during weaving. The thin and thick bar effects due to the variations in yarn. The appearance of broken weft is due to the breaking of filling yarn. The appearance of wrong weft is due to the process when weaving does not follow the

Fig. 1 Fully automated garment factory

rule of schematic design during weft insertion. There are different categories present in garments defect and various ways of classifying defects in colors and design.

In this research, grayscale garments defect detection techniques are introduced in order to get more efficiency in garment industries. The given proposed computer vision techniques are designed for the grayscale images and incorporating different feature extraction techniques to create discrimination between different defects. Besides this, the processing of grayscale image descriptors is more efficient than color descriptors. Further, the processing of grayscale image can significantly minimize the complexity in calculation. The fabric detection based on color concept has its own detection model and importance such as bad design of warp yarn garments [14–16].

Conventionally, various types of yarns are used for manufacturing of garments and different methods like felting, weaving, and knitting are used for this purpose. The fabric depends on fibers used, the fabric manufacturing method, tools applied for manufacturing, and finishing method. There exist primarily three issues linked to garments defect identification. (1) There are very wide areas of garments available in the market. The same technique cannot be useful in every type of garments that may lead to instability. (2) The different categories and characteristics of the defects in garments lead to various problems. (3) It is very complicated to manage huge quantity of defects samples, which creates data variance and failure of different machine learning algorithms. To overcome these issues, various researchers of this field make good efforts. The main findings of this paper are as follows:

(1) To propose a garments defect detection method.
(2) To utilize the benefits of edge detection method in image segmentation step.
(3) To utilize the methods of machine learning in detecting defects in garments.
(4) To improve and enhance the defect detection accuracies using TILDA datasets.
(5) To prove that our proposed system is best by comparing our results with existing results.

The rest of our paper is organized as follows: next section discusses the various related research papers. The proposed methodology has been described in Sect. 3, results and experiments have discussed in Sect. 4, and lastly concluded in Sect. 5.

2 Related Works

Texture-based inspection refers to representation of region in a given image by texture-based contents and is very useful in various applications of computer vision areas [17–19]. Liang et al. introduced an intelligent framework for assessment of fabric surface design [20]. They used the three feature extraction methods for their system. They used attention-driven technique for fault detection using saliency map method and different statistical measures, and wavelet texture property is used to represent fabric surface. They used fuzzy ARTMAP-based neural networks that is applied to learn properties from grading and identification of quality of fabric surface. The backpropagation NN, fuzzy ARTMAP, and SVM were compared to get the optimized classification performance (see Fig. 2). The assessment results illustrated the introduced technique using fuzzy ARTMAP that produced superior results to identify fabric surface.

Wen et al. explained that garments flaw is unexpected properties, and wavelet reaction for these properties and background is entirely different, and therefore, wavelet could accurately identify the issues [21]. The technique used for garments defect detection can be categorized into adaptive and non-adaptive. The non-adaptive method uses the combination of present bases with the different feature extraction

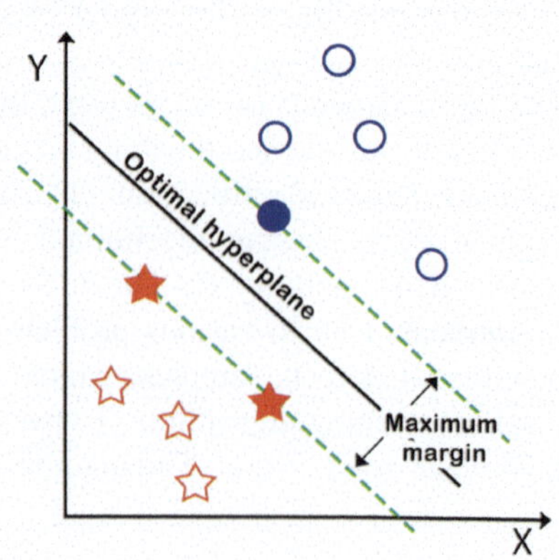

Fig. 2 Support vector machine

techniques for identification. Due to issues in present wavelet and non-adaptive cannot produce good results with all garments types in the industry. Wen et al. introduced a technique for identification of different garments types using adaptive wavelet method. Adaptive wavelet was used to identify fabric defect, and it primarily consists of training and classification stage. They used the combination of an optimal method and wavelet bases to deal with different types of fabrics. They used the cost function model to get the filter coefficient to identify the garments defects. The limitation of their work is that the normal images are used for the training of model for identification of the garments defects. The investigation results show that introduced technique efficiently achieved optimal performance in the identification of garments defects.

Karlekar et al. introduced a real-time and efficient framework that is capable of fabric defect detection [22]. They used wavelet concept because it identifies line properties with less complexity. The images are converted to grayscale image. They used median filter and histogram utilization for removal of unwanted things from image. Wavelet-based technique is used at various scales to extricate different features for identifying defects. The thresholding is further used for image segmentation, and morphological filter is used for removing extra noise. The introduced work achieved satisfactory results for segmentation and thresholding of garments defects using warp knitting tool.

Yapi et al. proposed a learning-based technique using statistical signatures for differentiating between defective free and defective garments [23]. The introduced framework is capable of managing various types of garments that range from simple to complex. The framework consists of three basic steps: preprocessing, training, and testing. Their framework is robust to illumination and noise. It is also invariant to scale translation and transformation. Their proposed work is able to get better results as compared to existing results.

Hamdi et al. introduced a technique based on unsupervised learning to identify defects in pattern garments images [24]. The introduced framework consists of three steps: preprocessing of images, texture filtering, and identification of defects. In preprocessing step, enhancing contrast of given image and extrication of fabric design takes place. The given image has been segmented to various regions depending on texture technique, and then it gets filtered to separate noises in second step. Further, the filtered image is partitioned into blocks, and then median of whole blocks is combined with k-means to differentiate defected blocks with maximum difference gain.

Wu et al. introduced dictionary-based technique to optimize the representation of texture garments and enhance the calculation speed which could be used in the detection of defects [25]. The combination of least squares techniques and discrete cosine techniques is applied to learn dictionary. The introduced research produced good outcomes in terms of error rate, calculation speed, and consistent results. The introduced framework can also be used to evaluate the effect of density in garments pattern representations and for the identification of garments defects.

Hanbay et al. introduced a vision-based framework for the identification of garments defects in circular knitting tool [26]. The introduced framework comprises two steps: feature extrication and classification. The shearlet transform is used in normalized images to extract important features from the images. The three-layered artificial neural networks are used to classify defects. The performance assessment has been done on knitted garments on a circular tool machine in textile industry. The five metrics have been used for quantitative evaluation. The introduced technique produces better results as compared to the existing work. The input of image has been done using high-density camera and Intel core i9 (3.5 GHz) with 8 GB RAM. They wind up that wavelet-based technique is primarily supreme fitted for real-time-based applications due to high calculation performance.

3 Proposed Methodology

The given Fig. 3 shows the proposed methodology used in garments defect detection using image processing. The steps involved in the proposed systems are: image segmentation, feature extraction, and classification (see Fig. 3).

3.1 *Image Segmentation*

This step is the basic advance in image investigation. Segmentation separates an image into various segment parts or regions. More precisely, edge detection is a method that is used in image processing to construct image segments and pattern recognition more effectively [27, 28]. It reduces the measure of information and removes the useless information. The accuracy of classification depends on the flawlessness of detecting important edges. It is used to detect strength discontinuities in image. We used Canny edge detection technique for the above purpose [29, 30].

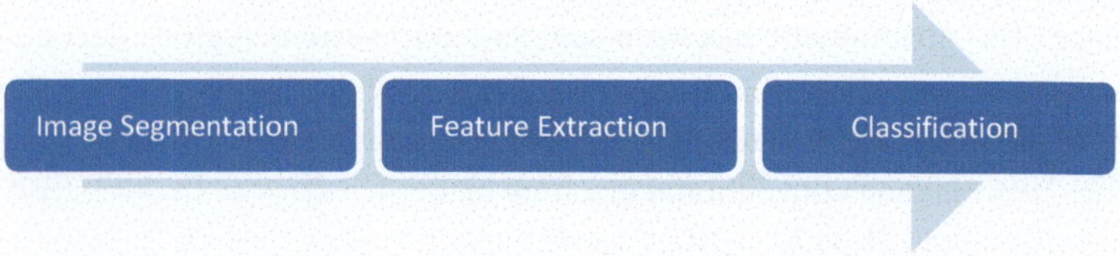

Fig. 3 Proposed methodology

3.2 Feature Extraction

The feature extraction method is applied to the fabric image acquired by the computer. The central point of local property is used to represent image part that is most important for the classification. The image purely depends on the local feature extractor which is achieved from different region called ROI [31]. Independent component analysis (ICA) is put in to garments image to extricate the important features. ICA technique is used to identify region of interest in the given image [32]. An ROI in image is a collection of pixel, which has position and could be identified. An ROI has high local data that seems to be repeatable in various images and produces more accurate classification [33–36].

3.3 Classification

Support vector machine (SVM) is a supervised machine learning technique which has been applied to identify defects in fabrics. We are using K-cross validation rule for the bias and variance error. The whole dataset is divided into n partitions. In each partition, n-1 is used for training and remaining 1 is used to test framework. The final result is the average of results achieved from n partition (see Fig. 2).

4 Experiments and Results

We simulate all the experiments in the computer system which has very good configuration of Intel i7 processor; 16 GB RAM; and Python 3.8.1 environment (see Table1). To understand the performance of our defect detection system in fabrics using TILDA dataset, we are comparing accuracies of existing work with our proposed system (see Table 2 and Fig. 4).

Table 1 Proposed methodology

Step 1	Edge detection method is used to improve the fabric images
Step 2	ICA is used to extract independent components from the images and reduce the dimensions of the image and store the important data only
Step 3	SVM is used to classify defects in the images

Table 2 Comparison of our work with existing work

Author	Accuracy (%)
Habib et al. [37]	82.97
Saeidi et al. [38]	78.4
Islam et al. [39]	85.57
Murino et al. [40]	77
Karayiannis et al. [41]	84
Kumar et al. [42]	85.9
Kuo et al. [43]	76.5
Proposed work	89.4

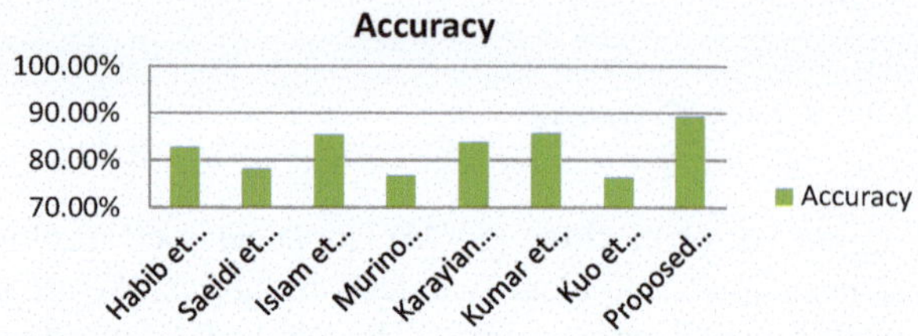

Fig. 4 Comparison of our work with existing work

5 Conclusion and Future Works

The identification of defects in garments plays a very important role in performance of any industries. The industry required real-time detection system which controls quality of the garments and works quickly, efficiently, and effectively. Manual monitoring is time-consuming and less efficient that produces great loss. The automated monitoring of garments takes less time and more efficient as compared to manual monitoring. The accuracy rate of all research in fabric detection system till now is very less. Our paper is able to resolve many issues regarding garments defect system and introduced innovative garments defect detection system using ICA and SVM. The combination of ICA and SVM is able to enhance the detection rate of 89.4% which is better than the existing systems. In the future, we will work on sensor-based fabric defect detection system and try to design optimal systems which are more accurate to detect defects in garment and, further, try to incorporate some more feature extractors and machine learning techniques.

References

1. Latif A, Rasheed A, Sajid U et al (2019) Content-based image retrieval and feature extraction: a comprehensive review. Math Probl Eng 2019. Article ID 9658350
2. Li Y, Luo H, Yu M, Jiang G, Cong H (2019) Fabric defect detection algorithm using RDPSO-based optimal Gabor filter. J Text Inst 110(4):487–495

3. Wong W, Jiang J (2018) Computer vision techniques for detecting fabric defects. In: Applications of computer vision in fashion and textiles. Elsevier, Amsterdam, Netherlands
4. Czimmermann T, Ciuti G, Milazzo M et al (2020) Visual-based defect detection and classification approaches for industrial applications-a survey. Sensors 20(5):1459
5. Dario P, Kolhe S, Patil P (2009) A review of automatic fabric defect detection techniques. Adv Comput Res 1(2):18–29
6. Ngan HYT, Pang GKH, Yung NHC (2011) Automated fabric defect detection-a review. Image Vis Comput 29(7):442–458
7. Ngai EWT, Peng S, Alexander P, Moon KKL (2014) Decision support and intelligent systems in the textile and apparel supply chain: an academic review of research articles. Expert Syst Appl 41(1):81–91
8. Hanbay K, Talu MF, Özgüven OF (2016) Fabric defect detection systems and methods-a systematic literature review. Optik 127(24):11960–11973
9. Wen J, Wong W (2018) Fundamentals of common computer vision techniques for fashion textile modeling, recognition, and retrieval. In: Applications of computer vision in fashion and textiles. Elsevier, Amsterdam, Netherlands
10. Siegmund D, Samartzidis T, Fu B, Braun A, Kuijper A (2017) Fiber defect detection of inhomogeneous voluminous textiles. In: Mexican conference on pattern recognition. Springer, Berlin, Germany
11. Nisha MF, Vasuki P, Roomi SMM (2017) Survey on various defect detection and classification methods in fabric images. J Environ Nanotechnol 6(2):20–29
12. Kumar A (2008) Computer-vision-based fabric defect detection: a survey. IEEE Trans Industr Electron 55(1):348–363
13. Niles SN, Fernando S, Lanerolle WDG (2015) A system for analysis, categorisation and grading of fabric defects using computer vision. Res J Text Appar 19(1):59
14. Zhang J, Wang J, Pan R, Zhou J, Gao W (2018) A computer vision-based system for automatic detection of misarranged warp yarns in yarn-dyed fabric: part I: continuous segmentation of warp yarns. J Text Inst 109(5):577–584
15. Zhang J, Wang J, Pan R (2019) A computer vision-based system for automatic detection of misarranged warp yarns in yarn-dyed fabric: part II: warp region segmentation. J Text Inst 110(9):1359–1367
16. Wang J, Zhang J, Wang L, Pan R, Zhou J, Gao W (2020) A computer vision-based system for automatic detection of misarranged colour warp yarns in yarn-dyed fabric: part III: yarn layout proofing. J Text Inst 1–9
17. Mei S, Wang Y, Wen G (2018) Automatic fabric defect detection with a multi-scale convolutional denoising autoencoder network model. Sensors 18(4):1064
18. Xie X (2008) A review of recent advances in surface defect detection using texture analysis techniques. ELCVIA Electron Lett Comput Vis Image Anal 7(3):1–22
19. Ping Tian D et al (2013) A review on image feature extraction and representation techniques. Int J Multimed Ubiquitous Eng 8(4):385–396
20. Liang Z, Xu B, Chi Z, Feng D (2012) Intelligent characterization and evaluation of yarn surface appearance using saliency map analysis, wavelet transform and fuzzy ARTMAP neural network. Expert Syst Appl 39(4):4201–4212
21. Wen Z, Cao J, Liu X, Ying S (2014) Fabric defects detection using adaptive wavelets. Int J Cloth Sci Technol 26(3):202–211
22. Karlekar VV, Biradar M, Bhangale K (2015) Fabric defect detection using wavelet filter. In: Proceedings of the 2015 international conference on computing communication control and automation, Pune, India, February 2015, pp 712–715
23. Yapi D, Allili MS, Baaziz N (2017) Automatic fabric defect detection using learning-based local textural distributions in the contourlet domain. IEEE Trans Autom Sci Eng 15(3):1014–1026
24. Hamdi AA, Sayed MS, Fouad MM, Hadhoud MM (2018) Unsupervised patterned fabric defect detection using texture filtering and K-means clustering. In: Proceedings of the 2018 international conference on innovative trends in computer engineering (ITCE), Aswan, Egypt, February 2018, pp 130–144

25. Wu Y, Zhou J, Akankwasa NT, Wang K, Wang J (2019) Fabric texture representation using the stable learned discrete cosine transform dictionary. Text Res J 89(3):294–310
26. Hanbay K, Talu MF, Özgüven ÖF, Öztürk D (2019) Realtime detection of knitting fabric defects using shearlet transform. Tekstil Ve Konfeksiyon 29(1):1–10
27. Kumar T, Sahoo G (2010) Novel method of edge detection using cellular automata. Int J Comput Appl 9(4):38–44
28. Zhu Q (1996) Efficient evaluations of edge connectivity and width uniformity. Image Vis Comput 14:21–34
29. Senthilkumaran N, Rajesh R (2009) Edge detection techniques for image segmentation – a survey of soft computing approaches. Int J Recent Trends Eng 1(2):250–254
30. Rizon M, Hashim MF, Saad P, Yaacob S (2006) Face recognition using Eigen faces and neural networks. Am J Appl Sci 2(6):1872–1875
31. Rosten E, Porter R, Drummond T (2010) FASTER and better: a machine learning approach to corner detection. IEEE Trans Pattern Anal Mach Intell 32:105–119
32. Wikipedia, Corner Detection. http://en.wikipedia.org/wiki/Corner_detection. Accessed 16 March 2011
33. Chang JY, Chen JL (2001) Automated facial expression recognition system using neural networks. J Chin Inst Eng 24(3):345–356
34. Jianli L, Baoqi Z (2007) Identification of fabric defects based on discrete wavelet transform and back-propagation neural network. J Text Inst 98(4):355–362
35. Rahul M, Yadav V (2023) Garment defect detection system based on histogram using deep learning. In: 4th doctoral symposium on computational intelligence, India, March 03, 2023
36. Shukla R, Yadav V (2020) Input data characterization using machine learning and deep learning. In: 1st international conference on computational research and data analytics, India, October 24, 2020
37. Habib MT, Rokonuzzaman M (2012) A set of geometric features for neural network-based textile defect classification. ISRN Artif Intell 2012, Article ID 643473, p 16
38. Saeidi RD, Latifi M, Najar SS, Ghazi Saeidi A (2005) Computer vision-aided fabric inspection system for on-circular knitting machine. Text Res J 75(6):492–497
39. Islam MA, Akhter S, Mursalin TE (2006) Automated textile defect recognition system using computer vision and artificial neural networks. In: Proceedings world academy of science, engineering and technology, vol 13, pp 1–7
40. Murino V, Bicego M, Rossi IA (2004) Statistical classification of raw textile defects. In: 17th international conference on pattern recognition (ICPR 2004), ICPR, vol 4, pp 311–314
41. Karayiannis YA, Stojanovic R, Mitropoulos P, Koulamas C, Stouraitis T, Koubias S, Papadopoulos G (1999) Defect detection and classification on web textile fabric using multi resolution decomposition and neural networks. In: Proceedings on the 6th IEEE international conference on electronics, circuits and systems, Pafos, Cyprus, September 1999, pp 765–768
42. Kumar A (2003) Neural network based detection of local textile defects. Pattern Recogn 36:1645–1659
43. Kuo CFJ, Lee C-J (2003) A back-propagation neural network for recognizing fabric defects. Text Res J 73(2):147–151

A Random Walk-Based Approach for Clustering of Food Items

Priyanka Das and Rupam Bhattacharya

Abstract Most of the recipe sites have a distinct page structure comprising a title, list of ingredients, preparation instructions, comments, and thus it is very challenging to cluster similar recipes. The work presented here initially starts with the formation of a recipe network and finding 'leading' nodes from the network. To the best of our knowledge, there doesn't exist any research that addresses the concept of finding 'leading' nodes from a recipe network and subsequently clustering similar recipes based on the ingredients. In order to fill this gap, we have proposed a method RWRC (Random Walk-based Recipe Clustering). To determine the 'leading' nodes, the proposed approach incorporates data from the breadth-first search tree. To create a breadth-first search (BFS) tree, the graph is explored in level-order, starting with each node as the root node. At each level of the traversal, we have computed the degree centrality, betweenness centrality and closeness centrality of each node. The 'leading factor' has been considered as the average of these centrality measures for all the nodes present in the network. A random walk-based graph clustering approach has been addressed that initiates from the root node of the original recipe network. If the 'leading factor' of the next visiting node in the random walk is equal to or greater than the average 'leading factor' of the whole network, the agent moves to its neighbours; otherwise, the stopping requirement is satisfied. Further, these nodes discovered by this random walk are organised into a cluster. When a walk comes to an end, the visited nodes from the initial graph are extracted, and the subsequent walk begins in the remaining graph. The random walk continues until there are no undiscovered node(s) left. These extracted nodes form the clusters of recipes.

Keywords Random walk · Clustering · Recipe network · Breadth first search

P. Das (✉) · R. Bhattacharya
Department of Computer Application Institute of Engineering and Management, Kolkata, West Bengal, India
e-mail: priyankadas700@gmail.com

R. Bhattacharya
e-mail: rupam.bhattacharya@iem.edu.in

S. Dutta et al. (eds.), *Cyber Intelligence and Information Retrieval*, Lecture Notes in Networks and Systems 1025, https://doi.org/10.1007/978-981-97-3594-5_32

1 Introduction

At present time, many cooking websites offer meal recipes using a variety of cultures and cooking techniques. Despite the abundance of recipe data, it has little to no organisation. But, there has been a lot of growing interest in knowledge extraction from recipe data as the availability of recipes has increased enormously. While working with recipes data, some researchers concentrated on clustering, some on classification and only a few on retrieving diverse patterns from the data. A novel technique for recommending healthier alternatives of daily meal has been proposed in [1]. This recipe recommendation system employs a similarity metric established using a user-centered methodology, which also takes into account regional culinary myths and realities. The research in [2] looked into the underlying characteristics of Chinese cuisine, and the recipes were modelled as graphs based on workflows such as cooking methods. Based on page structure and significant terms, a clustering method for user-generated recipe sites has been introduced in [3]. The result shows the efficiency of the strategy for classifying closely related recipes. Another research entails the creation of a web application named 'WishDish' [4], that helps in shopping food products, receiving recipe suggestions for those items, and arranging weekly meals on a calendar. In this research, based on weighted recipes, K-medoids clustering has been used for generating the recipe ideas. As the cooking recipes are mostly about ingredients and accompanying qualities, understanding the structure of the ingredient phrases is of utmost importance. A clustering method for user-generated recipe sites based on page structure and the main ingredient and seasoning of the cuisine in this research [5]. It allows users to sort their search results into categories based on what they're looking for. An experiment has been performed to determine the effectiveness of the proposed strategy. A recent study [6] discover patterns in 26 world cuisines using data from RecipeDB [7], an online recipe database, and investigate their interrelationships further.

Some of the frequent inquiries regarding clustering recipes are as follows: What similarities do two recipes share? What are the most frequently used ingredients? How may the recipes be grouped? Therefore, we are consequently interested in learning how to group comparable recipes based on ingredients to optimise the recipe search process. The proposed work describes how using the most frequent ingredients, one may easily group dishes into clusters and determine their resemblance. For serving this purpose, initially we have collected data from a recipe site and each recipe has a special name, an accompanying picture, a list of items that go with it, and instructions for how to prepare it. But only the recipe's name and ingredient list were taken into account for our analysis. Since the current method relies on grouping recipes based on ingredients, we must vectorise the ingredients in order to represent the recipes in order to identify the commonalities between them. In order to achieve this, binary vectors were produced by employing the One Hot Encoding method to convert numerical category variables. Next, the similarity among the recipes has been measured and based on that a recipe graph has been constructed. The nodes represent the vertices/nodes of the graph and the edges connecting the nodes represent the

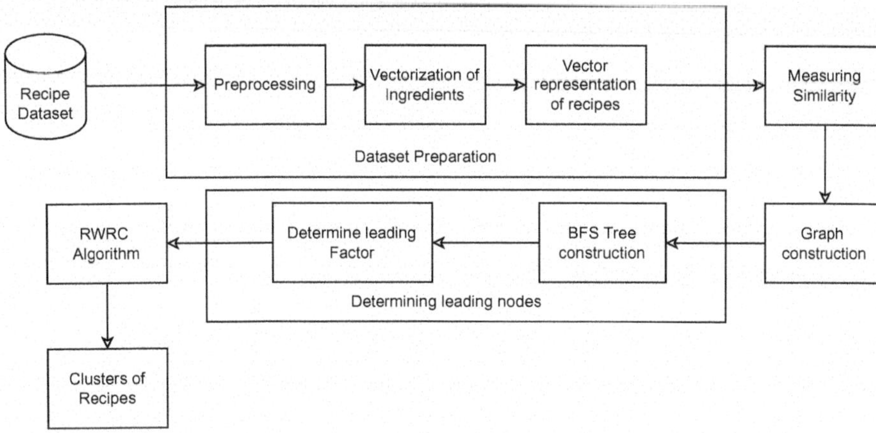

Fig. 1 Workflow diagram of the proposed approach

links between the recipes. By moving through each node in level-order, beginning at the root node, a breadth-first search tree has been constructed. Few centrality ratings have been obtained for each node at each level of the traversal. By averaging these centrality measurements over all network nodes, the 'leading factor' has been taken into account. At the recipe network's root node, a random walk-based graph clustering technique has also been used. Figure 1 shows the workflow diagram of the proposed approach.

The remainder of the paper is laid out as follows: The proposed methodology is discussed in Sect. 2. The experimental results are shown in Sect. 3, and the study is concluded with prospective future works in Sect. 4.

2 Proposed Methodology

The details of the method have been illustrated in the following subsections.

2.1 Data Collection

The data for this experiment has been collected from Cookpad, which is a collaborative website and users can post their own recipes. For the present work, we have considered near about 5000 unique recipes and each recipe has a unique name along with an image, the corresponding list of ingredients and the cooking procedure. But for our analysis, we have considered only the name of the recipe and list of ingredients.

Table 1 Text parsing rule for discarding quantity of ingredients

List of ingredients mentioned in the recipe description	Ingredients after discarding quantity
1 tablespoon sunflower oil, 1/2 cup milk,	Sunflower oil, milk
1/4 teaspoon salt, 1 teaspoon sugar, 3 cups flour	Salt, sugar, flour

2.2 *Preprocessing of the Experimental Data*

The majority of the recipes taken into account in this experiment were gathered from a website with a collaborative design and as a result, the dataset is filled with unreliable data. We have discarded the weights and measures of the ingredients mentioned in the recipes. This has been done with the assumption that as the proposed work is dealing with the ingredients only, the quantity of the ingredients will not be adding any value to our model. An example of this text parsing rule is shown in Table 1. Stemming often makes the words unrecognisable, and we need to look up for the unstemmed word to gain understanding on it. Therefore, only lemmatisation is carried out for obtaining the unique list of ingredients.

2.3 *Vectorisation of Ingredients*

Binary vectors have been created by converting numerical category variables using One Hot Encoding approach available in Scikit-Learn library. Say, the present work comprises recipe data having m categories. A one-hot vector v is then a binary vector of length m where only a single entry can be one, all others must be zero. We set the ith entry to 1, and all others to 0, to indicate that this v represents the categorical variable taking on the ith possible value. Here, in this work, we can represent the one-hot encoded vector of ingredients as $\left\{ \mathrm{v} \epsilon \{0, 1\}^m : \sum_{i=1}^{m} \mathrm{v}_i = 1 \right\}$.

2.4 *Measuring Similarity of Recipes*

Once all the ingredients have been vectorised, the objective is to create similar group of recipes. As a result, it's important to gauge how comparable the recipes are. For this purpose, cosine similarity, as specified in Eq. (1) has been employed.

$$Sim_{xy} = \frac{\vec{R}_x . \vec{R}_y}{||\vec{R}_x||||\vec{R}_y||} \tag{1}$$

where $\vec{R}_x$ and $\vec{R}_y$ are two m-dimensional vectors that symbolise two distinct recipes. The range of the cosine similarity value is [0, 1], with 1 showing the most similarity and 0 signifying the greatest dissimilarity between the recipes.

A complete weighted undirected graph (also known as a recipe network) termed $G_R = (N, E_R, W_R)$ is created based on the similarity score, where N is the set of nodes, E_R is the set of edges in G_R, and W_R is the set of weights assigned to the edges in E_R. The weights assigned to the edges are considered to be reciprocal of the cosine similarity score.

2.5 *Determining the Leading Factor*

In graph theory a vertex or a node is the fundamental unit based on which a graph is constructed. All the nodes present in a graph can not have equal significance. To foresee and appropriately govern the graph-based system (of any application), it is essential to precisely and effectively identify the important nodes. These important nodes hold special significance in any graph-based system. These significant nodes may also be referred to as influential nodes or as the leading nodes. Here, a graph-traversal framework as discussed in [8] has been followed for recognising the 'leading nodes' and subsequently a 'leading factor' has been measured for each node in the graph. The graph considered here is the original recipe network $G_R = (N, E_R, W_R)$, where N = set of nodes, E_R = set of edges and W_R is the set of assigned weights in the edges in G_R. The proposed method takes information from the breadth-first search tree into account when identifying the 'leading' nodes. The graph is investigated in level-order, with each node serving as the root node, in order to generate a breadth-first search (BFS) tree. We calculated each node's degree centrality $C_d(n)$, betweenness centrality $B(n)$, and closeness centrality $C(n)$ at each level of the traversal. The average of these centrality measurements across all network nodes has been referred to as the 'leading factor', Lf_{avg}. Figure 2 describes the procedure for determining the leading factor from the bfs tree construction. Figure 2a is the original recipe network and Fig. 2b shows the steps where each node has been considered separately for constructing the bfs tree followed by calculating the 'leading factor'. It is observed that as there are seven nodes present in the original recipe graph, seven separate bfs trees have been constructed. Algorithm 1 describes the whole process of determining the 'leading factor'. This part of the proposed work uses graph traversal approach for identifying important nodes from a recipe network, but the recognition of these nodes further helps in grouping similar recipes based on the ingredients.

2.6 *Random Walk-Based Recipe Clustering*

This section reflects the most significant part of the present work leading to the clustering of recipes based on random walk applied on the recipe network $G_R =$

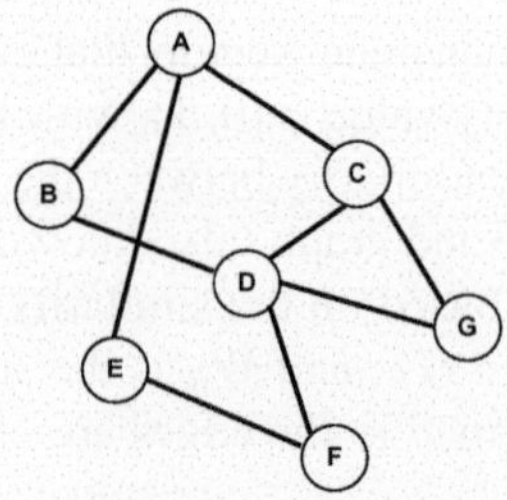

(a) Original Recipe Graph

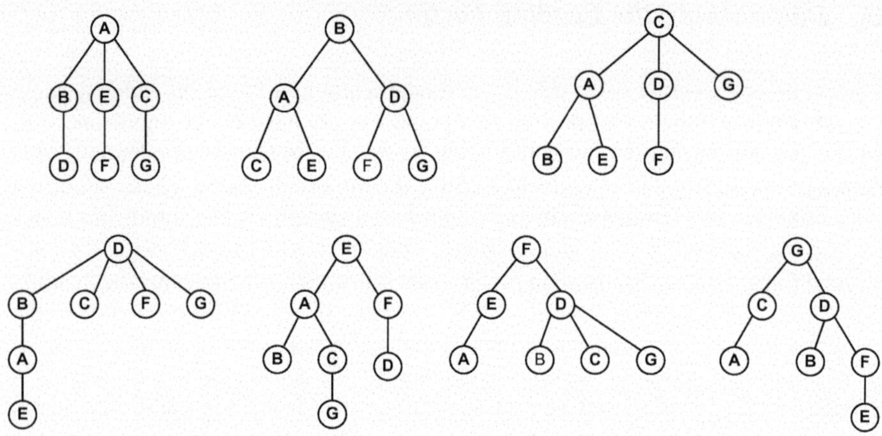

(b) Breadth first search trees generated from original recipe graph

Fig. 2 Level order tree traversal for determining the ‘leading factor’

(N, E_R, W_R). A probabilistic process of visiting nearby nodes is called a random walk in a graph. If there is connectivity through an edge, a walking agent that starts from a seed vertex can jump to any of its neighbours. The sole responsibility of the walking agent is to begin its journey from the seed vertex and then leap to any of its neighbours at random. The choice of the neighbour is sometimes determined by a particular criterion based on the problem domain. For the proposed work, the primary aim is to extract coherent clusters from the graph. Initially, for each node $n_i \in N$ in the network G_R, the leading factors $Lf(n_i)$ and Lf_{avg} are measured. Next, the walking agent for random walk starts at the root node $n_i \forall i = 1, 2, \ldots, r$. Here, the selection of neighbour depends on the importance of each node present in the recipe network. Algorithm 1 in Sect. 2.5 describes the process of finding ‘leading’ nodes and the ‘leading factor’ by which the significance of the nodes can be measured. The clustering technique described in this work is dictated by the fact that the walking agent goes to its neighbours if the next visiting node’s (neighbour) ‘leading factor’ in the random walk is equal to or higher than the average ‘leading factor’ of the whole

Algorithm 1: Leading factor calculation from bfs trees

```
/* This algorithm finds the leading factor from the bfs trees constructed by considering each
node as the root node*/
Input: Original recipe graph G_R = (N, E_R, W_R), where N = set of nodes, E_R = set of
       edges and W_R is the set of assigned weights in the edges in G_R
Output: Leading factor Lf_avg
begin
    Let r̂ and Q be the root node and queue respectively
    Q.enqueue( r̂) /*Inserting r̂ in queue until all its neighbour vertices are marked*/
    mark r̂ as visited
    repeat
        Calculate the centrality measures for all the visited nodes
        while Q is not empty do
            v = Q.dequeue( ) /* Remove that node from queue, whose neighbour will be
            visited now*/
            Process all the neighbours of v
            for all neighbours w of v in Graph G_R do
                if w is not visited then
                    Q.enqueue(w) /* Store w in Q to further visit its neighbour */
                end
                else
                    mark w as visited
                end
            end
        end
    until all the nodes are visited
    Calculate the average centrality measure for each visited node
    Lf_avg ← average centrality measure /* Lf_avg is the leading factor*/
end
```

network, i.e., for each node $n_i \in N$ in G_R, if $Lf(n_i) \geq Lf_{avg}$, then the node n_i is added to the set of visited nodes. Otherwise, it moves to the next neighbour. When this walk comes to an end, the visited nodes from the initial graph G_R are extracted, and the subsequent walk begins in the remaining graph. The nodes visited by the random walk are extracted separately along with their incident edges. Let g_i be the subgraph of G_R containing all the visited nodes $n_1, n_2, \ldots, n_r$ along-with their edges. Let $G_1 = \{g_1, g_2, \ldots, g_r\}$. Now, the remaining graph is $G_R = G_R - G_1$, where the random walk gets started again. However, for each subgraph $g_i \in G_1$, the leading factors $Lf(n_i)$ and Lf_{avg} are recomputed, and this process iterates until all nodes in the original graph are processed. Algorithm 2 describes the RWRC approach. The singleton clusters produced by this method are labelled as noise and thus eliminated. But there's a chance that several nodes could tie for first place. Any one of the nodes is chosen to deal with this circumstance. As opposed to hard coding the walk, the current approach provides the random walker with a priority list from which to choose. Depending on how many times the suggested technique is run, this can lead to various clustering structures. A clustering method based solely on random walks shares the same characteristics. However, because of the guidance factor incorporated in the current approach, the final clustering quality will be superior to a simple random walk. The random walking agent starts from the root node but chooses the neighbour based on the 'leading factor' and this guidance helps in extracting coherent clusters from the recipe network.

Algorithm 2: Random Walk based Recipe Clustering (RWRC)

```
/* This algorithm finds clusters of recipes from the recipe graph G_R = (N, E_R, W_R) */
Input: G_R = (N, E_R, W_R), where N = set of nodes {n_1, n_2, ..., n_r}, E_R = set of edges in
      G_R and W_R = set of weights assigned to the edges in E_R by cosine similarity
      measured among the recipes
Output: Cluster C_R of recipes
begin
 | C_R = {φ}
 | repeat
 |  | for each node n_i ∈ N in G_R do
 |  |  | Calculate the leading factors Lf(n_i) and Lf_avg as mentioned in Algorithm (1)
 |  | end
 |  | Execute the random walk starting from root node n_i where i = 1, 2, ..., r.
 |  | for each node n_i ∈ N in G_R do
 |  |  | if Lf(n_i) ≥ Lf_avg then
 |  |  |  | node n_i added to set of visited nodes
 |  |  | end
 |  |  | else
 |  |  |  | move to next neighbour
 |  |  | end
 |  | end
 |  | Let g_i be the subgraph of G_R containing all the visited nodes n_1, n_2, ..., n_r
 |  | along-with their incident edges.
 |  | G_1 = {g_1, g_2, ..., g_r}
 |  | G_R = G_R − G_1
 |  | repeat
 |  |  | for each subgraph g_i ∈ G_1 do
 |  |  |  | compute Lf(n_i) and Lf_avg followed by selecting the nodes from random walk
 |  |  | end
 |  | until G_1 is empty
 | until G_R is a null graph
 | C_R = C_R ∪ G_1
 | /* G_1 contains all subgraphs each of which is a cluster of recipes */
 | Return C_R
end
```

---

3 Experimental Results

The proposed work has been implemented in a machine running Ubuntu GNU/Linux version 16.04 on an Intel(R) Core i3-5005U CPU @ 2.00 GHz processor using Python 3.9 with its built-in multiprocessing libraries and a number of modules like numpy

Table 2 Details of the dataset

Dataset	No. of recipes
Cookpad dataset	4416
Recipe 5k	4826
Tarla Dalal	2840
Food.com	1610
RecipeDB	1250
Allrecipes.com	1430
Epicurious.com	1360

Table 3 Number of clusters obtained

Dataset	No. of clusters
Cookpad dataset	18
Recipe 5k	15
Tarla Dalal	16
Food.com	17
RecipeDB	16
Allrecipes.com	13
Epicurious.com	14

1.14, networkx 2.1, and matplotlib 2.2. The basic statistics of the experimental data is shown in Table 2. We can see that this dataset is unbalanced because the number of recipes collected from different cooking sites differs. This experimental dataset is highly imbalanced, which may have an impact on the performance of the proposed classification method. Therefore, we have used upsampling and downsampling with sklearn library from Python. Therefore, this data imbalance problem has been solved by upsampling and downsampling to make the number of recipes equal. The number of each recipe has been sampled to 1250. Another example has been given in Table 4 where the recipes along with their ingredients are shown. Figure 3 shows the original graph (in the left) as well as the clusters obtained by the proposed approach. The variation in colours represent different clusters. Table 3 shows the number of clusters obtained by the proposed RWRC approach against each dataset. We have used the proposed approach on data collected from Recipe 5K, Tarla Dalal, Food.com, RecipeDB, Allrecipes.com, and Epicurious.com.

Table 4 Summary of the dataset

Recipe	Ingredients
Rasgulla	Full fat milk, ice water, lemon juice, sugar, water, green cardamom pods, rose water
Raw Jackfruit Curry	Raw jackfruit, potato, tomato, onion, green chillies, garlic, ginger-garlic paste, turmeric powder
Kolkata style Egg Chicken Roll	Boneless breasts, yogurt, red chilli powder, cumin powder, coriander powder, ginger-garlic paste, salt, mustard oil

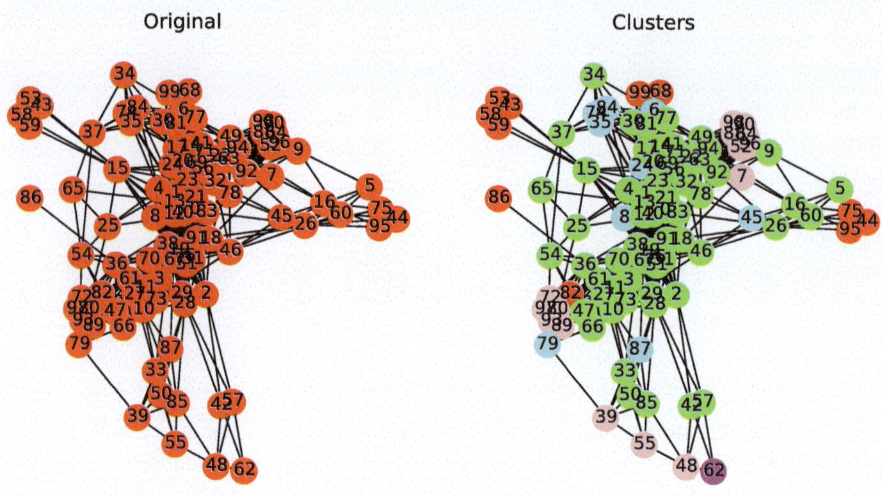

Fig. 3 Clusters obtained by RWRC approach on Cookpad data

3.1 Evaluation

Six clustering evaluation metrics like the Dunn index (DI), Davies–Bouldin index (DBI), Silhouette coefficient (S), and Calinski–Harabasz index (CH) are used to compare the clustering outcomes of the various methodologies. Table 5 shows the evaluation of the proposed work based on 4 different internal metrics. The evaluation has been performed on seven different cooking data. From this it is observed that the proposed approach provides the best Dunn score for Cookpad data, shown in bold in Table 5. As the low score is considered to be good in case of Davies–Bouldin index, the proposed work has provided the best score of 49 for Tarla Dalal data. Similarly, the best scores for Silhouette coefficient has been obtained for both Tarla Dalal and RecipeDB. Apart from these, RWRC provides the best score of 79 in Calinski–Harabasz index for data collected from Food.com.

Table 5 Evaluation of the proposed approach (RWRC) based on internal metrics (in %) on different data

Dataset	Indices			
	DI	DBI	S	CH
Cookpad dataset	**82**	53	76	75
Recipe 5k	79	51	73	78
Tarla Dalal	79	**49**	**78**	75
Food.com	78	55	75	**79**
RecipeDB	78	60	**78**	75
Allrecipes.com	79	57	74	76
Epicurious.com	78	53	68	72

4 Conclusion and Future Plans

This proposed work efficiently clusters various food items based on ingredients, but still it requires more in-depth research for considering the other factors involved for finding similar recipes. Thus in future, we wish to incorporate the uses of cooking devices along with the images of the recipes for clustering similar food items.

References

1. Hanai S, Nanba H, Nadamoto A (2015) Clustering for closely similar recipes to extract spam recipes in user-generated recipe sites. In: Proceedings of the 17th international conference on information integration and web-based applications and services, pp 1–5
2. Wang L, Li Q, Li N, Dong G, Yang Y (2008) Substructure similarity measurement in Chinese recipes. In: Proceedings of the 17th international conference on world wide web, WWW'08, pp 979–988
3. Youri van Pinxteren GG, Kamsteeg P (2011) Deriving a recipe similarity measure for recommending healthful meals. In: Proceedings of the 16th international conference on intelligent user interfaces, IUI'11, pp 105–114
4. Phirke T, Dighe P, Rathi D, Iyer J, Giri N (2016) Wishdish: recipe prediction using k-medoids clustering technique for big data analytics. Int J Comput Appl 153(4):53–56
5. Nadamoto A, Hanai S, Nanba H (2016) Clustering for similar recipes in user-generated recipe sites based on main ingredients and main seasoning. In: 2016 19th international conference on network-based information systems (NBiS), pp 336–341
6. Sharma T, Upadhyay U, Kalra JS, Arora S, Ahmad S, Aggarwal B, Bagler G (2020) Hierarchical clustering of world cuisines. CoRR. arXiv:2004.12283
7. Batra D, Diwan N, Upadhyay U, Kalra JS, Sharma T, Sharma AK, Khanna D, Marwah JS, Kalathil S, Singh N, Tuwani R, Bagler G (2020) Recipedb: a resource for exploring recipes. Database 2020
8. Liu Y, Wei X, Chen W, Hu L, He Z (2021) A graph-traversal approach to identify influential nodes in a network. Patterns 2(9)

Predicting Future Automobile Ownership Trends in Uzbekistan Using Linear Regression

Danish Ather, Malikhan Singh, Kavita Arora, Sonal Pathak, Suhail Javed Qurashi, and Anupam Singh

Abstract The rapid urbanization and economic growth of Uzbekistan over the past decades have engendered significant changes in automobile ownership trends. Understanding these trends is crucial for infrastructure planning, sustainable transportation strategies, and environmental considerations. This Paper seeks to employ Machine Learning (ML) techniques to forecast the number of cars per 100 households in Uzbekistan by 2025, based on historical data from 2010, 2015, and 2020. The dataset, derived from official statistics provided by the State Committee of the Republic of Uzbekistan on Statistics, shows a remarkable increase from 21 cars per 100 households in 2010 to 48 by 2020. Leveraging regression models, we aim to predict the expected growth for 2025. Preliminary analyses hint at a continuing upward trend, impacted potentially by factors such as Population growth, GDP growth, urban migration, and increasing global connectivity. The findings from this research will not only serve as a prediction tool but will also shed light on the socio-economic factors influencing car ownership, providing valuable insights to policymakers and urban planners. While the limited data points present a challenge for complex model utilization, the study underscores the potential of ML in forecasting and the need for more comprehensive data collection in future studies. The predictive insights drawn from this research can significantly aid in shaping transportation policies, mitigating potential environmental impacts, and planning urban infrastructures in Uzbekistan.

D. Ather (✉)
Amity University, Tashkent, Uzbekistan
e-mail: danishather@gmail.com

M. Singh
Computer Science and Engineering, College of Engineering & Technology, Mathura, India

K. Arora · S. Pathak · S. J. Qurashi
School of Computer Applications, Manav Rachna International Institute of Research and Studies, Faridabad, Haryana, India

A. Singh
Department of Computer Science and Engineering, Graphic Era Hill University, Dehradun, India

S. Dutta et al. (eds.), *Cyber Intelligence and Information Retrieval*, Lecture Notes in Networks and Systems 1025, https://doi.org/10.1007/978-981-97-3594-5_33

Keywords Automobile ownership trends · Machine learning forecasting · Uzbekistan transportation · Urban infrastructure planning · Socio-economic influences

1 Introduction

The nexus between urbanization, economic growth, population growth, and automobile ownership is one that has intrigued scholars, policymakers, and urban planners alike. As cities grow and economies flourish, the need for personal mobility often rises, leading to an upsurge in automobile ownership. Uzbekistan, a Central Asian nation, stands as a compelling case study in this respect, having experienced significant transformations in its socio-economic landscape over the past few decades. This evolution is nowhere more evident than in the automobile ownership patterns of its citizens.

According to statistics from the State Committee of the Republic of Uzbekistan on Statistics, there were roughly 21 vehicles per 100 homes in the nation in 2010 [1]. This number increased to 42 in 2015, a 100 percent increase in just five years. At a slightly slower rate, the trend persisted, reaching 48 cars per 100 homes in 2020 [1] (see Fig. 1). Such quick changes demand a careful analysis to comprehend the underlying causes and forecast future paths. Predicting automobile ownership trends is crucial for a variety of applications, from infrastructure planning to environmental sustainability issues, as metropolitan centers in Uzbekistan continue to grow and develop.

Fig. 1 Number of cars per 100 households [1]

Historically, a variety of factors have been blamed for the increase in car ownership in different countries. Top on the list are frequently economic growth and population growth, with a rising middle class having the means to purchase personal autos. It is possible to link the rising number of vehicles in Uzbekistan with the country's steady economic growth since the early 2000s, which has been fueled by exports of natural gas, cotton, and gold. Urbanization trends, which see rural people flocking to cities in search of better opportunities, further fuel this growth and put more strain on the transportation system [2].

However, merely understanding past and present trends is insufficient. The ability to predict future trends can equip policymakers and urban planners with the insights necessary to make informed decisions. Machine Learning (ML), with its prowess in detecting patterns and making predictions based on historical data, emerges as a potent tool in this endeavor. ML's capability in forecasting has been tapped in diverse domains, from stock market predictions to weather forecasting [3, 4]. In the realm of transportation and urban planning, leveraging ML techniques offers the promise of accurate forecasts, optimized infrastructure development, and targeted policy interventions.

The current study, thus, pivots on the intersection of automobile ownership trends in Uzbekistan and the predictive capabilities of ML. By harnessing data from the previous years and feeding it into sophisticated algorithms, the goal is to project the number of cars per 100 households in Uzbekistan by 2025. While the primary objective is predictive in nature, the study also delves into understanding the socio-economic variables influencing automobile ownership, weaving a comprehensive narrative around urban mobility in modern-day Uzbekistan.

2 Motivation

The evolution of the twenty-first century has brought about unprecedented technological progress and an accelerated pace of globalization. These developments have introduced myriad opportunities but have concurrently presented significant challenges, especially in sectors like urban planning and transportation. As societies advance, the demand for enhanced mobility and connectivity becomes paramount. This is most distinctly illustrated by the spike in automobile ownership in emerging economies, signifying socio-economic progression, and altered lifestyles.

Positioned at the heart of Central Asia, Uzbekistan offers a fascinating lens through which these global shifts can be observed. From its stance as a post-Soviet republic to its ascent as an emerging economic contender, the journey has been noteworthy. As its populace experiences increasing prosperity, their aspirations evolve, leading to a burgeoning demand for personal mobility. Data, as observed from the earlier section, vividly shows this growth trajectory from 2010 to 2020.

Yet, these numbers, while indicative of progress, also raise pertinent questions. What are the ramifications of such trends on the urban infrastructure of Uzbekistan's cities? Is the nation's transportation grid equipped to handle this influx? How will this affect the environment? Cities around the world provide critical insights. Proliferating car ownership often leads to urban challenges including congestion, exacerbated air pollution, and an overburdened infrastructure. Metropolises such as Delhi and Beijing have contended with deteriorating air quality, attributing a significant portion to vehicular emissions. Urban centers like Los Angeles are emblematic of the challenges arising from dense traffic [5]. Such global experiences underscore the necessity of not just understanding the present, but also adeptly anticipating the future. This forms the crux of our research motivation.

Harnessing the capabilities of Machine Learning (ML) appears to be a viable solution. Its transformative impact across diverse sectors, from healthcare to finance, is a testament to its potential in analyzing vast datasets, discerning patterns, and generating predictions. Given the automobile landscape's rapid shifts in Uzbekistan, there exists an imperative to deploy ML's prowess for forecasting future trajectories. This not only aids in preparedness but ensures that growth remains sustainable.

Moreover, a deeper dive into the cultural dimensions reveals that automobiles aren't merely transport mediums; they represent socio-economic aspirations, personal freedom, and a modern ethos. As the middle class in Uzbekistan expands, the allure of personal vehicles is anticipated to intensify. Grasping these nuances can provide a richer context to the projections made using ML algorithms. In essence, the study's motivation stems from the intricate tapestry of socio-economic evolution, impending urban challenges, and the transformative potential of ML. By charting out predictions for Uzbekistan's automobile ownership patterns by 2025, we aim to furnish stakeholders with invaluable insights. This knowledge can inform infrastructural investments, policy formulation, and strategies geared toward environmental preservation.

3 Literature Survey

The automobile industry has continuously undergone transformations since the inception of the first car. The rate of automobile ownership, particularly in emerging economies, serves as an essential indicator of socio-economic growth, urbanization trends, and infrastructure development. Given the importance of this topic, several studies have delved into the myriad factors influencing car ownership, the implications of rising ownership rates, and the use of predictive tools such as machine learning to anticipate future trends. The literature present in coming section aims to synthesize the findings of pertinent studies in these domains.

In an attempt, [6] examined the correlation between economic growth, urbanization, and car ownership in several Asian countries. The study highlighted that as GDP per capita increased, the demand for personal vehicles surged, primarily driven by the expanding middle class seeking better mobility options and viewing car ownership as a status symbol. Likewise, [7] investigated the urban–rural divide concerning automobile ownership. They postulated that urban areas, while having higher numbers of vehicles, might have lower ownership rates per household due to accessible public transport alternatives. In contrast, rural households, with limited public transport, relied more on personal vehicles.

The environmental ramifications of increased automobile ownership have been a significant concern for researchers and policymakers. For instance, [8] carried out a study on the environmental impact of rising car numbers in Mexico City. Their results pointed toward exacerbated levels of pollution and called for the introduction of sustainable urban transport solutions. In terms of infrastructure, conducted a study focusing on the strain on urban roads due to the surge in car ownership. Their research emphasized the necessity of substantial investments in transport infrastructure and the promotion of alternative transport modes.

A newer area of exploration is the use of predictive analytics, especially machine learning, to forecast car ownership trends. Zairova et al. [9] implemented ML algorithms to predict car ownership rates. They utilized variables like income, population density, and existing transport infrastructure to feed their models. Their findings offered reasonably accurate predictions for the subsequent five years, underlining the potential of ML in this domain.

Following a similar trajectory, combined socio-economic factors, urbanization rates, and prior trends in car ownership to predict the future landscape in Brazil. The machine learning models they utilized offered insights into not just the expected growth but also potential saturation points in car ownership. As noted in motivation section, automobiles aren't merely tools of transport; they signify much more in the socio-cultural context. In this context, [10] undertook a comprehensive study exploring the cultural dimensions of car ownership in the Middle East. They posited that beyond the utility factor, cars in many cultures symbolize freedom, status, and modernity. Such cultural nuances can significantly influence ownership trends, especially in societies experiencing rapid socio-cultural transformations.

The literature available on car ownership trends, their implications, and predictive models is expansive. There's a consensus on the intricate interplay of socio-economic, environmental, infrastructural, and cultural factors influencing these trends. The emergence of machine learning and predictive analytics offers a promising avenue to anticipate and address future challenges and opportunities in this domain.

4 Methodology

This section elucidates the methodological approach employed to predict the car ownership rate in Uzbekistan for 2025 using a Linear Regression model (see Fig. 2).

Data Collection: The data pertaining to the number of cars per 100 households in Uzbekistan was obtained from the State Committee of the Republic of Uzbekistan on Statistics for the years 2010, 2015, and 2020. We also collect the data on population of Uzbekistan from statistical agency under the president of the republic of Uzbekistan for the same years.

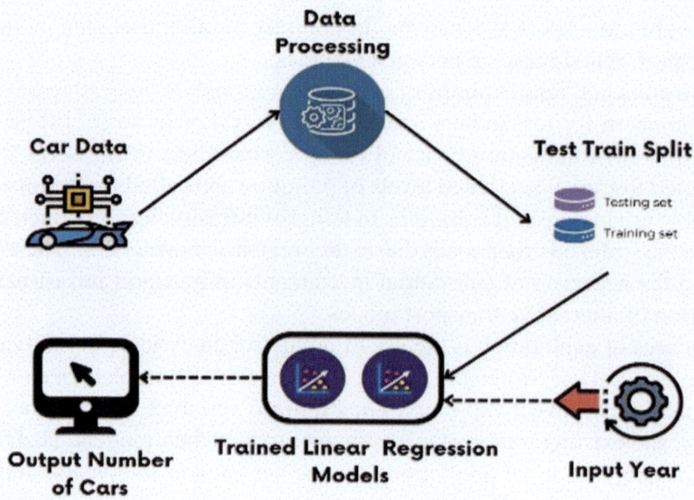

Fig. 2 Methodology

Data Preprocessing: Before applying regression models, the data was inspected for any missing values or outliers. Since the dataset was compact, there was no need for handling missing data. However, normalization of the data was carried out to ensure that the scales were consistent, aiding in convergence during model training.

Model Selection: For this study, two regression models were chosen:

Linear Regression: This is a fundamental predictive analysis technique that quantifies the relationship between one dependent and one independent variable by fitting a linear equation to the observed data. The simplest form of the equation with two variables is defined as:

$$Y = \alpha + \beta X + \varepsilon \quad (1)$$

where:

- Y is the dependent variable: Number of Cars per 100 Households.
- X is the independent variable: Population.
- α is the y-intercept.

Model Training: The Linear Regression and was trained on the dataset for the years 2010, 2015, and 2020. We used the years as independent variables and the number of cars per 100 households as the dependent variable.

Model Evaluation: The models' performance was evaluated using the Root Mean Square Error (RMSE). Given the limited data points, cross-validation was employed to ensure a comprehensive evaluation as shown in Eq. (2):

$$\mathrm{RMSE} = \sqrt{\frac{1}{n}\sum_{i-1}^{n} (y_i - \hat{y}_i)^2} \quad (2)$$

where

- y_i is the actual value.
- $\hat{y}_j$ is the predicted value.
- n is the number of observations.

Prediction for 2025: With the trained models, predictions for the year 2025 were made to estimate the number of cars per 100 households in Uzbekistan. For prediction we take the average growth rate of the population and forecast the car ownership for the year 2023, 2024, and 2025.

5 Results and Discussion

Our state-of-the-art AI and machine learning model has been employed to predict the number of cars based on population. Using linear regression, the results generated are as follows in Table 1.

Model's Statistical Summary:

- Average (Mean) Number of Cars Predicted: 37.000
- Standard Deviation of Predicted Number of Cars: 14.177
- Model Accuracy (R-squared): 0.914
- Total Data Points Analyzed: 3
- Overall Model Validity (F-test): 21.374 with a Probability > F of 0.036
- Model's Complexity Evaluation (AIC): 19.832 and (BIC): 18.029
- RSME: 1.75.

Legend for Significance:

- *** Highly Significant ($p < 0.01$)
- ** Significant ($p < 0.05$)
- * Marginally Significant ($p < 0.1$).

Following the results of the model, the regression equation can be written as follows:

Table 1 Linear Regression results

Parameter	Coefficient	Robust standard error	t-value	*p*-value	Significance
Population (β)	0.00000459	0.000000993	4.62	0.036	**
Constant (α)	−105.254	30.818	−3.42	0.081	*

$$Y = -105 + 0.00000459X \tag{3}$$

Using Eq. 2 we predict the number of cars as per following Table 2.

For the Linear Regression model, the evaluation revealed an RMSE (Root Mean Square Error) of 1.75. This metric gives us an idea about the prediction error magnitude, with lower values indicating better model performance. Following an exhaustive training and evaluation phase utilizing the data from 2010, 2015, and 2020, certain insightful outcomes surfaced from the application of Linear Regression model. Following figure depicts the actual and forecasted values.

In the Fig. 3 above the asterisks (*) next to the number of cars for 2023–2025 suggest that these values are predictions or estimates. Similarly, the double asterisks (**) next to the population for 2024 and 2025 also signify predictions or a different category.

To predict the number of cars for the year 2025, a Linear Regression models was employed. The Linear Regression model predicted that there will be 75 cars for every 100 households by 2025, 64 in 2024 and by the end of 2023, 60. The results suggest a potential plateauing or slowing down in the growth rate of car ownership in the coming years.

Considering the RMSE values and predictions, it's evident that the Linear Regression model, with its inbuilt robustness gives quite accurate prediction. It might indicate that not all provided years are equally influential in determining the car ownership trend, and model's ability to weigh them accordingly proved beneficial.

The predictions suggest a continuing upward trend in car ownership in Uzbekistan, albeit at a decelerating pace. Given that the increase from 2010 to 2015 was 21 cars and from 2015 to 2020 was 6 cars, the predicted growth from 2020 to 2025 is around 27 cars per 100 households, depending on the model's predictions.

Table 2 Prediction for number of cars

Year	Number of cars per 100 households#	Population##
2010	21 (actual)	28,001,443 (actual)
2015	42 (actual)	31,022,525 (actual)
2020	48 (actual)	33,905,242 (actual)
2023	60*	36,024,947 (actual)
2024	64*	36,956,713.78**
2025	75*	39,267,192.23**

* Forecasted as per Eq. 2
** Predicted based on 9% growth rate in population for the previous period
#*Source* State Committee of the Republic of Uzbekistan on Statistics
Source Stats.uz online at https://stat.uz/en/official-statistics/demography

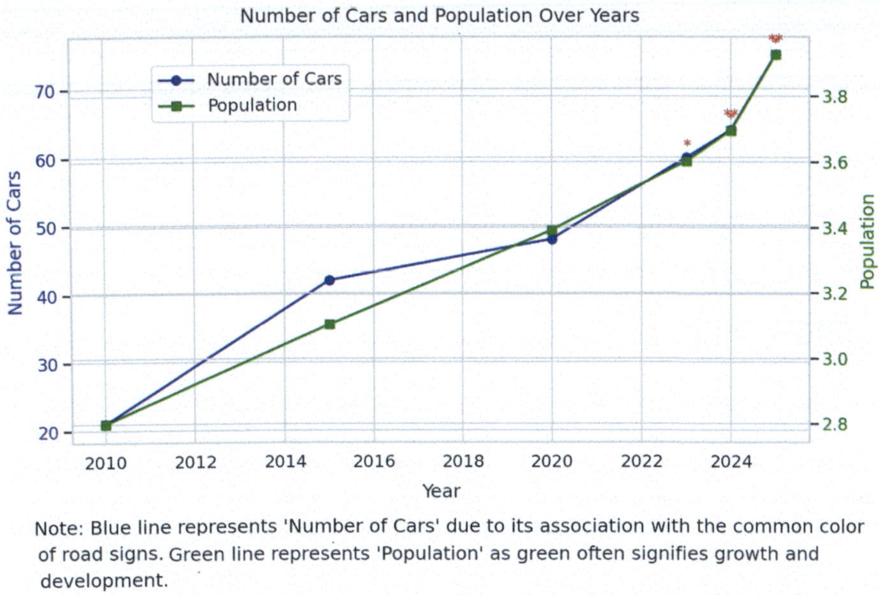

Fig. 3 Linear forecast for year 2025

6 Conclusion

In the realm of data-driven decision-making, predictive modeling plays a pivotal role, allowing stakeholders to forecast trends and make informed choices. This study focused on deciphering the growth pattern in the number of cars per 100 households in a specific region, considering data from 2010, 2015, and 2020. The overarching aim was to utilize this data to project the numbers for the year 2025 using a Linear Regression.

The progression in car ownership from 21 cars per 100 households in 2010 to 48 cars by 2020 elucidates a marked trend of increasing vehicle ownership. This growth can be attributed to a multitude of factors, such as rising incomes, urbanization, and possibly the evolution of consumer needs. While the past data presented an evident upward trend, forecasting the 2025 data became a challenging endeavor, which necessitated the application of robust predictive models.

Linear Regression, in its essence, establishes a linear relationship between the dependent and independent variables. This method was chosen due to its simplicity and interpretability. However, as with most real-world data, the risk of overfitting or capturing noise always lurks.

The results shed light on intriguing insights. The Linear Regression model returned an RMSE of 1.75 for its predictions. Drawing predictions for 2025, the Linear Regression projected 75 cars per 100 households.

To conclude, in the context of car ownership, this study not only offers a forecast for stakeholders but also underlines the relentless growth in vehicle acquisition, hinting toward larger socio-economic implications. Future studies could delve deeper into understanding the factors driving this growth, thereby paving the way for more nuanced policy-making and urban planning. As the world steers toward a more data-centric paradigm, the value of precision and adaptability in predictive models, as demonstrated in this study, will only amplify.

7 Future Scope

While this research has made strides in projecting car ownership rates for 2025, it opens doors to a multitude of further exploratory avenues. Delving deeper, future endeavors could benefit from more granular data, possibly on a monthly or weekly scale, capturing intricate patterns in the rapidly changing car market. A holistic understanding of car ownership could be further enriched by considering various external parameters such as economic fluctuations, fuel prices, public transport dynamics, and infrastructural developments.

Furthermore, as the machine learning landscape continuously evolves, embracing newer algorithms or even sophisticated models like neural networks might offer enhanced predictive precision. A regional perspective, distinguishing between urban and rural areas or varying geographical regions, could reveal unique trends and patterns essential for localized urban planning. Moreover, while this study leveraged regression-based model, the application of specialized time-series forecasting techniques might capture latent cyclical or seasonal trends.

Another promising angle of exploration lies in gauging the environmental ramifications of increased car ownership. In an era of heightened environmental consciousness, such insights could guide sustainable vehicle usage policies. Additionally, the qualitative dimensions of the rise in car ownership, driven by socio-cultural and behavioral aspects, remain largely untouched, offering a fertile ground for research through surveys or ethnographic studies.

Technological disruptions, such as the proliferation of electric and autonomous vehicles, are poised to redefine traditional car ownership paradigms. Monitoring and integrating these trends can lead to more contemporary and forward-looking predictions. Similarly, understanding the effects of government interventions, from tax policies to green vehicle incentives, could provide valuable information on how regulatory landscapes influence car ownership trends.

Lastly, situating this data within a global context, by comparing it with trends from other countries, can offer a broader perspective, helping us understand the global standing of the studied region. As we advance in this data-driven era, the confluence of technology, vast datasets, and real-world questions offers a myriad of opportunities for deeper, more nuanced exploration in the realm of car ownership and its implications.

References

1. State Committee of the Republic of Uzbekistan on Statistics, "Automobile Ownership Data 2010–2020". https://stat.uz/en/contacts1/181-ofytsyalnaia-statystyka-en/6372-macro-indicators. Accessed 10 July2023
2. Olma N (2021) Monotonous motorscapes: Uzbekistan's car industry and the consolidation of a post-socialist shortage economy. Centr Asian Surv 40(2). https://doi.org/10.1080/02634937.2020.1858756
3. Jain A, Sarkar A, Ather D, Raj D (2022) "Temperature based automatic fan speed control system using Arduino. SSRN 4159188
4. Jain A, Sarkar A, Ather D, Raj D (2022) Temperature based automatic fan speed control system using Arduino. SSRN Electron J. https://doi.org/10.2139/ssrn.4159188
5. Nuriddinov S, Avazov B, Hasanov F, Rakhmonova Y (2021) Analysis of the causes of traction electric failures of electric cargo cars operated on railways of the Republic of Uzbekistan. In: E3S web of conferences. https://doi.org/10.1051/e3sconf/202126405041
6. Gritsina MA (2019) The Caracal Caracal caracal Schreber, 1776 (Mammalia: Carnivora: Felidae) in Uzbekistan. J Threat Taxa 11(4). https://doi.org/10.11609/jott.4466.11.4.13470-13477
7. Galdini F (2022) 'Backward' industrialisation in resource-rich countries: the car industry in Uzbekistan. Compet Change. https://doi.org/10.1177/10245294221125850
8. Rahimov RV, Khatamov SA, Rakhmatov ZX (2017) Scientific substantiation of technical solutions for the improvement of the construction of the body of a hopper car for transportation of cement produced in the Republic of Uzbekistan. Eur Sci Rev. https://doi.org/10.29013/esr-17-3.4-13-17
9. Zairova D, Kadirov MA, Khojiev N, Khikmatov F, Shokuchkorov K (2021) Running quality assessment of a passenger car produced in the Republic of Uzbekistan. In: E3S web of conferences. https://doi.org/10.1051/e3sconf/202126405058
10. Makhkamdjanov BM (2006) Cars with hybrid drive and elektrocarsa way to decision of the ecological problems in Uzbekistan. Renew Energy 31(5). https://doi.org/10.1016/j.renene.2005.08.010

Adversarial Attacks on Neural Networks

Pranjal, Vaishnavi, Divyansh, Dharm Raj, Vishal Jain, and Ambuj Kumar Agarwal

Abstract Adversarial attacks on neural networks are unplanned and skillfully produced inputs that are intended to influence the network's output or predictions in a negative way. Adversarial attacks represent a serious threat to the security and dependability of neural networks since they can be exploited to deceive or manipulate the system. A purposeful input that aims to influence neural networks' predictions is called an adversarial attack. The security and dependability of neural networks are seriously threatened by these assaults because they can be exploited to trick or control the system. Adversarial attacks come in a variety of forms, including physical, black-box, gradient-based, and transferability attacks. Several defense mechanisms, such as adversarial training and input sanitization, have been developed to lessen the effects of these attacks.

Keywords Autoencoder · Neural network · Long short-term memory (LSTM) · Generative adversarial networks

Pranjal · Vaishnavi · Divyansh · D. Raj (✉) · V. Jain · A. K. Agarwal
Department of Computer Science & Engineering, Sharda School of Engineering & Technology, Sharda University, Greater Noida, India
e-mail: dharmraj4u@gmail.com

Pranjal
e-mail: pranjalthakur252003@gmail.com

Vaishnavi
e-mail: vaishnavi.kv2@gmail.com

Divyansh
e-mail: divyanshtri12vedi@gmail.com

V. Jain
e-mail: drvishaljain83@gmail.com

A. K. Agarwal
e-mail: ambuj4u@gmail.com

S. Dutta et al. (eds.), *Cyber Intelligence and Information Retrieval*, Lecture Notes in Networks and Systems 1025, https://doi.org/10.1007/978-981-97-3594-5_34

1 Introduction

In recent years, adversarial attacks on neural networks have gained significant attention, especially as deep learning models gain popularity across a range of applications, including computer vision, natural language processing, and speech recognition. The inaccurate output of a neural network as a result of these attacks can be used by attackers to accomplish a variety of harmful objectives, such as getting around security measures, changing data, or harming people or organizations. Numerous factors make adversarial attacks on neural networks conceivable. Due to their inherent complexity and difficulty in comprehension, deep learning models are vulnerable to assault. Additionally, as neural networks are frequently trained on huge datasets, biases, and vulnerabilities may be introduced.

Huang et al. [1] suggests several protection strategies have been created by researchers to lessen the effects of hostile attacks. By including hostile samples into the training data, adversarial training can increase the neural network's defense against these assaults. Defensive distillation involves training a secondary model to detect and exclude adversarial examples from the input data, whereas input sanitization entails filtering out or altering inputs to remove hostile perturbations. In other words, situations that are only slightly different from instances that are correctly identified and picked from the distribution of data are wrongly classified by these machine learning models. Many different models with diverse architectures trained on distinct portions of the training data commonly misclassify the same adversarial sample. This implies that fundamental flaws in our training methods are exposed by adversarial samples. It was unclear what caused these adversarial cases, but speculative explanations have indicated that it may be related to deep neural networks' excessive nonlinearity, maybe in combination with inadequate model averaging and inadequate regularization of the purely supervised learning.

A distributed denial-of-service (DDoS) attack is a purposeful attempt to obstruct a server, service, or network's regular traffic by saturating the target or its surrounding infrastructure with an overwhelming amount of Internet traffic.

1.1 Reinforcement Learning

Kaelbling et al. [2]—A subset of machine learning is reinforcement learning. It requires acting prudently to maximize benefit in a particular circumstance. Many computer systems and applications use it to determine what to do in a certain circumstance. The reinforcement agent chooses how to complete the job; there is no right or incorrect answer in reinforcement learning. Unlike supervised learning, where a model is trained utilizing the solution key given in the training data, this method does not require any outside assistance. It is necessary for it to draw lessons from its past experiences in the lack of a training dataset.

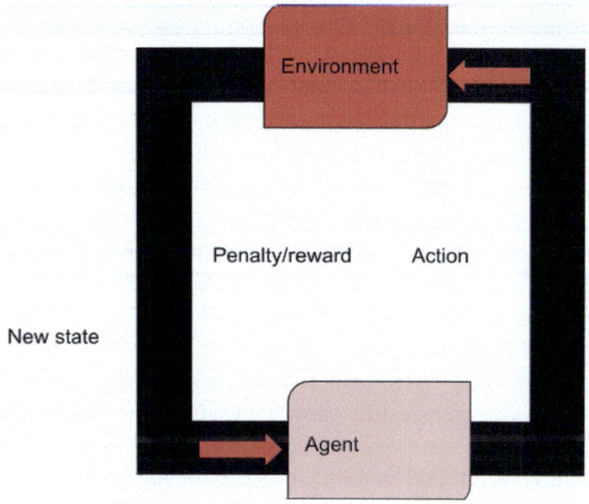

Fig. 1 Reinforcement learning block diagram

The discipline of reinforcement learning (RL) (Fig. 1) focuses on decision-making. It entails knowing how to react in a circumstance to benefit the most. Machine-learning systems that use a trial-and-error method collect the data for RL. Data are not an input for either supervised or unsupervised machine learning.

Algorithms used in reinforcement learning determine the next step based on the results. The algorithm receives input after each step that aids in determining whether the decision it made was excellent, bad, or neutral. It is a useful strategy for computers that must make numerous little judgements without human oversight.

2 Literature Survey

In recent years, adversarial attacks on neural networks have gained significant attention, especially as deep learning models gain popularity across a range of applications, including computer vision, natural language processing, and speech recognition. The inaccurate output of a neural network as a result of these attacks can be used by attackers to accomplish a variety of harmful objectives, such as getting around security measures, changing data, or harming people or organizations. Huang et al. [1]—this paper presents attack taxonomy against networked machine-learning algorithms, explains the specific application-related limitations on an adversary's capabilities, and suggests two models for modelling an adversary's capabilities. Investigate machine-learning system weaknesses and discuss defenses against attackers. Goodfellow et al. [3]—Examples of adversary are inputs created by applying small, purposeful worst-case mutations to explanation from the dataset, wherein the model's perturbed input results give an incorrect response with an elevated level of confidence. Carlini et al. [4] with an input x and a desired categorization t, it is feasible to create an additional input x' that is identical to x but classed as t, making neural networks vulnerable to adversarial scenarios. As a result, it is challenging to deploy neural networks in safety-sensitive circumstances. Eykholt et al. [5]—Recent research shows that small-magnitude perturbations applied to the input make modern deep neural networks (DNNs) vulnerable to adversarial situations. Since new physical systems use DNNs in safety–critical contexts, adversarial instances could mislead them and cause dangerous situations.

Ilyas et al. [6]—Even in the black-box condition, where the attacker just has query access to the model, current neural network-based classifications are vulnerable to adversarial circumstances. Madry et al. [7]—Recent studies have demonstrated that adversarial instances, or inputs that are nearly identical to real data but are misclassified by the network, can affect neural networks. In fact, some of the most recent research results show that adversarial attacks against deep learning models may already be possible. Zügner et al. [8]—When applied to the node classification issue, neural network models for networks have shown great performance. It provides the first examination of adversarial assaults on attributed networks with a special emphasis on models based on graph convolution theory. Huang et al. [9]—It is generally known that classifiers based on machine learning are susceptible to inputs that have been nefariously crafted by adversaries to produce misclassification. The use of such adverse events in applications involving computer vision has been thoroughly studied. Santhanam et al. [10]—Recent studies show that modern models are particularly sensitive to adversarial perturbations of the input. Which provides a technique for identifying and defending against adversarial attacks that utilizes a GAN's discriminator and generator that have both been trained on the same dataset.

Miyato et al. [11]—Contrary to adversarial training, which allows supervised learning algorithms to be regularized, virtual adversarial training enables the extension of supervised learning algorithms to the semi-supervised setting. The presented method produces cutting-edge outcomes for a number of benchmark purely supervised and semi-supervised tasks. Madry et al. [7] —Recent studies have shown that deep neural networks are susceptible to adversarial instances, or inputs that are nearly identical to real data but are misclassified by the network. This study investigates the neural networks' resilience to adversarial attacks using robust optimization in order to solve this issue. Chen et al. [12] upstream procedures are typically sped up or made more accurate by using pre-trained models from self-supervision. Increasing the final model's resilience and decreasing processing costs are two convoy benefits of using adversarial fine-tuning. Several pre-trained self-supervised models are also shown to have a number of adversarial weaknesses. Li et al. [13] Deep learning has lately significantly improved visual recognition. Using a cascade classifier, an efficient adversarial detection system was produced. Additionally, after being trained on a specific adversary producing mechanism, the final classifier may accurately identify adversaries from a completely other mechanism. Papernot et al. [14]—By employing large datasets and computationally effective training methodologies, deep learning beats traditional approaches for a number of machine-learning challenges. However, adversarial samples—inputs created by rivals with the intention of making deep neural networks misidentify—are a threat for deep neural networks because of flaws in the training process. Table 1 shows the literature survey of previous researches.

Table 1 Literature survey of previous researches

Year	Author	Algorithm used	Objective
2010	Huaping JiaJun LiuMin ZhangXiaohu HeWeixi Sun	Deep belief neural network (DBN)	The development of the technique for machine learning generated and created from the ANN is known as deep learning. The traits of the human eye and brain can be extracted using these methods. Massive interconnections of neurons and stage-by-stage ANN are used to build DL algorithms, which can perform high-level
2020	D. Li, D. Kotani and Y. Okabe	Generative adversarial networks	One of the most important models for deep learning in recent years has been GAN. It is made up of two models: a discriminative and generative model that may be used to create a fluid game system. GAN has been used in a variety of applications, including intrusion detection, voice processing, data augmentation, and image production
2016	Yajie Wang, Yu-an Tan, Wenjiao Zhang, Yuhang Zhao, Xiaohui Kuang	Deep neural networks (DNNs)	Convolutional neural networks (CNNs), which are a representation of different deep learning models, are already widely used in a variety of computer vision tasks, including recognizing objects in images, image segmentation, etc. Deep neural networks (DNNs) have grown quickly and gained transcendent progress
2020	Dan Tang, Liu Tang, Wei Shi, Sijia Zhan & Qiuwei Yang	Convolution neural network (CNN)	In the area of deep learning, the CNN model is a great classification technique for image recognition. It has the ability to recognize the differences between map features and find feature maps that have LDoS attacks
2018	Yunsheng Fu; Fang Lou; Fangzhi Meng; Zhihong Tian; Hua Zhang; Feng Jiang	Recurrent neural networks (RNNs)	By learning neural networks with a variety of features, RNN may produce classifiers that distinguish between attack and non-attack data while preserving the attack features of the input vectors
2020	Kun Yang; Junjie Zhang; Yang Xu; Jonathan Chao	AutoEncoder	A DDoS assaults detection framework (AE-D3F) based on AutoEncoder that only uses regular traffic to develop the detection model and has the ability to update it automatically over time

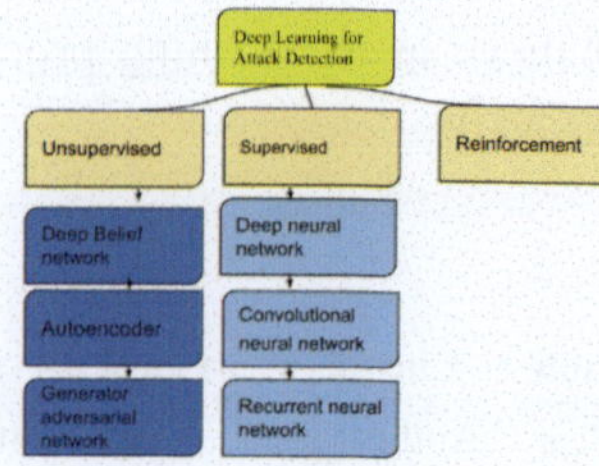

Fig. 2 Classification of the available deep learning attack detection techniques

Lakhdhar [15]—To defend networks from security threats, cyber defenders must identify vulnerabilities that a criminal could exploit and repair them. However, vulnerabilities are constantly growing and altering. Furthermore, allocating the time and resources required to fix all the stolen assets and patch every vulnerability found is not always possible. Due to these reasons, the defender must be in possession of a collection of metrics that can be used to automatically link newly discovered vulnerabilities to likely attack methods. Security tools will be better able to react in real time to all vulnerabilities exploitation attempts by selecting and ranking relevant response tactics. Figure 2 shows the classification of the available deep learning attack detection techniques.

3 Autoencoder-Based Methods for Attack Detection

AE may be viewed as a method for compressing data and has a structure that resembles a neural network. In fact, after compressing the input to a feature space representation, it is feasible to reassemble the representation form the output. AE has applications in decreasing dimension and outlier identification even though it may be thought of as a standard representation learning approach. The condensed feature space of AE is used by researchers in the field of cybersecurity to capture abnormal behavior, which has the advantage of providing dynamic representation for attack types that have not yet been defined.

Long-Short-Term Memory (LSTM)-Based Intrusion Detection Systems

An intrusion detection system (IDS) is a device or piece of software that monitors a network for hostile activity or rules being breached. It looks for malicious activity or security weaknesses on a system or network (Fig. 3). IDS protects networks (network-based intrusion detection system, NIDS) or computers (host-based intrusion detection system, HIDS) by either looking for signs of known assaults or deviations from usual

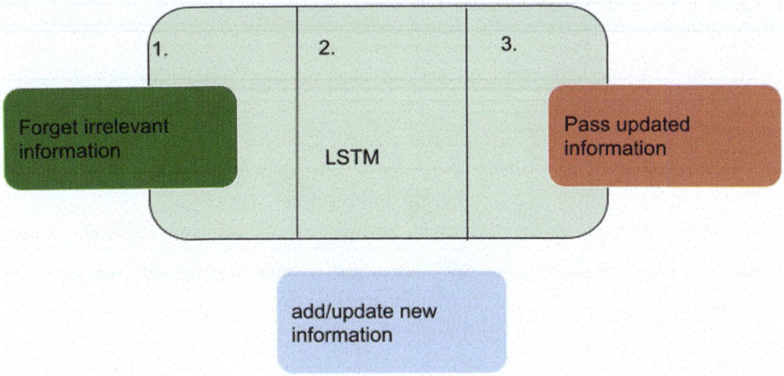

Fig. 3 LSTM architecture

activity. Algorithms based on deep learning have been shown to be more effective in intrusion detection than conventional machine learning methods. In this paper, Long Short-Term Memory (LSTM) will be used to recognize attacks using deep learning approaches.

Other IDS include classification methods such K-nearest, SVM, or decision trees in addition to feature selection in some cases. Superficial learning (single neural networks), which depends on using feature engineering in hand to generate the features of the ML model, is used by nearly all IDS that utilize machine-learning methodologies [16]. In addition, because of the overwhelming volume of data submissions, shallow learning highlighted his incapacity to tackle immediate environmental issues. Deep trying to learn-based models have become more and more well-liked in recent years. A few examples are recurrent neural networks (RNN), variable autoencoders (VAE), and long short-term memory (LSTM) (Fig. 4).

In assaults utilizing autoencoders, the attacker teaches a model of an autoencoder to discover the distribution of the initial input data, such as photographs. After the autoencoder has been trained, the attacker may use it to create adversarial examples by converting the input picture into a reduce-dimensional representation and then reversing the process to produce an image that has been significantly altered in order to deceive the target model.

Applications of Deep Learning Structures for Attack Detection

Because deep learning has so much potential for developing security applications, it has been frequently employed in cybersecurity. Only a few examples of connected applications include traffic analysis, spam detection, virus detection, intrusion detection, and phishing. Through these illustrations of successful applications, it is considered that the innovation brought forth by deep learning frameworks may be leveraged to assess consumer needs. Figure 5 shows the graph shows model's accuracy between

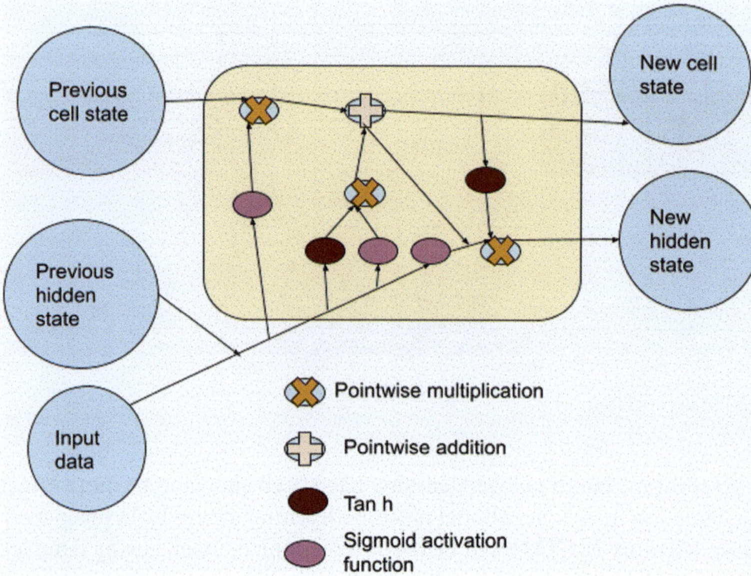

Fig. 4 Long short-term memory (LSTM) networks

training and testing dataset. Figure 6 shows the graph shows the objective of a model to find a give of weights and biases which have low loss.

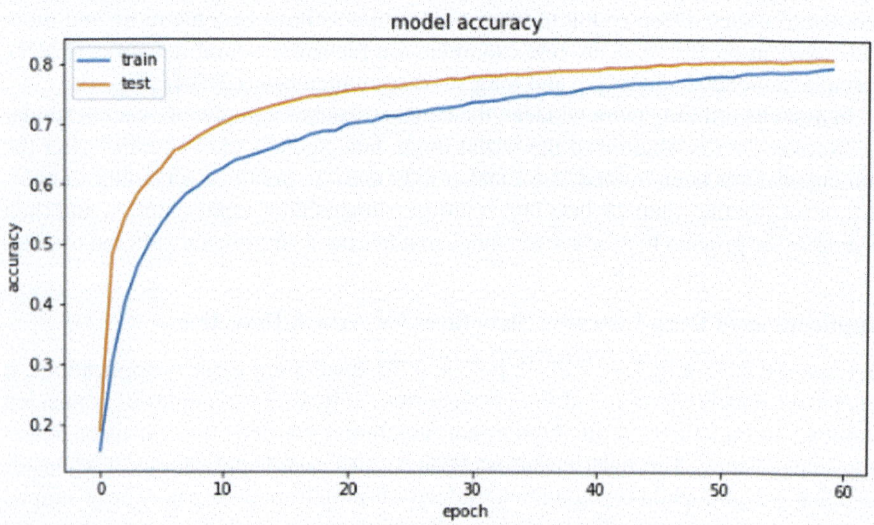

Fig. 5 Graph shows model's accuracy between training and testing dataset

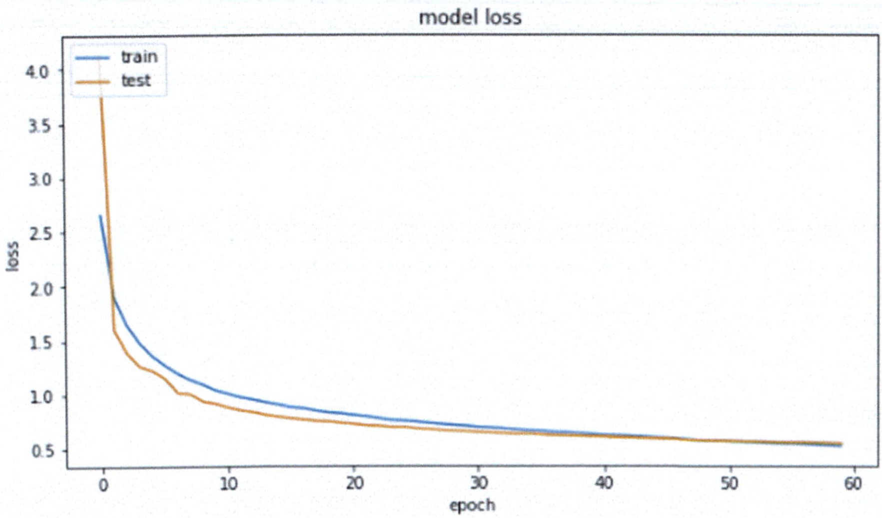

Fig. 6 This graph shows the objective of a model to find a give of weights and biases which have low loss

4 Implementation Results

Botnet Detection with an Autoencoder.

Dataset used: N-BaIoT Dataset [17].

```
0
Shape of data: (9548, 115)
Detected anomalies: 0.2429491919298616%
1
Shape of data: (107685, 115)
Detected anomalies: 100.0%
2
Shape of data: (102195, 115)
Detected anomalies: 100.0%
3
Shape of data: (122573, 115)
Detected anomalies: 100.0%
4
Shape of data: (237665, 115)
Detected anomalies: 100.0%
5
Shape of data: (81982, 115)
Detected anomalies: 100.0%
```

Table 2 The result detection the output

Sl.no	Shape of data	Anomalies detected
1	(9548,115)	0.245
2	(107,685,115)	100.0
3	(102,195.115	100.0

Result (Table 2) showing different anomalies of the different dataset where the autoencoder detection for botnet is applied and showing the accuracy loss that has occurred.

References

1. Huang L, Joseph AD, Nelson B, Rubinstein BI, Tygar JD (2011) Adversarial machine learning. In: Proceedings of the 4th ACM workshop on security and artificial intelligence, pp 43–58
2. Kaelbling LP, Littman ML, Moore AW (1996) Reinforcement learning: a survey. J Artif Intell Res 4:237–285
3. Goodfellow IJ, Shlens J, Szegedy C (2014) Explaining and harnessing adversarial examples. arXiv:1412.6572
4. Carlini N, Wagner D (2017) Towards evaluating the robustness of neural networks. In: 2017 IEEE symposium on security and privacy (sp). IEEE, pp 39–57
5. Eykholt K, Evtimov I, Fernandes E, Li B, Rahmati A, Xiao C, Song D et al (2018) Robust physical-world attacks on deep learning visual classification. In: Proceedings of the IEEE conference on computer vision and pattern recognition, pp 1625–1634
6. Ilyas A, Engstrom L, Athalye A, Lin J (2018) Black-box adversarial attacks with limited queries and information. In: The international conference on machine learning. PMLR, pp 2137–2146
7. Madry A, Makelov A, Schmidt L, Tsipras D, Vladu A (2017) Towards deep learning models resistant to adversarial attacks. arXiv:1706.06083
8. Zügner D, Akbarnejad A, Günnemann S (2018) Adversarial attacks on neural networks for graph data. In: Proceedings of the 24th ACM SIGKDD international conference on knowledge discovery and data mining, pp 2847–2856
9. Huang S, Papernot N, Goodfellow I, Duan Y, Abbeel P (2017) Adversarial attacks on neural network policies. arXiv:1702.02284
10. Santhanam GK, Grnarova P (2018) Defending against adversarial attacks by leveraging an entire GAN. arXiv:1805.10652
11. Miyato T, Dai AM, Goodfellow I (2016) Adversarial training methods for semi-supervised text classification. arXiv:1605.07725
12. Chen T, Liu S, Chang S, Cheng Y, Amini L, Wang Z (2020) Adversarial robustness: from self-supervised pre-training to fine-tuning. In: Proceedings of the IEEE/CVF conference on computer vision and pattern recognition, pp 699–708
13. Li X, Li F (2017) Adversarial examples detection in deep networks with convolutional filter statistics. In: Proceedings of the IEEE international conference on computer vision, pp 5764–5772
14. Papernot N, McDaniel P, Jha S, Fredrikson M, Celik ZB, Swami A (2016) The limitations of deep learning in adversarial settings. In: 2016 IEEE European symposium on security and privacy (EuroS&P). IEEE, pp 372–387

15. Lakhdhar Y, Rekhis S (2021, May) Machine learning based approach for the automated mapping of discovered vulnerabilities to adversarial tactics. In 2021 IEEE Security and Privacy Workshops (SPW) (pp. 309–317). IEEE
16. Salem AMG, Bhattacharyya A, Backes M, Fritz M, Zhang Y (2020) Updates-leak: data set inference and reconstruction attacks in online learning. In: The 29th USENIX security symposium. USENIX, pp 1291–1308
17. https://www.kaggle.com/datasets/mkashifn/nbaiot-dataset
18. Biggio B, Corona I, Maiorca D, Nelson B, Šrndić N, Laskov P, Roli F et al (2013) Evasion attacks against machine learning at test time. In: Machine learning and knowledge discovery in databases: European conference, ECML PKDD 2013, Prague, Czech Republic, September 23–27, 2013, Proceedings, Part III 13. Springer, Berlin, Heidelberg, pp 387–402
19. Li D, Kotani D, Okabe Y (2020) Improving attack detection performance in NIDS Using GAN, 2020 IEEE 44th Annual Computers, Software, and Applications Conference (COMPSAC), Madrid, Spain, pp. 817–825. https://doi.org/10.1109/COMPSAC48688.2020.0-162

Recycled Car Price Extrapolation by LASSO and Linear Regression

Soumen Ghosh, Shneha, Rituja, Sabyasachi Samanta, and Tanmay Sinha Roy

Abstract Given the fluidity of the used automobile market, appraising the value of a recycled car may be difficult and entail a number of variables. Predicting the price of recycled car by using machine learning can have positive effects on the economy, the environment, and society. It can also increase market transparency and open up new business options. The used car market in India is a booming industry, it was valued at \$32.14 billion, and it is expected to reach \$74.70 billion in the coming years. Though COVID-19 has affected a number of industries, the used car market has been minimally impacted by it. A huge amount of people wanting a personal vehicle of their own and the increase in car prices leads more people to buy used cars. As a result, the market for old cars will expand. We have attempted to forecast used car prices with this statistical model by utilizing machine learning techniques such as LASSO (least absolute shrinkage and selection operator) regression and linear regression. This model is based on information gathered from past customers and a variety of automotive attributes. We attempted to determine a certain percentage of the anticipated price's accuracy with the aid of this model.

Keywords Car price prediction · LASSO regression · Linear regression · R2 score · Machine-learning application

1 Introduction

With the increase in the price of cars and the high amount of money spent on maintenance and insurance of cars, pre-owned cars are much sought after. The market of used cars is having a hike because of the quick growth of new car production and sales. While a brand new car might have lots of new technologies incorporated in it,

S. Ghosh (✉) · Shneha · Rituja · S. Samanta
Haldia Institute of Technology, Haldia, WB, India
e-mail: soumenghoshcse2812@gmail.com

T. S. Roy
Swami Vivekananda University, Kolkata, India
e-mail: tanmoysinha.roy@gmail.com

S. Dutta et al. (eds.), *Cyber Intelligence and Information Retrieval*, Lecture Notes in Networks and Systems 1025, https://doi.org/10.1007/978-981-97-3594-5_35

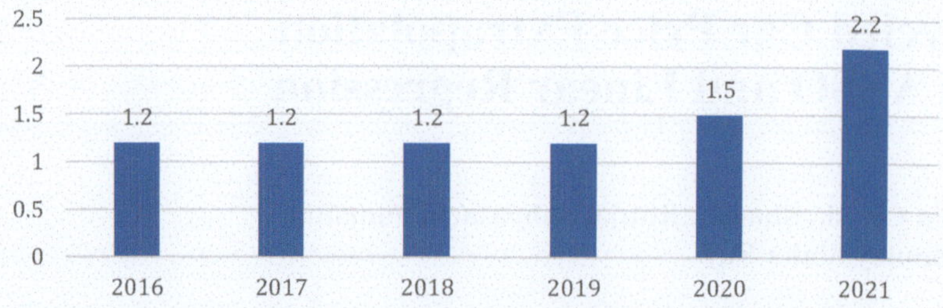

Fig. 1 Market ratio of pre-owned to new car across India from FY 2016–2021 [6]

it is seen many times that Indian middle-class families go for a pre-owned car over a new one. Thus, pre-owned cars come to the rescue. While buying a car predicting the price is important for the customer as well as the seller. Even insurance companies are interested in the proper valuation of a car. Previously, the new-to-used car ratio was 1:1.2, but now it has become 1:2.2. When ten new cars are sold, the market has 22 used cars available for purchase. Similarly, the average holding time of cars has reduced from 5–6 years to 3 years these days [1, 2]. With the boom in the used car market, as more Indian middle-class families and young population are opting for used cars over new ones, the unorganized setup of the used car industry from roadside garages has moved to well-organized shops and shifted more toward online shopping. In FY 2021–22, an estimated 4.4 million used cars were sold in India. The other factor for the rise in the used car industry is because of the demand for luxurious vehicles. It has been noted that the sale of luxury cars in India has seen a hike of 20% [3, 4].

Nowadays, online shops such as 'CARDEKHO', 'QUIKR', 'OLX', 'CARWALE', etc., have made the car buying experience so much easier for customers still the consumers and sellers should be informed about the trends and patterns that help in finding out the value of a car. Figure 1 shows the market ratio of pre-owned to new car across India from financial year 2016 to 2021. Various algorithms can be used to predict the value of a car. These algorithms work using the data of some of the features of a car [5]. In this model, we have used the dataset from 'CARDEKHO'. The dataset consists of nine parameters, each parameter used in this analysis of finding the used car price accuracy. The parameters on which this analysis is based are as follows: Car name, year of purchase of a car, selling price of a car, present price of a car, kilometers driven, fuel type, seller type, transmission, number of owners.

2 Literature Review

Alhakamy et al. [7] studied the data of used car prices and developed an understanding through executing various machine learning and big data tools. They exhibited in a way of answering the questions on different features used for predicting car prices to make them more pocket friendly by using linear regression model based on the significance of independent features on dependent variables, the efficiency of using this model, and how used cars can lead to overall economic and environmental growth. Bukvić et al. [8] presented how in developing nations there is a huge demand for cars with economic challenges which leads to the need for research and development in this field. They offered insight into data of car models in the Croatian market for predicting the prices and provided solutions based on supervised machine-learning techniques comparing its accuracy to a classification algorithm. Mammadov [9] focused on using a linear regression model to predict car prices in the U.S. His goal was to provide, how car prices are determined in the U.S. automobile market. The obtained result was studied involving five major parameter like data visualization, feature scaling and engineering, feature selection, and model building. Ahtesham et al. [10] proposed a machine learning based model to help new entrants understand the important pricing factors in the automobile industry. Their predictions was based on different factors like mileage, and year of production. They also employs three different machine learning algorithms like linear regression, gradient boosting, and decision tree and implemented by using Apache Spark and Big Data tool with a prediction of accuracy of 89%. Amik et al. [11] created a tool to assist prospective purchasers in determining the cost of used automobiles in Bangladesh. With a 91% accuracy rate, they used a variety of methods on the dataset, including random forest, decision trees, LASSO regression, extreme gradient boosting, and linear regression.

Liu et al. [12] also proposed a used car prediction model with the significant improvements in mobile internet for traditional offline used car trading which is no longer able to meet the needs of consumers. They predicted by using PSO-GRA-BPNN, by filtering feature variables with grey relational analysis and optimizing the traditional BP neural network using particle swarm optimization algorithm. The proposed PSO-GRA-BPNN model has a significantly prediction effect with a smaller MAPE and MAE, and higher R and R2. Pattabiraman et al. [13] proposed a consistently rising of past decade, resulting in the emergence of a flourishing second-hand car industry. By using ML techniques like LASSO regression, multiple regression, and regression trees, they developed a statistical model that utilizes previous customer data.

3 Methodologies

3.1 Data Acquisition

Here, we have used the dataset of 'CARDEKHO'. 'CARDEKHO' is one of India's leading online platforms for car-related ventures, either buying new cars or buying and selling used cars [14]. After lots of searching and sorting, we obtained this dataset of used cars of 'CARDEKHO' from the 'KAGGLE' platform [6]. The dataset consists of 301 features and 9 instances including the target class. The features of the dataset are demonstrated as follows. Table 1 represents a sample dataset for car prediction with different attributes. In Tables 1 and 5, 'the fuel type 'P' for Petrol, 'D' for Diesel and price 'L' for 'Lakh' are taken. Seller type as 'Dealer' and Transmission as 'Manual' are taken as 'Yes' for positive cases.

Table 1 Sample dataset for car prediction

Sl. no	Car model name	Year of production	Selling price (in Lakh)	Present price (in Lakh)	Kms driven	Fuel type (petrol/ diesel)	Seller type (dealer)	Transmission (manual)	Owner
0	MarutiRITZ	2014	3.35L	5.59L	27,000	P	Yes	Yes	0
1	Maruti SX4	2013	4.75 L	9.54 L	43,000	D	Yes	Yes	0
2	Maruti CIAZ	2017	7.25 L	9.85 L	6900	P	Yes	Yes	0
3	Maruti WAGNOR	2011	2.85 L	4.15 L	5200	P	Yes	Yes	0
4	Maruti SWIFT	2014	4.6 L	6.87 L	42,450	D	Yes	Yes	0
…	…	…	…	…	…	…	…	…	…
296	Honda CITY	2016	9.5 L	11.6 L	33,988	D	Yes	Yes	0
297	Honda BRIO	2015	4 L	5.9 L	60,000	P	Yes	Yes	0
298	Honda CITY	2009	3.35 L	11 L	87,934	P	Yes	Yes	0
299	Honda CITY	2017	11.5 L L	12.5 L	9000	D	Yes	Yes	0
300	Honda BRIO	2016	5.3	5.9 L	5464	P	Yes	Yes	0

3.2 Data Preprocessing

Many times, it has come to notice that the dataset we have found is not suitable for direct use in machine-learning (ML) algorithms. Sometimes, there are missing values, invalid data, duplicate data, or noise. For further processing and analyzing the data, certain pre-processing is required to make the data suitable for further use inspection. Data pre-processing is defined as cleaning and transforming the data to better quality to make it suitable for further tasks. First, we checked for the number of unique values per feature and output shows as follows. Table 2 shows a sample with the number of unique values per feature.

From the above result, the car name column has 98 unique values. Since there are lots of unique values and some of them have 1 or 2 instances, we drop this column in our pre-processing. Further, we check for any null values that are present or not. After that, we get the result as shown in the output below. Since no null value is there so no more processing of data is required. Thus, now the dataset is ready for work (Table 3).

Table 2 The number of unique values per feature

ds.nunique()	
Car_Name	98
Year	16
Sellingprice	156
Present_Price	147
Fuel_Type	3
Seller_Type	2
Transmission	2
Owner	3
dtype: int64	

Table 3 Revised data without NULL value

Car_Name	0
Year	0
Selling_Price	0
Present_Price	0
Kms_Driven	0
Fuel_Type	0
Seller_Type	0
Transmission	0
Owner	0
dtype: int64	

Table 4 Finding basic properties like count, mean, std, min, max

ds.decribe()

	Year	Selling price	Present price	Kms driven	Owner
Count	301	301	301	301	301
Mean	2013.627	4.661296	7.628	36,947	0.04
std	2.891554	5.082812	8.644	38,889	0.25
min	2003	0.1	0.32	500	0.0
25%	2012	0.9	1.2	15,000	0.0
50%	2014	3.6	6.4	32,000	0.0
75%	2016	6.0	9.9	48,767	0.0
max	2018	35	92.6	500,000	3.0

3.3 Data Analysis

Analyzing data involves examining and interpreting data to extract insights and draw conclusions. This process typically involves using statistical methods and tools to identify patterns, trends, and relationships within the data. Data analysis aims to extract meaningful information from raw data, here we are analyzing the obtained dataset and finding its basic properties like count, mean, std., min, max, etc. These properties help in understanding the dataset and deriving its usefulness. Table 4 shows the finding basic properties like count, mean, std, min, max.

We have evaluated count for different features as given below.

Further we have plotted the count plot for the above data items for Fuel_Type, Seller_Type, and transmission and Owner columns. Figures 2 and 3 shows the count for different features and plot for Fuel_Type, Seller_Type, transmission, and owner columns, respectively. Given below are the plots for each of them.

3.4 Data Encoding

Data encoding is the process of converting data from one format or representation to another. This is done to ensure that the data is compatible with a particular system or application or to make it more efficient for storage or transmission. There are different types of data encoding techniques, including text encoding, image encoding, audio encoding, and video encoding, also it is an important process in computing, as it ensures data can be stored, transmitted, and processed efficiently and effectively. In our dataset, we are encoding the Fuel_Type column, Seller_Type column, and transmission column to numerical values to make our analysis more practical and efficient for transmission. Given below are the steps involved in this process.

Table 5 Dataset after encoding

Sl. no	Car name	Year of production	Selling price (in Lakh)	Present price (in Lakh)	Kms driven	Fuel type (petrol/ diesel)	Seller type (dealer	Transmission (manual)	Owner
0	MarutiRITZ	2014	3.35 L	5.59 L	27,000	P	Yes	Yes	0
1	MarutiSX4	2013	4.75 L	9.54 L	43,000	D	Yes	Yes	0
2	MarutiCIAZ	2017	7.25 L	9.85 L	6900	P	Yes	Yes	0
3	MarutiWAGNOR	2011	2.85 L	4.15 L	5200	P	Yes	Yes	0
4	MarutiSWIFT	2014	4.6 L	6.87 L	42,450	D	Yes	Yes	0
…	…	…	…	…	…	…	…	…	…
296	Honda CITY	2016	9.5 L	11.6 L	33,988	D	Yes	Yes	0
297	Honda BRIO	2015	4 L	5.9 L	60,000	P	Yes	Yes	0
298	Honda CITY	2009	3.35 L	11 L	87,934	P	Yes	Yes	0
299	Honda CITY	2017	11.5 L	12.5 L	9000	D	Yes	Yes	0
300	Honda BRIO	2016	5.3 L	5.9 L	5464	P	Yes	Yes	0

```
print(ds.Fuel_Type.value_counts())

Petrol    239
Diesel     60
CNG         2
Name: Fuel_Type, dtype: int64
```

```
print(ds.Seller_Type.value_counts())

Dealer        195
Individual    106
Name: Seller_Type, dtype: int64
```

```
print(ds.Transmission.value_counts())

Manual       261
Automatic     40
Name: Transmission, dtype: int64
```

```
print(ds.Owner.value_counts())

0    290
1     10
3      1
Name: Owner, dtype: int64
```

Fig. 2 Count for different features

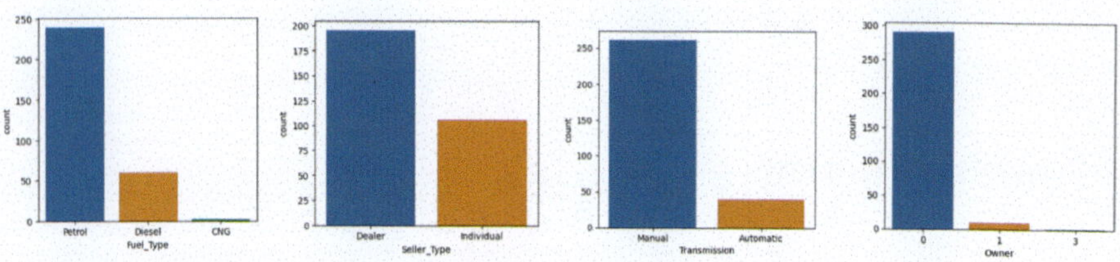

Fig. 3 Plot for Fuel_Type, Seller_Type, transmission, and owner columns

```
#encoding "Fuel_Type" Column
ds.replace({'Fuel_Type':{'Petrol':0,'Diesel':1,'CNG':2}},inplace=True)

# encoding "Seller_Type" Column
ds.replace({'Seller_Type':{'Dealer':0,'Individual':1}},inplace=True)

# encoding "Transmission" Column
ds.replace({'Transmission':{'Manual':0,'Automatic':1}},inplace=True)
```

In this case, Fuel_Type is changed to represent CNG, diesel, and petrol as the numerical values 0, 1, and 2, respectively. The values 0 and 1, respectively, are assigned to Seller_Type Dealer and Individual. Lastly, manual and automatic are changed to the numerical values 0 and 1, respectively, under the Gearbox feature. Table 5 represents the dataset after encoding with different parameters. The obtained dataset after encoding.

3.5 Data Splitting

Data splitting is a process in which a dataset is divided into two or more separate subsets or partitions. This is done primarily in machine learning and data analysis to separate data into different sets for training, testing, and validation purposes. The primary objective of data splitting is to prevent over-fitting, when a machine learning model performs exceptionally well on the training data but fails to perform well on new or unseen data. By splitting the dataset into separate training and testing sets, we can evaluate the performance of a machine-learning model on new data and ensure that it is not over-fitting to the training data. In summary, data splitting to ensure that models are correctly evaluated and can generalize well to new data. For this model, we have divided the dataset into 85:15 ratio.

3.6 Algorithms Used

3.6.1 Linear Regression

Linear regression is used for predictive modeling, the goal is to estimate or predict the value of the dependent variable based on the value of the independent variables. The basic idea is to find the best-fit line through the data which explains the best relationship between dependent and independent variables. This line is called the regression line or the best-fit line [15]. There are two main types of linear regression: Simple linear regression and multi-linear regression. Simple linear regression is modeled using a straight line with the equation:

$$Y = b_0 + b_1 * X \tag{1}$$

Here, Y is the dependent variable, X is the independent variable, b_0 is the intercept, and b_1 is the slope of the line. The intercept represents the value of Y when X is zero, and the slope represents the value of Y for a one-unit increase in X. The equation for multiple linear regression is as follows:

$$Y = b_0 + b_1X_1 + b_2X_2 + \cdots + b_nX_n \tag{2}$$

Here, Y is the dependent variable, $X_1, X_2, \ldots, X_n$ are the independent variables, and $b_0, b_1, b_2, \ldots, b_n$ are the intercept and the slopes for each of the independent variables.

3.6.2 LASSO Regression

Regression analysis using shrinkage is used in LASSO (least absolute shrinkage and selection operator), a sort of procedure that yields more accurate results than linear regression. Shrinkage means the shrinking of data values toward the central point or mean. The goal of this algorithm is to minimize the sum of the squared errors between the predicted values and the actual values, subject to the constraint that the sum of the absolute values of the regression coefficient is less than or equal to a certain value.

$$\sum_{i=1}^{n}\left(y_i\left[-\sum_{j}\left[x_{ij}\beta_j\right]\right]\right)^2+\lambda\sum_{j=1}^{p}\left|\beta_j\right| \quad (3)$$

Here, λ is the amount of shrinkage.

4 Result Analysis

The dataset was evaluated using two regression algorithms as mentioned above. For both of these methods, we calculated the respective R-squared score and RMSE (Root Mean Squared Error). Using these metrics, we determined the accuracy of the two aforementioned algorithms. R-squared score statistics are used to represent the proportion of the variance for a dependent variable that the independent variables explain collectively. R-squared generally ranges between 0 to 1 and is often stated in percentage. So if R-squared is 100% that means dependent variables are completely explained by independent variables. How well the data match the regression line is indicated by the R-squared score. The mean-squared error (MSE) and R-squared are fairly close. The formula to find R-squared:

$$R^2 = 1 - \frac{unexplained\ \ variation}{total\ variation} \quad (4)$$

The next matrix that we used for this model is RMSE (root-mean-squared error). RMSE is the square root value of what we have obtained from MSE (mean-squared error). It is the standard deviation of the residuals. Residual is the measure of how far are data points from the regression line [5]. This tells us how densely distributed are the data points around the best-fit line. To calculate RMSE, we use the following formula mentioned below.

$$RMSE = \sqrt{\left(1/N\sum\nolimits_{I=1}^{N}(p_i - a_i)^2\right)} \quad (5)$$

Here, N is the sample size,p_i is predicted value and a_i is the actual value

After splitting the data into train and test sets we checked the R2 score and RMSE to determine the performance of the linear and LASSO regression. The R-squared score and root-mean-squared error of test data for both the regression algorithm are demonstrated below in a table.

Algorithm	R^2 score	RMSE
Linear regression	0.844	1.531
LASSO regression	0.894	1.265

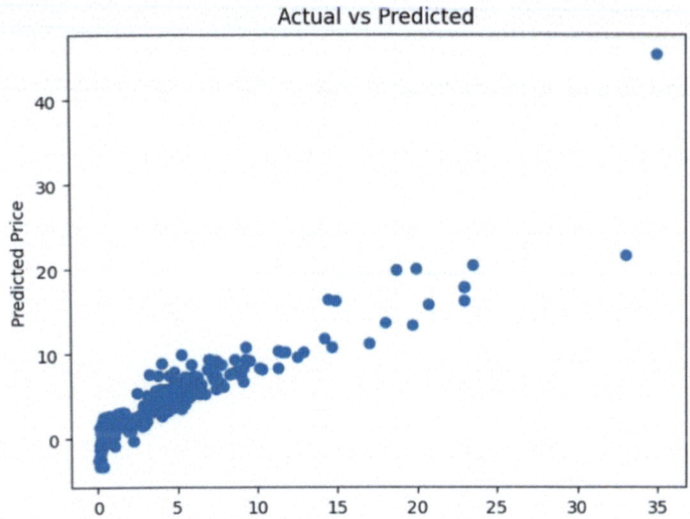

Fig. 4 Scatter plot actual price versus predicted price graph

From the above table after comparing the values for both types of regression we conclude, using LASSO regression, we get better accuracy than Linear regression. It is visible that not only is the R2 score more in LASSO regression but also the root-mean-squared error is reduced in it.

After performing linear regression and LASSO regression calculating metrics like R-squared score and root-mean-squared error, we plotted some scatter plots for a better result of the model.

The above scatter plot shows (Fig. 4), the actual price versus predicted price graph obtained from the training data after performing the linear regression algorithm on the dataset. From the graph, it is visible that the data points concentrate around the regression line. Similarly, we plotted the graph for test set data and got a similar kind of result data sensing around the best-fit line, the result is shown below.

The second algorithm that we have used is LASSO regression. For this again, we calculated the R-squared score and root-mean-squared error. As mentioned earlier using this model resulted in a better R2 score and reduced root-mean-squared error. Thus, this model (Fig. 5) resulted in better accuracy and was proven to be better than linear regression. For better understanding, scatter plot was plotted on train set data.

Similarly scatter plot was done for test set data which is exhibited above (Fig. 6). It shows that for both the test and train set the data gives around the best-fit line which is what we needed. Also when we compare the scatter plot of test data for linear regression with the test data for LASSO regression it comes that the data in the LASSO regression is much more densely concentrated and near to the regression line than the data in linear regression (as in Fig. 7). Thus, this proves that LASSO regression results in better accuracy than the linear regression model.

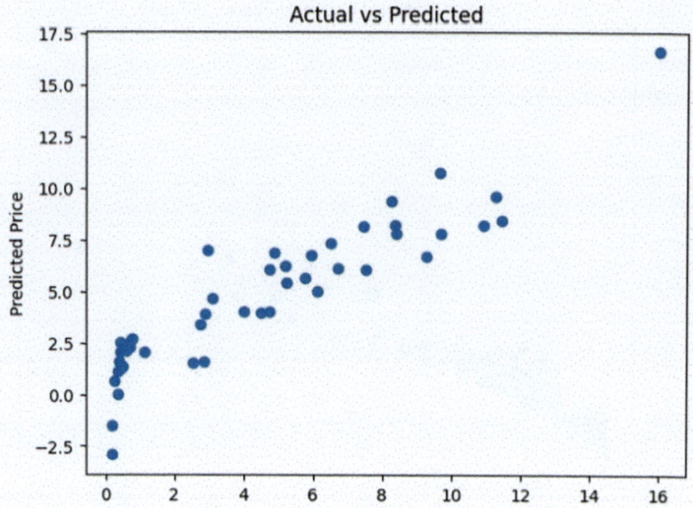

Fig. 5 Scatter plot actual price versus predicted price graph (graph for test set data)

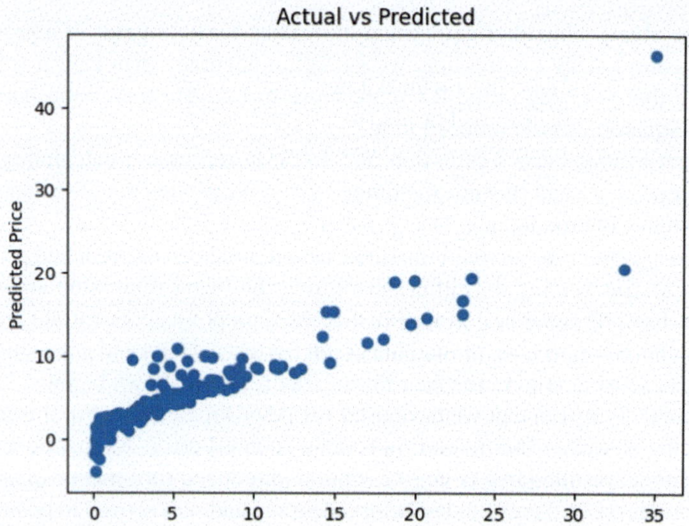

Fig. 6 Scatter plot actual price versus predicted price graph (LASSO regression)

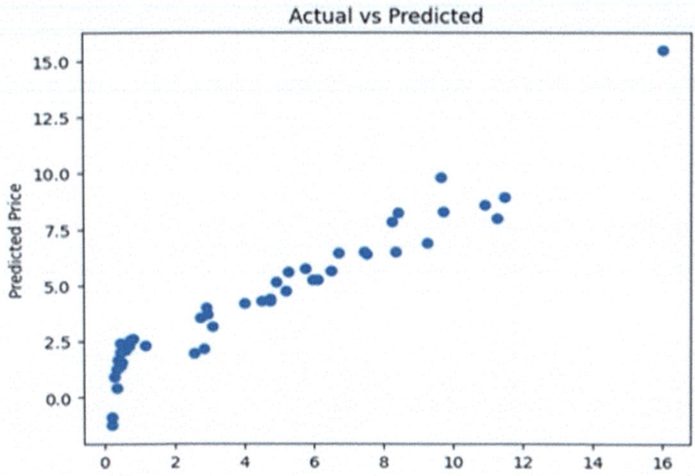

Fig. 7 Scatter plot actual price versus predicted price graph (test set data and LASSO regression)

5 Conclusion

When dealing with high-dimensional data with multicollinearity, LASSO is helpful for feature selection, model interpretability, and managing high-dimensional data; nevertheless, linear regression may be chosen when you think all characteristics are necessary for reliable predictions and when model simplicity is not your main priority. This study explored the use of linear regression and LASSO regression to predict used car prices using the data taken from 'CARDEKHO' (using 'KAGGLE' database). The result of the analysis showed that both linear regression and LASSO regression can be effective in predicting used car prices, with LASSO outperforming linear regression in terms of accuracy. We have got the value of the R2 score as 0.844 in the case of linear regression while the value of the R2 score is 0.894 in the case of LASSO regression. The study also found that the most important predictors of used car prices were the present price of the car, kilometers driven, fuel type, and owner. Both buyers and sellers can utilize these factors to help them decide on a fair price because they played a big part in the forecasting process. This further emphasizes the predictive capacity of machine learning methods like LASSO regression and linear regression. All things considered, this advances the topic of used car pricing and shows how machine learning can be applied to solve practical issues in the auto sector.

References

1. Fathalla A, Salah A, Li K, Li K, Francesco P (2020) Deep end-to-end learning for price prediction of second-hand items. Knowl Inf Syst 62(12):4541–4568. Springer Science and Business Media LLC. https://doi.org/10.1007/s10115-020-01495-8
2. Akhtar M, Moridpour S (2021) A review of traffic congestion prediction using artificial intelligence. In: Bazant M (ed) J Adv Transp 2021:1–18. Hindawi Limited. https://doi.org/10.1155/2021/8878011
3. Kumar K, Pande BP (2022) Air pollution prediction with machine learning: a case study of Indian cities. Int J Environ Sci Technol. Springer Science and Business Media LLC. https://doi.org/10.1007/s13762-022-04241-5
4. Krishnan JR, Selvaraj V (2022) Predicting resale car prices using machine learning regression models with ensemble techniques. In: 2nd international conference on mathematical techniques and applications, ICMTA2021. AIP Publishing. https://doi.org/10.1063/5.0108560
5. Das Adhikary DR, Sahu R, Pragyna Panda S (2022) Prediction of used car prices using machine learning. In: Biologically inspired techniques in many criteria decision making. Springer Nature Singapore, pp 131–140. https://doi.org/10.1007/978-981-16-8739-6_11
6. "India Used Car Market Analysis - Industry Report - Trends, Size & Share." Mordor Intelligence. https://www.mordorintelligence.com/industry-reports/india-used-car-market
7. Alhakamy A, Alhowaity A, Alatawi AA, Alsaadi H (2023) Are used cars more sustainable? Price prediction based on linear regression. Sustainability 15(2):911. MDPI AG. https://doi.org/10.3390/su15020911
8. Bukvić L, PašagićŠkrinjar J, Fratrović T, Abramović B (2022) Price prediction and classification of used-vehicles using supervised machine learning. Sustainability 14(24):17034. MDPI AG. https://doi.org/10.3390/su142417034
9. Mammadov H (2021) Car price prediction in the USA by using liner regression. Int J Econ Behav (IJEB) 11(1). https://doi.org/10.14276/2285-0430.3049
10. Ahtesham M, Zulfiqar J (2022) Used car price prediction with Pyspark. In: Digital technologies and applications. Springer International Publishing, pp 169–179. https://doi.org/10.1007/978-3-031-01942-5_17
11. Amik FR, Lanard A, Ismat A, Momen S (2021) Application of machine learning techniques to predict the price of pre-owned cars in Bangladesh. Information 12(12):514. MDPI AG. https://doi.org/10.3390/info12120514
12. Liu E, Li J, Zheng A, Liu H, Jiang T (2022) Research on the prediction model of the used car price in view of the PSO-GRA-BP neural network. Sustainability 14(15):8993. MDPI AG. https://doi.org/10.3390/su14158993
13. Venkatasubbu P, Ganesh M (2019) Used cars price prediction using supervised learning techniques. 9(1):S3. ISSN: 2249-8958. https://doi.org/10.35940/ijeat.A1042.1291S319
14. "Vehicle dataset." Kaggle. https://www.kaggle.com/datasets/nehalbirla/vehicle-dataset-from-cardekho
15. Liverani C (2020) A beginner's guide to stepwise multiple linear regression. Medium, 10 June 2020. https://medium.com/analytics-vidhya/a-beginners-guide-to-stepwise-multiple-linear-regression-e81df82b9f9e. Accessed 10 March 2023

A Socio-Legal Perspective on Gender-Based Discrimination in Machine Learning Algorithm

Reena Roy, Anil Kumar Dixit, S. K. Chaturvedi, Sagar Saxena, Gagandeep Kaur, and Ravi Kant

Abstract The problem of gender-based prejudicialness began the moment women joined professional workplaces, businesses, and other institutions. Regardless of stringent regulations and laws, almost all organizations exhibit bias against women. However, its varieties and density may vary based on a country's location, industries, or level of prosperity. Due to this AI-ML algorithms are also getting affected producing biased results while recruiting women or at the time of promotion or appraisal because the data that ML is learning or getting trained is not equitable or fair. In this paper, we discussed those women in managerial roles face prejudicialness from their peers. This shows the problem's apex. Therefore, the prejudice against women must be addressed through coordinated efforts of every pertinent organization. Further in this paper, we identified the common algorithmic errors that cause biases and prejudicialness against women in the workplace. Thus, this paper shows the relationship between Artificial Intelligence, Machine Learning, and Gender Discrimination. In conclusion, the present paper proposes several gender discrimination mitigations through ML algorithms and legal techniques.

R. Roy · A. K. Dixit · S. Saxena (✉) · R. Kant
Law College Dehradun, Uttaranchal University Dehradun, Dehradun, India
e-mail: adv.sagarsaxena@gmail.com

R. Roy
e-mail: reenaroy2709@gmail.com

A. K. Dixit
e-mail: anilkdixit44@gmail.com

R. Kant
e-mail: ravi9kant@outlook.com

S. K. Chaturvedi
School of Law, H.N.B Garhwal Central University, Tehri Garhwal, India
e-mail: chaturvedilaw@gmail.com

G. Kaur
School of Law, University of Petroleum & Energy Studies, Dehradun, India
e-mail: gkaur@ddn.upes.ac.in

S. Dutta et al. (eds.), *Cyber Intelligence and Information Retrieval*, Lecture Notes in Networks and Systems 1025, https://doi.org/10.1007/978-981-97-3594-5_36

Keywords Machine learning · Artificial intelligence · Convolutional neural networks · United Nations · Convention on the Elimination of All Forms of Discrimination against Women · Office of the United Nations High Commissioner for Human Rights

1 Introduction

The fast-growing use of machine learning and artificial intelligence techniques opens new horizons in terms of accuracy, efficiency, and rapidity in everyday life applications. Applications of artificial intelligence play a progressively significant part in the structuring of society as a whole, governing a vast array of social networks and structures, and even influencing interpersonal relationships and desires [1].

As mankind keeps evolving, the growing presence of AI in our daily existence is becoming crucial for the transformation of our community and social well-being. A growing number of occupations are being computerized and supplanted by AI. Yet, the substitution of jobs is contingent upon the nature of the employment and its duties. As simple, repetitive duties that can be coded and utilized, low-skill occupations such as business, office duties, and administrative assistance have a greater probability of being taken over by AI. Employees in such roles may receive a promotion if they are qualified, or they might get reassigned to work from home on other duties. This form of substitution and change can happen in any type of organization [2].

As the variety of jobs declines progressively as a result of AI replacement, gender parities, and prejudicialness are among the issues that arise too. Globally, gender parity is an inherent right and value; however, threats to the premise of sex continue to exist today. AI creates opportunities while presenting issues for equal treatment of women and equality of opportunity. Due to the confluence of numerous technological advances and the impact of these technologies on our professional and personal lives, it is necessary to focus on the relationship between AI and equal rights for women [3].

AI's emergence paved the way for the development of ML, which trains machines to learn on their own. The "learning" procedure occurs by offering data from previous actions that accomplished the same duties or by giving data through the Internet. The machine will learn an operation depending on how it is instructed; if the machine had been either deliberately or inadvertently designed to prefer a particular race or gender for job opportunities, for example, it will carry out so. The machine will acquire a predilection based on its training. Kate Crawford accurately described the genesis of the increasing incidence of prejudice against women within machine intelligence; "Just like every technological innovation before it, AI will reflect the principles of its developers" [4].

Integrating gender theory, specifically feminist linguistic theory, into a method relating to ML from written information might avoid the emergence of discrimination based on gender and reduce the need for altering algorithms. Integrating AI with deployment, it is anticipated that AI will help minimize prejudice against women in companies; additional research is required to confirm the hypothesis. Additionally, with AI and the quest for gender parity, there is still an enormous amount of work to be done. A few standards for addressing issues and possibilities may become obsolete over time, while others may emerge. Therefore, nation-states should regularly exchange these cutting-edge techniques to inspire others, even if certain techniques have been modified to a specific context in a given state [5].

As a result of global requirements, the proportion of female employees in various companies and institutions is rising consistently. Nevertheless, despite the initiatives of governments and the grievances or objections of distressed women, discrimination could not be eradicated from the workplaces of various industries. In most developing countries, the number of violations of laws and statutes is increasing. Women may experience discrimination in the form of uneven employment opportunities, disparities in monetary compensation, communication, conduct, assigned roles in employment, promotions, preconceived notions, and the varying importance accorded to males over women in various organizations. However, in advanced countries like the USA, female employees are paid less than their male counterparts. The numerous women promoted to managerial positions have always been considerably below average. The majority of supervisors believe women cannot execute difficult tasks. Thus, the level of trust in female employees is not of the same level as that of fellow male employees [6].

As a result of this reason, superiors routinely assign more essential tasks to men. Although cases of gender-based assault in the workplace have fallen in developed nations, they remain extremely prevalent in developing nations. Societies and organizations that are advanced, improved, and well-developed should not tolerate gender-based discrimination. This indicates that there is still an abundance of opportunities for achieving gender equality. Women employees should receive the same respect, responsibility, reward, and appraisal as their male counterparts. Explicitly apparent egalitarianism in every company and organization requires effort, stringent regulations, and a mental shift on the part of management and executives. Supervisors and other partners may require instruction, training, and on-the-job courses to effectively handle and reduce this major issue facing the modern sophisticated world.

Thus, this review study offers an organized and exhaustive study of the relevant interdisciplinary academic literature to clarify existing information and comprehend the connection between AI & ML and gender-based equality in the workplace. In the following pages, the authors aspire to accomplish the following objectives will be discussed:

A. The present research focuses on the rise of gender-based discrimination and how ML algorithmic development is affected by this social misery, as well as a comparison of different technologies.
B. In addition, this study focuses on global aspects and methods that can be used to curb the issues related to gender-based discrimination crime using A.I. and machine learning.

This study is structured as follows. Section 2 states the algorithmic discrimination mechanisms. Section 3 describes the technical background of ML algorithmic creation. Section 4 states four common algorithmic hiring errors. Then Sect. 5 states some recommendations for mitigating discrimination against gender biases through legal & technological ways. Further Sect. 6 introduces the proposed methodology of the ML algorithm. Section 7 states the international women's rights concept. Further Sect. 8 provides recommendations. Finally, the paper ends with Sect. 9 with the conclusion.

2 Algorithmic Discrimination Mechanisms

2.1 AI-ML-Based Algorithms

Given that AI is an extensive subject, it can be generally defined as a collection of technologies that aim to "render computers do actions better than human brains are capable of". To accomplish this, machines utilize sets of principles known as algorithms. Algorithms are computational commands designed for resolving issues, answering queries, and carrying out specific functions. It can be very straightforward, such as stating that if criteria A is satisfied, then result B ought to take place, or it can be highly complicated, such as integrating numerous sets of standards to construct an entire system. A complicated set of algorithms conducts the process of ML, which is defined as "the capacity of machines to acquire their unique understanding by deriving patterns from unstructured datasets". The functioning of algorithms based on machine learning, however, is of much more serious concern. Technology may bring challenges in and of its own by contributing toward the expansion of discriminatory norms on a wider scale compared to what would be possible with humans [7]. As a result of ML's capacity to discover and understand patterns in any specific dataset, the risk of prejudice is increased. Even if they are not commanded to discriminate, ML's algorithms produce discriminatory results. In addition to automating discrimination, ML techniques bring the unforeseen risk of unintentionally reproducing existing structures of disparity. An ML algorithm designed to determine job-related promotions and relying on previous successful applicants' data could be made gender-blind and still be prejudiced against female applicants, for example, if requested to sort out applicants based on an evaluation of their performance at employment that includes their usual hours of work. In the above instance, women might feel disadvantaged because based on conventional gender roles in society at large, they bear a disproportionate share of the burden of parenting and therefore work fewer hours or even part-time. The difficulty with ML is that algorithms are not only susceptible to transmitting human prejudices in their functioning but also serve as a mechanism for perpetuating disparity patterns that are inherently embedded in the data they compute [8].

2.2 In What Ways May ML algorithms Discriminate

Algorithmic decision-making has been defined as the new "regulatory realm" for equal protection of law and equal opportunity. Numerous entry points exist for discrimination to infect ML algorithms. An algorithm requires two components to function: a code that codifies an issue in arithmetic and data, which is an array of variables that are input from which the machine can acquire knowledge. Many classifications have been suggested to define how both the code and the data can be prejudiced, thereby introducing discrimination into algorithmic functions. Kleinberg recognized discriminatory risks in the result of the choice given to the algorithm, the inputs of data used to teach it what they name "candidate determinants", and the learning process, including the data used to train it [9].

Barocas and Selbst also suggest a five-tiered classification of discrimination's causes [10]:

- The first point of entry relates to the description, or what the researchers refer to as "issue definition", using the construction of "target variables", which are further subdivided into "class labels".
- The second entry point is the instruction data used to teach the algorithm, which may be prejudiced because of how it was collected or the data itself.
- Third entry point is the selecting of appropriate algorithm attributes, i.e., which attributes should be regarded as pertinent for the algorithmic framework.
- The fourth entry point is "proxy prejudiced", in which discrimination is caused by known as "redundant encodings", or "situations in which participation in a protected group appears to be embedded in different data", which the algorithm considers to be associated with the final result. 18 utilizing a non-protected class such as a residence or postal code as a substitute for a protected class such as ethnicity is a standard illustration.
- The fifth and final entry point they describe is the concealment of deliberate discrimination within the algorithm's architecture.

From the perspective of discrimination theory, algorithmic bigotry's origins can be classified into two broad categories of disparity-producing processes. On the one hand, bias, and stereotyping undermine the equal participation of individuals in society. On the other hand, systemic forms of disparity reflect previous prejudice that has been established and codified throughout history. Algorithmic prejudice can further entrench both pervasive and figurative disparities through sustaining prejudiced access to social amenities, e.g., employment, medical services, societal advantages, the use of rights, etc., and by deceiving or obscuring particular sections of the population. Thus, algorithmic prejudice both results from and reinforces hierarchical status beliefs and preconceived notions, as well as established patterns of disparity. From the vantage point of discrimination statute, these two major axes can be used to classify the processes through which discrimination may enter algorithms. The effects of ML algorithms concerning prejudice and data-driven disparities in terms of technology, development, as well as the performative behavior of discrimination are therefore serious and it is necessary to assess whether existing anti-discrimination laws are adequate to avert and recourse them [11].

3 Technical Background

The term "recruiting algorithms" refers to a broad range of computational instruments that are capable of assessing a candidate's suitability for a specific position in the company and provide an index that serves as input to make subsequent recruiting choices. We authors use the terms "ML-based algorithm Algorithms" collectively, even though "algorithms" encompass a far wider range of varied techniques for solving problems. In the case of computerized recruiting, "algorithms" primarily refer to automated ML algorithms [12].

In technical phrases, autonomous ML is an ensemble of algorithms that can modify how they make decisions according to exposure to previous data. In other words, the algorithm "learns" how to arrive at more accurate choices or form more accurate inferences as it is confronted with previously identified and labeled data, a procedure known as learning. Multiple techniques are subsequently employed to make sure that learning produces an Algorithm that can extend to freshly collected data, allowing the algorithm to use patterns learned from training to infer absent characteristics from fresh data that arrive (see Fig. 1). This is referred to as data mining, the technique of discovering correlations and generating statistical understanding from massive data sets [13].

Figure 1 describes the ML algorithm development process in the following three ways: First is a specified algorithm whose underlying reasoning can adapt and "learn" from encounters with annotated data. Then a learning set of annotated data is fed to the Algorithm, which then discovers complicated relations based on the data that are stipulated. Lastly, the Algorithm can then deduce the identifier of data that is unlabeled by drawing from previous data points.

Fig. 1 Machine learning algorithm creation process

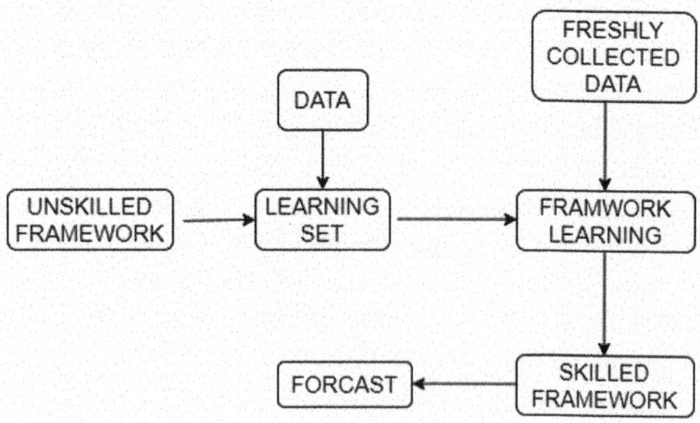

4 Four Common Algorithmic Recruiting Errors that Affect Decisions

4.1 Datasets

Because the majority of ML algorithms are constructed by identifying mathematical trends in data that has been collected, their ability to forecast is constrained by the relationship between variables in the learning dataset. Significantly, modern ML cannot differentiate between cause and effect. This is unfavorable when protected traits, such as gender, have a strong association with adverse results such as unemployment, because it impedes social transformation. If, for example, the data were collected during a time when women's employment was socially disapproved of, the algorithms learned from these data would reflect the gender-based discrimination of that period. Because these trends are completely embedded in the learning data, appropriate dataset utilization is essential to ensure that the predictions are chosen equitably [14].

4.2 Algorithm Framework

Algorithms of ML have particular algorithmic designs and utilize distinct "learning" and "forecasting" techniques. For example, a network of convolutional neural networks [CNN] may be used for detecting images, but distinct structures may be preferred for vocal recognition. Because the Algorithm's requirements and architecture essentially characterize the procedure for making choices, errors in the framework of the Algorithm may result in unanticipated outcomes [15].

4.3 Metrics

In addition to the Algorithm's framework and the dataset used for learning it, we think about its actual efficacy. The term metrics generally relates to the collection of data used to evaluate the efficacy of an Algorithm. In general, the accuracy of an Algorithm is assessed using a variety of error metrics such as false affirmative and false adverse rates used to measure efficiency error. An ML algorithm may be deemed suitable for application in social contexts if a few of these metrics fall within a given margin or if some of these metrics are comparable across various groups of individuals. As with prejudices, however, impartiality has proved to be a challenging concept in algorithmic learning. Equity has been described as a subjective, non-standardized concept. As there is no agreement on the definition of justice, there is a lack of consensus. By establishing a particular definition of equity, however, algorithm developers create a false sense of impartiality, resulting in the "formalism paradox". Extremely challenging to codify in the syntax of mathematics is the complicated geopolitical and historical terrain; as a result, we frequently observe algorithms that neglect essential social realities [16].

4.4 Applications

The Algorithm may also fail if it is deployed in a manner that justifies prevailing prejudices in the recruitment procedure. Due to the unprecedented complexity of obtaining this data, there currently aren't any statutes enforcing a public disclosure of tool use in recruiting. Rather, we have to trust that these recruiting algorithms have been trained with adequate data, can draw the inferences they purport to be equipped to make, and are utilized appropriately [17].

5 Advanced and Legal Techniques for Mitigating Discrimination Against Women

The solution to women's discrimination problems must be clarified and bolstered using a suitable legal system as well as using advanced technologies like AI & ML algorithms in this context. Also, cognitive and psychological shifts are crucial to achieving this objective. In this regard, the alteration of the cultural traits of a community and its structure is crucial. Supervisors who have received specialized training to fight against gender-based prejudice in this discipline can effectively manage gender-based discrimination. The state's rules and laws must be respected and adhered to in every way and at every opportunity by the companies. Everybody should guarantee equitable recruitment and appraisal based on the doctrine of equity and good conscience. For this purpose, special training sessions may be organized. Increased awareness of the issue is required. Several state and international laws prohibit discrimination based on gender as these laws provide numerous protections for women's rights in the workplace [18]. Also, concerning common algorithmic Recruiting errors that affect the decision: Dataset, Algorithm Framework, Metrics, and Applications few questions to guide the hiring software company that uses the automated recruiting AI-ML algorithms technologies are given below [19, 20].

5.1 In the Respect of Dataset

It is extremely challenging to "correct" an algorithm trained on prejudiced data. Consequently, it is crucial to incorporate the general traits of the dataset. This section aims to describe the approaches in which the dataset can reflect human prejudices, as well as the efforts made to reduce this:

- How were data regarding training gathered?
- If labels are linked with the data, what categories of labels exist?
- Was there any test set developed to verify the data?
- Was the distribution of the test set and the learning set comparable?

5.2 *In Respect of Metrics*

To detect the existence of prejudices such as the algorithm's lack of accuracy and its efficacy for distinct groups, it is essential to analyze the metrics encircling the data. This section seeks to demonstrate the efficacy of the algorithm on various protected categories and substantiate efforts to combat prejudice:

- How was the efficacy of the algorithm assessed?
- Why were these metrics selected over others?
- What precautions have been utilized to make sure that algorithms are not prejudiced based on ethnicity, color, religion, sexuality, country of origin, or one's age?
- Which definition of equity or fairness have you chosen to employ in your endeavor to reduce prejudice in the algorithm?

5.3 *Applications*

The majority of possible prejudice arises from recruiting supervisors' selective use of algorithmic outputs while selecting applicants. This section attempts to clarify how the algorithm is implemented in practice and how it interacts with other models:

- Is the algorithm utilized in practice regularly?
- How are the algorithm's outputs utilized in decision-making?
- Are there additional algorithms that are used together with this one?

6 Universal Women's Rights Law

Creating and utilizing technologies rooted in normative conceptions of female actions, personalities, and social positions may promote prejudice against women. CEDAW entered into force in Sep. 1981 to eliminate prejudice against women. Since it came into power, 189 member states have signed it. Article 2 of the Convention outlines the fundamental obligations of the member states regarding the elimination of discrimination. The eliminating of hatred and prejudice based on adverse perceptions of women is addressed in Article 5 of the Convention, requiring Member States to adopt appropriate steps to alter the societal and cultural norms of conduct of males and females, to accomplish the eradication of biases and customary and all other behaviors centered on the notion of being inferior or supremacy of both of the genders or on preconceived notions of roles for male and female [21].

Also, the OHCHR is the primary human rights agency at the UN, and its responsibility includes the treaty-based supervision bodies, which include the CEDAW Committee. The elevation of discrimination based on gender to an issue of concern for the OHCHR is therefore related, indicating that it is an important concern among the different UN human rights conventions such as the Committee on the Rights of the Children and the Committee on the Rights of Individuals with Disabilities. The OHCHR report on gender biases defines preconceived notions as follows: "A stereotype about gender is a broad assumption or assumption about the features or traits that males and females hold or ought to have, or the roles males and females play or should represent. It also notes that "there is an increasing acceptance that gender prejudice represents an important, yet unresolved challenge to the acknowledgment, movement, and pleasure of female rights as human beings" and that gender prejudice is harmful and discriminatory "when it limits women's or men's ability to nurture their skills, pursue their professions, and make decisions about their careers and life". It also observes that both hostile, negative, and appearing prejudices can be detrimental for both males and females [22].

Indeed, the principles set forth by the UN provide the private sector with additional direction on their respect for human rights duties, emphasizing that private players must consider the possibility of humanitarian rights consequences of their offerings and actions [23].

7 Recommendations

After analyzing the different predictive algorithmic models and studies of researchers we propose the following:

a. Cognitive and psychological shifts are crucial in society among all individuals as well as in organizations in the respect of women candidates for achieving the goal of equity and good conscience. More awareness and training programs are needed to be conducted in every organization.
b. Further Advanced research in the field of AI-ML algorithm is required to make sure that ML algorithm should not be corrupted with the prejudiced data that AI is taking for learning. There must be Technology should be used that will automatically eliminate negative data.

8 Conclusion

In the present study, we have outlined the various ways in which the application of machine learning algorithms may lead to discrimination based on gender. Gender-based discrimination in the AI era is a big issue we need to consider with utmost priority. Also, CEDAW & OHCHR agencies are working under the supervision of the UN's guiding principles and standards which we authors noted with respect to international non-discrimination laws. Integrating gender theory, specifically the gender equality concept, into the approach to ML from mathematical codes and data may cause the learning of gender biases; however, the algorithms must be modified otherwise ML will learn corrupted data only; more impartiality and equity are required while serving data to ML algorithm. Also, developers and businesses must reconsider their use and development of ML algorithms. Regarding digital forms of discrimination, the absence of a coherent and exhaustive equality law framework regarding the various grounds of prejudice, such as ethnicity, gender, faith, sexual orientation, and years of age, also appears problematic. In addition, the intricate nature of the human–machine connection may cause the mathematical distinction between "direct and indirect prejudice to obscure". While it is true that ML algorithms increase the likelihood of seeing certain forms of prejudice flourish, it is also important to note that while they do increase the likelihood of seeing certain forms of discrimination thrive, they also decrease the likelihood of seeing other forms of discrimination occur. As a result, the emergence of new technology poses both a threat and an opportunity to advance the goal of equality. When AI & ML algorithm is deployed, it is anticipated that AI will lower instances of gender discrimination in organizations as well as at workplaces for women; however, additional research is required to validate this claim.

References

1. Rahwan I, Cebrian M, Obradovich N, Bongard J, Bonnefon JF, Breazeal C, ... Wellman M (2019) Machine behavior. Nature 568(7753):477-486
2. Loebbecke C, Picot A (2015) Reflections on societal and business Algorithm transformation arising from digitization and big data analytics: a research agenda. J Strateg Inf Syst 24(3):149–157
3. Heilman ME, Caleo S (2018) Gender discrimination in the workplace. The Oxford Handbook of workplace discrimination, pp 73–88
4. Crawford K, Calo R, There is a blind spot in AI research. Nature 538(7625):311–313
5. Zerbes I (2020) A Union that strives for more. EuCLR European Crimi Law Rev 10(3):253–254
6. Leavy S (2018) Gender bias in artificial intelligence: the need for diversity and gender theory in machine learning. In: Proceedings of the 1st international workshop on gender equality in software engineering, May 2018, pp 14–16
7. Nuseir MT, Al Kurdi BH, Alshurideh MT, Alzoubi HM (2021) Gender discrimination at the workplace: do artificial intelligence (AI) and machine learning (ML) have opinions about it? In: The international conference on artificial intelligence and computer vision, May 2021. Springer International Publishing, Cham, pp 301–316

8. Heaton JIG, Bengio Y, Courville A (2016) Deep learning: The MIT Press, 800 pp, ISBN: 0262035618. Genetic programming and evolvable machines, vol 19(1–2), pp 305–307 (2018)
9. Kim PT (2016) Data-driven discrimination at work. Wm Mary L Rev 58:857
10. Kleinberg J, Ludwig J, Mullainathan S, Sunstein CR (2018) Discrimination in the age of algorithms. J Legal Anal 10:113–174
11. Barocas S, Selbst AD (2016) Big data's disparate impact. California Law Rev 671–732
12. Sloane M, Moss E (2019) AI's social sciences deficit. Nat Mach Intell 1(8):330–331
13. Naim I, Tanveer MI, Gildea D, Hoque ME (2015) Automated prediction and analysis of job interview performance: the role of what you say and how you say it. In: 2015 11th IEEE international conference and workshops on automatic face and gesture recognition (FG), vol 1. IEEE, May 2015, pp 1–6
14. El Mrabet MA, El Makkaoui K, Faize A (2021) Supervised machine learning: a survey. In: 2021 4th international conference on advanced communication technologies and networking (CommNet). IEEE, Dec 2021, pp 1–10
15. Obermeyer Z, Emanuel EJ (2016) Predicting the future—big data, machine learning, and clinical medicine. N Engl J Med 375(13):1216
16. Küster D, Krumhuber EG, Steinert L, Ahuja A, Baker M, Schultz T (2020) Opportunities and challenges for using automatic human affect analysis in consumer research. Front Neurosci 14:400
17. Selbst AD, Boyd D, Friedler SA, Venkatasubramanian S, Vertesi J (2019) Fairness and abstraction in sociotechnical systems. In: Proceedings of the conference on fairness, accountability, and transparency, Jan 2019, pp 59–68
18. Stevenson M (2018) Assessing risk assessment in action. Minn L Rev 103:303
19. Alyammahi A, Alshurideh M, Kurdi BA, Salloum SA (2020) The impacts of communication ethics on workplace decision making and productivity. In International conference on advanced intelligent systems and informatics. Cham: Springer International Publishing, Sept 2020, pp 488–500
20. Alameeri K, Alshurideh M, Al Kurdi B, Salloum SA (2021) The effect of work environment happiness on employee leadership. In: Proceedings of the international conference on advanced intelligent systems and informatics 2020. Springer International Publishing, (2021)
21. Ludwig J, Mullainathan S (2021) Fragile algorithms and fallible decision-makers: lessons from the justice system. J Economic Perspect 35(4):71–96
22. Session TS (2005) Convention on the elimination of all forms of discrimination against Women
23. Adams R, Loideáin NN (2019) Addressing indirect discrimination and gender stereotypes in AI virtual personal assistants: the role of international human rights law. Cambridge Int Law J 8(2):241–257

IoT-Based Health and Safety System for Manual Scavengers

Vasundhara Bhatt, Poonam Rawat, Anil Kumar Dixit, Sagar Saxena, Gagandeep Kaur, and Monika Kothiyal

Abstract The purpose of this research study is to investigate the potential applications of Internet-of-Things (IoT) technology in the process of creating a health and safety system for manual scavengers, with a particular emphasis on the efforts taken by the United Nations Development Programme (UNDP). Manual scavenging, a profession that is both dangerous and degrading, is still practiced in many regions of the globe, even though it poses significant threats to the health and safety of the people who engage in it. In this article, we will study the difficulties that are encountered by manual scavengers and evaluate how IoT-based solutions might successfully solve these difficulties. To be more specific, it will investigate the incorporation of Internet-of-Things devices, such as environmental monitoring systems, wearable sensors, and real-time data analytics, to improve occupational well-being, reduce the risk of health hazards, and offer prompt assistance in the event of an emergency. In addition, the paper will conduct an analysis that is critical of the UNDP's efforts in promoting sustainable development and combatting manual scavenging. It will do this by emphasizing effective case studies and policy frameworks that have been implemented in various locations. This research aims to contribute to the development of comprehensive and effective IoT-based health and safety systems for manual scavengers by synthesizing research findings, technological advancements, and UNDP

V. Bhatt · P. Rawat · A. K. Dixit · S. Saxena (✉)
Law College Dehradun, Uttaranchal University Dehradun, Dehradun, India
e-mail: adv.sagarsaxena@gmail.com

V. Bhatt
e-mail: vasundharabhatt9@gmail.com

P. Rawat
e-mail: deanlaw@uumail.com

A. K. Dixit
e-mail: anilkdixit44@gmail.com

G. Kaur
School of Law, University of Petroleum & Energy Studies, Dehradun, India
e-mail: gkaur@ddn.upes.ac.in

M. Kothiyal
ICFAI Law School Dehradun, Dehradun, India

S. Dutta et al. (eds.), *Cyber Intelligence and Information Retrieval*, Lecture Notes in Networks and Systems 1025, https://doi.org/10.1007/978-981-97-3594-5_37

interventions. The end goal is to improve the quality of life for these marginalized communities and promote inclusive sustainable development.

Keywords IoT · Health and safety system · UNDP · Manual scavenger

1 Introduction

Scavenging human waste by hand involves physically cleaning, transporting, disposing of, or managing waste from dry latrines and sewers. This activity is known as manual scavenging. Even though the practice has been illegal in India since 1993, it is nevertheless carried out there, and workers are used to physically clean things like septic tanks, sewers, and so on. These workers are exposed to a harsh and dangerous environment on the job, and as a result, their lives are in danger of being cut short by disease, accidents, or even death. Researchers have suggested Internet-of-Things (IoT)–based health and safety solutions for manual scavengers as a solution to this problem [1].

These systems make use of Internet-of-Things devices to obtain data about the health and safety of workers and then wirelessly communicate this data to control canters for effective monitoring and control. The control centers gather data on the workers' health metrics as well as environmental factors and then inform both the workers and the authorities about any possible hazards [2]. Because of this, the frequency of accidents and the number of lives lost are both reduced, as are the costs of company responsibility and insurance [3]. One example of this kind of system that has been suggested is the Internet-of-Things -based health and safety system for manual scavengers and miners, which was developed by Deepa R, an assistant professor at the Chaitanya Bharathi Institute of Technology (A) in Hyderabad, Telangana, India [4]. The system gathers and sends data from a variety of health and environmental sensors to the control room wirelessly (point-to-point connection), and it also sends the data to the Thing Speak IoT Cloud so that it may be monitored and reported. On the other hand, the control room is the location that is in charge of receiving the data from the sensors and then sending out SMS warnings to the relevant authorities so that they can maintain control over the whole operation [5].

In today's world, every industry is changing as a direct result of developments in technology. Concurrently, industrial control and safety systems are transitioning from isolated standalone systems to ubiquitous systems that are intercommunicable and share a significant amount of data. As a result of networking all of the things and devices that are engaged in a given activity or company, the Internet of Things (IoT) is having an exponentially increasing influence throughout the whole business spectrum. The Internet of Things (IoT) refers to the collection of people, things, times, locations, services, and networks that can communicate with one another [6]. Thus, in the pursuit of integrating national value creation with sustainable development to attain free from manual scavenging practices with the IoT, we aim to explore the following research objectives:

A. Examining the IoT devices is necessary to monitor manual scavenging workers in adherence to the SDGs and international laws and ensure accountability for its progress.
B. What are the legal and policy frameworks for mitigating the practices of manual scavenging and how effectively are IoT devices facilitating a value-creation process that is in line with the achievement of the SDGs?

The paper consists of seven main sections. The first part begins with a short introduction about manual scavenging, SDG, and IoT. The second section states SDGs which are working toward the elimination of manual scavenging practices and the rights of humans. The third section investigates manual scavenging as a critical examination. The fourth section describes IoT technology and its application in occupational safety. The fifth section states policy framework legal perspectives and environmental sustainability prohibiting manual scavenging. The sixth section provides recommendations from the author's side. Lastly, the seventh section concludes with the conclusion.

2 The Sustainable Development Goals (SDGs)

The UN created the (SDGs) in 2015 to address social, economic, and environmental issues. These are the "Global Goals for Sustainable Development". The third Sustainable Development Goal (SDG) promotes healthy lifestyles and well-being for all ages. Goal 6 of the SDGs says that everyone deserves clean water and drainage. These goals are especially important for manual scavengers' health and safety systems because the activity is cruel and risky [7]. Manually cleaning, executing, and discarding human waste are waste extraction. This is called "manual scavenging", and marginalized groups do it to survive. Though illegal in most of these countries, this practice is popular in many developing nations. Manual scavengers risk exposure to harmful substances, illnesses, and gases. They often work without hygiene or protective gear, increasing their risk of injury and disease [8]. To achieve the (SDGs) for manual scavenger health and safety, coordinated measures and interventions at all levels are needed. Following Sustainable Development Goal goals are especially relevant to this issue [9]:

I. SDG 3: Good Health and Well-Being [10]
Target 3.3 strives to eliminate fatalities that may have been avoided and lessen the impact of existing illnesses. This goal may be connected to the need for better health outcomes for those who engage in manual scavenging, such as the reduction of occupational dangers and disease exposure.
II. SDG 6: Clean Water and Sanitation [11]
The attainment of universal access to sanitation and hygiene services that are both sufficient and equitable is the primary emphasis of Target 6.2. This objective directly addresses the need to offer safe and dignified alternatives to

manual scavenging, such as the development of mechanical methods for waste management and the improvement of sanitary infrastructure.

III. SDG 8: Decent Work and Economic Growth [12]
Protecting workers' rights and fostering workplaces that are safe and secure are the primary focuses of Target 8.8. This objective is critical for those who engage in manual scavenging because it helps them advocate for the abolition of dangerous activity and ensures that communities who are impacted have access to alternative employment options.

IV. SDG 10: Reduced Inequalities [13]
Target 10.2 emphasizes the need to encourage the social, economic, and political inclusion of all people while also working to empower them. To address the health and safety problems of manual scavengers, active measures to eliminate inequality and guarantee that they are fully integrated into society are required. Manual scavengers often belong to groups that are disadvantaged and socially excluded.

V. SDG 16: Peace, Justice, and Strong Institutions [14]
Target 16.3 emphasizes the need to promote the rule of law and ensure that all people have equitable access to justice. To put an end to the practice of manual scavenging, there has to be strict enforcement of laws that ban the activity, and individuals responsible for continuing the practice need to be held accountable.
To make progress toward these Sustainable Development Goals (SDGs), it is essential to address the concerns of manual scavengers on their health and safety comprehensively. This involves bringing awareness to the issue, offering education and training, promoting other economic options, upgrading sanitary facilities, and implementing stringent legal measures to prevent the practice of manual scavenging. Securing the health and well-being of manual scavengers and making progress toward the Sustainable Development Goals (SDGs) will need concerted efforts on the part of governments, organizations representing civil society, and members of the international community working together.

3 Manual Scavenging: A Critical Examination

Many countries still practice manual scavenging, which involves cleaning septic tanks, sewers, and open drains by hand. This degrading job poses serious health and safety dangers. Manual scavenging involves hand-cleaning sewage tanks, drainage systems, and uncovered sewers. Scavenging by hand is examined in this section, including its meaning, history, health and security issues, and cultural and moral consequences. Highlighting these elements may raise awareness of the challenges that manual scavengers confront and the urgent need for new solutions to this significant problem. Solving this serious issue requires success [15]. Manual scavenging involves removing waste and trash from toilets and restrooms. Minorities often do this duty. Manual scavenging's roots can be traced to deeply ingrained cultural and societal practices; therefore, understanding its past is crucial. Understanding hand-scavenging's cultural and historical roots is crucial. Throughout human history, certain populations were discriminated against and forced to scavenge by hand [16]. The health and safety risks of manual scavenging are one of the biggest challenges. Manual scavengers risk exposure to harmful chemicals and infections, as well as frequent mishaps. These people suffer dangers because of a lack of protection machinery, personal safety equipment, and training. Manual scavengers face immediate and future physical hazards, including respiratory disorders, skin problems, and mental agony. Manual scavenging affects culture and morality. Society often discriminates against those who labor in this sector, marginalizing and excluding them. Laborious garbage digging violates fundamental human rights like life, dignity, and equal opportunity. The gender-specific aspects of manual scavenging must be examined because women bear the brunt of obstacles and face extra overlapping impediments [17].

3.1 Definition and Historical Context

Defining manual scavenging: Human excreta that is cleaned, transported, disposed of, or handled manually in an unsanitary lavatory, exposed drain, sewer, septic tank, or pit constitutes manual scavenging in other words simply scavenging by hand and its numerous manifestations, including the manual cleaning of sewage tanks, drainage systems, and uncovered drainage. Historical background: Examine the beginnings of the practice of manual scavenging and its past prominence in various areas and civilizations. Cultural and caste-based dimensions: Examine the socioeconomic and cultural variables that have led to the continuing practice of manual scavenging in certain regions [18].

3.2 Health and Safety Hazards Faced by Manual Scavengers

Occupational health risks: Examine the mental disorders, and breathing health risks linked to manual scavenging, including exposure to toxic fumes, infectious agents, and injuries to the body. Long-term health consequences: Study the lasting medical consequences that manual scavengers experience, such as breathing illnesses infections of the skin, and mental disorders. Lack of protective measures: Emphasize the lack of adequate security supplies, protective apparel, and instruction, worsens the dangers that manual scavengers experience [19].

3.3 Social and Ethical Implications

Stigma and social exclusion: Understand the cultural shame and prejudiceness that manual scavengers face., which leads to their being excluded and marginalized from the general population.

Violation of human rights: Manual scavengers' assault on fundamental freedoms, such as their entitlement to exist, respect, and equal chances, is looked at. Gender dimensions: Examine the gendered elements of scavenging by hand, such as the disproportionate strain placed on women and the intersecting problems they experience. Ethical considerations: Examine the legal consequences of employing people in dangerous and dehumanizing jobs and the moral obligation that society has to tackle this problem. By analyzing the definition, context of history, health and safety risks, and the societal and moral consequences of scavenging by hand, this part aims to offer a comprehensive comprehension of the variety of difficulties encountered by people in this work [20].

4 IoT Technology and Its Application in Occupational Safety

The creation of IoT technology has led to the introduction of new technologies with a chance to enhance occupational safety in an array of industries, such as manual scavenging. The IoT is a network of interconnected devices that are equipped with the software, devices, and connectivity features required to collect and exchange data in real-time. Regarding workplace security for manual scavengers, the IoTs have enormous potential for improving working conditions while minimizing health risks to workers. For most IoT-based solution implementations, integrating ubiquitous detectors, surveillance systems, and immediate analysis of data is essential [2]. It is possible to utilize wearable sensors for tracking the bodily functions of manual scavengers. This category includes heartbeat rate, the temperature of the body, and levels of oxygen. These updates are capable of helping identify indicators of anxiety, exhaustion, and an array of illnesses, allowing for immediate intervention and guidance. Moreover, wearables can independently track the motions and positions of scavengers, guaranteeing their security and providing them to reach out for guidance in the event of a crisis or accident. Sewage tanks, sewage lines, and drainage systems may contain dangerous gases that can be monitored by IoT-enabled ecological monitoring systems. These gases include noxious vapors and asphyxiating gases. By providing early warnings and alarms, these systems can prevent possible medical issues by enabling prompt evictions and remedies [5].

Real-time evaluation of data is a crucial element of IoT-based workplace security systems. Through the real-time analysis of enormous amounts of data, it is possible to identify patterns and trends, allowing for the prevention of safety risks at work. This entails anticipating potentially hazardous situations, optimizing work schedules, and implementing preventative measures so that manual scavengers can operate without risk to their well-being. The incorporation of IoT technology into the traditional scavenging process offers an array of advantages, including improved well-being and security, higher efficiency, decision-making based on data, and the promotion of private authority through the availability of data and assets. However, for IoT to be successfully implemented in manual scavenging, multiple challenges must be overcome. These issues include cost and availability, technological infrastructure requirements, data confidentiality and safety, and involvement of stakeholders [21]. Regarding safety at work for these individuals, efforts to leverage IoT technology should prioritize inclusion, affordability, as well as cooperation between government departments, entities, and the manual scavenger society. Utilizing the potential of the IoT can significantly improve the safety at work of manual scavengers, which will contribute to their mental health as well as their freedoms and sense of worth.

4.1 Outline of IoT

The IoT is an internet of connected gadgets equipped with software, sensor technology, and connectivity features that enable them to gather and share data. By facilitating continuous tracking, analysis of information, and automated processes, IoT technology has transformed numerous industries, including safety at work. In the wake of manual scavenging, IoT can significantly enhance safety at work and the psychological health of those involved in this dangerous work [22].

4.2 IoT-Based Safety at Work Systems

In the context of manual scavenging, IoT-based methods to improve safety at work comprise an array of devices and applications. Typically, these techniques incorporate ubiquitous detectors, surveillance systems, and real-time information analytics [2]:

- ***Wearable Detectors***:
  The bodily functions of manual scavengers, such as pulse rate, temperature of the body, and levels of oxygen, can be tracked using sensors worn on the body. These sensors collect statistics in real-time that can be examined to identify indications of tiredness, stress, or medical issues. In addition, ubiquitous instruments can monitor the location and movement of manual scavengers, allowing for immediate help in the event of injuries or situations of emergency [23].

- ***Ecological surveillance systems***:
  Ecological surveillance systems equipped with IoT detectors are capable of identifying and measuring hazardous fumes and deadly gases in sewage tanks, sewage

lines, and drainage. These detectors may supply manual scavengers and management with preliminary warnings and signals, allowing for opportune escape or assistance to avoid adverse health effects [24].

- ***Real-time data analytics***:
  IoT technology allows the gathering and continuous evaluation of enormous quantities of data. By utilizing sophisticated data analysis methods, developments and patterns can be recognized, allowing preventative steps to be taken to minimize risk at work. To guarantee the protection of manual scavengers, data analysis in real-time can aid in predicting dangerous situations, improving work schedules, and instituting safeguards [25].

4.3 IoT in Manual Scavenging: Benefits and Drawbacks

There are numerous advantages associated with using IoT technology in the form of manual scavenging, but there are also particular obstacles that must be resolved [26]:

- ***Benefits of IoT implementation***:

A. IoT-based systems can substantially improve the protection and health of manual scavengers by offering continuous surveillance, alerts in advance, and prompt assistance during situations of emergency.
B. Improved efficiency: IoT technologies may maximize workflows, minimize the need for human involvement, and increase the general efficiency of traditional scavenging activities.
C. Statistics-driven decisions: IoT-generated information and analyses may offer useful information for choice-making, allowing for improved utilization of resources, evaluation of risks, and avoidance.
D. Empowering and consciousness: IoT solutions may encourage manual scavengers by granting them access to the required data, instruction, and resources for their safety at work and freedom.

- ***Problems with IoT execution***:

A. Costs and inclusivity: To assure both affordability and availability for the manual scavenger society as a whole inexpensive IoT solution must be created.
B. Private and data protection: The acquisition and use of confidential private information through IoT devices necessitate stringent security and confidentiality precautions.
C. Stakeholders' participation: Effective implementation of IoT-based workplace security systems requires the participation of government departments, entities, and the manual scavenger society.

Utilizing IoT technology, manual scavengers' occupational safety can be significantly enhanced. The deployment of IoT-based solutions must be accessible, and economical, and emphasize the health and rights of those who collect waste manually.

5 Policy Framework and Legal Perspectives

The regulatory and legal implications of manual scavenging must be examined to solve the problem and protect workers. National efforts are underway to criminalize hand scavenging, punish those who do it, and encourage alternate survival methods. International conventions like the ILO urge efforts to end forced labor, including hand scavenging. The UNDP helps end manual scavenging by advancing legislation, skills, and cooperative arrangements. Despite this, enforcement limitations, social barriers, rehabilitative requirements, information gaps, and global coordination remain [10]. Operational policy creation must address the above issues, implement and enforce current regulations, raise awareness, and promote global collaboration to end manual scavenging while protecting the rights and equality of those affected. Legal Actions, International Agreements were made by the all the signatories of UN are in mission to completely eradicate this practice.

5.1 Legislation and International Conventions are Needed to Solve Manual Scavenging and Ensure Worker Safety. This Section Examines Manual Scavenging Laws and International Agreements [27]

- State law: Detailing the rules and laws enacted by legislatures that forbid manual scavenging, punish perpetrators, and encourage restoration and alternate sources of income for those affected.
- International standards: Discuss pertinent international agreements, which include the ILO conventions, which promote the abolition of any kind of enslaved labor, including manual scavenging.
- The UNDP's Role in Policy Advocacy and Execution: The UNDP is instrumental in advocating for changes in policy and carrying out measures that ban manual scavenging. This section analyzes UNDP's unique efforts:
- Legislative advocacy: Assessing the UNDP's attempts to raise consciousness about the problem of manual scavenging and encourage changes in policy at all levels of government.
- Training and skill creation: Discuss the UNDP's efforts aimed at enhancing the abilities and knowledge of manual scavengers so that they may explore other sources of income and break the downward spiral of scavenging by hand.

- Coordinated relationships: Emphasizing the UNDP's partnerships with governments, nonprofit groups, and other parties in carrying out initiatives and programs aimed at eradicating the practice of manual scavenging and assuring safety at work.

5.2 Obstacles and Suggestions for Policy Formation

While policy structures have made progress in resolving manual scavenging, numerous obstacles remain. This section examines the obstacles and gives suggestions for effective policy construction:

- Analyzing the difficulties related to ensuring the compliance and execution of the current regulations and laws, such as tracking, logging, and responsibility.
- Societal and traditional barriers: confronting the societal and cultural obstacles that impede the successful execution of regulations, such as caste-based stigma and prejudice.

5.3 Restoration and Alternate Source of Income

Examine the challenges of providing manual scavengers with complete restoration and reliable alternate sources of income, such as training for skills, opportunities for schooling, and monetary assistance.

5.4 Consciousness and Indoctrination [28]

Highlighting the significance of education efforts and training initiatives directed at all participants, including legislators, police agencies, and members of the public. International cooperation: Suggested for increased foreign interaction and partnership for sharing efficient procedures, exchanging knowledge, and collaboratively approaching the international problem of manual scavenging. This section offers insights into the regulatory landscape encircling the practice of manual scavenging by examining legislation, international treaties, and the contribution of the UNDP in advocating for and implementing policy. In addition, it identifies the obstacles that must be surmounted and makes proposals for the creation of efficient strategies that thoroughly address the issue while guaranteeing the protection, respect, and rights of manual scavengers.

6 Recommendations

After examining the study or research of distinguished scholars or researchers:

A. IoT gadget utilizing various sensors, such as an ultrasonic detector for determining the level of waste, a gas detector to identify toxic gases, a pulse detector, a cardiac detector, a notification system, and a cooling device must be provided as a solution for a variety of problems replacing manual scavenging.
B. Implementing or enforcing government legislations or laws guided by SDGs must be more strictly implemented by administration agencies in the remote areas of the countries so that scavenger people will see another source of income.

7 Conclusion

The study concluded that IoT-based safety and health solutions might benefit manual scavengers, focusing on UNDP efforts. This study explores how IoT technology could reduce manual laborers' work-related dangers and improve their health. Wearable sensors can follow manual scavengers' biological functions using IoT technology, delivering real-time data for early medical diagnosis and action. Ecological surveillance systems with IoT devices can detect harmful gases. These gadgets protect volunteers by warning and notifying them. Database-driven real-time data analytics optimizes operations and asset allocation. The analysis stressed the UNDP program's role in policymaking. UNDP has helped raise awareness of manual scavenging's dangers and promote revolutionary norms, functions, and partnerships. Their efforts have supported hand-scavengers, given them other means of existence, and protected their liberties and humanity. However, further obstacles must be overcome to implement IoT-based wellness and security solutions for hand-scavengers. Cost-effectiveness, technology infrastructure, data confidentiality and safety, and stakeholder involvement are some of the obstacles. Different levels of government, organizations, and manual dumpster divers must work together to solve these issues. Conclusion: IoT-based health and safety solutions have many untapped potential for increasing manual scavengers' workplace safety. Positive progress has been aided by the UNDP's broad program development and implementation. By addressing the challenges and ensuring thorough and sustainable solutions, we can eliminate manual scavenging and create a safer, cleaner environment for everyone.

References

1. Ramadhin SPK, Anand S, Aishwarya R, Yogitha R (2021) Smart drainage and health monitoring system of manual scavenger using IoT. In: 2021 5th international conference on intelligent computing and control systems (ICICCS), pp 412–416. IEEE. J Clerk Maxwell, a treatise on electricity and magnetism, 3rd ed, vol 2. Oxford: Clarendon, 1892, pp 68–73, May 2021

2. Shanthi KG, Sesha Vidhya S, Keerthi B, Niviya BR, Manimegalai S, Nisha A (2006) IoT-based portable artificial electronic olfactory system for the safety of manual scavengers
3. Mohanraj S, Aravindh R, Arun A (2023) IoT based system for manhole monitoring and management. In: 2023 9th international conference on advanced computing and communication systems (ICACCS), Mar 2023, vol 1. IEEE, pp 1357–1361
4. Deepa R (2020) IoT-based health and safety system for manual scavengers and miners. Int Res J Eng Technol 7:6401–6407
5. Nandini Y, Lakshmi KV, Srujan TIS, Yasheswi M, Jagadish KS (2023) Design of real-time automatic drainage cleaning and monitoring system using IoT. In: 2023 7th international conference on computing methodologies and communication (ICCMC), Feb 2023. IEEE, pp 1291–1296
6. Rajalakshmi B, Saranya S, Surya S, Bathool AS (2022) Electronic nose system for protecting workers in the sewage system. In: 2022 second international conference on artificial intelligence and smart energy (ICAIS), Feb 2022. IEEE, pp 1689–1692
7. Hák T, Janoušková S, Moldan B (2016) Sustainable development goals: a need for relevant indicators. Ecol Ind 60:565–573
8. Allen C, Metternicht G, Wiedmann T (2018) Initial progress in implementing the Sustainable Development Goals (SDGs): a review of evidence from countries. Sustain Sci 13:1453–1467
9. Howden-Chapman P, Siri J, Chisholm E, Chapman R, Doll CN, Capon A (2017) SDG 3: ensure healthy lives and promote wellbeing for all at all ages. A guide to SDG interactions: from science to implementation. International Council for Science, Paris, France, pp 81–126
10. Bennich T, Weitz N, Carlsen H (2020) Deciphering the scientific literature on SDG interactions: a review and reading guide. Sci Total Environ 728:138405
11. Ortigara ARC, Kay M, Uhlenbrook S (2018) A review of the SDG 6 synthesis report 2018 from an education, training, and research perspective. Water 10(10):1353
12. Rai SM, Brown BD, Ruwanpura KN (2019) SDG 8: Decent work and economic growth–A gendered analysis. World Dev 113:368–380
13. Úbeda F, Forcadell FJ, Aracil E, Mendez A (2022) How sustainable banking fosters the SDG 10 in weak institutional environments. J Bus Res 146:277–287
14. McDermott CL, Acheampong E, Arora-Jonsson S, Asare R, de Jong W, Hirons M, ... Setyowati A (2019) SDG 16: peace, justice and strong institutions–a political ecology perspective
15. Shahid M (2015) Manual scavenging: Issues of caste, culture and violence. Soc Change 45(2):242–255
16. Kumar K, Deepalatha R (2022) Indian inclusion efforts relevance to scavenger community: a cross-sectional study in Haryana. Contemp Voice Dalit 14(1):36–52
17. Wankhede A (2021) The legal defect in the conditional prohibition of manual scavenging in India. Contemp Voice Dalit 2455328X211047730
18. Ziyauddin KM (2022) Unread history of an erstwhile manual scavenging community: from being an urban to a marginalized. J Soc Incl Stud 8(1):115–133
19. Rangamani S, Bheemappa KANNAMEDI, Obalesha RG (2015) Health issues of sanitation workers in a town in Karnataka: findings from a lay health-monitoring study. Natl Med J India 28(2):70–73
20. Pradhan S, Mittal A (2020) Ethical, health, and technical concerns surrounding manual scavenging in urban India. J Public Health 28:271–276
21. Chandrasekaran R, Mylsami G, Prasath M (14) Smart septic tank cleaning machine. Int J Early Childhood Special Ed 14(5)
22. Sankari S, Shwetha S, Sandhiya S (2021) Advanced sewer clearance and toxin indicator
23. Singh AR, Jalva MHS, Vanmathi B, Vyshali S (2014) An automated sewage line block detection and removal system
24. Tentzeris MM, Georgiadis A, Roselli L (2014) Energy harvesting and scavenging [scanning the issue]. Proc IEEE 102(11):1644–1648
25. Costa F, Genovesi S, Borgese M, Michel A, Dicandia FA, Manara G (2021) A review of RFID sensors, the new frontier of Internet of things. Sensors 21(9):3138
26. Ng IC, Wakenshaw SY (2017) The internet-of-things: review and research directions. Int J Res Mark 34(1):3–21

27. Stankovic JA (2014) Research directions for the Internet of Things. IEEE Internet Things J 1(1):3–9
28. Albreem MA, Sheikh AM, Alsharif MH, Jusoh M, Yasin MNM (2021) Green Internet of Things (IoT): applications, practices, awareness, and challenges. IEEE Access 9:38833–38858

Detection of Deceptive Hotel Reviews Through the Application of Machine Learning Techniques

Md Jakir Hossain, Md Tanvir Chowdhury, Habibur Rahman, Md Shohrab Hossain, Shabrina Akter Shara, Tashfia Choudhury, Maleha Israt Chowdhury, Tasnim Israk Synthia, and Abu Talha

Abstract In the contemporary digital landscape, online reviews play a pivotal role in shaping tourists' decisions, leveraging the pervasive influence of the internet. However, discerning the authenticity of reviews has become an intricate challenge, blurring the line between genuine and deceptive evaluations. This ambiguity provides fertile ground for individuals seeking to exploit the uncertainty, as witnessed in the proliferation of false reviews and the spread of unfavorable rumors through social media platforms. The consequences are profound, with travelers and hosts alike navigating a landscape where traditional verification systems prove insufficient, leaving consumers vulnerable to misinformation. To address this pressing issue, we propose

M. J. Hossain (✉) · M. T. Chowdhury · H. Rahman · S. A. Shara · T. Choudhury · M. I. Chowdhury · T. I. Synthia · A. Talha
Department of Computer Science and Engineering, East West University, Dhaka, Bangladesh
e-mail: mdjakirhossen13@gmail.com

M. T. Chowdhury
e-mail: mdtanvirchowdhury015@gmail.com

H. Rahman
e-mail: habibresearchmail@gmail.com

S. A. Shara
e-mail: shabrinashara98@gmail.com

T. Choudhury
e-mail: tashfia1998@gmail.com

M. I. Chowdhury
e-mail: maleha.ick@gmail.com

T. I. Synthia
e-mail: sinu02114@gmail.com

A. Talha
e-mail: mtalha147258@gmail.com

M. S. Hossain
Department of Textile Engineering, Ahsanullah University of Science and Technology, Dhaka, Bangladesh
e-mail: mdshohrab1991@gmail.com

S. Dutta et al. (eds.), *Cyber Intelligence and Information Retrieval*, Lecture Notes in Networks and Systems 1025, https://doi.org/10.1007/978-981-97-3594-5_38

a novel machine learning-based approach. Our strategy employs classifiers such as Random Forest, Support Vector Machine, and Decision Tree, each exhibiting robust validation accuracy values of 92.19%, 87.19%, and 86.56%, respectively. This innovative framework serves as a powerful tool to distinguish between authentic and fraudulent reviews, providing a more reliable foundation for decision-making in the travel sector. By leveraging the capabilities of machine learning, we aim to create a system that not only identifies deceptive practices but also contributes to restoring trust in online reviews. As the travel industry contends with the challenges posed by the digital era, this approach offers a promising avenue for mitigating the impact of false information and enhancing the overall integrity of the review ecosystem.

Keywords Online reviews · Fake reviews · Tourism · Machine learning · Classifier · Validation accuracy

1 Introduction

Fake reviews' detection refers to reviews that are full of alteration and deception of information and spread without having any authenticated source that is needed to be detected. When a customer tries to book a hotel, reviews are the main source for their information. They read the reviews for the opinion of another guest on the hotel service. Depending on that review, they decide as to the book this room or not. If they saw a positive review, they probably book the room. But sometimes they saw this fake review, they become confused, and they can't figure out if this hotel will be good for them or not. Since reviews are important for sharing authentic feedback about negative and positive service. But in that case, when someone tries to mislead this information, we considered these reviews be labeled as fake. This misleading information costs certain services of a nation. Many ways have been used to detect such reviews to avoid spreading and misleading. This could not be done as it is still dependent on human manual detection. Moreover, social media has become a fast-growing thing. It has become quite easy to spread any fake reviews rapidly. It has become quite impossible to detect which reviews are real and which are fake. In such a scenario, machine learning approach to detect fake reviews and labeling is the only approach that can reduce this problem. Initially fake review detection was introduced by Jinal et al. [1]. The purpose behind the proposed method lies in addressing the pervasive issue of fake reviews, which undermines the integrity of online information, particularly in the hotel-booking domain. Because decision-makers are depending more and more on consumer reviews, these platforms' credibility is under risk due to the proliferation of false information. The goal is to create a powerful machine learning-based system that can distinguish between real and fraudulent evaluations, improving the trustworthiness of online reviews and protecting the interests of customers and service providers alike.

2 Literature Review

In [8] research paper presents a novel approach that employs machine learning techniques to detect fraudulent online reviews. The proposed method incorporates a combination of review content analysis and reviewer behavior analysis as key features. The study utilizes an authentic dataset of restaurant reviews from Yelp to showcase the superior performance of the K-Nearest Neighbors (KNN) algorithm compared to other classifiers. The KNN algorithm achieves an impressive F-score of 82.40%. The incorporation of behavioral characteristics results in a 3.80% increase in F-score, providing significant insights for addressing the issue of deceptive reviews in the realm of E-commerce. This study [9] aims to examine the escalating problem of fraudulent reviews in the realm of electronic commerce. It proposes the implementation of an advanced system that leverages n-grams and sentiment scores to effectively identify and flag such deceptive reviews. The study utilized a conventional dataset of hotel reviews and applied supervised machine learning methods. The results were notable, as the Support Vector Machine (SVM) and Random Forest models achieved accuracies of 93% and 95%, respectively. These performances surpassed the effectiveness of previously employed approaches. In [10] research paper explores the significant impact of online reviews on the success of businesses and the obstacles presented by fraudulent reviews. Using natural language processing techniques, this research investigates a set of 15 linguistic characteristics to differentiate between fraudulent and authentic reviews. The results of the study suggest that counterfeit reviews exhibit a higher degree of redundancy, increased frequency of pauses, and longer sentence structures. Machine learning algorithms demonstrate a notable level of precision in the identification of fraudulent reviews through the utilization of linguistic features. In the contemporary era of digital connectivity, the practice of online shopping has experienced a substantial upsurge, wherein the evaluations and assessments of products by consumers have emerged as a pivotal factor in shaping their purchasing decisions. The identification of fraudulent reviews is a significant field of study, which can be classified based on the data elements involved, such as the review text and metadata, as well as the various approaches employed for detection, including rule-based, graph-based, machine learning-based, and deep learning-based methods. Academic researchers frequently utilize widely used datasets in their investigations within this particular field [11]. This study examines the utilization of machine learning methodologies to analyze consumer sentiment in online reviews within the hospitality and tourism sector. By undertaking a systematic literature review, this study assesses the current body of research, identifies areas that have not been adequately explored, and provides valuable insights into the potential of this collaboration. This study enhances the comprehension and potential advancements of sentiment analysis within this particular context [12]. This study examines the significant impact of online reviews on business performance. The system utilizes natural language processing techniques and examines 15 distinct linguistic characteristics to differentiate between counterfeit and authentic reviews. The results of the study indicate that fraudulent reviews demonstrate a higher degree of redundancy, longer pauses, and increased sentence length. Machine learning algorithms demonstrate a notable level of accuracy in discerning fraudulent reviews by leveraging linguistic cues [13]. This [14] study shows the how CNN network works in characters and we also implemented it in our research.

3 Methodology

As detecting fake hotel reviews is so important for any country, we are working with some different perspectives here. First of all, we need to collect data for preprocessing. Then we train those data with different models. Figure 1 shows the method.

Data Collection: The initial step involves gathering a substantial dataset of hotel reviews. This dataset is crucial for subsequent analysis and model training. The data should ideally encompass a wide range of reviews, both genuine and potentially fake, from various sources and platforms.

Data Preprocessing: Before feeding the data into machine learning models, it undergoes a series of preprocessing steps. This includes tasks such as text cleaning, tokenization, and removing irrelevant information like HTML tags or special characters. The goal is to prepare the text data for analysis.

Model Training: The preprocessed data is then used to train different machine learning models. These models are designed to learn patterns and features within the reviews that can help distinguish between genuine and fake ones. The figure suggests that various models are employed in this phase, which could include algorithms like Random Forest, Support Vector Machine, or Decision Trees.

Model Evaluation: To assess the effectiveness of each model, rigorous evaluation is performed. This involves using metrics such as accuracy, precision, recall, F1-score, and more to determine how well the models can differentiate between real

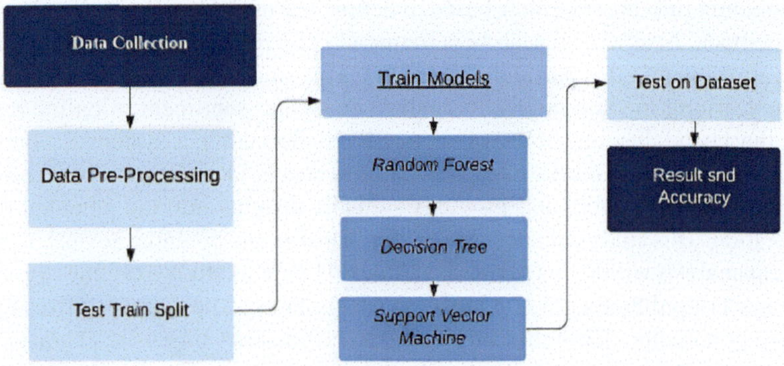

Fig. 1 Comprehensive approach to addressing the critical issue of detecting fake hotel reviews

and fake reviews. The results are essential for selecting the most accurate model for deployment.

Model Deployment: Once the most effective model is identified, it can be deployed in real-world scenarios. This could involve integrating the model into online review platforms, hotel-booking websites, or apps used by travelers. The deployed model would help users make more informed decisions by flagging or filtering potentially fake reviews.

3.1 Dataset Description

Data collection is the most important step. As most of the effort to detect deceptive reviews gets limited because of data, there are many ways to collect data such as fast checking, online platform kaggle, github. Our data are combined of both real and deceptive hotel reviews from different sources. The dataset we have used is stored into the 'dataset.csv' file which is available in Kaggle [4]. There are two sorts of dataset, one is an X, and another one is a Y set. The full dataset has four features. The first feature contains a hotel for every review. Then, the second feature contains the polarity of that review. Then next feature is a source, which declares the source of that review and the last feature is text. The reviews of this file have been classified as positive and negative. X contains only two features, source and text. The body part of the article comes under the text feature. It contains all the information on the reviews. The most important feature of the Y dataset is the deceptive label. The label indicates if a review is reliable or not. If a review is reliable, then the label is indicated as truthful, and if not, the label is deceptive indicating that the review is fake. We convert the label using a label encoder. We split the dataset as train and test, here 80% training and 20% test. We have a total test dataset of 320.

3.2 Dataset Preprocessing

The dataset collected for fake reviews detection is a raw dataset. This dataset contains noise such as special characters, uppercase, number, commonly used words. Such a dataset needs to be preprocessed. As the performance of any machine learning depends on the input data set, the data need to be clean for better results by removing unnecessary and irrelevant features that may be deleterious to classification performance [5]. So, preprocessing plays an important role. We have followed some steps in our project for preprocessing.

RegEx: To clean the dataset, we used re and string. Here, re is a build-in package in Python. It can be used to clean the dataset using regular expressions. It helps to remove texts that are misleading the data. By using the Python library, we have made the dataset clean and preprocessed from unnecessary words. After cleaning the data, we make a new data frame ‘complete text’ using ‘source’ and ‘text’ columns, where ‘complete text’ defines our independent variable x and ‘deceptive’ defines dependent variable y dataset. After that, we used our training algorithms on it.

Removed Number: In a particular text, any number does not change the meaning of a sentence. Therefore, if it is removed from the article, then it will minimize the noise in the data. We have used a Python library to keep all alphabet values in the dataset and remove all numerical values. It effectively removes all the numeric values minimizing noise in our dataset.

Remove Lower Case: It is difficult to check dataset when sentences are written in uppercase and lowercase both, so we have used text. Lower() to convert all the words in lower case.

Remove Text in Square Brackets: Removing square brackets[] is necessary. Without having square brackets in dataset is considered to be good dataset, so we used re.sub to remove square brackets.

Remove Special Character: Having special character in dataset gives error or improper result for the specific dataset. So we used resub to remove all the special to remove the special character.

Remove Links: Dataset having links are not consider as good dataset to work with, so we decided to use re.sub to remove links.

Ignore Not Needed Line: The text which exists between < > ; we need to ignore them so that we can get a proper test and train dataset.

Remove Punctuation: Punctuation mark should be ignored while working with dataset so we used re.sub to reduce punctuation mark.

Remove Newline: We need to remove the newline so that the dataset will be considered fresh and working will be easy. So we used re.sub to remove newline.

Remove Digit: Digit will be ignored to get a proper dataset, so we used re.sub to remove the digit remained in the dataset.

Removed Punctuation and Special Characters: We have only kept the textual data in our research. In this case, we have removed all characters that are special characters or not alphabets. We used the Python library for removing all specialized characters. All these characters do not create any meaning to the sentence so by removing it. We didn’t change any meaning of any article or news. After removing special characters, we removed punctuation that is not necessary to interpret a meaning full sentence. After removing special characters and punctuation, we only kept the word of the text in our data. We made all the texts into lowercase and make sure that we only have text with English words and no extra symbol is on the dataset.

4 Implementation

In this section, we will first discuss the dataset, then secondly the necessity of preprocessing to get a clean dataset will be also mentioned, and thirdly model development using the dataset and result based on classification implementation on the dataset.

Implementation: Count Vectorizer is the only model that we have used in our research. We have implemented Deceptive Hotel Reviews Detection using machine learning taken dataset by applying feature extraction technique named Count victor. This is one of the effective algorithms which is in the case of the pre-training vectorization dataset. The entire occurrence of words is considered by this algorithm. We have using Count Vectorizer for our dataset. It divides the whole sentence into words which is called tokens [6]. After that it calculates the number of instance of each word in the given data. It makes the data simpler for the classification to give an accurate result. After processing the dataset, only important words will be retained for the next step [7]. Then in the next step, we have implemented three known machine language classifications such as Random Forest, Decision Tree, and Support Vector Machine and got the best result out from them.

Random Forest: Here, we have chosen Random Forest as our first classifier. Random Forest is one of the popular supervised machine learning algorithms from classification. It generates good predictions which are easily understandable. This algorithm provides higher accuracy in predicting the output. By using attribute randomness, it will create many Decision Trees which give a more accurate result. In our code, we have used n_estimators which is 1000, here n_estimators means number of trees in the forest, and we have written random state which is 0, where random state is used by the random number generator.

In this model, we follow these processes and generate 1000 numbers of trees to predict the higher accuracy of result (Fig. 2).

Decision Tree: Decision Tree (DT) is also a supervised learning method used for classification and regression. Decision Tree classifier is based on the building of a

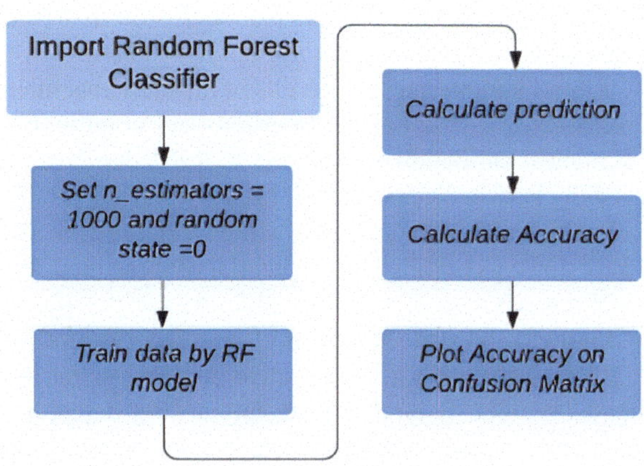

Fig. 2 Random Forest classifier process

tree which represents a decision of the training data. The goal is to conduct a model that learns simple decision rules from the data features and then predicts the value of a target variable. It takes input from the learning data. Based on the best possible division of the features, this classifier begins to build the tree iteratively. The process depends on some predefined functions such as entropy, information gain, Gini index. The leaf node contains the child node which determines the decision whether the review is truthful or deceptive.

SVM Support Vector Machine: Support Vector Machine is also a supervised machine learning algorithm that can be used for both classification and regression [2]. It used for both linear and nonlinear data. The main goal of SVM classifier is to create a decision boundary that can create isolated n-dimensional spaces in the classes. SVM classifier basically separates the given data into classes and finds the best divisible hyperplane which classifies the given training data. These hyperplanes set target variable values for future predictions. This process is designed to make it easier to classify new upcoming data points in the future [3]. SVM is one of the most popular and broadly used classifiers. First of all, we collect the dataset from Kaggle and the dataset has total five columns in total. As we do not need the hotel column in this project for this reason we remove the column. Then, we just remove all the numbers, special characters, links, text in square brackets from those text reviews. We give priority to those words; those are helping to keep the sentences meaningful. Then, we convert the lines to individual vectors and change them to numerical. When we apply multinominal functions to the numerical datasets, it generates some accuracy rate and confusion matrix. Then we apply logistic regression and Support Vector Machine, and this thing generates another accuracy rate and confusion matrix.

5 Experimental Result

After implementing all the three classification algorithms, we have got an accurate result. We prepared a confusion matrix and a table for comparing all three machine learning algorithms. The confusion matrix for all the algorithms we have implemented is given below in Fig. 3.

Table 1 summarizes the results of different machine learning models for detecting fake hotel reviews. The Count Vectorizer combined with Random Forest achieved the highest validation accuracy at 92.19%, demonstrating its effectiveness in distinguishing between genuine and fake reviews. Support Vector Machine also performed well with an 87.19% accuracy, while Decision Tree achieved an accuracy of 86.56%. These findings highlight the potential of these models in addressing the issue of fake hotel reviews.

Figure 4 shows accuracy result of all algorithms.

Table 2 presents a comparative analysis of the proposed model's performance against previously employed techniques for detecting fraudulent online reviews. In [8], the proposed model utilizing Random Forest achieved an impressive accuracy

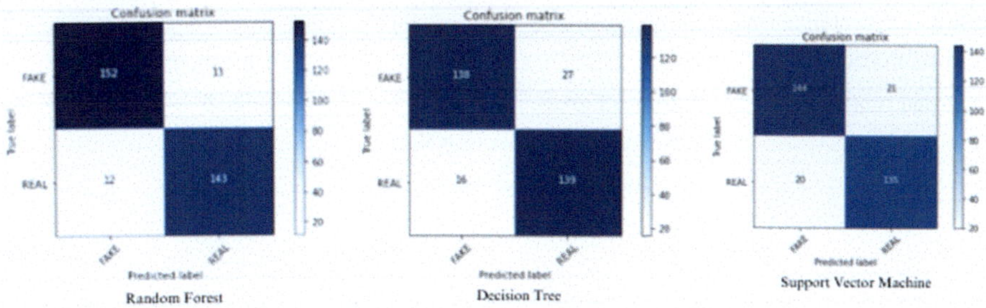

Fig. 3 Confusion matrix for all the algorithms

Table 1 Evaluation and interpretation of experimental outcomes

Models	Classifier	Validation accuracy (%)
Count vectorizer	Random Forest	92.19
Count vectorizer	Support Vector Machine	87.19
Count vectorizer	Decision Tree	86.56

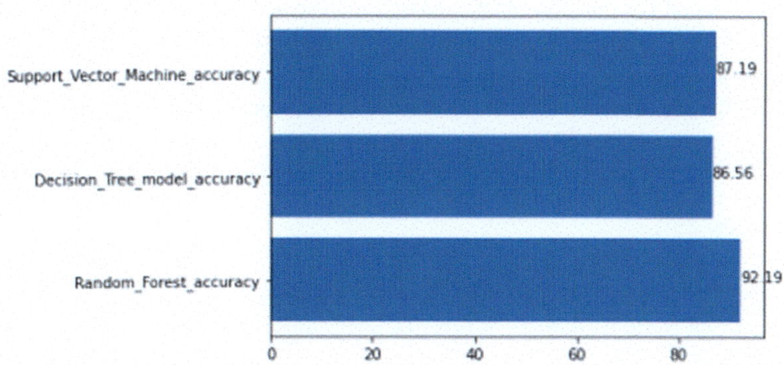

Fig. 4 Accuracy result of all algorithm

of 92.19%, outperforming all other methods. However, in [9], the use of K-Nearest Neighbors (KNNs) resulted in a lower accuracy of 82.40%, indicating its relative performance compared to the proposed model. Additionally, Support Vector Machine (SVM) in [9] achieved an accuracy of 93%, which, while high, did not surpass the proposed model's accuracy. These results highlight the effectiveness of the proposed Random Forest-based approach in addressing the challenge of identifying deceptive reviews in the realm of online reviews.

Table 2 Comparative analysis with previously used techniques

Paper reference	Model	Accuracy (%)
Proposed model	**Random Forest**	**92.19**
[8]	K-Nearest Neighbors (KNNs)	82.40
[9]	Support Vector Machine (SVM)	93

6 Conclusion

When the popularity of social media is increasing day by day, straightly it is growing its consumers a lot faster. The fact is that more people collect hotel reviews from that hotel page which is available on social media. So, it is quite tough to find authentic reviews for any hotel. As, social media is also a carrier of fake reviews, which has a serious negative impact on society. In our project, we collected data from an authentic source and preprocessed those by cleaning and feature extraction by Count Vectorizer. Then, we use three algorithms. The algorithms we used were SVM, Random Forest, and Decision Tree. Then, we just on the holdout set and after getting the result we just picked the model (Random Forest) which gave us the best result. We worked for our default database which was declared in our system. After working with this process, we wanted to work with manual input of data and that was not working fluently. So, there is a big scope to work on if the system can take input from the user and analysis any manually given data.

References

1. Jindal N, Liu B (2007) Review spam detection. In: Proceedings of the 16th international conference on World Wide Web—WWW '07, Banff, Alberta, Canada, p 1189. https://doi.org/10.1145/1242572.1242759
2. Elmurngi E, Gherbi A (2017) An empirical study on detecting fake reviews using machine learning techniques. In: 2017 seventh international conference on innovative computing technology (INTECH), Aug 2017, pp 107–114. https://doi.org/10.1109/INTECH.2017.8102442
3. Kashti RP, Prasad DPS, Analysis of classifiers for fake review detection, vol 6, no 9, p 5
4. Deceptive Opinion Spam Corpus, Kaggle. https://www.kaggle.com/datasets/rtatman/deceptive-opinion-spam-corpus
5. Guyon I, Elisseeff A (2003) An introduction to variable and feature selection. J Mach Learn Res 3, no null:1157–1182
6. Bansode* M, Pardeshi S, Ovhal S, Shinde P, Birajdar A (2021) Fake review prediction and review analysis. IJITEE 10(7):143–151. https://doi.org/10.35940/ijitee.G9042.0510721
7. Etaiwi W, Al-Naymat G (Sep 2017) The impact of applying different preprocessing steps on review spam detection. Procedia Comput Sci 113:273–279. https://doi.org/10.1016/j.procs.2017.08.368
8. Elmogy AM, Tariq U, Ibrahim A, Mohammed A (2021) Fake reviews detection using supervised machine learning. In: Proceedings of [(IJACSA) international journal of advanced computer science and applications, vol 12, No 1, 2021], College of computer engineering and sciences, Prince Sattam Bin Abdulaziz University, KSA

9. Alsubari SN, Deshmukh SN, Alqarni AA, Alsharif N, Aldhyani THH, Alsaade FW, Khalaf OI, Data analytics for the identification of fake reviews using supervised learning. Comput Mater Continua. https://doi.org/10.32604/cmc.2022.019625
10. Abri F, Gutierrez LF, Namin AS, Jones KS, Sears DRW, Linguistic features for detecting fake reviews. Texas Tech University
11. Tang H, Cao H, A review of research on detection of fake commodity reviews. Northwest Minzu University
12. .Jain PK, Pamula R, A systematic literature review on machine learning applications for consumer sentiment analysis using online reviews. Indian Institute of Technology (Indian School of Mines), Dhanbad, India
13. Abri F, Gutierrez LF, Namin A, Jones KS, Sears DRW, Fake reviews detection through analysis of linguistic features. Texas Tech University
14. Hossain MJ, Zaman SS, Akash FR, Alam F, Reza AW, Arefin MS (2023) Deciphering handwritten text: a convolutional neural network framework for handwritten character recognition. In: Vasant P et al (eds), Intelligent computing and optimization. ICO 2023. Lecture notes in networks and systems, vol 729. Springer, Cham. https://doi.org/10.1007/978-3-031-36246-0_18

Detecting Crop Pests and Diseases Through Deep Learning Techniques for Improved Yields

Md Tanvir Chowdhury, Md Jakir Hossain, Omar Rafat Adnan, Monjurul Islam Sumon, Md Sabbir Hossain, Habibur Rahman, Mahir Tagwar, Farid ishraq Zarif, Sajida Islam Ipty, Momena Khatun Methila, and Raifa Runa

Abstract To optimize agricultural operations and guarantee food security, crop categorization relies heavily on the precise identification of a wide range of agricultural goods. This study offers a thorough assessment of a crop classification model, concentrating on the classification performance of four different crop categories: cashew, maize, tomato, and cassava. While the model successfully classifies cashew

---

M. T. Chowdhury (✉) · M. J. Hossain · O. R. Adnan · M. I. Sumon · M. S. Hossain · H. Rahman · M. Tagwar · F. Zarif · S. I. Ipty · M. K. Methila · R. Runa
Department of Computer Science and Engineering, East West University, Dhaka, Bangladesh
e-mail: mdtanvirchowdhury015@gmail.com

M. J. Hossain
e-mail: mdjakirhossen13@gmail.com

O. R. Adnan
e-mail: omarrafatadnan@gmail.com

M. I. Sumon
e-mail: 123monjurul@gmail.com

M. S. Hossain
e-mail: sabbirhossain1338@gmail.com

H. Rahman
e-mail: habibresearchmail@gmail.com

M. Tagwar
e-mail: onontaj1999@gmail.com

F. Zarif
e-mail: ishraq12583@gmail.com

S. I. Ipty
e-mail: islamsanjida17@gmail.com

M. K. Methila
e-mail: methilaraiyan@gmail.com

R. Runa
e-mail: raifa97runa@gmail.com

S. Dutta et al. (eds.), *Cyber Intelligence and Information Retrieval*, Lecture Notes in Networks and Systems 1025, https://doi.org/10.1007/978-981-97-3594-5_39

(79.89%), maize (61.60%), and tomato (43.28%) with outstanding accuracy, it has substantial difficulties with cashew. With a classification accuracy of barely 14% and unusually high loss values, cassava is the most significant challenge. These findings underline the difficulty of correctly differentiating certain crops, notably cassava, and emphasize the need to improve and fine-tune the model. Future research endeavors should prioritize improving the model's accuracy and generalization, aiming to create a robust and reliable crop classification system that can effectively support agricultural communities and enhance food security.

Keywords Agricultural · Food security · Classification · Deep learning · Generalization

1 Introduction

Bangladesh is characterized as an agrarian nation, whereby a significant proportion of its people, around 50%, is mostly engaged in agricultural activities. Moreover, over 70% of the country's land is allocated for the cultivation of various crops. The primary agricultural commodities farmed by farmers in this region include a variety of major crops such as rice, jute, wheat, tea, pulses, oil seeds, vegetables, and fruits. Currently, crops are cultivated via the use of current scientific advancements and technological innovations. Technical. With technical backing, it can be enhanced. Close observation is necessary for the management of perennial rice crops, notably for the control of diseases that can have a considerable impact on production and, consequently, the post-harvest life. Any condition that interferes with a plant's normal physiological processes and causes recognizable symptoms is referred to as a disease in the context of plants. Observable phenomena known as symptoms can be used to confirm the presence of a disease. Pathogens, the agents responsible for diseases in plants, are frequently the cause of these illnesses. The leaves and stems of plants are common sites for the appearance of these pests and diseases. As a result, identifying plants, leaves, and stems, determining the presence of pests or diseases, measuring the level of pest or disease incidence, and recognizing the symptoms of pest or disease attacks, play a key part in the effective cultivation of crops. To maintain healthy plants and guarantee a harvest, this process is essential. A single experiment in biological science might generate thousands of photos. The research pictures might be necessary for other research projects like categorizing lesions, rating quantitative features, estimating the area of an insect's diet, etc. Almost the majority of these chores are completed by hand or using various software programs. It is not only a large lot of work, but it also has two big problems: long processing times and subjectiveness brought on by many people. Plant biologists therefore require effective computer tools to automatically extract and analyze important content to conduct high throughput studies. In this case, image processing is crucial.

2 Literature Review

Sharma et al. [1] discuss technological advancements in automated crop pest and disease detection. Their study examines the use of AI and computer vision in crop protection. The authors provide insights into the challenges and ongoing research in this area, emphasizing the importance of AI in modern agriculture. Jaware et al. [2] focus on crop disease detection using image segmentation. Although the publication year is not specified, their work is noteworthy for its application of image processing techniques in identifying crop diseases. This approach is critical for early detection and intervention. Selvaraj et al. [3] present research on AI-powered banana diseases and pest detection. While the publication year is not mentioned, their work underscores the significance of AI in identifying diseases and pests in specific crops like bananas. Boissard et al. [4] propose a cognitive vision approach to early pest detection in greenhouse crops. Although the publication year is not specified, their research explores the use of cognitive vision techniques, which could provide valuable insights into greenhouse crop protection. Ouhami et al. [5] offer a comprehensive survey on computer vision, IoT, and data fusion for crop disease detection using machine learning. Their recent study discusses the integration of various technologies and highlights ongoing research in this field. Badage [6] discusses crop disease detection using machine learning in the context of Indian agriculture. While the publication year is not specified, this work sheds light on the specific challenges and opportunities for AI-based crop protection in the Indian context. In [8–10] authors applied different CNN techniques for performing different tasks.

3 Methodology

Designing an effective model for crop disease classification involves several key steps and considerations. Here's a proposed design for a crop disease detection model Shown in Fig. 1.

In this section, we have explained the workflow of crop disease detection. Image pre-processing: The acquired images are then preprocessed to correct for any distortion or noise, and to enhance the visibility of the lesion. This might include techniques such as image normalization, histogram equalization, or denoising.

3.1 Dataset Description

We have collected this dataset from Mendeley Data [7]. This dataset comprises two essential components. First, it includes a collection of raw images, encompassing a total of 24,881 images representing four key crops: cashew (6,549 images), cassava (7,508 images), maize (5,389 images), and tomato (5,435 images). Second, the

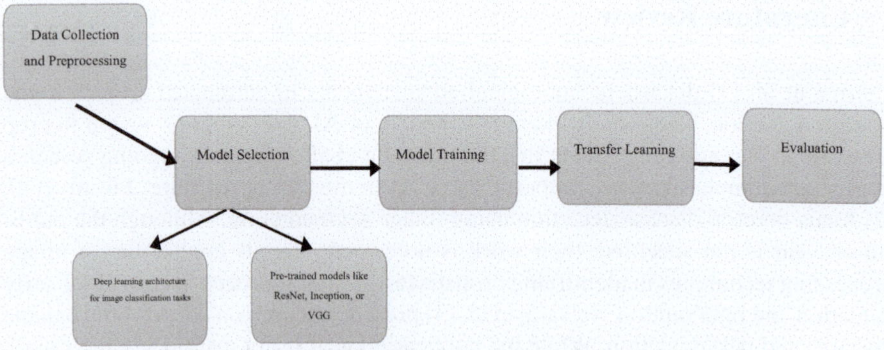

Fig. 1 Proposed design for a crop disease detection model

Fig. 2 Examples (tomato: Septoria leaf spot and healthy) of data used for analysis/training

dataset contains augmented images, further divided into train and test sets, resulting in a total of 102,976 images. These images are categorized into 22 classes, corresponding to various crop pests and diseases commonly found in Ghana. Figure 2 show the visualization of datasets used for analysis.

4 Implementation

This part gives a thorough explanation of how the study was carried out, how the experiments were set up, and how the performance of the model was assessed. It includes each stage and technique utilized to carry out the study and assess its efficacy.

Implementation: In this research, we used convolutional neural network (CNN) models. The implementation of a CNN model for crop disease classification involves several key steps, as described below:

Data Generation: To create the training, validation, and test datasets, the flow from directory method is employed. This method allows for specifying various parameters, including image size, batch size, and classification mode.

CNN Architecture: The architecture of the CNN is constructed through Keras' Sequential API, featuring a combination of convolutional layers, max-pooling layers, and fully connected layers. The convolutional layers employ the Rectified Linear Unit (ReLU) activation function, enhancing non-linearity in feature extraction. For the final layer dedicated to multiclass classification, the softmax activation function is applied, facilitating the assignment of probabilities to different classes. To optimize and train the model, the Adam optimizer is chosen, which adapts learning rates for each parameter individually. Categorical cross-entropy serves as the loss function during training, particularly suitable for scenarios involving multiple classes. Evaluation of the model's performance is conducted using precision as the metric, reflecting the accuracy of positive predictions. Figure 3 visually represents the intricate architecture of the CNN, showcasing the interconnected layers responsible for learning hierarchical features from input data. This comprehensive configuration, encompassing activation functions, optimizer, and evaluation metrics, contributes to the model's efficacy in addressing multiclass classification challenges.

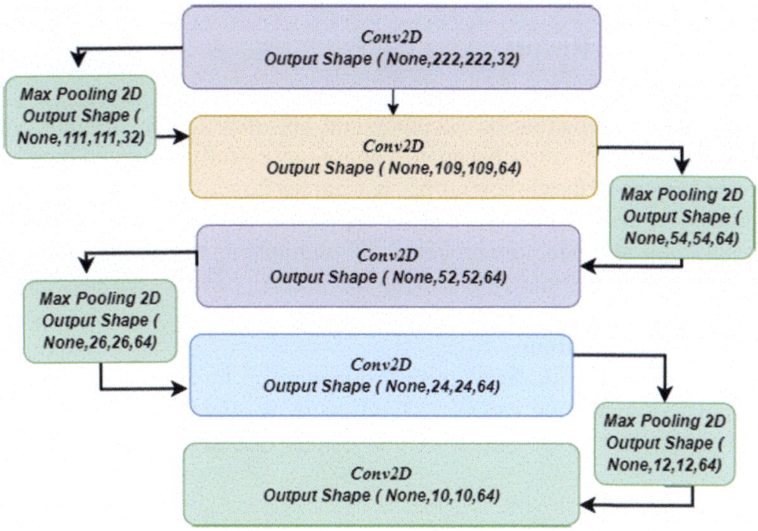

Fig. 3 Architecture of sequential CNN

Early Stopping: The Early Stopping callback is configured to monitor validation accuracy and halt training if there is no significant improvement over a specified number of epochs, controlled by the patience parameter. This helps prevent overfitting.

Model Training: The CNN model is trained using the training data generator. The validation accuracy is monitored during training, and early stopping criteria are applied. The training history, including loss and accuracy measures, is saved for visualization.

Visualizing Training Progress: To assess the model's convergence and detect overfitting, training and validation loss curves are plotted over epochs. Additionally, training and validation accuracy curves are visualized to evaluate generalization performance.

Model Saving: The trained CNN model is saved to a file, such as 'cnn.h5', for future use or deployment.

Model Evaluation: The test dataset is evaluated by making predictions using the trained model. Confusion matrix calculations provide insights into classification performance. An in-depth classification report, including F1-score, precision, recall, and accuracy, is generated to assess the model's performance across multiple classes.

5 Experimental Result

Table 1 provides information on accuracy and loss metrics for different types of crops. The proportion of crops that are lost or not successfully harvested is called loss accuracy. For example, the cashew crop loss accuracy is measured at 0.5499, meaning that around 54.99% of cashew crops experience loss. This measure guarantees more efficient agricultural production processes by illuminating the effectiveness of crop management approaches and emphasizing the necessity of focused loss mitigation strategies.

Performance Evaluation

Tables 2, 3, and 4 provide detailed confusion matrices for crop diseases, including cashew, maize, tomato, and cassava. These matrices offer insights into precision, recall, F1-score, and support metrics for various disease categories. The tables

Table 1 Image detection accuracy of models for each crop category

Crop	Test		Validation	
	Loss	Accuracy	Loss	Accuracy
Cashew	0.5499	0.7966	0.5623	0.7989
Maize	0.9727	0.6263	1.0236	0.6160
Tomato	1.7821	0.5028	1.4217	0.4328
Cassava	4,172,452.2500	0.1598	22,834,672.0000	0.1400

present a comprehensive evaluation of the classification performance, aiding in the assessment of disease detection models in agricultural applications.

Table 2 Confusion matrix of cashew

	Precision	Recall	F1-score	Support
Anthracnose	0.78	0.60	0.68	1838
Gumosis	0.87	0.97	0.92	425
Healthy	0.65	0.93	0.77	1336
Leaf miner	0.81	0.73	0.77	1487
Red rust	0.94	0.91	0.93	1815
Accuracy			0.80	6901
Macro avg	0.81	0.83	0.81	6901
Weighted avg	0.81	0.80	0.80	6901

Table 3 Confusion matrix of maize

	Precision	Recall	F1-score	Support
Fall armyworm	0.48	0.38	0.43	284
Grasshopper	0.55	0.78	0.65	411
Healthy	0.45	0.64	0.53	211
Leaf beetle	0.90	0.85	0.87	950
Leaf blight	0.55	0.72	0.62	1004
Leaf spot	0.55	0.37	0.44	1261
Streak virus	0.64	0.63	0.63	1004
Accuracy			0.62	5125
Macro avg	0.59	0.62	0.60	5125
Weighted avg	0.63	0.62	0.62	5125

Table 4 Confusion matrix of tomato

	Precision	Recall	F1-score	Support
Healthy	0.57	0.92	0.71	500
Leaf blight	0.37	0.17	0.23	1309
Leaf curl	0.00	0.00	0.00	532
Septoria leaf spot	0.52	0.80	0.63	2340
Verticillium wilt	0.43	0.27	0.33	764
Accuracy			0.51	5445
Macro avg	0.38	0.43	0.38	5445
Weighted avg	0.43	0.51	0.44	5445

6 Conclusion

Evaluation of our crop categorization model has provided important information about how well it works for a variety of crop types. Notably, the model achieves roughly 79.89% and 61.60% validation accuracy, respectively, indicating excellent accuracy in cashew and maize identification. It does, however, have significant difficulty correctly categorizing tomatoes and cassava, with validation accuracies of around 43.28% and 14%, respectively. The notably elevated loss values linked to cassava highlight the intricacy of this classification work, disclosing the intricacy required in accurately classifying certain crops. These results highlight the necessity of thorough model improvement, particularly about difficult crop categories such as tomatoes and cassava. The accuracy constraints that have been found highlight the need for focused efforts to maximize the model's performance. Subsequent investigations ought to focus on improving the model's precision and generalizability to guarantee its resilience over a wider range of crops. These improvements are critical to strengthening crop classification systems' overall effectiveness in agricultural settings, where precise classification is critical to crop management and decision-making procedures.

References

1. Sharma V, Tripathi AK, Mittal H (2022) Technological advancements in automated crop pest and disease detection: a review & ongoing research. In: 2022 International conference on computing, communication, security and intelligent systems (IC3SIS). IEEE, pp 1–6
2. Jaware TH, Badgujar RD, Patil PG (Year not specified) Crop disease detection using image segmentation
3. Selvaraj MG, Vergara A, Ruiz H, Safari N, Elayabalan S, Ocimati W, Blomme G (2019) AI-powered banana diseases and pest detection
4. Boissard P, Martin V, Moisan S (2008) A cognitive vision approach to early pest detection in greenhouse crops
5. Ouhami M, Hafiane A, Es-Saady Y, El Hajji M, Canals R (2021) Computer vision, IoT and data fusion for crop disease detection using machine learning: a survey and ongoing research. Remote Sens 13:2486
6. Badage A (Year not specified) Crop disease detection using machine learning: Indian agriculture
7. Mensah Kwabena P, Akoto-Adjepong V, Adu K, Abra Ayidzoe M, Asare Bediako E, Nyarko-Boateng O, Boateng S, Fobi Donkor E, Umar Bawah F, Songose Awarayi N, Nimbe P, Kofi Nti I, Abdulai M, Roger Adjei R, Opoku M (2023) Dataset for crop pest and disease detection. Mendeley Data 1. https://doi.org/10.17632/bwh3zbpkpv.1
8. Hossain MJ, Zaman SS, Akash FR, Alam F, Reza AW, Arefin MS (2023) Deciphering handwritten text: a convolutional neural network framework for handwritten character recognition. In: Vasant P et al (eds) Intelligent computing and optimization. ICO 2023. Lecture notes in networks and systems, vol 729. Springer, Cham. https://doi.org/10.1007/978-3-031-36246-0_18
9. Hossain MS et al (2023) Performance evaluation of intrusion detection system using machine learning and deep learning algorithms. In: 2023 4th international conference on big data analytics and practices (IBDAP), Bangkok, Thailand, pp 1–6. https://doi.org/10.1109/IBDAP58581.2023.10271964

10. Banerjee D, Kukreja V, Hariharan S, Jain V, Dutta S (2023) An intelligent framework for grassy shoot disease severity detection and classification in sugarcane crop. In: 2023 2nd international conference on applied artificial intelligence and computing (ICAAIC), pp 849–854

A Novel Framework for Automatic Plant Disease Detection Using Convolutional Neural Networks

Ayan Sar, Anvi Goel, Tanupriya Choudhury, Ketan Kotecha, and Abhishek Bhattacharya

Abstract This research aims to introduce a groundbreaking framework for the automated detection of plant diseases leveraging the power and potential of Convolutional Neural Networks (CNNs). The productivity of agriculture was substantially affected by the rapid spread of plant diseases, which necessitated more efficient and timely detection methods. This proposed framework employed the state-of-the-art CNN architecture, which could be adept at learning intricate patterns within different plant images. This model was trained on a comprehensive dataset that encompassed various diverse plant diseases and healthy specimens. For the enhancement of generalization, the usage, and employment of various transfer learning techniques were also necessitated. This framework demonstrated superior accuracy, with sensitivity and specificity in the identification of various plant diseases across different plant species. Moreover, after the model's efficiency was validated through its real-time testing on many diverse crops. This research aims to contribute to the advancement of precision agriculture by providing a more robust and scalable solution for early disease detection, which could empower farmers to adopt preventive measures and mitigate losses in yield.

A. Sar · A. Goel
School of Computer Science, University of Petroleum and Energy Studies (UPES), Dehradun, Uttarakhand 2480007, India

T. Choudhury (✉)
Research Professor, CSE Department, Graphic Era Deemed to be University, Dehradun, Uttarakhand 248002, India
e-mail: tanupriya.choudhury@sitpune.edu.in

Adjunct Professor, CSE Department, Symbiosis International (Deemed University), Symbiosis Institute of Technology, Pune, Maharashtra 412115, India

K. Kotecha
Symbiosis Center for Applied Artificial Intelligence, Symbiosis International (Deemed University), Symbiosis Institute of Technology, Pune 412115, India
e-mail: Director@sitpune.edu.in

A. Bhattacharya
Sister Nivedita University, Chakpachuria, West Bengal, India
e-mail: abhishek.bh@snuniv.ac.in

S. Dutta et al. (eds.), *Cyber Intelligence and Information Retrieval*, Lecture Notes in Networks and Systems 1025, https://doi.org/10.1007/978-981-97-3594-5_40

Keywords Plant disease detection · Convolutional neural networks · Automated agriculture · Precision farming · Transfer learning

1 Introduction

In this present realm of precision agriculture, where technological advancements are always reshaping traditional farming practices, the identification and timely management of different plant diseases stand as the critical components for ensuring and making a step forward toward global food security. This research introduced a groundbreaking framework for automatic plant disease detection, along with leveraging the power of Convolutional Neural Networks (CNNs), which would revolutionize the way of monitoring and safeguarding crops. The prevalence of plant diseases will always be a significant threat to crop yields and consequently, impact the agricultural ecosystem. The traditional methods of disease identification would often rely on manual inspection, which may lead to delays in diagnosis and subsequent treatment. This proposed framework would address this challenge by harnessing the real potential of CNNs, which is a class of deep learning algorithms known for their exceptional ability to extract intricate patterns from visual data. This novel framework was integrated with a sophisticated CNN architecture, which was further trained on an extensive dataset comprising diverse images of healthy and diseased plants. Through the usage of transfer learning techniques, the knowledge acquired by pre-trained networks was leveraged which enhanced the model's ability to generalize and adapt to the various plant species and disease types. The robustness of the system was further bolstered by the employing of data augmentation strategies for the simulation of real-world scenarios and improving the network's resilience to diverse environmental conditions.

2 Problem Identification

Agriculture plays a pivotal role in global food production, and in plant diseases which significantly could impact crop yield and quality. The traditional methods of plant disease detection are very time-consuming, labor-intensive, and very importantly, lack accuracy. The integration of modern technologies, such as Convolutional Neural Networks (CNNs), offered a very promising avenue for automation in plant disease detection.

Existing manual inspection methods are very much prone to human error and subjectivity, which leads to unreliable results. Also, early detection of plant disease is very crucial, but traditional techniques often fail to identify symptoms at an early stage. The limited scalability of the traditional methods which are not easily applicable to large-scale agricultural settings, is unable to make timely detection.

Increasing global population and climatic change would necessitate improved agricultural efficiency for the ensuring of food security. Automated plant disease detection could also facilitate early intervention, which in turn, would minimize crop losses and reduce the need for excessive pesticide usage. The latest advent of deep learning, particularly CNNs, has demonstrated very significant potential in image-based tasks, including object recognition. While some of the automated systems exist, they may not be able to fully exploit the capabilities of the advanced and latest deep learning techniques. The existing models also find it struggling to cope with the variations in environmental conditions, plant types, and disease manifestations. Some models may not find them generalizing well to diverse datasets, which might hinder their real applicability across different regions and crops. Despite much progress in automated plant disease detection, the need for a novel framework that will address the limitations of current approaches is bigger. This framework aimed to enhance the accuracy, robustness, and scalability by leveraging the power of CNNs and incorporating innovative features.

3 Literature Review

Strange et al. in [1] examined the global threat to food security posed by different plant diseases, which highlighted the need for more advanced detection methods. Uthayakumar et al. in [2] proposed an algorithm focusing on lossless compression in the field of networks working under wireless sensors, which presented a working application for wind plant data. Sladojevic et al. in [3] utilized deep neural networks for plant disease detection along with its recognition through different leaf image classifications. Sharma et al. in [4] conducted a performance analysis related to deep learning, specifically in CNN models for plant disease detection, with a focus on image segmentation techniques. Lakshmanaprabhu et al. in [5] introduced a clinical decision support system, completely online, that employed optimal deep neural networks in the field of medical applications. Barbedo et al. in [6] investigated the impact of the dataset size and its variety on the effectiveness of deep learning and transfer learning for plant disease classification. Barbedo et al. in [7] explored the different plant disease identification from its lesions and spots using different deep-learning techniques. Guo et al. in [8] proposed a different kind of plant disease identification system which was based on deep learning algorithms for its effective use of smart farming applications. Selvaganesan et al. in [9] designed and analyzed the pepper leaf disease detection system using the concept of deep belief networks. Khatoon et al. in [10] presented an effective image-based automatic diagnostic system for tomato plants using different deep-learning techniques. Chen et al. in [11] applied deep transfer learning for image-based plant disease identification, showcasing its efficacy of transfer learning in the context of CNNs for plant disease detection. Bharali et al. in [12] developed a plant disease detection system that utilized leaf image classification with CNNs, ultimately contributing insights into the image-based approaches for plant disease identification. Xie et al. in [13] proposed a real-time detector for grape leaf diseases using improved CNNs, which illustrated advancements in deep learning-based techniques for the rapid and accurate identification of plant diseases. Zhang et al. in [14] improved object detection for tomato diseases using deep learning while presenting enhancements in the application of CNNS for its precise identification of plant diseases. Janarthan et al. in [15] implemented deep metric learning for citrus disease classification with sparse data while addressing the complex challenges related to limited dataset availability in this context. Barman et al. in [16] compared various CNNs for smartphone image-based real-time classification of citrus leaf diseases, ultimately providing insights into the suitability of different CNN architectures for their practical applications. Kang et al. [17] proposed a multi-objective piglet image segmentation method which was further based on an improved noninteractive GrabCut algorithm, ultimately highlighting different segmentation techniques for the agricultural image analysis. Houssein et al. in [18] introduced a novel Black Widow optimization algorithm for multilevel thresholding image segmentation, contributing to the field of image segmentation techniques applicable to plant disease detection. Howard et al. in [19] presented MobileNets, as an efficient CNN designed for mobile vision applications, providing a solid technological foundation for lightweighted yet effective CNNs in this domain. Gaurav et al. in [20] introduced BEPO, as a binary emperor penguin optimizer, which demonstrated its effectiveness in the automatic feature selection. Xiao et al. [21] proposed a multiple hidden layer extreme machine learning method, ultimately showcasing its application in the context of plant disease detection. Singh et al. in [22] presented PlantDoc, as a dataset tailored for visual plant disease detection. This resource contributed to the advancement of plant disease research for algorithm evaluation. Rangarajan et al. in [23] focuses on tomato crop disease classification using pre-trained deep learning algorithms. Karthik et al. in [24] proposed an attention-embedded residual CNN for disease detection in tomato leaves, which introduced an innovative approach that incorporated attention mechanisms to improve the network's ability. Rakesh et al. in [25] deep transfer learning in conjunction with an optimal kernel extreme machine learning model for plant disease diagnosis and classification.

4 Methodology

Here, as shown in Fig. 1, the methodology began with an introduction that highlighted the significance of automated plant disease detection using the potential of Convolutional Neural Networks (CNNs) and similarly outlining the research objectives. The dataset collection and preprocessing steps involved one of the major parts of this research of suitable dataset selection of plant images with disease labels along with the implementation of necessary transformations. Next, the model architecture was designed, with the incorporation of CNN along with potential transfer learning from pre-trained models. The core of this methodology lies in the design of an effective CNN architecture. The training process involves the setup of parameters and fine-tuning of the model for optimal performance. The evaluation metrics, including

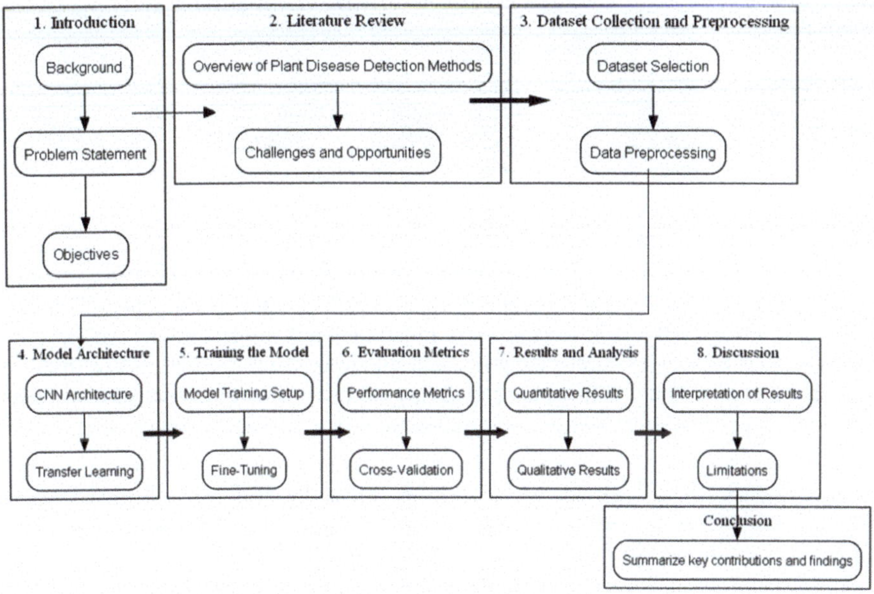

Fig. 1 Proposed methodology for the framework on paddy disease detection

accuracy and precision, were also defined and cross-validation was applied for the assessment of the model's robustness. The results, including both qualitative and quantitative, were also presented and analyzed, which was followed by a discussion of findings and limitations and potential future scope.

5 Datasets and Algorithms Discussed

The dataset that was used for this novel framework is the Plant Village dataset, which is comprised of a complete collection of images designed for mainly disease detection in plants. It consisted of a diverse set of 61,486 images under 39 unique disease categories. These categories comprised of total range of 16 plant varieties and multiple backgrounds while providing a rich and varied dataset for both training and evaluation.

The dataset is referenced from this link: https://data.mendeley.com/datasets/tywbtsjrjv/1. It had been made perfectly for the efficient and accurate support of this research and development in this diverse field of agriculture, besides providing particular attention to the automated identification and proper diagnosis of diseases.

For better performance and to increase the efficiency of the dataset, various data augmentation techniques were also applied such as Image Flipping, Noise injection, PCA Color Augmentation, and various other techniques.

Fig. 2 Overview of dataset images

This dataset served as the foundation for the training and evaluation of the CNN in this proposed framework. Images were pre-processed using common techniques available like resizing, center cropping, and conversion to tensors. These preprocessing steps played a crucial role in ensuring that the input data was standardized and well-compatible with the model architecture. The dataset can be visualized in Fig. 2.

The framework is built by leveraging the potential of Convolutional Neural Network (CNN) as the core algorithm for automatic plant disease detection. The CNN architecture was implemented using the PyTorch library.

The Convolutional Layers consisted of a full series of convolution operations, each followed by the Rectified Linear Unit (ReLU) activation and its batch normalization. The max-pooling layers with a kernel size of 2 were also employed to downsample the spatial dimensions of the feature maps. The convolution operation between the input image I and a filter (kernel) K can be defined as in Eq. (1).

$$(I * K)(i, j) = \sum_{m} \sum_{n} I(m, n) K(i - m, j - n)$$

$$\text{Convolution operation} \tag{1}$$

The output from this convolutional layer was flattened and passed through a dense layer consisting of 50,176 input features (which were derived from the flattened shape) and 1024 output features, along with ReLU activation and dropout for its regularization.

The final dense layer produced the output with a size of 'K', where 'K' is the number of disease categories (targets).

The model was then trained using the custom training loop defined in the 'batch_gd' function. Here, iterations were done over the training data in mini-batches. Then zeroing the gradients was done to avoid accumulation. Then the model predictions were computed and calculated for cross-entropy loss. It was followed by backpropagation for computing gradients and updating the model parameters using the Adam optimizer.

The batch normalization normalized the input of each layer, which helped in the stabilization and acceleration of the training process. Given input x, mean μ and standard deviation σ, batch normalization can be calculated as in Eq. (2).

$$BN(x) = \gamma\left(\frac{x-\mu}{\sigma}\right) + \beta$$

Batch normalization function (2)

where γ is the scaling factor, and β is the shifting factor.

The architecture of the CNN algorithm used here can be visualized in Figs. 3 and 4.

Layer (type)	Output Shape	Param #
Conv2d-1	[-1, 32, 224, 224]	896
ReLU-2	[-1, 32, 224, 224]	0
BatchNorm2d-3	[-1, 32, 224, 224]	64
Conv2d-4	[-1, 32, 224, 224]	9,248
ReLU-5	[-1, 32, 224, 224]	0
BatchNorm2d-6	[-1, 32, 224, 224]	64
MaxPool2d-7	[-1, 32, 112, 112]	0
Conv2d-8	[-1, 64, 112, 112]	18,496
ReLU-9	[-1, 64, 112, 112]	0
BatchNorm2d-10	[-1, 64, 112, 112]	128
Conv2d-11	[-1, 64, 112, 112]	36,928
ReLU-12	[-1, 64, 112, 112]	0
BatchNorm2d-13	[-1, 64, 112, 112]	128
MaxPool2d-14	[-1, 64, 56, 56]	0
Conv2d-15	[-1, 128, 56, 56]	73,856
ReLU-16	[-1, 128, 56, 56]	0
BatchNorm2d-17	[-1, 128, 56, 56]	256
Conv2d-18	[-1, 128, 56, 56]	147,584
ReLU-19	[-1, 128, 56, 56]	0
BatchNorm2d-20	[-1, 128, 56, 56]	256
MaxPool2d-21	[-1, 128, 28, 28]	0
Conv2d-22	[-1, 256, 28, 28]	295,168
ReLU-23	[-1, 256, 28, 28]	0
BatchNorm2d-24	[-1, 256, 28, 28]	512
Conv2d-25	[-1, 256, 28, 28]	590,080
ReLU-26	[-1, 256, 28, 28]	0
BatchNorm2d-27	[-1, 256, 28, 28]	512
MaxPool2d-28	[-1, 256, 14, 14]	0
Conv2d-29	[-1, 256, 14, 14]	590,080
ReLU-30	[-1, 256, 14, 14]	0
BatchNorm2d-31	[-1, 256, 14, 14]	512
Dropout-32	[-1, 50176]	0
Linear-33	[-1, 1024]	51,381,248
ReLU-34	[-1, 1024]	0

Fig. 3 CNN architecture

```
     BatchNorm2d-31        [-1, 256, 14, 14]             512
         Dropout-32              [-1, 50176]               0
          Linear-33               [-1, 1024]      51,381,248
            ReLU-34               [-1, 1024]               0
         Dropout-35               [-1, 1024]               0
          Linear-36                 [-1, 10]          10,250
================================================================
Total params: 53,156,266
Trainable params: 53,156,266
Non-trainable params: 0
----------------------------------------------------------------
Input size (MB): 0.57
Forward/backward pass size (MB): 145.11
Params size (MB): 202.78
Estimated Total Size (MB): 348.46
----------------------------------------------------------------
```

Fig. 4 CNN architecture (Contd. from Fig. [3])

The utilization of Convolutional Neural Networks (CNNs) in the automated framework enhanced the automated detection of plant diseases by leveraging the potential of their unique ability to learn hierarchical representations of image data. CNNs are particularly very effective in the capturing of spatial dependencies and local patterns within different images, which plays a crucial role in the identification of subtle visual cues associated with plant diseases. The convolutional layers in CNNs are capable of automated extraction of relevant features, which enables the model to discern intricate patterns that are indicative of various diseases affecting plants. This hierarchical feature learning allowed the model to generalize well to diverse datasets and eventually contribute to the improved accuracy that was observed in our results.

5.1 Potential Benefits and Limitations

Implementing this novel framework in the field of agriculture and crop management offered several potential benefits:

Early disease detection. The framework for automated detection enabled the system for early identification of plant diseases, which further allowed for timely intervention and disease management. This, in turn, potentially prevented the widespread damage of crops and the improvement of overall crop yield.

Precision Agriculture. By the accurate identification and localization of diseases, farmers could apply targeted treatments, such as the application of pesticides or disease-specific irrigation, which would reduce the need for the widespread application of resources.

Resource Optimization. Efficient disease detection would aim for optimal use of resources, with minimization of unnecessary chemical treatments and reduction of environmental impact. This could lead to further cost savings for farmers and a more sustainable approach to agriculture.

5.2 Limitations and Challenges Associated

With the proposed algorithm of CNN offering very promising results, the following limitations and challenges were also discussed here:

Limited Data Representativity. The effectiveness and potential of CNNs are heavily dependent on the diversity and representativity of the dataset. The incomplete or biased datasets could result in the model being less robust than it could be, making it less adaptable to the variations in plant diseases. This could hinder the generalizations to real-world scenarios.

Interpretability. CNNs were often considered as the "black-box" models, which made it challenging to interpret their decisions. The main understanding of the rationale working behind the model's predictions was crucial for gaining trust from end-users, especially in diverse application as critical as agriculture.

Computational Requirements. The training of deep CNNs could also be computationally intensive, which would require significant resources. This might pose challenges, particularly for small-scale farmers or regions with limited access to high-performance computing infrastructure.

6 Results and Discussions

The system on which the deep neural networks were implemented and the dependencies used for the successful training and implementation of the model:

Processor: 11th Gen Intel(R) Core (TM) i5-11400H @ 2.70 GHz 2.69 GHz.
Memory: 24.0 GB (23.7 GB usable).
Disk: NVMe Micron_2450_SSD (512 GB).
Network: MediaTek Wi-Fi 6 MT7921 Wireless LAN Card (100 Mbps up and down).
GPU 1: Intel® UHD Graphics 6000.
GPU 2: Nvidia GeForce RTX 3050 Ti Laptop GPU.
Dependencies and packages: TensorFlow, Keras, CUDA Toolkit, matplotlib, and cuDNN libraries should be configured as per the GPU available. In my case, I used TensorFlow version 2.10.0 with CUDA version 11.2 and cuDNN library 8.1.

The CNN exhibited an overwhelming and promising performance during the training phase, achieving a high accuracy of 93.070% as in Fig. 5 along with the loss graph in Fig. 6. This indicated that this model effectively learned the intricate patterns and features which was present in the training dataset. This also showcases its ability to capture the underlying characteristics of various diseased and healthy plant images. The validation accuracy, which was also a crucial metric for assessing the model's generalization capabilities, also reached an impressive accuracy of 91.530% as in Fig. 7 along with the loss function graph in Fig. 8. This result also indicated that the model not only memorized the training data but also learned how to generalize well to unseen samples. This high validation accuracy underscored the robustness of this proposed framework in the differentiation between different plant diseases. The ultimate evaluation of this model's performance was also conducted on the previously unseen test dataset. The model also achieved a commendable test accuracy of 90.940% as in Fig. 9 along with the loss function graph in Fig. 10, which confirmed its ability to accurately classify plant diseases in various real-world scenarios. This result also suggested that this framework is reliable and effective in practical applications, where accurate disease detection plays a crucial role in accurate crop management.

For the benchmarking of the effectiveness of this proposed framework, this was compared with other existing state-of-the-art methods in plant disease detection. The model demonstrated competitive and superior accuracy, which showcased its potential as a very advanced solution in this field.

From Fig. 5, the training accuracy graph illustrates the model's ability for the improvement of performance over epochs. A steady increase in accuracy was observed, which indicated its effective learning. Simultaneously, the training loss

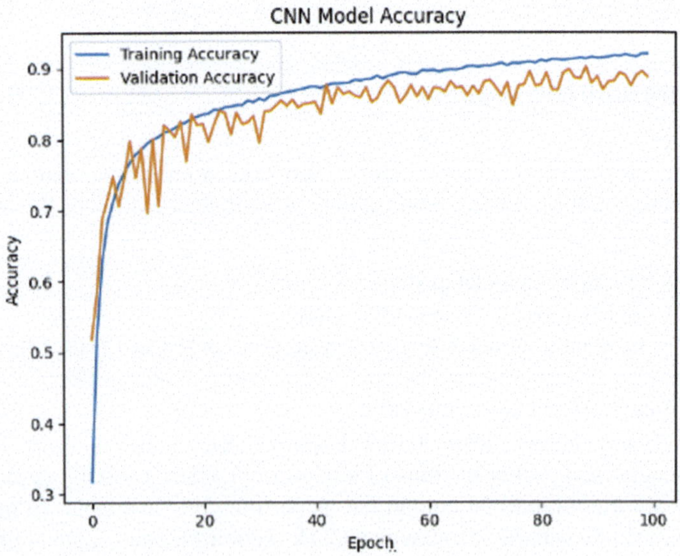

Fig. 5 Training accuracy graph

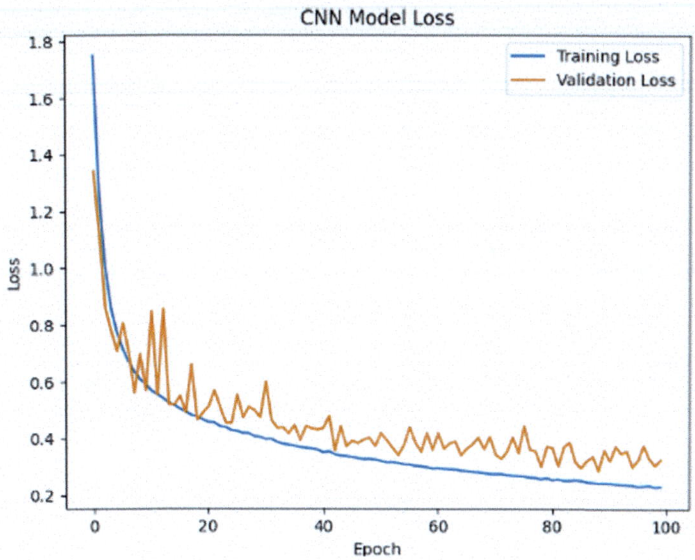

Fig. 6 Training loss graph

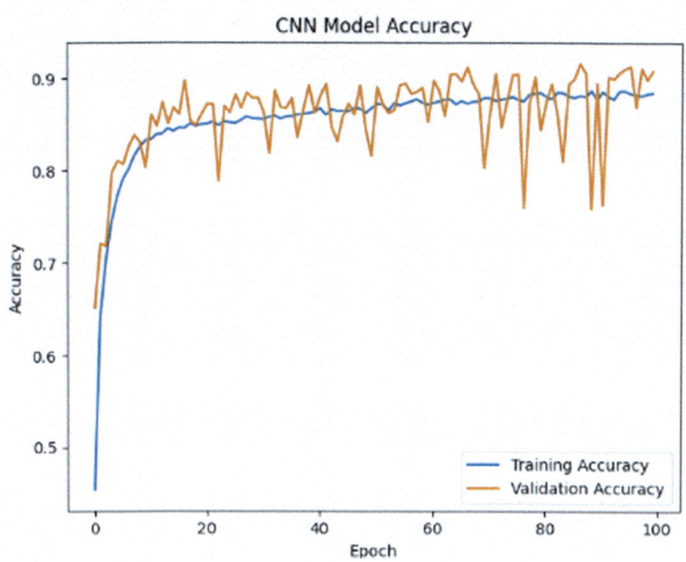

Fig. 7 Validation accuracy graph

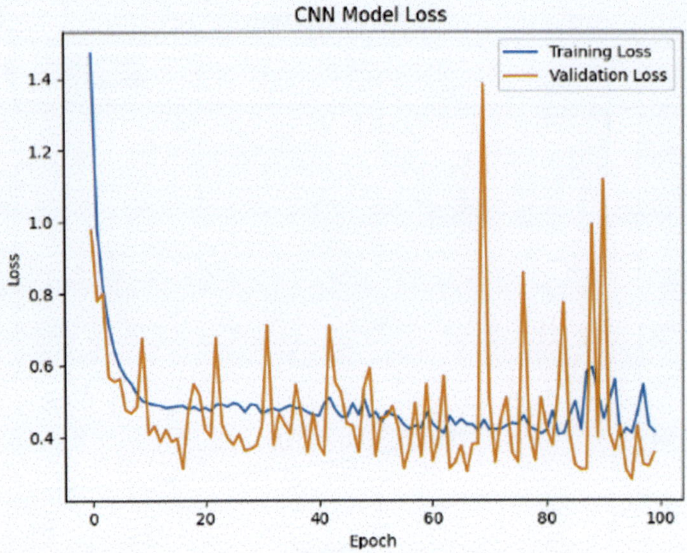

Fig. 8 Validation loss graph

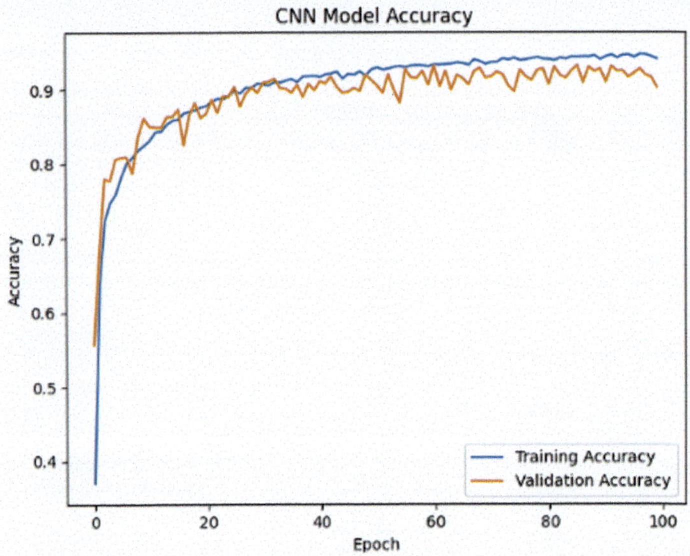

Fig. 9 Test accuracy graph

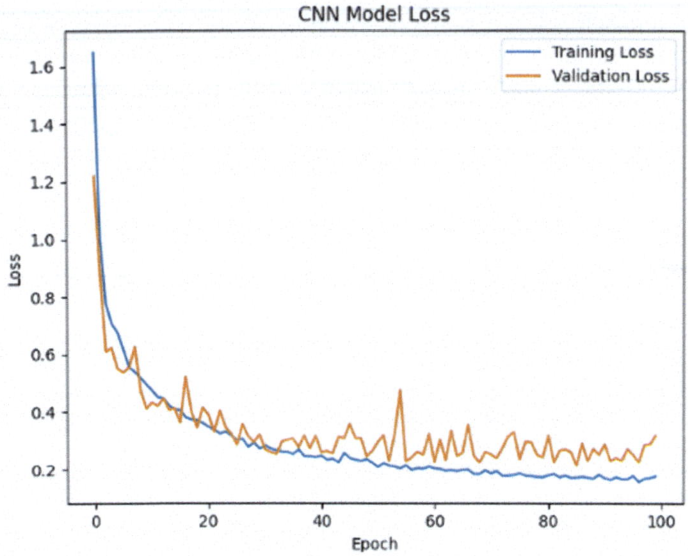

Fig. 10 Test loss graph

graph in Fig. 6 demonstrated a decreasing trend, which signified that the model can minimize errors during the training process. The validation accuracy graph in Fig. 7 helped to understand the ability of the model to generalize unseen data. The steady increase in its accuracy, along with the decreasing validation loss in Fig. 8 also confirmed that the model was not overfitting to the training data and was capable of making very accurate predictions on new instances. The test accuracy graph in Fig. 9 evaluated the model's real-world performance. The stable accuracy and low-test loss in Fig. 10 signified that the framework generalized well to previously unseen data, besides providing confidence in the practical scenario.

7 Conclusion

This research introduced a novel framework for automatic plant disease identification by leveraging Convolutional Neural Networks (CNNs). The model achieved a high training accuracy of 93.070%, validation accuracy of 91.530%, and test accuracy of 90.940%. This demonstrated the resilience and capability of the proposed model for generalization. The effective diagnosis of plant diseases with the help of this framework posed great potential while providing a very useful tool for crop management and precision agriculture. The competitiveness of the proposed model in the actual field was further portrayed by the accurate analysis with other state-of-the-art methods. After these effective findings and experimentation, there is a strong future scope where additional research and improvement could be done on the dataset to handle a larger range of scenarios, along with coping with different kinds of environmental conditions. With the high accuracy of the model along with its accuracy, robustness, and computational efficiency, it stood as a practical and valuable solution for the aiding of farmers and agronomists in timely disease detection and identification, ultimately contributing wholly to sustainable management.

References

1. Strange RN, Scott PR (2005) Plant disease: a threat to global food security. Annu Rev Phytopathol 43(2005):83–116
2. Uthayakumar J, Vengattaraman T, Amudhavel J (2017) A simple lossless compression algorithm in wireless sensor networks: an application of wind plant data. IIOAB J 8(2):281–288 (ESCI-Web of Science Thomson Reuters)
3. Sladojevic S, Arsenovic M, Anderla A, Culibrk D, Stefanovic D (2016) Deep neural networks based recognition of plant diseases by leaf image classification. Comput Intell Neurosci
4. Sharma P, Berwal YPS, Ghai W (2020) Performance analysis of deep learning CNN models for disease detection in plants using image segmentation. Inf Process Agric 7(4):566–574
5. Lakshmanaprabu SK, Mohanty SN, Krishnamoorthy S, Uthayakumar J, Shankar K (2019) Online clinical decision support system using optimal deep neural networks. Appl Soft Comput 81:105487. https://doi.org/10.1016/j.asoc.2019.105487
6. Barbedo JGA (2018) Impact of dataset size and variety on the effectiveness of deep learning and transfer learning for plant disease classification. Comput Electron Agric 153:46–53
7. Barbedo JGA (2019) Plant disease identification from individual lesions and spots using deep learning. Biosyst Engl 80:96–107
8. Guo Y, Zhang J, Yin C, Hu X, Zou Y, Xue Z, Wang W (2020) Plant disease identification based on deep learning algorithm in smart farming. Discr Dyn Nat Soc
9. Selvaganesan S, Jana S, Begum AR (2020) Design and analysis of pepper leaf disease detection using deep belief network. Eur J Mol Clin Med 7(9):1724–1731
10. Khatoon S, Hasan MM, Asif A, Alshmari M, Yap YK (2021) Image-based automatic diagnostic system for tomato plants using deep learning
11. Chen J, Chen J, Zhang D, Sun Y, Nanehkaran YA (2020) Using deep transfer learning for image-based plant disease identification. Comput Electron Agric 173:105393. https://doi.org/10.1016/j.compag.2020.105393
12. Bharali P, Bhuyan C, Boruah A (2019) Plant disease detection by leaf image classification using convolutional neural network. Springer, Singapore, pp 194–205
13. Xie X, Ma Y, Liu B, He J, Li S, Wang H (2020) A deep-learning-based real-time detector for grape leaf diseases using improved convolutional neural networks. Front Plant Sci 11
14. Zhang Y, Song C, Zhang D (2020) Deep learning-based object detection improvement for tomato disease. IEEE Access 8:5660756614
15. Janarthan S, Thuseethan S, Rajasegarar S, Lyu Q, Zheng Y, Yearwood J (2020) Deep metric learning based citrus disease classification with sparse data. IEEE Access 8:162588–162600
16. Barman U, Choudhury RD, Sahu D, Barman GG (2020) Comparison of convolution neural networks for smartphone image based real time classification of citrus leaf disease. Comput Electron Agric 177:105661. https://doi.org/10.1016/j.compag.2020.105661
17. Kang F, Wang C, Li J, Zong Z (2018) A multi-objective piglet image segmentation method based on an improved noninteractive GrabCut algorithm. Adv Multimed
18. Houssein EH, Helmy BE, Oliva D, Elngar AA, Shaban H (2021) A novel black widow optimization algorithm for multilevel thresholding image segmentation. Expert Syst Appl 167:114159. https://doi.org/10.1016/j.eswa.2020.114159

19. Howard AG, Zhu M, Chen B et al (2017) Mobilenets: efficient convolutional neural networks for mobile vision applications. https://arxiv.org/abs/1704.04861
20. Dhiman G, Oliva D, Kaur A, Singh KK, Vimal S, Sharma A, Cengiz K (2021) BEPO: a novel binary emperor penguin optimizer for automatic feature selection. Knowl Based Syst 211:106560. https://doi.org/10.1016/j.knosys.2020.106560
21. Xiao D, Li B, Mao Y (2017) A multiple hidden layers extreme learning machine method and its application. Math Prob Eng
22. Singh D, Jain N, Jain P, Kayal P, Kumawat S, Batra N (2020) PlantDoc: a dataset for visual plant disease detection. In: Proceedings of the 7th ACMIKDDCoDS and 25th COMAD, 2020, pp 249–253
23. Rangarajan AK, Purushothaman R, Ramesh A (2018) Tomato crop disease classification using pre-trained deep learning algorithm. Procedia Comput Sci 133:1040–1047
24. Karthik R, Hariharan M, Anand S, Mathikshara P, Johnson A, Menaka R (2020) Attention embedded residual CNN for disease detection in tomato leaves. Appl Soft Comput 86:105933. https://doi.org/10.1016/j.asoc.2019.105933
25. Rakesh S, Sudhakar P (2020) Deep transfer learning with optimal kernel extreme learning machine model for plant disease diagnosis and classification. Int J Electr Eng Technol (IJEET) 11(9):160–178

Author Index

S. Dutta et al. (eds.), *Cyber Intelligence and Information Retrieval*, Lecture Notes in Networks and Systems 1025, https://doi.org/10.1007/978-981-97-3594-5

The manufacturer's authorised representative in the EU is Springer Nature Customer Service Centre GmbH, Europaplatz 3, 69115 Heidelberg, Germany. If you have any concerns regarding our products, please contact ProductSafety@springernature.com

Printed and bound by CPI Group (UK) Ltd, Croydon, CR0 4YY
30/09/2026
02225232-0009